Comments from ...

"While the popularity of CCD's for astronomical imaging has grown exponentially in the last few years, there is a surprising dearth of informative or helpful literature on this exciting new field of amateur astronomy. This book fills that niche admirably, providing a complete "how-to" for the beginning to intermediate CCD astronomer. It removes the "black art" from CCD imaging and allows anyone to easily learn what it takes to begin maximizing the use of their CCD camera and get started down the path to acquiring wonderful images. A must-have that will become the bible of many CCD astronomers."

- Jeff Hapeman

"Ron has the knack of explaining and then demonstrating processing techniques in an understandable manner. For me as a Photoshop novice, I was able to become productive in image processing very quickly. Also, this book is chock full of practical insights on both hardware and software. This book will become the essential handbook for CCD Imaging."

- John Smith, Tucson, AZ

"An excellent resource and reference guide for the amateur and experienced astrophotographer which provides a thorough description of the issues, rationale and processes involved for each step in the CCD imaging of the universe and its countless wonders. Many valuable illustrations are also supplied to support each underlying theme and issue as well as an analysis of imaging software and hardware. A complete reference guide which is a must for any astrophotographer's library."

- Anthony Ayiomamitis, Athens, Greece

"I attempted ccd imaging in the mid 1990's and gave up on it after some pretty dismal results. After learning about this book, and utilizing the information provided here, I took up the hobby again and was producing good images after just a few nights out (and excellent images after just a few months!)"

- Randy Nulman

"At long last! A superb volume of information that helped me transition 25 years of film photography experience into the new realm of CCD imaging."

- John Gleason, www.celestialimage.com

"If you want to learn CCD imaging (and skip the PhD in mathematics!), this is your guide to success."

- John Polhamus

"The most complete and helpful guide to CCD imaging available anywhere. A must have for both the beginner and seasoned CCD imager."

- Rob MacKay, www.darkhorizons.org

More Comments from Readers

"As I strive to achieve the highest level of quality possible in Astronomical CCD imaging, The New CCD Astronomy has been there every step of the way. It is without a doubt, the most comprehensive and authoritative 'how to' book ever written for this most rewarding hobby."

- Mark Jenkins
Beloit, WI

"Mr. Wodaski gives an excellent coverage of methods to be used in taking CCD images with equipment available to any amateur. This is a practical book, full of comprehensive advice, which rescued me from the muddle of conflicting, confusing literature previously available. The New CCD Astronomy is a classic that all neophyte and experienced imagers should own."

- Morgan S. Wilson, MD

"As a newcomer to CCD imaging, I was reading everything I could find. I ordered Ron's book after reading the first online drafts."

- Frank Hainley
Moraga, California

"When I pack for the dark sky site, The New Astronomy gets packed first...even before my scope! A scope, a CCD camera, a little starlight, and The New Astronomy are a recipe for success. Your images will improve dramatically almost overnight."

- Mark R. Holbrook
www.ccdastronomy.com

"The New CCD Astronomy sets the bar a notch higher than anything else in print today. It will become the defacto standard for CCD-based astronomy."

- Tom Skinner
Physics teacher and longtime amateur astronomer

Bubble Nebula

The New CCD Astronomy

Ron Wodaski

How to capture the stars
with a CCD camera
in your own backyard.

NEW ASTRONOMY PRESS

Published by New Astronomy Press
PO Box 1766, Duvall, WA 98019 USA

New Astronomy Press
http://www.newastro.com

© 2002 Ron Wodaski

No part of this publication may be reproduced, stored in a retrieval system, or transmitted in any form or by any means, electronic, mechanical, photocopying, recording, scanning, or otherwise, excepted as permitted under Sections 107 or 108 of the 1976 United States Copyright Act, without the prior written permission of the copyright owner. Requests for permission should be address to Ron Wodaski, PO Box 1766, Duvall, WA 98019 Email: permissions@newastro.com.

First published 2002

Published in the United States of America

Printed by Edwards Brothers, Inc. Ann Arbor, MI

Typeset in Adobe Garamond and Adobe Helvetica

Library of Congress Control Number: 2001119349

ISBN 0-9711237-0-5

10 9 8 7 6 5 4 3 2 1

*For my wife Donna Brown.
Without her support, none of this
would have been possible.*

Acknowledgements

I would like to extend a hearty thank you to the many online readers who supported me in a wide variety of ways during the time I wrote this book. This includes everything from contributing images to the book, to volunteering to review chapters and making suggestions. My readers have been a major motivation to me during the time I've been writing and researching this book. Among my best helpers were those who took the time to ask insightful and sometimes desperate questions about the art of CCD imaging.

Various companies have supplied hardware or software to make it easier for me to review their products. These include SBIG, Finger Lakes Instruments, Software Bisque, Hutech, and Custom Scientific, who provided products and took time to explain sometimes complex hardware or software. Many other companies provided assistance in a variety of ways, including Anacortes Telescope & Wild Bird, Diffraction Limited, Astro-Physics, Axiom Research, Excelsior Optics, Optec, and Tom Osypowski.

Product names occur often in the book. These names are protected by patents, copyrights, trademarks, and so on. These names are the property of their respective owners.

Unless otherwise noted, images were taken by the author. Images contributed by other individuals are attributed individually. They are copyrighted by their respective authors who retain all rights.

About the cover

The cover image is built from two different images taken by Tony Hallas. The lower half is a time exposure of the observing field at Sunglow Ranch in Arizona, where Tony, myself, and a number of other imagers and observers spent a wonderful week under the stars in May, 2000. Tony walked around to the individual observers and "painted" their equipment with a red flashlight during the exposure. The white glowing objects are laptop screens; mine is the second one from the right. The upper half of the image is the Sagittarius star cloud, an area full of imaging opportunities.

Table of Contents

Part One: Getting Started with CCD

Chapter 1: Using a CCD Camera 1
Section 1: About CCD Cameras 2
CCD Cameras Are Cool(ed) 3
A Typical CCD Session 4
CCD Exposures 6
Section 2: Acquiring Images 8
The Imaging Process 8
Image Control (Brightness and Contrast) 15
Section 3: What Can I Do with a CCD Camera? 19
CCD Imaging Basics 19

Chapter 2: Practical Focusing 31
Section 1: Focusing Fundamentals 32
Focusing a CCD Camera 32
Diffraction Effects 33
Optimal Focus Position 36
Focusing Explained 37
Automated Focusing 39
The Implications of Focal Ratio 39
Section 2: Acquiring Images 41
The Focusing Process 43
Moving Primary Mirror Issues 46
The Zen of Focusing 46
Section 3: Software-Assisted Focusing 48
Brightest Pixel Focusing 48
FWHM Focusing 50
Subframe Focusing 53
Focus Analysis 54
Focuser Issues 57
Section 4: Aids to Focusing 58
Focusing with Diffraction Spikes 58
Focusing with a Mask 61
Other Focusing Aids 64
The JMI NGF-S Focuser 65
Section 5: Alternative Focusing 65
The Optec TCF-S (Temperature Compensating Focuser) 66
RoboFocus 67
Automatic Focusing 67
Using @Focus 68
Focuser types 70
Keys to Successful Automated Focusing 70
Start Close to Focus 71
Setting @Focus Parameters 72
Setting Step Sizes 74
A Sample @Focus Run 75
Verify @Focus Results 77
Section 6: Other Focusing Aids 78
Rings for Parfocal Eyepieces 78
Flip Mirrors and Off-Axis Guiding 78

Chapter 3: Practical Imaging 80
Section 1: Setting Up Your Telescope and Mount 82
Collimation: First, Last, and Always 82

    Collimating a Cassegrain's Secondary 84
    Free Play (Backlash) Adjustments 86

Section 2: Signal versus Noise 89
    Signal to Noise Ratio 89
    Reducing Noise 91

Section 3: Imaging the Sun, Moon, and Planets 94
    Choosing and Using Filters 94
    Solar Imaging. 95
    Lunar Imaging. 98
    Planetary Imaging 101

Section 4: Imaging the Deep Sky. 107
    Taking Longer Exposures 107
    Unguided Imaging 109
    Stacking (Combining) Images 110
    Dealing with Light Pollution 110
    Principles of Astrometry and Photometry 112

Section 5: Fun Science with a CCD Camera 112
    AutoAstrometry for a Single Image. 115
    Troubleshooting AutoAstrometry 116
    AutoAstrometry for Multiple Images in a Folder. 116
    Searching for Minor Planets and Supernovae 119

Part Two: Taking Great Images

Chapter 4: The Hardware Explained. 127

Section 1: Start with a Solid Mount 130
    Types of Mounts 130
    German Equatorial Mounts 130
    Dob with Equatorial Platform 134
    Fork Mounts 135

Section 2: Selecting a Telescope. 137
    Focal Length Issues 137
    Telescope types for CCD imaging. 142

Section 3: Choosing Camera and Software 152
    The Blooming Facts. 152
    Which Should You Choose?. 159
    One-Shot Color Cameras. 161
    Cameras by Manufacturer 161
    Camera and Image-Processing Software 163

Section 4: Matching Camera, Telescope, & Mount 166
    Image Scale Explained 166
    The CCD Calculator 168
    Focal Ratio is King. 171
    The Bottom Line 174
    Setting up for Imaging 177
    Casual Imaging 177

Section 5: Imaging Options. 177
    The One Night Stand 178
    Good-Weather Setup 180
    Remote Control. 181
    An Imaging Road Show 183
    Observatory Imaging 185

Chapter 5: Taking Guided Exposures 185

Section 1: What Does Guiding Do?. 188
    How Guiding Works 188
    How Mounts Move 191

Section 2: Autoguiding Hardware 192
- Self-Guiding 194
- Dedicated Guiding Cameras 198

Section 3: Mount Calibration 201
- Connect the Hardware 202
- Choose a Good Guide Star 202
- Calibration for Autoguiding with CCDSoft 203
- Calibration using MaxIm DL 210
- Calibration Tips 213

Section 4: Autoguiding in Action 215
- Resuming Guiding after Downloading 215
- A Typical Autoguiding Session 215
- Recalibration and Guide Errors 217
- Auoguiding Possibilities 218
- Adjusting and Tuning Your Mount 221
- Polar Alignment 222
- Examining Autoguiding Data in a Spreadsheet ... 224

Chapter 6: Increasing Image Quality 225

Section 1: Image Reduction 228
- CCD Chips Explained 230
- Dark Frames Explained 232
- Flat fields explained 238
- Bias frames explained 244

Section 2: Using Dark Frames 248
- Taking a Dark Frame 248
- Applying a Dark Frame 256

Section 3: Using Flat-Field Frames 263
- About Flat Fields 263
- Taking a Flat Field 264
- Applying a Flat Field to an Image 275

Section 4: Image Reduction in Action 282
- Taking a Bias Frame 282
- Using Reduction Groups 283
- Creating Reduction Groups 284

Section 5: Other Tips on Cutting Down Noise 287
- Aligning Images 287
- Combining Images (Maxim DL) 290
- Combining Images (CCDSoft V5) 290
- Aligning and Combining with Registar 291
- Track & Accumulate 295

Section 6: Dealing with Light-Pollution Gradients 297
- Gradients Are a Problem 299
- Using a Light Pollution Suppression Filter 300
- Removing Gradients with Software 303
- Removing Gradients in Photoshop 305
- Removing Gradients with Subtraction 314

Part Three: Advanced Image Processing

Chapter 7: Color Imaging 327

Section 1: Principles of Color Imaging 322
- Color filters 322
- Response Curves 324
- Color Filters Explained 325
- Luminance Layers 326
- CMY Filters 328

Section 2: Using a Filter Wheel 330
- Selecting a Filter Wheel 330
- Automating Color Imaging (CCDSoft) 330
- Automating Color Imaging (MaxIm DL) 332
- The Color Imaging Process...333
- Color Imaging Steps (CCDSoft)...336
- Color Imaging Steps (MaxIm DL) 336
- Color Imaging Guidelines 336

Section 3: Color Combining 338
- Color Combining in CCDSoft...339
- Color Combining in MaxIm DL 347

Section 4: Advanced Color Combining 354
- Normalizing with MaxIm DL..354
- Color Combine in Photoshop 355
- Making an LRGB image..359
- Color Processing Guidelines..362

Chapter 8: Image Processing Fundamentals 371

Section 1: Selecting Data to Display 366
- Data versus Image 366
- Histogram Adjustments 367
- Histogram Changes in Photoshop 376
- Digital Development 386
- Digital Development in Astroart 393

Section 2: Improving Image Clarity 395
- Sharpening..395
- Smoothing...400
- Deconvolution...403

Chapter 9: Image Processing for Celestial Objects 419

Section 1: Processing Sun, Moon, and Planets 414
- Seeing Conditions 414
- Sharpening..416

Section 2: Globular Clusters 421
- Resolution Requirements 422
- Digital Development in Maxim DL 422
- Levels and Curves in Photoshop..424
- How Much Sharpening?..428

Section 3: Galaxies.. 429
- Exposure Guidelines...431
- Getting Better Signal-to-Noise Ratios...................................432
- Dynamic Range 435
- Galaxy Processing Tips...436
- Digital Development 449

Section 4: Nebulae.. 450
- Exposure Guidelines...450
- Throwing Out the Pixel Laws..452
- Nebula Processing Tips 453
- Narrow-Band Filters...460

Section 5: Making Mosaics 462
- Taking the Images 463
- Setting Up the Mosaic Images 465
- Aligning Images...467

Index 471

Tips for readers

I have spent over a year researching and writing the book you hold in your hands. It is chock full of practical advice for everyone interested in imaging the objects that populate the skies, night and day. CCD imaging has given me more pleasure than I can possibly describe, and I can only hope that this book will help you do the same.

Unlike many books, this one is more than the paper it is printed on. There is a book web site, and you can also order a CD-ROM with a copy of the web site and the book content for a small fee. The book stands well on its own, but you will get the most out of your hard-earned dollars by checking out the suggestions below.

- Be sure to visit the book web site at **http://www.newastro.com/newastro/book_new**.
- You can download files from the web site that will allow you to follow along with many tutorials. Look for links at the start of the tutorials.
- The complete text of the book is available online. Many of the images in the book are shown in color in the downloadable Adobe Acrobat files. This is especially useful for the content in chapter 7, but many images throughout the book will reveal more information when seen in color. Be sure to take advantage of the free one-year web subscription that comes with the book, and download the Acrobat files!
- Your one-year subscription to the web site also includes other benefits. These include additional tutorials, discussion groups moderated by the author, a searchable database of CCD imaging targets, and more. If you purchased the book online, you recevied a username and password when you ordered the book. If you bought the book in a bookstore, follow the instructions below to obtain your usename and password.

Thank you for trusting me to lead you on the journey to CCD imaging. The trip has been a delight to me, and I hope that I have managed to convey my sense of excitement and wonder in these pages.

If you did not buy this book online

Online purchasers automatically receive a username and password for the book web site. Other purchasers need to fax the following information to the publisher to get their username and password:

- A copy of your receipt
- Your name, phone, and address (in case there are any questions about your request)
- Your email address (very important; your username and password will be emailed to you)

If you have any comments or suggestions, we'd love to hear them. You can find our fax number on the New Astronomy Press web site at **http://www.newastro.com**.

If you would like to send comments about the book, or ask questions about the book or CCD imaging in general, please join the Yahoo discussion group that has been set up for readers of this book at **http://groups.yahoo.com/group/ccd-newastro**.

1 Using a CCD Camera

PART ONE: GETTING STARTED WITH CCD

CCD imaging involves capturing some well-traveled photons. Many of those photons have crossed incredible distances to glide down the barrel of your telescope and strike the light-sensitive pixels of your camera.

The photons knock some electrons loose. The CCD camera counts and digitizes the data, and sends the results to your computer, where you see a picture.

Section 1: About CCD Cameras

FIGURE 1.1.1. AN EXAMPLE OF A DEEP, LONG-EXPOSURE CCD IMAGE.

Anyone can make the CCD magic happen. Figure 1.1.1 shows what you can expect if you master the process of CCD imaging. This image of the Crescent Nebula involves a total of three hours of exposure time through various filters.

Not every image has to take that long, of course. Figure 1.1.2 on the next page is a 10-second image of M42, the Great Nebula in Orion. The Crescent Nebula in figure 1.1.1 is a relatively dim object, so you need long exposures to get details like you see in the figure. M42 is relatively bright, especially the core, and you can get reasonable results with much shorter exposures. Longer exposures, however, will still bring out more faint detail.

Shorter exposures are easier when starting out, and they are a good way to get familiar with the processes involved in CCD imaging.

To get good at CCD imaging, you'll need to learn a lot of new things. That challenge is part of what makes CCD imaging something special. There are some things that you might expect to be easy, like focusing, that turn out to be a significant challenge. Chapter 2 is dedicated to teaching you everything you might ever want to know about focusing.

There are other things that you might expect to be hard, such as determining the exposure time, that turn out to be easy. This book will take you through the learning process one step at a time, and tell you what to expect, and how to evaluate your results, so that you can get up to speed and taking images as soon as possible.

The secrets to taking good CCD images are not really secrets. I can think of five things that will make for the best possible CCD images. Why these five things matter will become clear as you read on.

- Long exposures
- Precise focus
- A steady mount
- Precise polar alignment
- High-quality optics

And, if you are a beginning CCD imager, you might as well add number six to the list:

- A fast focal ratio and a short focal length make it easier to image.

It takes nine chapters and hundreds of pages to explore these areas, but it won't be long before you are putting these rules into practice and having a blast with CCD imaging.

FIGURE 1.1.2. AN EXAMPLE OF A SHORT CCD EXPOSURE OF A BRIGHT OBJECT (M42).

CCD Cameras Are Cool(ed)

You might be wondering why you would need to buy a specialized CCD camera for astrophotography in the first place. Why not use a film camera? Why not use a digital camera or a video camera?

The short answer is that a specialized CCD camera has some distinct advantages over other technologies. There are some specific situations where other types of cameras do a better job than a CCD camera. Your best choice depends on what you want to accomplish.

A CCD camera is better than film because you can be successful with shorter exposures, and you get instant feedback on your technique because you can see the image right on your monitor.

BUT: CCD chips are very small, and film covers a much wider field of view. And some people prefer the visual appearance of film images.

A CCD camera is better than a digital camera because CCD cameras have dramatically less noise. The CCD chips in cameras intended for astronomical use are typically cooled to 25 to 40 degrees Centigrade below the ambient temperature. This allows you to take time exposures with dramatically less noise.

BUT: Digital cameras take superb solar, lunar, and planetary images, especially through large, fast telescopes because they capture more light for shorter exposures.

A CCD camera is better than a video camera because video cameras are limited to extremely short exposures -- 1/60th of a second. A CCD camera can take exposures of an hour or more for deep images.

BUT: Video cameras are great for live shots of the sun, moon and planets, and for sharing those images with an audience in real time.

Film, digital cameras, and video cameras all have their place in the realm of astrophotography. But CCD cameras, with their superb quality, instant feedback, and low noise, are a cut above the others for most purposes. CCD is much easier than film in many ways. I have the utmost respect for successful film astrophotographers. I rely totally on the CCD camera's ability to help me focus, and to show me immediately if I've made a mistake. Film imagers have no such luxury.

If you have already bought a CCD camera, or are in the market for one, you can rest assured that it's one of the easiest ways to get incredible images of galaxies, clusters, planets, nebulae, and all the other cool things out there in the universe.

A Typical CCD Session

If you haven't used a CCD camera, you might be wondering what a typical imaging session is like. It starts the way any visual observing session would start: setting up the telescope in the usual manner. Most CCD imaging is done with some form of equatorial mount. Such a mount must be aligned to the north celestial pole. Imaging requires a more accurate polar alignment, but there are tricks for getting that done. Most types of telescopes are suitable for CCD imaging, but the mount must track very accurately to be suitable.

The CCD camera attaches to the focuser. Most CCD cameras have a 1.25" or 2" nosepiece that you insert into the focuser like an eyepiece. A Newtonian has the eyepiece far up on the tube, and that is where you would mount a CCD camera. Schmidt-Cassegrains, refractors, and many other types of telescopes have the eyepiece holder at the back of the scope, and that's where you put the CCD camera. You attach the camera the same way you would attach an eyepiece. You can mount CCD cameras in other ways on some telescopes, such as using a motorized focuser or a more secure connection between telescope and camera. Each camera and telescope manufacturer offers different options, so there are many ways to attach a CCD camera. If you need help deciding how best to mount your camera, visit the discussion group for the book. All web links can be found on the home page of the book web site at **http://www.newastro.com**.

The camera has cables that connect to your computer. Once the camera is on, you connect to the camera with software that controls the camera's functions. Examples include CCDSoft, MaxIm DL, and Astroart. Using the camera control software, you choose settings such as the amount of cooling to use, whether to image with the full CCD chip or just a portion of it, whether to bin (join) pixels to increase sensitivity, and so on. You may vary these choices during the course of the night as needed.

Focusing is the next step. There are many ways to rough focus a CCD camera. For example, you can use a parfocal eyepiece to get close to focus, and then do critical focusing with the CCD camera in place. Most CCD cameras come with software that allows you to rapidly download a small portion of the image. You can use this visual feedback from the camera control program to evaluate focus as you make changes. The real trick is learning how to do critical focusing. As mentioned earlier, there's an entire chapter to help you learn how to focus. Even a small error in focus position can affect the image, so it's worth taking the time to focus accurately. The more you do it, the better you get at it. You refocus periodically during the night because focus changes with temperature, and sometimes with the physical movement of the telescope.

Once you are focused, it's time to point the telescope at the object you want to image. You can do this using a finder scope, or you can use digital setting circles or a goto mount to aim the telescope. The smaller your CCD chip and the longer your focal length, the more of a challenge this will be. Goto scopes are very popular for CCD imaging because they let you put objects on the chip more easily. However, all goto scopes are not created equal. To put objects "on the chip" reliably, you'll need a first class mount. A little hunting around is common at longer focal lengths. If you have star-hopping skills from your visual observing, you will find them useful for CCD imaging.

Some CCD imagers use an autoguider. This is a second CCD chip or camera that is aimed at the same area of the sky as the imaging camera. The main purpose of autoguiding is to allow you to take the long exposures needed for better quality images. The autoguider takes images at regular intervals, and measures the position of a guide star. The autoguider software then adjusts the mount to keep the guide star centered. If you are using an autoguider, the key step is to find a suitably bright guide star. If necessary, you take a few minutes to perform a guiding calibration. This allows the autoguider software to learn the speed at which your mount moves to make guide corrections. You then initiate the autoguiding process. Like focusing, autoguiding is a skill that will take a little time to master.

FIGURE 1.1.3. A RAW CCD EXPOSURE, AS IT APPEARS WHEN DOWNLOADED TO A COMPUTER (NGC5566).

bining them later during image processing. This will give you results similar to taking a single long exposure. A person taking single images might take images of 5-10 minutes, although this can vary quite a bit. A person taking multiple images might take a bunch of 1-3 minute exposures totaling 6-15 minutes to get roughly the same results.

When the image is done, the camera shutter closes, and the camera reads out the image from the CCD chip, converts it to digital format, and the camera control software downloads it to your computer. Figure 1.1.3 shows what the raw frame can look like. It looks like a disaster! Fear not; there is a good image lurking under all that noise.

If you are taking an unguided exposure, you typically take some test exposures to find out how long of an exposure your mount and polar alignment will allow you to take. This is the longest exposure you can take of a given object, and it varies with declination. The closer you are to the celestial equator, the greater the drift that results from poor polar alignment.

At this point, you are focused, you are pointing at the object you want to image, and now all you have to do is take a long enough exposure to get the image. Various factors can limit the longest exposure -- skyglow, mount capabilities, whether or not you are guiding, etc. -- but longer is almost always better. The images are digital, so you might wind up taking more than one image and com-

FIGURE 1.1.4. THE SAME IMAGE AS ABOVE, AFTER IMAGE REDUCTION.

FIGURE 1.1.5. A FULLY-PROCESSED IMAGE OF NGC5566.

There are enough variables in CCD imaging that you will often need to take a moment to check the image. Is it well focused? Is the object of interest centered properly? Did the autoguiding work OK? Did the mount track accurately? If anything goes wrong, you can see it right away and fix the problem. You can then take another image if necessary. When you have a satisfactory image or sequence of images, you find another object and do it all over again.

Either before or after your imaging session, you will take some special images. These are called bias frames, dark frames, and flat-field frames. These images record the noise characteristics of your camera and optics. Later, during image processing, you will use the special frames to clean up your regular images in a process called image reduction. Figure 1.1.4 shows the result of image reduction. The noise is largely gone, and what remains is a pretty cool image.

Finally, you will use various image processing tools to brighten, darken, sharpen, smooth, crop, sum, average, resize, colorize and otherwise process the images to make them look their best. Figure 1.1.5 shows the result of some fancy processing, using the techniques described in this book. CCD imaging really is magic!

CCD Exposures

An image can take anywhere from a few milliseconds to a few hours to capture. Bright objects like planets, the moon, and the sun are often imaged with the shortest possible exposure times. Some cameras may not have short enough exposure times, and you can use a filter to cut the brightness, just as you would use a moon filter to cut the brightness for visual observing of the moon.

SECTION 1: ABOUT CCD CAMERAS

Many deep-space objects are reasonably bright, and require exposures of a minute or less to preserve detail in the bright areas. These include bright nebulae like M42 and Eta Carinae and galaxies like M31. However, many of these bright objects also have very dim details, so it also pays to take long exposures of such objects. You will need to use special techniques to preserve both the brightest and dimmest details of the object. This is explained in detail in chapters 3, 6 and 9.

FIGURE 1.1.6. A COMPARISON OF A LONG EXPOSURE (LEFT) AND A SHORT EXPOSURE (RIGHT).

When it comes to imaging the dimmest deep-sky objects, long exposures are best. Images of 5, 10, even 30 to 60 minutes can be used for such objects to reveal subtle details clearly. To take such long exposures, you need a way to guide the mount during the exposure, and you need a mount that moves smoothly and accurately. Some CCD cameras are self-guiding (e.g., SBIG ST-7E), while others require an external autoguider (e.g., FLI CM-10E). Chapter 4 covers mounts in detail, and chapter 5 has details on autoguiding.

Figure 1.1.6 shows one of my first images of a Messier object: M65, the small image at right. You might find it hard to recognize because the exposure is too short to show much detail. The much longer exposure at left reveals more detail. Longer exposures require more careful attention, but they are the key to getting beautiful images.

The best results come with the longest exposures, through the best optics, on the most stable mount. When you can bring all of these to bear on an image, the results can be stunning as shown in figure 1.1.1 and 1.1.5.

If you image in color, most cameras require a filter wheel with red, green, and blue filters. Other filter combinations are available, but red/green/blue is by far the most common approach to color. You take an image through each color filter, and then use software to combine them to create a color image. This is called an RGB image. You can also combine a white-light (luminance) image with the red, green, and blue images. These are called LRGB images. There are also other color techniques, such as CMY and false color, but RGB and LRGB are the most commonly used by amateurs. Chapter 7 will get you going with color imaging.

When you are starting out, you will probably begin with short exposures. As you learn more about how to control the mount, the camera, and the software involved, you will enjoy the many improvements that come from taking longer and longer exposures. This can take the form of long single exposures, or combined exposures that are the sum (or median) of many shorter exposures.

For more information about the relative advantages of long exposures and combined exposures, please see chapter 3.

CHAPTER 1: USING A CCD CAMERA

Section 2: Acquiring Images

Many camera control programs, such as CCDOPS, MaxIm DL, Astroart, etc. use basically the same process for acquiring images.

- Connect to and set up the camera
- Set the exposure parameters
- Set automatic or manual image reduction
- Select a filter if you have a filter wheel
- Turn automatic save on or off
- Set the number of images
- Take the image

This section will walk you through the imaging process as it exists in CCDSoft version 5, a camera control program from Software Bisque. This is the software I use myself when imaging, and I think it does a great job of organizing all the tasks and features involved in camera control. It doesn't yet support every camera, but new cameras are being added. If CCDSoft doesn't support your camera, look into Maxim DL or Astroart, in that order.

The Imaging Process

The following example shows how to take an image using CCDSoft version 5. I use CCDSoft for my own imaging because the camera control features of CCDSoft are well organized and comprehensive.

One of the first questions that comes to mind for CCD imaging is: How long should I expose? The simple answer is that it varies a lot. Jupiter or the moon might require an exposure of 0.05 second with one setup, and .1 second with another. A distant galaxy might need 10 or 20 minutes of exposure to reveal dim details. Deciding how long to expose for various objects gets easier with experience. How deep you can go with a given imaging setup depends on the sensitivity of the camera and the focal ratio of your telescope; see chapter 4 for full details.

TIP: Because of the large dynamic range typical of most CCD cameras, very long exposures are possible for many objects. Various factors, however, can stop you from taking long exposures. These range from the limitations of your mount to an object passing behind a tree. The key to success is to take exposures that are long enough to suit your purposes. If you are hunting for minor planets, you want exposures that go deep enough to find the minor planets, but are short enough to avoid showing motion. If you want a beautiful image of a galaxy, long or multiple exposures will reveal the details in the dim areas of the galaxy. Whatever type of imaging you do, you need a certain level of signal that will overcome the noise inherent in the process of collecting light from the distant reaches of space. Whether you find yourself taking long exposures or taking shorter exposures and combining them, the more photons you collect of a given astronomical subject, the better your results will be.

You set exposure times on the Take Image tab of the Camera Control panel, shown in figure 1.2.1.

The settings in figure 1.2.1 are for one full-frame three-minute exposure with an automatic dark frame. For info about dark frames, see chapter 6. The bin mode is set to 1x1 (full resolution). Binning allows you to control image scale; see chapter 4.

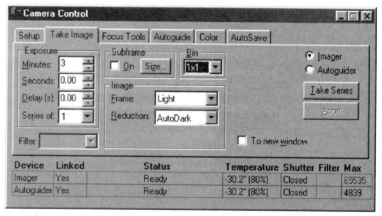

FIGURE 1.2.1. EXPOSURE SETTINGS ARE SET ON THE TAKE IMAGE TAB OF THE CAMERA CONTROL PANEL.

Once the telescope is rough focused, it's always a good idea to take a test exposure to verify focus quality. I suggest taking an unbinned, subframe test exposure to monitor how good your focus position is. Various factors can change the best focus position, such as temperature changes. (The focus position shifts slightly with changes in temperature.) See chapter 2 for details on focusing.

> **TIP:** A good method for checking exposure is to take a very short full-frame exposure with the Focus Tools tab, using the maximum binning mode available (that is, the lowest resolution), such as 3x3. Find an area with unsaturated stars, and click and drag a subframe around them. Now take an unbinned exposure (set bin mode to 1x1). Examine the resulting image for focus quality. To get a sense of how much the focus is affected by seeing conditions, take a sequence of 3-10 such images and observe the change in the Sharpness graph. The greater the fluctuation in the Sharpness value, the greater the impact of seeing on focus. The longer your focal length, the more of an issue the seeing will be with respect to image quality.

How long should you expose your image? The dominant factor in determining exposure for a given CCD camera and telescope combination is focal ratio. This is the ratio between the aperture and the focal length. The slower your focal ratio (e.g., f/10), the longer your exposures must be. The faster your focal ratio (e.g., f/4), the shorter your exposures can be. This is true no matter what the aperture of your telescope -- all f/8 telescopes require the same exposure duration for the same camera. Chapter 4 shows examples illustrating why this is so.

For example, an f/1.95 system such as a Fastar-equipped Celestron Schmidt-Cassegrain telescope might saturate the sky background using an ST-237 camera in two minutes. An f/5.6 refractor, on the other hand, would require almost four times the exposure to saturate, and an f/10 SCT would require even longer exposures.

The brightness of your subject also affects exposure times if you are using a non-antiblooming (NABG) CCD camera (see chapter 4 for details on antiblooming and non-antiblooming cameras). A bright star can cause blooming (see figure 1.2.4) if the exposure is long enough. On the other hand, if you are imaging a dim galaxy, exposure times of 10 to even 60 minutes might have little or no blooming. It all depends on the brightness of the stars and the galaxy core. Experiment with various exposure times for each type of object you image to learn what works best for your camera and telescope. Blooming can be cleaned up manually in an image editor, such as Photoshop or Picture Window Pro. Larger blooms are harder to clean up, and they can mask some of the object you are trying to image.

The bottom line is that NABG cameras are limited in the length of exposure you can take by blooming. You can take multiple short exposures and combine them to overcome this. Antiblooming cameras allow you to take longer single exposures, but they are less sensitive and *require* longer exposures. The better your mount, the more you can take advantage of an antiblooming camera. A good mount will track and guide accurately, thus allowing those long exposures.

> **TIP:** There are two kinds of exposures: unguided and guided. In an unguided exposure, the mount tracks at a rate that is as close as possible to the rate at which the stars appear to move (the sidereal rate). There is no feedback to tell the mount how accurately it is following the stars. At some point in an unguided exposure, the difference between the sidereal rate and the mount's tracking rate and/or polar alignment error causes stars to leave a linear trail on the image. The greater the difference, the larger the trailing effect will be. In a guided exposure, there is a feedback mechanism which adjusts the tracking rate of the mount.

During an unguided exposure, the mount is doing its best to track the motion of the stars relative to the earth. Some mounts are more capable in this regard than others. CCD imaging is very demanding of a mount, requiring tracking and corrections at extremely high levels of precision. As you image, you will learn how far you can push your particular mount, as well as how to get the best possible performance out of it.

Polar alignment has a major impact on tracking accuracy. If you have a very good polar alignment, you will be able to take longer unguided exposures. This assumes that the mount itself tracks well. If your polar alignment is only approximate, you will see star trails on relatively shorter exposures even if your mount tracks accurately. At a focal length of 1000mm, you should be able to polar align well enough for one minute exposures without star trails on a regular basis. At 500mm, two minutes should be practical. At 2000mm, 20 to 30 seconds may be all you can get with a simple polar alignment. For more precise polar alignment, consider using a drift alignment (see the section "Manual Drift Alignment" in chapter 4).

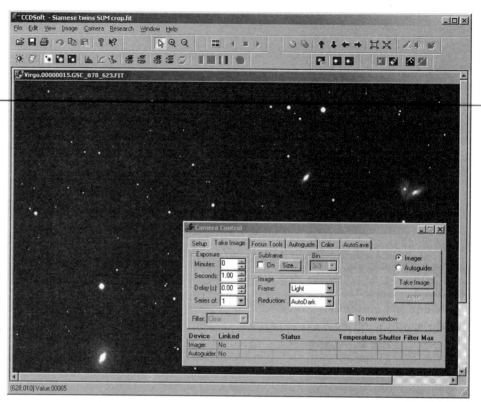

FIGURE 1.2.2. THE DOWNLOADED EXPOSURE APPEARS IN A WINDOW IN CCDSOFT.

The quality of an unguided exposure depends on the accuracy of your mount. A mount that tracks poorly (i.e., it has a large periodic error as described in chapter 4) will not be as effective for astrophotography as one that has a very steady tracking speed (small periodic error). Generally speaking, if the periodic error is small enough it will hide errors within the width of the smallest stars in your images. Some mounts have periodic error correction (PEC) that compensates for periodic error. The longer the focal length of your telescope and the smaller the pixels in your CCD detector, the more accurately your mount must track for successful imaging.

To take an exposure, follow this checklist until you are familiar with the Take Image tab and are ready to experiment with additional options. Each of these items will be covered in depth shortly:

1. Make the Take Image tab active on the Camera Control panel.
2. Select an appropriate bin mode.
3. Select the exposure duration.
4. Set Frame type to "Light."
5. Set an appropriate reduction setting. If you aren't familiar with image reduction, use AutoDark to automatically take and subtract a dark frame from each image. Other camera control programs have a similar feature.
6. If you have a color filter wheel, select the appropriate filter. Use the Clear filter for your first images to keep things simple.
7. If you want to take more than one image, select that number in the "Series of" drop-down list.
8. Click the Take Image/Series button. Wait for the image to download; it will appear in the CCDSoft window after the download (see figure 1.2.2). If AutoSave is active, the filename is built automatically. If TheSky is running the name of the object at the center of the image is included in the filename. Coordinates are stored in the image header. They are based on what the telescope reports and will not be precisely accurate if your polar alignment is off.

The following sections contain more information about these steps. Other camera control programs will have similar features, and the information below also applies to many other software programs.

Select Bin Mode

Binning combines groups of pixels together, in effect giving you different pixel sizes with the same camera. This allows you to match pixel size to telescope focal length more flexibly. For a given telescope, small pixels require longer exposures but provide higher resolution (subject to the limitations of the seeing conditions). Binned pixels allow shorter exposures but provide lower resolution.

For any given combination of camera, telescope, and seeing conditions, one bin mode or another will provide the optimal compromise between exposure duration and resolution. Image scale, which is the amount of sky covered by one pixel (or one binned super-pixel), measures the relationship between camera and telescope. Image scale is described in detail below.

Figure 1.2.3 shows how binning works. Binning 2x2 groups four pixels together, giving you virtual pixels made up of 4 actual pixels. Binning 3x3 yields virtual pixels made up of nine actual pixels. The higher the bin number, the larger the chunk of sky covered by a single virtual pixel.

Figure 1.2.4 shows the relative sizes of images taken with 1x1 binning (large image) and 2x2 binning (small image). Both images cover exactly the same area of sky, about 1 by 1.5 degrees in this example. The 2x2 binned image shows that the binned detector is more sensitive for the short exposure time used on this bright object. There is more detail visible in the dim areas of M42 in the binned image.

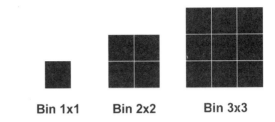

Figure 1.2.3. Bin modes group pixels together for greater sensitivity, at a cost of lower resolution.

Note: If you are using a non-antiblooming camera, binned pixels bloom at the same rate as unbinned pixels. It is the individual pixels that bloom, not the binned ones. So the full well capacity -- the bucket size, the point where overflow occurs -- doesn't change when you bin.

Generally speaking, you will most often use a bin mode that will deliver between 2 and 3 arcseconds of sky coverage per pixel. This is the most common range for general-purpose CCD imaging. When working within this range, each pixel in the image "sees" or covers 2 to 3 arc seconds of the sky. The seeing conditions

Figure 1.2.4. The relative sizes of images taken in different bin modes. Inset is binned 2x2. Large image is binned 1x1.

on most nights allow approximately 2 to 3.5 arc seconds of detail, and that is the reason for this range of values. If you have better or worse seeing conditions, select a bin mode based on your local conditions.

You get the most benefit from high-resolution imaging on nights when the seeing is exceptionally good. High-res is often used for planets, but it also yields very fine detail on galaxies and other deep-sky objects. The key is to wait for exceptional seeing conditions that support high resolution.

Low-resolution imaging (more than 3 arc-seconds per pixel) allows you to image a wide area using camera lenses or telescopes with short focal lengths (under 700mm). Seeing is unlikely to affect low-resolution wide-field images, so you can do this kind of imaging on almost any clear night.

You can use various tools and web pages to determine arc seconds per pixel for your camera and telescope combination, or use the following formula:

$$\frac{(205 * pixel\_size\_in\_microns)}{telescope\_focal\_length\_in\_mm}$$

If you are binning 2x2, be sure to double the pixel size to get the correct value. For example, if you have an ST-7E camera (9-micron pixels) and are using it unbinned on a Meade LX200 8" at f/10, this is

$$\frac{(205 * 9)}{2000}$$

or 0.93 arc-seconds per pixel. That's very high resolution. Since a typical night offers 2-4 arcsecond seeing conditions, you would rarely be able to use all that resolution. Binning 2x2 yields 18 micron pixels, which gives you 1.86 arc-seconds per pixel. This is just under two arc seconds per pixel, and offers a good compromise on nights of typical seeing.

TIP: I created a program for determining image scale for hundreds of camera/telescope combinations. You can download the program from **http://www.newastro.com/newastro/book_new/camera_app.asp**. See chapter 4 for details.

You can also approach image scale from another direction: using focal reducers. These are available for specific telescopes, and in general-purpose models which will fit a wide variety of telescopes. For example, Meade and Celestron sell focal reducers for their SCTs that reduce the focal ratio to 63%. A 63% focal reducer (often called an f/6.3 reducer) brings the focal length of an 8" SCT down to 1260mm, which provides 1.47 arc seconds per pixel unbinned with an ST-7E.

TIP: Binning also increases the sensitivity of your camera for short exposures, as shown in figure 1.2.4. Smaller pixel size means higher resolution, while larger pixel size means greater sensitivity. You are making a tradeoff between resolution and sensitivity whenever you select a bin mode for your image. Try an experiment for yourself. Take a one-minute exposure of a galaxy using 1x1, 2x2, and (if available) 3x3. Note the differences in resolution and depth of detail for different bin modes.

Select exposure length

Apart from issues around blooming and saturation, your maximum exposure length is determined by the accuracy of your polar alignment, by how accurately your mount tracks, and by whether or not you are doing guided exposures. If you see stars trailing into lines in unguided exposures, shorten your exposure until this goes away (or take some time to improve your polar alignment or the accuracy of your mount's tracking). If you see wiggly lines on long exposures, then your mount's periodic or random tracking error is too large for the current focal length. Use PEC, get a focal reducer, or switch to a scope with a shorter focal length that better fits the capabilities of your mount.

Table 1.1 includes some recommended minimum exposure times. The table assumes a focal ratio in the range of about f/6 to f/8. Use longer exposures for slower focal ratios, and shorter exposures for faster focal ratios. If you get blooming on a particular object, use a shorter exposure. If you don't get much detail in the image, or if it seems washed out or grainy, go to a longer exposure. Generally speaking, unless you run into blooming, saturation, or other problems, longer exposures are generally better. Experiment to find the best exposures for your setup. You might want to keep a written record of successful exposures which you can use as a guide for future imaging sessions.

Table 1.1: Suggested Exposures

Object	Suggested Exposure range
Sun	Use a visual solar filter, plus any additional filters needed to reduce the incoming light. Polarizing filters, moon filters, or other specialized filters such as hydrogen-alpha filters can reduce the light throughput so that your camera's shortest exposure will be sufficient for solar images. Do not use so-called photographic solar filters. These are intended for film, which is much less sensitive than your CCD detector. Never attempt to image or view the sun without a proper solar filter!
Bright planets	Planets such as Venus, Jupiter, Saturn and Mars require very short exposures, as little as 5 to 10 thousandths of a second. However, if you use a Barlow or eyepiece projection to increase the focal ratio to f/20 or slower, longer exposures (up to a full second) may be required.
Moon	The moon is also very bright, especially around full moon, and may require some effort on your part to attenuate the light sufficiently for imaging. Polarizing filters and moon filters, or both, will get the job done. I've also heard reports of successful imaging with a solar filter, but haven't tried this myself. As with solar images, the shortest possible exposure of your camera should be tried first. Then add additional filtering if that isn't short enough, or increase the exposure if it's not long enough.
Open clusters, bright globular clusters	These require relatively short exposures, and exposure length is usually limited by blooming if you don't have an anti-blooming camera. If you do have an anti-blooming camera, you can take longer exposures, even long enough to image background galaxies. For clusters with very bright members, or globulars with very bright cores such as M13, exposures under a minute should be enough to show good detail, but take multiple images and combine them to reduce noise. Dimmer clusters may require exposures longer than a minute. Very bright clusters, such as the Pleiades or the Beehive, usually only work well with anti-blooming cameras. For color imaging, be careful not to saturate bright stars so you can get truer star colors. Anti-blooming cameras consistently deliver better star color in all images because they do not saturate as easily an NABG cameras do. Saturated star images lead to white stars.
Galaxies	Galaxies come in a huge range of brightness levels. M31 will show up in an exposure of a few seconds; M101 may require 5 to 10 minutes or more to get details. Edge-on spirals and elliptical galaxies are brighter and you can get by with shorter exposures. Face-on spirals are usually the dimmest galaxies and require long exposures. When using a scope with a very long focal length, large pixels or binning may be required to get reasonable exposure times. For many galaxies, the core is much, much brighter than the arms, and with an NABG camera, the core may bloom before you get the desired detail in the arms. Take multiple exposures if this is the case.
Nebulae	Nebulae come in an even wider range of brightness than galaxies . Some, like M42, will yield excellent results in less than 20 seconds. Others, like the Rosette Nebula, may require 20-minute exposures to get good detail. A few trial exposures will help you determine the best exposure for any given nebula. Even a bright nebula like M42 has extremely faint detail in it, so capturing and representing the full range of detail can be a challenge. See the Nebula section in chapter 9 for some processing tips.

Note: If you are using an IR blocking filter, it will affect your exposures times. Various CCD cameras are more or less sensitive to IR light, and blocking infrared will require longer exposures. Blocking infrared is often desirable on refractors to reduce star bloating from chromatic shift (the inability of a refractor to bring all colors of light to the same focus).

Generally speaking, longer exposures are better in most cases because they provide better-quality data. However, blooming, skyglow, the risk of passing satellites or airplanes, and other factors usually limit the longest exposure time you can use. You can take multiple exposures and combine them in various ways to increase the quality of your images. See chapter 8 for details on aligning and combining images. Longer exposures improve the signal-to-noise ratio of your images, reducing grain and increasing dim detail. Combining exposures is nearly as effective.

For example, a single 30 minute exposure will have a slightly better signal-to-noise ratio than six five-minute images, but 8 five-minute images will have a better S/N than one 30-minute exposure.

Of course, three 30-minute exposures would be better still, and this is the approach I often take. But you'll need an anti-blooming camera to do that for most objects.

Set Frame type to "Light"

The next step is to make sure that the Frame is set to "Light." The Frame drop-down also includes settings for taking image reduction frames, which is explained in detail in chapter 6. A CCD detector generates a certain amount of noise, and image reduction removes a great deal of that noise.

Briefly, the frame types are:

Light - A normal image, taken with the shutter open.

Bias - A frame of the shortest possible exposure, taken with the shutter closed. It represents the minimum noise in the CCD detector and camera circuitry. This is subtracted from Dark frames so that the dark frames can be scaled when there is a difference in exposure length between the light and dark frames.

Dark - A frame taken with the shutter closed. It is in effect a picture of the electronic noise in the camera. This noise can be subtracted from a Light image to create a cleaner image. For best results, always use dark frames that are the same exposure length and chip temperature as your light frames. If you ever have to use different exposure lengths, make sure you take a bias frame so the software can scale the dark frame properly.

Flat Field - An image of an evenly illuminated field with the shutter open. Think of it as an image of the optical noise in the system, such as dust motes on glass surfaces or reflections off of the inside of the telescope. The Flat Field is applied to the Light frame to remove this source of noise.

> **TIP:** When you are taking an image, always remember to set the frame type to "Light." There is nothing more annoying than taking a long exposure of an object only to wind up with a dark frame instead! Actually, there is one thing that is equally annoying: taking an image with the subframe set to something really small. This typically happens when you use a subframe for focusing, and forget to turn it off before you take your image.

Choose an Appropriate Reduction Setting

When you select Light as the frame type, you can also choose the type of reduction to apply to the image. Reduction is the process of applying bias, dark, and flat-field frames to your image to reduce system noise. Full coverage of image reduction is in chapter 6.

For your first images, chose the "AutoDark" reduction setting. After your exposure is finished, the software will automatically take a dark frame with the same exposure settings, and subtract it from the image. If AutoSave is on, both the raw and reduced images will be saved to disk. AutoDark gives you a cleaner image with less thermal noise. When you gain more experience, you can explore the full range of image reduction options:

None - The software does nothing about image reduction. Use this setting when you wish to manually apply your own bias, dark, and flat-field images later, or when you simply want a quick image without any reduction.

AutoDark - This will follow the first exposure with a single dark frame. The dark frame is saved in memory and will be applied to all subsequent exposures with the same duration. If you change the exposure duration, a new AutoDark frame will be taken.

Bias, Dark, Flat - If you have previously taken bias, dark, and flat-field images and added them to a reduction group (see chapter 6), you can select which group to automatically apply to your images. The group names are hidden until you choose "Bias, Dark, Flat."

Select the Appropriate Filter

When you take a color exposure with a filter wheel or bar, select which filter to use. If you do not have a filter wheel, this setting will be disabled. For your first exposures, start with the Clear filter setting. The Filter drop-down on the Take Image tab is designed for taking one or more images using a single filter. If you plan to take color images with the red, green, and blue filters, use the Color tab instead.

> **TIP:** If you aren't sure if the filters in your filter wheel are set up in the default order, you can examine them visually without taking anything apart. Simply set the filter you want to test on the Take Image tab, and take a short exposure (one to three seconds). With a reasonably strong light behind you, look into the front of the camera. As the shutter opens, you will see the CCD detector. The color of the detector is the color of the current filter. Ignore any reflections off of the filter; the color of the CCD detector will give you the true color of the active filter.

Note: The filter setting, like the Bin and Frame Type settings, also affects exposures taken with the Focus and Autoguider tabs.

Set the number of images

If you want to take more than one image, select that number in the "Series of" drop-down list. You can then combine images to improve the signal to noise ratio, which will make your images less grainy. Take at least three images if you plan to use median combine, as that is the minimum number required to perform the math used in that combination method. Averaging and adding require only two images for proper operation. For most situations, four images provides the most obvious improvement over a single image, but you will continue to get small incremental improvements if you take larger numbers of images and combine them. After about 8 images, the differences become small, but they are still there.

Click the Take Image/Series Button

Once the Take Image tab options are set, click the Take Image/Series button. (The text on the button will be different if you choose one or multiple exposures.) The software will signal the camera to begin the exposure. At the end of the exposure, the image will be downloaded and displayed. The software will automatically adjust contrast using the black point and white point (Background and Range settings). You can fine-tune these settings if you aren't satisfied with the automatic results. A brief description of the contrast settings follows shortly. See chapter 8, Image Processing, for more information about the histogram tool and contrast settings.

Image Control (Brightness and Contrast)

Once you have taken an image, CCDSoft and other camera control programs can automatically balance the brightness and contrast of the image. You may have to turn on auto contrast in some programs to make this happen. In CCDSoft, automatic contrast adjustments are controlled by a checkbox on the Setup tab of the Camera Control panel. In MaxIm DL, automatic adjustments are, well, automatic.

You can also adjust contrast manually to suit your needs.

> **TIP:** CCDSoft and most other camera control programs (Mira, Astroart, MaxIm DL) include various image processing tools that allow you to do more than adjust the brightness and contrast of your images. Please see chapter 8 for information about other types of image processing tools.

You will also find that adjusting brightness and contrast helps with such things as determining best focus, and finding out whether your exposure was long enough to bring out the fainter details in the image. Although camera control programs include automatic contrast adjustments, the automatic setting isn't always (or even often) the setting that will give you the most useful information about your image.

For example, when imaging the California Nebula, the default presentation of the image might look like figure 1.2.5. The nebula is practically invisible, and you might question whether or not the exposure was long enough to be useful. Blooming has occurred, so it would be nice if you didn't have to use a longer exposure. If the exposure is long enough, you know you can continue imaging. If not, then you will have to try a longer exposure. Figure 1.2.5 looks dim, but is it really dim? Adjusting the brightness and contrast will tell us.

FIGURE 1.2.5. THE CALIFORNIA NEBULA, WITH AUTOMATIC BRIGHTNESS AND CONTRAST.

TIP: The vertical streaks above and below the brightest stars are called blooming. An anti-blooming CCD detector would have made it possible to take long exposures without blooming. The blooming spikes can be fixed by manual editing in an image editor. Most camera control software doesn't include sophisticated enough editing tools to handle blooming effectively. Image editing software such as Photoshop, Paint Shop Pro and others offer good options for dealing with blooming.

There are two approaches you can use to adjust an image's brightness and contrast:

Change the background and range settings - This method uses numeric values, and most programs will set them automatically if you turn that feature on. Use the Image | Brightness & Contrast | Background & Range menu selection to open the Background & Range dialog in CCDOPS. In MaxIm DL, use the View | Screen Stretch window to display the histogram and Minimum and Maximum settings. The minimum and maximum settings are the black and white points. Minimum is the black point. Maximum is the white point; in CCDSoft, the white point is equal to the background plus the range. See below for more on these terms.

Modify the image histogram - A histogram graphs the brightness values in an image, and allows you to adjust the contrast settings (and often other things as well) interactively. You can see the results of your changes in real time, which helps you choose the best settings. In CCDSoft, right click on an image and choose Histogram (see figure 1.2.6), or use the Image | Brightness & Contrast | Histogram menu selection.

FIGURE 1.2.6. THE CCDSOFT HISTOGRAM TOOL.

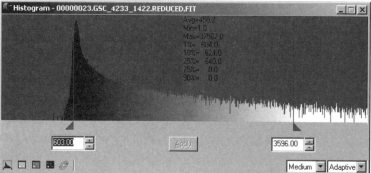

The buttons at bottom left and the drop-down lists at bottom right are presets that allow you to quickly find a histogram setting suitable for most images. See chapter 8 for details of how to use the CCDSoft Histogram tool. Other camera control programs offer their own versions of histogram tools. In MaxIm DL, for example, you do this with the Screen Stretch tool. CCDSoft's Histogram tool, however, is among the most flexible and is easy to use.

Whichever method you use, the background value sets the black point -- all pixels darker than the background setting will appear black. The range or maximum determines the white point -- all pixels brighter than the white point will appear white. The pixels darker and brighter than these values are still stored in the image, and you can use different black and white piont settings later if you wish to do so. Only pixels between the black point and the white point are displayed in various shades of gray. The higher the black point, and the lower the white point, the fewer pixels there will be sharing the gray levels, and the more detail you can see in the dim portions of the image. Of course, you lose very dim and bright detail when you do this, and that is why you will often find yourself making manual adjustments to these settings.

The automatic values calculated by the Background & Range dialog for the image in figure 1.2.5 were:

Background: 1946

Range: 3907 (White point: 1946+3907=5853)

These values are fine for the stars in the image, but the California nebula is very dim, and hardly shows up in the image. You can enhance how well dim pixels show up by using a slightly higher background setting (raise the black point a bit), and a shorter range (lower

FIGURE 1.2.7. THE CALIFORNIA NEBULA, WITH MANUAL ADJUSTMENT OF BRIGHTNESS AND CONTRAST.

the white point considerably). Figure 1.2.7 shows the result of these settings:

Background: 2380

Range: 650 (White point: 2380+650=3030)

CCDSoft and many other image processing programs also provide visual tools for adjusting the black point and white point. They provide a graph (the histogram) which you use to adjust the black point and white point. Figure 1.2.8 shows the MaxIm DL Screen Stretch window. The left triangle under the histogram

FIGURE 1.2.8. THE MAXIM DL SCREEN STRETCH WINDOW.

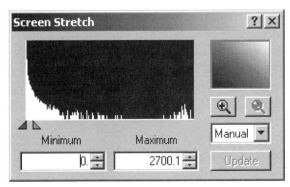

sets the black point, and the right triangle sets the white point. The drop-down list at lower right chooses among various histogram presets, or you can adjust manually.

In Mira, you would use the Stretch Palette Tool (see figure 1.2.9) to make histogram adjustments. Click the tool to activate it, and press the left mouse button down and hold it down while you move the mouse. This changes contrast and brightness. Or hold down the shift key while you move the mouse to change gamma and brightness.

See chapter 8 for detailed information on working with histograms.

TIP: Stars are very bright, and show up best with a large range setting. Nebulae and the arms of galaxies are dim, and have short brightness ranges. A low white point emphasizes these dim features, but at the cost of possibly "burning out" bright areas in the image by making them pure white.

In chapter 8, you will learn about non-linear histogram stretches. These allow you to balance dim and bright portions of the image more aggressively.

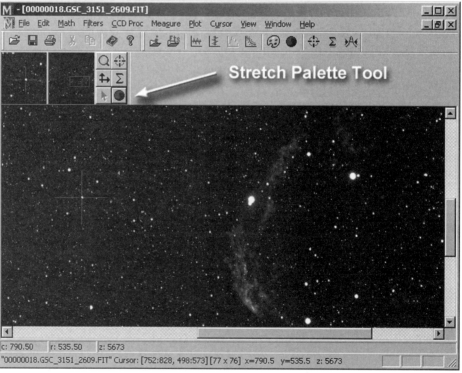

FIGURE 1.2.9. ADJUSTING BRIGHTNESS AND CONTRAST IN MIRA.

Section 3: What Can I Do with a CCD Camera?

If only it were so simple: attach a camera to a telescope and take incredible images of the objects that populate the night sky. Figure 1.3.1 shows an image of M16, the Eagle Nebula. It was taken with a ST-7E CCD camera, made by SBIG (Santa Barbara Instruments Group), through a 5" Takahashi refractor (FC-125). The sharpness, clarity, and detail of the image are typical of what you can expect under excellent seeing conditions using a competent CCD camera, an accurate mount, and quality optics. The purpose of this book is to show you how to take images that will look this good.

The image in figure 1.3.1 looks great because a number of things happened together:

- The air was stable.
- Light pollution was minimal.
- The optics were superb,
- The mount was rock steady and tracked accurately
- The focal length was appropriate to the subject and seeing conditions
- The exposure was long enough to bring out interesting details and suppress noise.

If any one of these things isn't true, the image quality will go down a bit but will still be pleasing. The further you get from these ideals, the less pleasing your images will be.

CCD Imaging Basics

The conditions listed above make up a laundry list of what matters most in CCD imaging. You'll learn a lot about each area in later sections of the book. What follows is a primer that will help orient you to the new world of CCD imaging.

Seeing Conditions

The steadier the air is, the better the results you can expect. When you go out in the evening to decide whether or not to image, the quality of the seeing is

FIGURE 1.3.1. AN IMAGE TAKEN WITH A 5" REFRACTOR AND AN ST-7E CCD CAMERA.

likely to be one of your top concerns. The steadier the air, the more likely you'll get an exceptional image.

For any given telescope, you will find that there is a limiting seeing condition. Focal length is the major determining factor. Short focal lengths are rarely limited by seeing. The longer your focal length, the more likely that seeing will be your limiting factor.

Generally speaking, the longer the focal length of your telescope, the more patient you must be to image successfully. With a focal length under 600-700mm you can image on just about any clear night. Short focal lengths are nearly invulnerable to seeing problems. At progressively longer focal lengths, the number of nights you can image successfully gets smaller and smaller. Depending on your location, focal lengths beyond 2000mm might preclude imaging on most nights.

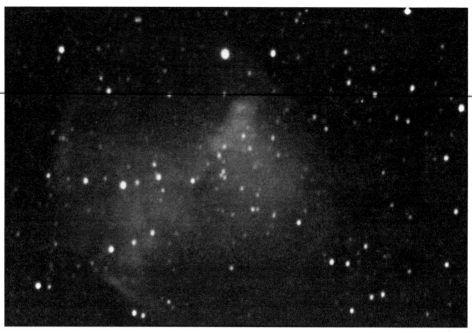

FIGURE 1.3.2. AN IMAGE OF M27 THAT SHOWS THE SOFTNESS TYPICAL OF POOR SEEING CONDITIONS.

The focal length of the telescope you use for imaging has a big impact on how often and how successfully you can image. If you want to image a lot, and like wide fields of view, a short focal length telescope will work well for you. If you want high magnification and have the patience to wait for those perfect nights, a longer focal length will better meet your needs.

Poor seeing makes your images look like they are out of focus. Figure 1.3.2 shows an image of the Dumbbell Nebula taken in poor seeing. Note that the stars are kind of fuzzy, and the overall image has a soft appearance. This is characteristic of imaging under slightly poor seeing conditions. If

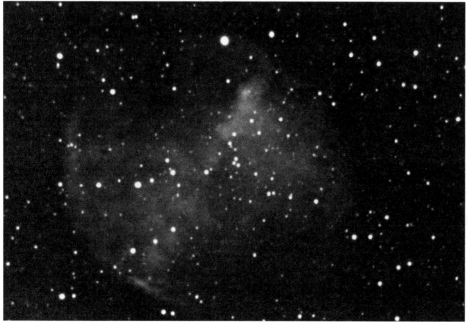

FIGURE 1.3.3. AN IMAGE OF M27 TAKEN UNDER BETTER SEEING CONDITIONS.

the seeing is really poor, the image will be even softer. It is easy to confuse poor seeing with poor focus. But when you simply can't seem to focus, it's likely that poor seeing is the cause of your trouble.

Figure 1.3.3 shows an image of M27 taken on a different night with the same equipment. The seeing was better than average on the second night, and the result is that the stars are more sharply defined, and more details are visible in the nebula. In fact, a number of dim stars are visible in the new image that are completely hidden in the first image. When the seeing is bad, details disappear.

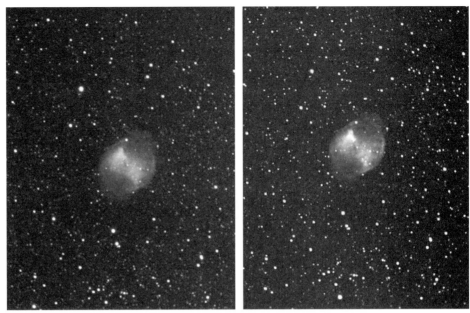

FIGURE 1.3.4. WIDEFIELD IMAGES ARE LESS AFFECTED BY POOR SEEING CONDITIONS.

Figure 1.3.4 shows how a short focal length "beats the seeing." The images in figure 1.3.4 were taken under poor seeing conditions, but at a focal length of 640mm. The level of detail is not as rich because the nebula is smaller, a natural consequence of using shorter focal lengths. But under poor seeing conditions, the detail isn't visible anyway. By imaging at a shorter focal length, a pleasing image can still be obtained. In fact, these images show more of the dim areas of the nebula than the images taken at longer focal lengths. A wider field of view often comes with a faster focal ratio, and you will get more dim details for a given exposure time with a fast focal ratio.

Bottom line on seeing conditions: The longer your focal length, the greater your dependence on the quality of the seeing and the larger your magnification. The faster your focal ratio, the faster you can image dim details, but with a lower magnification and wider field of view.

Light Pollution

Light pollution can make life miserable for astrophotographers using film. The brightness of the sky fogs the film, and limits the exposure time. The result is not very satisfying because you can't take deep exposures that would show all the cool stuff that's out there. You have to have a pretty dark sky to get excellent images with film.

Light pollution has an impact on CCD imaging, but you can take steps to overcome some of the limitations imposed by light pollution. The CCD equivalent of fogging is called saturation. Most CCD cameras don't saturate easily, so you can take long exposures even in a light polluted back yard. You can subtract the light pollution from the image, leaving you with an image. The image won't be as clear as one taken under a dark sky, but at least you have an image! The greater the light pollution, the less you will get, of course. But you can take good images with a CCD camera in light polluted conditions that preclude use of film.

The ability to cut through light pollution means that anyone with an average suburban backyard can image with a CCD camera. With extra effort, you can even image successfully from the center of a city!

The key to imaging through light pollution is to take long exposures. You can also take a large number of shorter exposures and combine them. Figure 1.3.5 shows what happens when you image from a light polluted location using a relatively short exposure time:

the light pollution nearly overwhelms the object.

The object is always brighter than the light pollution. The longer your exposure, the more readily you can leverage that small difference in brightness and make it work for you.

Figure 1.3.6 shows what happens when you subtract the light pollution from the image. The object (M27 in this example) stands out a little better, but it's not as sharp as in an equivalent image taken from a dark site. Even after subtracting the light pollution from a short exposure, you still don't get as much detail as you would like. The important point to note, however is that you can in fact subtract most of the light pollution.

FIGURE 1.3.5. LIGHT POLLUTION SHOWS UP ALL TOO CLEARLY IN A SHORT EXPOSURE.

When you take a longer exposure, and then subtract the light pollution, you can experience dramatic improvements. Figure 1.3.7 shows just how powerful the light pollution removal techniques in this book are. Half of the image shows the original image with all of the horrors of light pollution, and half shows the result of carefully removing the effects of the light pollution. (The bright vertical line is blooming.) The green coloring in the top half of the image is the result of light pollution from a nearby small town.

In addition to removing light pollution, you can do a little preventive work, too. Filters are available to cut out some of the worst light pollution, so that you have less work to do in cleaning up the image afterwards.

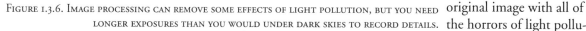

FIGURE 1.3.6. IMAGE PROCESSING CAN REMOVE SOME EFFECTS OF LIGHT POLLUTION, BUT YOU NEED LONGER EXPOSURES THAN YOU WOULD UNDER DARK SKIES TO RECORD DETAILS.

FIGURE 1.3.7. LIGHT POLLUTION CAN BE REMOVED FROM CCD IMAGES IF YOU KNOW HOW!

Bottom line on light pollution: Unlike film cameras, CCD cameras allow you to deal more effectively with the effects of light pollution. This won't eliminate the effects of light pollution, but you can reduce them significantly. Chapter 6 contains several options for removing the effects of light pollution from your images.

Optical Quality

It probably seems obvious to state that optical quality will make a difference in your images. However, I have heard many times on newsgroups that optical quality doesn't make a big difference. Speaking from my own experience, optical quality remains important, just as it is for visual and film use. The better the optical quality of your telescope, the better your images can be.

Note my careful use of the phrase "can be." If you don't optimize the rest of the system, you may not be able to see much difference between a good telescope and a great telescope. But once you learn your way around CCD imaging, you'll appreciate better optical quality. What CCD can do for you is make the most out of whatever level of optical quality you have. One of the first things to go with reduced quality is contrast. You can use image processing to enhance contrast, and get the most out of your telescope.

You can find optical quality at a variety of price ranges, but generally speaking the most convenient form of optical quality -- a high-end refractor -- is also the most expensive. By choosing carefully, you can get a scope with excellent optical quality without spending a huge amount of money.

For example, a high-end 6" APO refractor could run you anywhere from $9,000 to $16,000. You'll get superb optics, with virtually perfect correction. In addition, the optics will be extremely smooth, which improves contrast, which improves your ability to resolve subtle details. Since CCD cameras are especially good at detecting very, very small contrast differences, smooth optics are very important for the careful CCD imager.

But you don't have to spend that kind of money to get good optics. For example, a 7" or 8" Newtonian or Maksutov-Newtonian built around similar high-quality, highly-polished and smooth primary mirror might run you $1000 to $2000 (Newtonian) or $2500-4000 (Mak-Newt), less if you buy a high-quality mirror set and build the scope yourself. On the down side, the Newtonian design doesn't have the long back focus of a refractor, and the secondary mirror creates an obstruction in the optical path. But by jumping from 6" to 7" or 8", and going for high-quality mirrors, you can get wonderful images with the Newtonian design. The key is to aim for quality optics that fit your budget limits. Whatever telescope design you eventually wind up with -- and most designs work to some degree with CCD cameras -- optical quality always wins.

You will often here the phrase "aperture wins" in discussions of telescopes. This applies mostly to the visual realm. I have been imaging for over a year with 4" and 5" instruments of the highest possible quality, and I can assure you that such small instruments are capable of delivering stunning images. Aperture primarily impacts image scale -- bigger apertures tend to allow you zoom in on small objects. Small apertures make it easier to image wide fields of view. Figure 1.3.8 shows an image taken with a 4" refractor (Takahashi FSQ-106) of the Rosette Nebula. The image has exquisite detail because the refractor used had extremely high optical quality and a very flat field of view (not to be confused with the flat-field frame described in chapter 6). A flat field of view means that the telescope brings the view into focus along a flat plane, rather than a curved surface. The eye can accommodate differences in focus along a curved surface, but a CCD is flat. The larger the CCD chip, the more critical it is to get a large flat field.

The important components of optical quality are:

Sharpness - Do the optics deliver a sharp image? Use high power under steady skies on globulars, for example, to see how well the scope can resolve the stars in the cluster. Note: larger apertures will be naturally better at resolving detail; compare scopes of like aperture when comparing sharpness.

Figure 1.3.8. Even a small high-quality scope can deliver incredible images.

Contrast - Can you see subtle details visually and in your images? The full moon is a good test of contrast -- the ray structure will be more readily visible if the scope has good contrast. Compare an 8" SCT and a 5" APO refractor, perhaps at a star party, to see a range of contrast from average to superb. A secondary mirror may hurt contrast unless the design of the scope is very carefully thought out. Some superb imaging scopes, such as the Takahashi BRC-250, have huge obstructions but still manage superb contrast. It's harder to do it with a large secondary mirror, but it can be done. The bigger the secondary, the more carefully you should evaluate the effect on contrast. Don't rule out a scope for imaging because of a large secondary unless that large secondary hurts the contrast too much.

Smoothness of optics - Rough optics will scatter light and reduce contrast. The smoother the optics, the better able they are to resolve subtle differences in brightness. Smoothness takes time to do well, and is therefore most often found on expensive optics.

Color correction - Refracting telescopes must correct for the inherent color inaccuracies of their design. If you want a refractor for its smooth optics or convenient size, make sure it has superb color correction. Otherwise, colors will come to focus at different positions, and the CCD camera will faithfully record this flaw as halos around bright objects. Reflecting telescopes have zero problems with color correction, and as a result can be less costly for a given aperture than a refractor. Color correction is very costly to do well. Unfortunately, many reflecting telescopes sacrifice contrast, sharpness, or other things to keep their cost even lower. If you pick a reflecting telescope, pay a little more attention to its optical quality so you can be sure that it will be one that is suitable for CCD imaging.

Flatness of field - Your eye can accommodate a range of focus aberrations and still deliver a good image to your brain. A camera, however, records every focus problem faithfully. Many telescope designs deliver a sharp image at the center of the field of view, but a not-so-sharp image away from the center. This happens because the zone of critical focus is curved, and the camera's CCD chip is flat. The size of the area of sharp focus varies with the telescope type and the intent of the designer within a given type. In many cases, a field flattener (sometimes also called simply a corrector, and not to be confused with the corrector plate found on scopes such as Schmidt-Cassegrains) is available that will flatten the field for you. For example, the TeleVue Paracorr works well to flatten the field of Newtonian telescopes. Figure 1.3.9 shows an image taken through a high-quality 12.5" reflector without a corrector. Notice that the stars away from the center are elongated along a line from the center of the frame. This is called coma, and it is the result of a curved field. Figure 1.3.10 shows an image using a Paracorr. Coma is greatly reduced. There is still a bit of coma in this particular Paracorr image, but that is due to not getting the distance between the chip and the Paracorr exactly

FIGURE 1.3.9. IF YOU IMAGE WITH TOO LARGE OF A CCD CHIP FOR THE AVAILABLE FLAT-FIELD SIZE, YOU'LL GET COMA OUTSIDE THE FLAT FIELD AREA. THE INSET SHOWS A 2X BLOWUP OF THE TOP RIGHT CORNER.

right. A CCD camera with a smaller chip would image within the flat field at the center even with poor adjustment. Therefore the larger your chip, the more attention you must pay to such things. The larger the physical size of the CCD chip in a camera, the greater your concern should be about the size of the flat field. For any telescope, ask how flat the field is, and whether a field flattener is available for that telescope. An astrographs is a type of telescope that is specifically designed for imaging, and such scopes usually have the flattest fields available.

FIGURE 1.3.10. USING A CORRECTOR (A PARACORR IN THIS CASE) REDUCES FIELD CURVATURE AND THE STARS ARE SMALL, ROUND POINTS.

Accurate geometry/projection - Some telescopes will provide a flat field with good focus, but they won't project the sky evenly onto that field. For example, a doublet APO refractor is designed with superb color correction as the number one goal. The design may sacrifice the accuracy with which it projects the sky onto your CCD chip in order to achieve that superb color correction. You won't notice this until you try to assemble a mosaic from multiple images, and find that they don't quite line up the way you want them to. For the most accurate geometry in a refractor, go with a triplet or Petzval design unless you know that the doublet provides accurate projection.

Some eyepieces make similar compromises with respect to geometry. Such eyepieces sacrifice accurate geometry for the ability to provide a very wide field of view. Naglers are an example of this type of eyepiece, and you can see the geometry errors if you move the telescope while looking through the eyepiece.

Bottom line on optical quality: Whatever your budget, there is likely to be a type and size of telescope available that will have higher than average quality. That's the scope you should buy (see chapter 4 for more information).

Mount Quality

Of all the important components -- mount, telescope, and camera -- the mount is the primary key to success. Give me a superb mount, an average camera, and an average telescope, and I can give you good images. Granted, I can improve those images by getting a better camera and/or a better telescope. But without an adequate mount, you can't take good images at all. Perfect optics and a perfect camera would simply record the shortcomings of any mount perfectly.

A telescope mount that is fine for visual use may not work for imaging. The human eye can tolerate vibrations, inaccurate tracking, and a host of other flaws that mounts are heir to. The CCD camera, on the other hand, will faithfully record everything that is wrong or poorly adjusted in a mount. Everything.

The ways that a mount can go wrong make a long list. All of them interfere in one way or another with the mount's ability to track stars accurately, or the mount's ability to adjust its tracking speed accurately during guiding. Looseness can occur in the motors that drive the mount, and in the gears that transmit the motion from motor to axis. Individual gears can fail to mesh with their neighbors. Flexure can occur, so that the telescope isn't pointing where it should be pointing. The mount can fail to damp vibrations, or it might even amplify vibrations from a variety of sources (including the mount's drive motors). The details of how mounts can fail, and what to do about it, are covered in chapters 4 and 5.

> **TIP:** Many mount problems are non-fatal. You can adjust your mount to eliminate or reduce many common problems. Chapters 4 and 5 provide tips on how to get the most out of a mount. There are web sites for improving the more commonly used mounts.

It takes a lot of effort to design and built a mount that will track with the accuracy required for CCD imaging. The key areas of mount quality are:

- The worm and worm gear. These need to be cut with extreme precision for good tracking. They also should have a smooth finish to reduce the risk of random tracking errors.
- Big, high-quality bearings on RA and Dec axes. These support the load, and they must be smooth and move with the utmost precision.
- Good finishing of all surfaces that move. Burrs on gears, rough finish on bearings -- these kinds of things are what make many mounts unsuitable for serious astrophotography.
- Tight mesh between moving parts. The better made the mount is, the more closely the parts can mesh together. This is especially important for gears. If the gears are slightly out of round or irregular, they will have to be set further apart. This creates backlash, and the worse the backlash, the worse the performance of the mount. A little backlash is essential -- the gears must be free to move -- but too much is deadly.
- Lack of flexure. If the mount bends under typical loads, then it won't point accurately. Flexure can also affect polar alignment; a mount that flexes too much will not stay polar aligned, and will be unsuitable for serious imaging. It is also difficult to polar align a mount that flexes, since the measurements you take at various positions are not consistent with each other.

If budget is a concern, and it almost always is, you can reduce your need for the highest levels of mount accuracy by:

- Keeping the total weight of your imaging setup as low as possible (small, light telescope and camera).
- Using a short focal length telescope (under 650-750mm).

If you want to put a heavier load on your mount, or use longer focal lengths, you'll need to invest more heavily in a good mount. Speaking of "investing," it's not uncommon to spend 50% or more of your total equipment budget on the mount.

Bottom line on mount quality: It's not possible to overstate the need to have a stable mount for imaging.

Focal Ratio

You might think that aperture is the most important consideration in choosing a telescope. However, with the extreme sensitivity of most CCD cameras, it's not as important as you expect it to be. Focal length, mentioned earlier, plays an important role because it determines the image scale. But focal ratio (the ratio between aperture and focal length) plays a dominant role in choosing a telescope for CCD imaging.

Let's look at the focal ratio of some typical telescopes. Many SCTs (Schmidt-Cassegrains) have focal ratio of f/10. That is, the focal length is ten times the aperture. For example, an f/10 8" SCT has an aperture of 200mm, and a focal length of 2000mm. A 4" f/6 refractor has an aperture of 100mm, and a focal length of 600mm. You might expect that the 8" SCT would capture more light, and provide shorter exposure times. In fact, the opposite is true: the f/6 refractor will require shorter exposure times.

Telescopes with CCD cameras are just like regular cameras in this regard. A camera with a lens set to an f-stop setting of f/2 will require a shorter exposure than one set at f/5.6. The exposure time is based on the focal ratio, and the focal ratio alone.

Focal ratios are described as fast and slow. A faster focal ratio allows more light to enter for a given aperture. Focal ratios from approximately f/1 to f/5 are considered fast. You won't find telescope much faster than about f/2, however. Focal ratios of f/8 and beyond are considered slow. Focal ratios between f/5 and f/8 are middle-of-the-road. With any given telescope, adding a Barlow will make for a slower focal ratio. Adding a reducer will make for a faster focal ratio.

There is a limit to how much you can change the focal ratio with Barlows and reducers. There is always some price to pay for altering the native focal ratio. Barlows can magnify flaws as well as the image, for example. And focal reducers can cause vignetting (darkening) in the outer edges of your images. As with many things, better quality Barlows and reducers do a better job at avoiding these kinds of problems, or at least minimizing them when they are unavoidable.

What this means is that two telescopes with the same focal ratio will require the same exposure times, even if there is a difference in aperture. The scope with the larger aperture will provide more magnification. The section "Focal Ratio Is King" in chapter 4 has examples that show how and why this is true. Here are some ways that focal ratio can influence your choice of telescope for CCD imaging:

A telescope with a fast (f/2 to f/5) focal ratio is better for suburban locations, or any location that has significant light pollution. You will get more light for a given exposure. When you subtract the light pollution, you will be left with more of what you are interested in. You can still choose a slower focal ratio (f/6 or more), but you will need longer and longer exposure times. The greater your light pollution, the longer the exposure time will have to be for a give focal ratio to compensate.

A telescope with a fast focal ratio will deliver a wider field of view than one with a slow focal ratio. If you want the ease-of-use of a wide field instrument, choose a small apertures and fast focal ratio. If you want magnification, choose larger apertures and slower focal ratios.

A telescope with a slow focal ratio will provide a higher level of magnification for a given aperture. When the seeing conditions are good, such a telescope will provide stunning levels of fine detail with sufficiently long exposures. It will also require a better mount to hold it steady.

A telescope with a fast focal ratio is harder to make, and it is harder to make with a large flat field. Some fast telescopes, in fact, don't make the grade for CCD imaging at all. Verify the size of the flat field with the manufacturer of a scope with a fast focal ratio. See if a field flattener is available that would make the scope suitable for imaging.

On the other hand, many astrographs feature a wide field and a fast focal ratio, and include the necessary field flattening as part of the design. Astrographs are telescopes that are designed specifically for imaging. Many (but not all) astrographs provide a combination of a wide field of view, a fast focal ratio, and a large flat field. This combination makes them ideal for imaging, and they are a lot of fun to use. However, they usually cost more than visual-only telescopes because they require additional manufacturing time and costs.

Bottom line on focal ratio: The faster the focal ratio, the shorter your exposure times. This is true no matter what the aperture of the telescope is; exposure time is totally dependent on focal ratio alone. The slower your focal ratio, the greater the magnification and the longer the required exposure.

Exposure Duration

With a film camera, exposure is a critical choice. If the exposure is too short, details will be lost. If the exposure is too long, the film could become over-exposed. Exposure is less critical with a CCD camera.

First, a CCD chip is more sensitive than film. Exposures of just 5-20 seconds will usually show the presence of galaxies and nebulae. Such exposures are too short to show good detail, but they do demonstrate the incredible sensitivity of the CCD chip compared to film.

Figure 1.3.11 shows four short exposures using different CCD cameras and telescopes. Clockwise from top left:

- M51 - a 45-second exposure with a 4" refractor at f/5, ST-8E camera.
- M46 - a 10-second exposure with a 4" refractor at f/5, ST-9E camera.
- M82 - a 5-second exposure with a 16" Newtonian at f/5, ST-9E camera.
- M42 - a 10-second exposure with a 4" refractor at f/5, ST-8E camera.

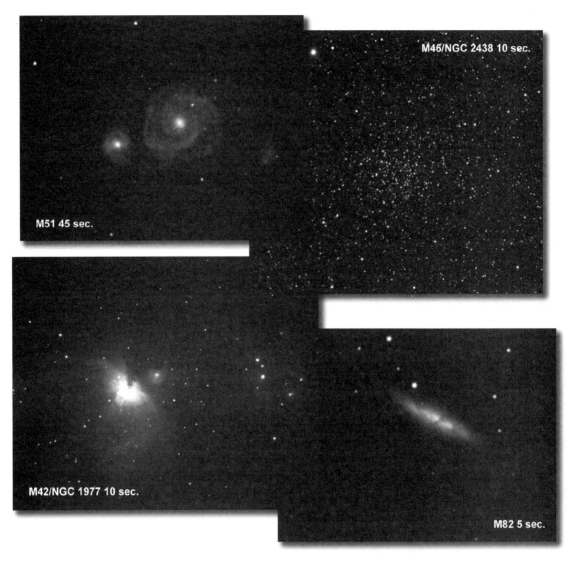

FIGURE 1.3.11. FOUR SHORT EXPOSURES AT F/5.

Telescopes with slower focal ratios require longer exposures to get the same results, and those with faster focal ratios would show even more detail in such ultra-short exposures. The main point is that, even with very short exposures, you will get results with a CCD camera.

That said, the reality is that longer exposures are almost always better. A long exposure will reveal more details than a short exposure will. Dim areas that seem empty in short exposures will show up as having interesting details in a long exposure. Long exposures are also less noisy because signal increases faster than noise.

As a result, longer exposures have less graininess, especially in the dimmer details.

Figure 1.3.12 shows images of M42 and M51 that involve much longer exposures. Note that the longer M42 image reveals a much greater extent of nebulosity around M42, as well as intriguing details in the nearby nebula NGC 1977. The M51 image not only reveals more details in the galaxy itself, but it also reveals the existence of dim streamers of stars around the smaller of the two colliding galaxies.

Figure 1.3.12 makes a key point about CCD exposures: you can get results with short exposures, but you can get superb results with long exposures. The natural question to ask is: how long? The answer is simple: as long as possible.

There are some things that will limit the length of your exposures. If you have a camera that features antiblooming, exposures are limited only by sky glow and your patience. If you have a non-antiblooming camera, your exposure length is limited by the time it takes for stars to bloom objectionably. You may also find that cosmic ray hits, satellite tracks, and other hazards are more common on longer exposures. You can limit your exposure length to reduce your risk.

If you use a non-antiblooming camera, you can take exposures limited by the level of blooming you are willing to tolerate. You can then combine those exposures to get results similar to (but not identical to) what you can get with single long exposures.

Non-antiblooming cameras accumulate light more evenly (this is called a linear response), and should always be used for scientific measurements.

As a practical matter, you can get good images with, and enjoy using either type of camera. Antiblooming cameras are easier to use, but if you want to do science with your camera (astrometry and photometry), you should get a non-antiblooming camera. Not all cameras are available in both versions. The antiblooming cameras will require increased exposure times, especially from light polluted locations, but they are also better able to take long exposures to overcome light pollution.

Bottom line on exposure duration: Longer is almost always better. When you can't do long, do a large number of multiple exposures and combine them using your camera control software. Long exposures require guiding; see chapter 5. If you can't decide between antiblooming and non-antiblooming, the antiblooming camera is the safer choice overall.

FIGURE 1.3.12. LONG EXPOSURES OF M42 AND M51 SHOW MUCH MORE DETAIL.

This might make it sound like you should always get an antiblooming camera, but its not that simple. You can find more details than you ever thought possible about blooming and antiblooming cameras in the next section, but the short version goes like this:

Antiblooming cameras allow long single exposures. They are 30% less sensitive than non-antiblooming cameras, but the convenience and quality of single long exposures mostly balances this out.

2 Practical Focusing

CCD cameras represent some pretty fancy technology, but in some ways they are just like ordinary cameras. As with a traditional film camera, the difference between a snapshot and a great photograph lies as much with the photographer as with the equipment.

CCD focusing is done in an iterative fashion, taking test exposures and then examining the results. The various camera control programs allow you to speed up the process so you can get focusing done efficiently.

Section 1: Focusing Fundamentals

FIGURE 2.1.1. AN IMAGE WITH GOOD FOCUS CAN BE STRIKING, SHOWING A HIGH LEVEL OF DETAIL.

You may have heard that focusing is difficult, or time consuming, or just plain annoying. It doesn't have to be that way. You do need to be extremely careful when focusing to get the best possible results. A little extra time spent focusing is well worth the effort. Good focusing is one of the single most important ingredients in taking a good image, and it is totally under the control of the operator.

In this chapter, you will learn everything you need to know to focus effectively. You'll learn:

- Why focusing is critically important.
- How to achieve best focus every time.
- Software aids for achieving best focus.
- Hardware aids for achieving best focus.
- Alternative focusing techniques.

Focusing a CCD Camera

The human eye is a marvelous instrument. It is capable of focusing over a wide range of distances. In addition, because our eyes function cooperatively with our brain, they are capable of accommodating to rapidly changing conditions. When you use your eye to look at a celestial object through a telescope, your eye can adjust to slight focus errors with little or no trouble. It can even adjust to rapid variations in focus caused by turbulence in the atmosphere, if the turbulence isn't too rapid or severe.

A camera, on the other hand, has no ability to accommodate even a slightly out-of-focus image. CCD cameras are no different from film cameras in this regard. If an image is out of focus, the photons are scattered over a wider area, and dim stars and faint details are lost.

Precise, accurate focusing is critical to success. Unfortunately, focusing any astro camera requires a little more work than focusing an ordinary camera. When you are taking a snapshot of the Grand Canyon, you have plenty of light to work with while you focus, and so does the automatic focuser found in most of today's cameras. Professional film cameras allow you to focus by looking at a ground glass screen that is conveniently located exactly the same distance from the camera lens as the film. The scene is bright, and focusing is relatively easy. Your camera might even have some optical tricks installed to make it even easier, such as split-ring focusing.

Working at night under the stars, there's not as much light to work with. The whole idea of using a camera is to take long exposures to capture as many of those scarce photons as possible. Focusing a film camera for astrophotography is a challenging task, quite unlike the daytime equivalent. Automatic focusers don't have enough light to work with, and a ground glass screen is suddenly too dim to focus with easily. Film-based astrophotographers have developed special devices to help them achieve good focus.

Instead of a ground glass screen, a CCD camera displays its results right on your computer screen. This means you aren't limited to the available light. You can collect photons over time while you are focusing, and this is a definite advantage over film. Still, there is significant work to be done in achieving good focus. No matter what type of object you wish to image with a CCD camera, most of the time you use a star for focusing. It is more challenging to focus on extended objects such as nebulae and galaxies because of their lower contrast. Solar system objects are a special case; see chapter 3 for details.

Stars are point sources of light. It would be ideal if stars became true points at perfect focus, but that's not the case. The image of a star on a CCD chip is actually a small disc. A number of factors contribute to this spreading out of the light from distant stars:

- The laws of optics describe how a point source of light gets spread out because of diffraction. See the section Diffraction Effects below for more information about diffraction.
- Turbulence in the atmosphere scatters light. The amount of scattering varies from night to night, even from hour to hour and minute to minute.
- Every telescope has at least some level of optical aberration. This can interfere with achieving perfect focus.
- Many telescopes require user collimation. If collimation isn't perfect, you won't be able to achieve best focus no matter how hard you try. Chapter 3 provides detailed help for collimating the telescopes most commonly used for CCD imaging.

On any given night, one or all of these factors will dominate. So "best focus" becomes an elusive prospect. And because the seeing is always tossing your star image around (a little or a lot), it takes some know-how to tell when you've got best focus. In this section, you will learn what it takes to achieve optimal focus for CCD imaging.

Diffraction Effects

The image of a star in a telescope eyepiece is in no way indicative of its actual diameter. All but a few stars are simply too far away to see as a disk even with fancy professional equipment. Yet if you magnify any star image, you will see a small disk. This is the result of diffraction effects. A photo, on film or with CCD, is a time exposure. That introduces a few other considerations, such as air turbulence, that increase the size of this disk. Those topics are covered elsewhere in this chapter.

Light is made up of photons. They aren't particles in the way we think of particles at our human scale. They also behave like little waves. This dual nature is common to the stuff you find at the subatomic level. It gives rise to phenomena that are sometimes non-intuitive. Diffraction is one of those phenomena. Wrap your mind for a moment around the idea of very small particles moving in a wavelike manner.

Waves are able to reinforce and cancel one another. If the peaks of two waves coincide (called reinforcement), the resulting wave is the combined height of the two peaks. If two troughs meet, the result is an extra-deep trough. Similarly, if a peak and a trough meet, they will cancel each other out, and the result is a flat spot.

These rules about waves also apply to photons. When two photons have waves that reinforce each other, you get a bright spot. When the waves cancel each other out because they coincide peak to trough, you get a dark spot.

When light passes through an aperture, the light waves are diffracted, changing the way the peaks and troughs interact at the focal plane. The net effect of these interacting photons results in a diffraction pattern caused by light waves reinforcing and canceling each other to varying degrees. In the case of the round objective of a telescope, this is a bright center surrounded by alternating dark and bright rings.

Photons have a very small wavelength, and they combine in a variety of ways as they come to focus at the back of the telescope. For a star, most of the light waves will combine to form a central bright spot. There will be some slightly off-axis photons resulting from diffraction, however, and they will travel a slightly different path to the focal plane. These will meet on-axis photons and depending on the distance from the optical axis, they will cancel or reinforce each other. This creates the bright and dark rings around the bright central spot. Each ring is less bright than the previous one because most of the light energy is toward the center of the pattern.

Figure 2.1.2 shows three examples of diffraction rings. The image on the left is an enlarged view of a star in an optically perfect refractor. The image is simulated with Aberrator 3.0, a software program that shows the effects of many types of optical defects on the image of a star or planet. You can download Aberrator from **http://aberrator.astronomy.net**.

The left image shows a bright center disk surrounded by successively dimmer diffraction rings. Most of the rings are too dim to show up clearly. The image in the middle shows the effect of a large central obstruction, such as the secondary mirror of a Schmidt-Cassegrain. The disk is slightly dimmer and the diffraction rings are slightly brighter. This reduces contrast, and explains why high-end refractors are desirable for imaging. The image at right shows what poor optics do to the diffraction pattern. Poor optics put more light into the diffraction rings, and they disturb the even distribution of that light in and around the central disk.

The central disk is called the Airy disk, named after a 19th century mathematician, George Airy, who established the mathematics behind the effect.

With perfect optics, 84% of the light from the star will be in the bright center, another 7% in the first ring, 3% in the second ring, etc. A telescope with a central obstruction, will spread the light out into the diffraction rings as illustrated by figure 2.1.2. Poor optics will cause a similar effect.

When the central spot is not as bright as the diffraction rings, it is because more of the light energy winds up in the rings. This is why a star test is such a useful way to analyze the quality of optics in a telescope: the diffraction rings tell the story of how good or bad the optics are. An experienced observer can often analyze a slightly out of focus diffraction pattern and determine the cause(s) of the non-standard pattern.

Air turbulence will also affect the appearance of the diffraction pattern. Figure 2.1.3 shows in-focus stars with the diffraction rings muddled by various degrees of turbulence. The images at left are the least affected by turbulence, and the images on the right are the most affected. The top images simulate a refractor, and the bottom images simluate an SCT. The obstructed telescope's image (SCT) breaks down with less turbulence. The same is true for optical quality: the poorer the optical quality, the more susceptible the telescope will be to loss of contrast and detail from turbulence.

FIGURE 2.1.2. THEORETICALLY PERFECT DIFFRACTION RINGS (LEFT); DIFFRACTION THROUGH A SCOPE WITH A 35% CENTRAL OBSTRUCTION (CENTER); DIFFRACTION THROUGH POOR OPTICS (RIGHT).

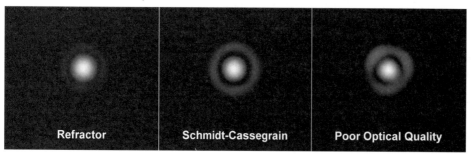

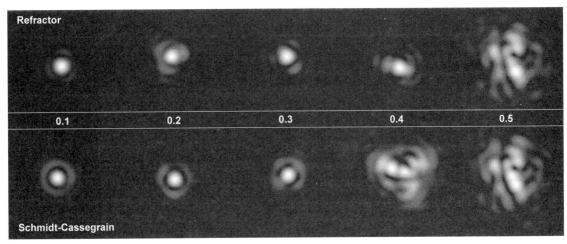

FIGURE 2.1.3. THE EFFECT OF INCREASING TURBULENCE ON DIFFRACTION PATTERNS IN DIFFERENT TELESCOPES.

No optical system, no matter how perfect, can ever do away with diffraction to create point-like images of stars; the wave nature of light won't allow it. Even perfectly focused, there will always be some spreading out of the light into a diffraction pattern. Thanks to turbulence, the image gets spread out even more. On any given night, turbulence (seeing) will limit the potential image quality for both film and CCD imaging. The shorter the focal length of your telescope (or the larger your camera's pixel size), the lower the magnification factor and the less impact turbulence will have.

Figure 2.1.4 shows a small section of an image captured with a CCD camera. You don't see any diffraction rings around the stars because two things conspire to hide them. One, the diffraction rings are very small, and would only show up at very high magnification. Most CCD imaging is not done at such high levels of magnification. Two, turbulence is nearly always the dominant factor, and it smears diffraction rings (and

FIGURE 2.1.4. STARS DON'T SHOW DIFFRACTION RINGS IN IMAGES.

the star's image) into a larger circle. The brighter the star (and the longer the exposure), the larger the circle. Why does brighter equal larger? The longer the exposure or the brighter the star, the larger the number of scattered and diffracted photons that will be recorded.

Optimal Focus Position

How important is optimal focus? Figure 2.1.5 shows an image of the globular cluster M13 in excellent focus. Notice how the individual stars are tiny points of light, and you can easily see darker spaces between the stars, even near the core of the cluster. There are large numbers of both bright and dim stars.

FIGURE 2.1.5. THE GLOBULAR CLUSTER M13, SHOWING AN EXAMPLE OF EXCELLENT FOCUS.

Figure 2.1.6, on the other hand, shows a globular cluster with less than perfect focus. Star sizes are larger, and they are not nearly as crisp and attractive. Fewer dim stars are present. Poor focus spreads out the light, and for dim stars this is enough to make them disappear entirely.

FIGURE 2.1.6. THE GLOBULAR CLUSTER M92 WITH POOR FOCUS.

The difference between these two images is a very, very tiny amount of focuser movement, yet the difference in appearance in enormous. Nothing you do on any given night will have more impact on the quality of your images than focusing. Taking the time to get perfect focus will make a big difference in the appearance of your images.

The care you must take to achieve perfect focus changes with the characteristics of various types of telescopes. The type of focuser in use, the arrangement of mirrors and/or lenses, and the focal ratio of the telescope all affect how hard you will have to work to achieve optimal focus.

One of the difficult challenges in CCD imaging is making very small focus changes. Some focuser designs make this easy, but many do not. For example, refractors typically use rack-and-pinion focusers. These are effective for visual use, but they are often too coarse for CCD imaging. What works well for the eye doesn't necessarily work well with a CCD camera. Schmidt-Cassegrain telescopes present a different kind of focusing problem. They use a moving primary mirror to achieve focus. The mirror can shift after you focus, or when you change the direction of focuser travel. This isn't a big problem for the eye, but it can be frustrating when focusing with a CCD camera.

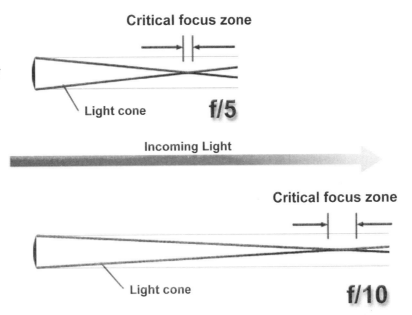

FIGURE 2.1.7. TELESCOPE FOCAL RATIO DETERMINES THE LENGTH OF THE CRITICAL FOCUS ZONE.

TIP: CCD imaging requires extremely fine control over focus position. Many telescope designs do not have an optimal focusing mechanism for CCD imaging. In most cases, there are alternative focusing mechanisms or techniques available that will help you achieve perfect focus. The details are covered later in this chapter.

Focusing Explained

The human eye is a marvelous instrument. Not only can it follow rapidly moving objects, but it can also adjust focus in real time, compensating for minor focus drift, curved fields, or minor out of focus conditions. The eye is a very forgiving focuser.

A CCD camera, on the other hand, is very unforgiving. You will need very precise focus to get sharp, detailed images from your CCD camera. Precise focus is also important for research (astrometry and photometry) so that you can accurately measure the brightness of objects.

Camera control programs have a variety of built-in features that will allow you to achieve the best possible focus. But there are also mechanical considerations that play a role in how effectively you can achieve focus.

"Critical focus" is actually a range of focuser positions, not just a single, exact spot along the focuser's range of travel. Since turbulence, diffraction, and other factors spread out the star's light into a disk, any focus position that provides the smallest possible disk will suffice. In other words, "critical focus" is actually a range of focus positions. Any position in the critical focus zone provides the best possible focus.

The actual size of this zone is based on the focal ratio of your telescope. The faster your focal ratio, the shorter the zone of critical focus is. For the purposes of focusing, a fast focal ratio is f/5 or lower; a slow focal ratio is greater than f/8. Scopes with a fast focal ratio will be more challenging to focus, and scopes with slow focal ratios will be a little more forgiving.

TIP: The focal ratio is the ratio of the telescope focal length to the aperture. For example, a telescope with an aperture of 100mm and a focal length of 500mm has a focal ratio of f/5. Lower-numbered focal ratios are said to be "fast" because they require shorter exposure times.

Figure 2.1.7 shows light cones for fast (f/5) and slow (f/10) refracting telescopes. The front of the hypothetical telescopes is on the left. Light enters from the left through the telescope objective, and comes to a focus where the converging lines cross. Other types of telescopes will have similar critical focus zones, but the light path be shaped differently. The f/5 light cone is steep; as you move away from the front of the scope, the light converges to focus very quickly. The f/10 light cone is shallow, and does not converge as rapidly as you move away from the front of the telescope. For all telescopes, the light diverges again past the focal point.

The telescope's focuser, manual or motorized, moves the CCD camera back and forth. The CCD detector inside the camera must sit within the critical focus zone to get a sharp image. To bring the CCD camera to focus, the focuser moves the camera so that the CCD detector is within the critical focus zone, as shown in figure 2.1.8.

Since telescopes with fast focal ratios have a smaller critical focus zone, you can focus more effectively if your focuser allows very small adjustments. You can improve your ability to focus a fast telescope by adding a motorized focuser.

TIP: Not all motorized focusers are ideal for CCD imaging. The best motorized focusers provide very fine control over focus position, sometimes using gears or a Crayford style add-on focuser to give you a high degree of control. Motorized focusers that simply move your existing focuser may or may not have fine enough control to give you an advantage. This is especially true with the rack and pinion focusers of refractors, which are relatively coarse without some kind of reduction gearing.

CCDSoft includes a focusing utility, @Focus, which can use a variety of motorized focusers to automatically find a position in the critical focus zone. The best focusers for use with @Focus are capable of making extremely small movements (the minimum movement is called the step size). Telescopes are susceptible to various kinds of focus drift, from looseness in the focuser to physical shrinking as temperatures fall during the night. A motorized focuser can help you adjust focus during a long imaging session.

When the seeing is very good (i.e., the atmosphere is stable), star images hold steady and are smaller and sharper. As the seeing deteriorates, star images begin to bloat up and lose their hard edge. This makes it more difficult to measure focus. CCDSoft's automatic focusing tool, @Focus, can average multiple images to help it converge on the best focus position even under poor seeing conditions. Manual focusing also becomes harder under poor seeing conditions. You wind up going through focus repeatedly, never quite sure where best focus is.

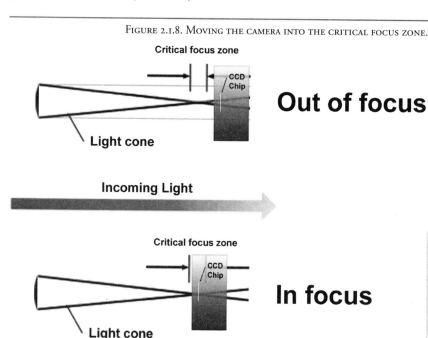

FIGURE 2.1.8. MOVING THE CAMERA INTO THE CRITICAL FOCUS ZONE.

TIP: If the zone of critical focus is very small, you may find it difficult to achieve critical focus reliably or easily using a manual focuser. There are several different types of motorized focusers available.

SECTION 1: FOCUSING FUNDAMENTALS

A motorized focuser can provide smaller adjustments in many cases than you can do by hand, and is thus better at making the small adjustments needed to position the CCD detector inside the critical focus zone. To get useful results from a motorized focuser, it should enable you to make movements that are no larger than one-half the size of your zone of critical focus. For optimal focusing, especially on nights of really good seeing, look for a focuser that will move in increments that are one-quarter or less the size of your critical focus zone.

The equation below computes the size of the critical focus zone (CFZ) in microns for a "perfect" optical system that is perfectly collimated:

$$CFZ = focal\_ratio^2 * 2.2$$

So for a C-14 at f/11:

$$CFZ = 11^2 * 2.2 = 266$$

266 microns is 0.266 millimeters or 0.01 inches. Here's an example with an f/5 focal reducer on the C-14, making it effectively an f/5.5 system:

$$CFZ = 5.5^2 * 2.2 = 66.5$$

66.5 microns is 0.0665 millimeters or 0.0026 inches. Note that the critical focus zone of the f/5.5 scope is one quarter that of the f/11 scope. This means that if you reduce the focal ratio by half, you reduce the close focus zone (CFZ) to one quarter.

Note: real world values for the CFZ are approximately 10%-30% greater than the theoretical values above.

Automated Focusing

CCDSoft includes two methods of focusing: automatically using @Focus, and manually using the Sharpness graph on the Focus Tools tab. To use @Focus, you must have a computer-controlled focuser (e.g., RoboFocus, Optec TCF-S), or a motorized focuser controlled by your mount (e.g., JMI NGF-S or motofocus connected to a Paramount GT-1100s, LX200, or Astro Physics GTO).

To access CCDSoft's focusing features, click on the Focus Tools tab of the Camera Control panel. Figure 2.1.9 shows the layout of the Focus Tools tab.

For a step-by-step example of using @Focus, see the Alternative Focusers section later in this chapter.

The Implications of Focal Ratio

Focal ratio describes the relationship between the aperture of a telescope and the focal length. For example a telescope with an aperture of 100mm and a focal ratio of f/8 would have a focal length of 800mm. The shorter the focal length, the wider the field of view. The longer the focal length, the greater the magnification for prime focus photography.

Telescopes of similar focal ratio have characteristics in common, even if the apertures of the two telescopes are different. For CCD imaging, a fast focal ratio allows you to take shorter exposures. Because a fast

FIGURE 2.1.9. THE FOCUS TOOLS TAB OF THE CAMERA CONTROL PANEL.

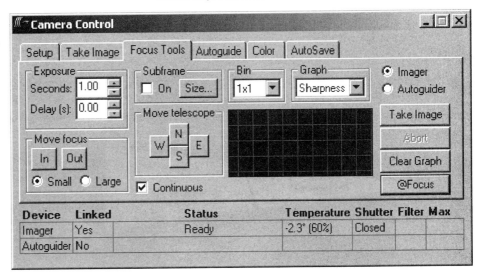

focal ratio also means a wide, diverging light cone, it will provide a wide-angle view of the sky. This allows you to image large objects, such as the moon. Or you can image multiple objects at once, or large nebulae such as the North American Nebula.

To get serious magnification with a fast focal ratio, you will need to use a scope that has a large aperture. For example, a 4" f/5 refractor equipped with an ST-8E camera will provide a field of view that is 1 x 1.5 degrees. A 16" f/5 Newtonian, on the other hand, will provide a field of view that is much smaller: 23.3 x 15.5 arcminutes.

A slow focal ratio telescope, on the other hand, will have a gently sloping light cone, and will cover a small area of sky. Small objects will fill the field of view. This is ideal for close-up images of a galaxy like M51 or a nebula like M16, the Eagle. If you want a wider field of view with a slow focal ratio, you'll need a small aperture. For example, an 8" f/10 SCT with an ST-9E has a field of view that is 17.6 x 17.8 arcminutes. A 3.5" f/10 telescope would have a field of view of 40 x 40 arcminutes with the same camera. Figure 2.1.11 shows some examples of different fields of view.

FIGURE 2.1.10. EXAMPLES OF CELESTIAL OBJECTS WITH DIFFERENT ANGULAR DIAMETERS.

Section 2: Acquiring Images

Focusing for CCD imaging involves multiple steps. It's an iterative process that involves gradually refining focus until you've got the best possible focus. Many times, you will go back and forth through best focus so you can find out where it is. The trick is to then move back to that perfect focus position.

You can determine focus in a variety of ways, and we'll look at these methods in this chapter:

- Visual focusing (quick but not always the most reliable)
- Software-assisted focusing (not as quick, but can be very reliable with the right focuser)
- Hardware-assisted focusing (slowest, but very reliable)

Figure 2.2.1 shows some typical star images in and out of focus. The telescope used for the example on the left is a refractor. The well-out-of-focus image is a broad circle with faint diffraction rings visible between the center and the edge. (Out of focus is about the only time you will encounter diffraction rings in your imaging.)

The example on the right shows a focused image that is typical of a telescope with a secondary mirror supported by a spider, such as a Newtonian. The length and thickness of the diffraction spikes will vary from one telescope model to another. Other types of telescopes will have their own characteristic out-of-focus appearance. Schmidt-Cassegrains, for example, will show the shadow of the secondary mirror if you are far enough out of focus.

TIP: Focusing by eye is a challenge. Always choose a bright star to do your visual focusing. Exactly how bright a star to choose depends on many factors, including the focal ratio of your telescope, the sensitivity of your camera, the seeing conditions, etc. You want a bright star that will not saturate your camera. See chapter 6 for details on calculating the saturation level of your CCD camera.

Figure 2.2.2 shows a succession of star images, ranging from moderately out of focus at left to accurately focused on the right. Notice that dim stars are invisible when you are out of focus, and as you get to critical focus more and more dim stars appear in the image.

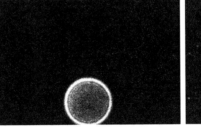

FIGURE 2.2.1. IN A REFRACTOR (LEFT), THE OUT-OF-FOCUS STAR IS CIRCULAR. IN MANY REFLECTORS (RIGHT), SPIDER VANES ADD DIFFRACTION SPIKES.

The simplest (but not necessarily the easiest) way to visually focus for CCD imaging is to observe a star image in your camera control software while making adjustments to the telescope's focuser. There are several attributes of a star image that you can use to determine when you are in focus, including the size of the star image and the nature of its border.

FIGURE 2.2.2. THE CHANGING APPEARANCE OF STARS AS FOCUS IMPROVES.

The size of the star image – As you get closer to the best focus point, the star image will shrink to a smaller size, as shown in figure 2.2.2. Although the star is a point source, you will not be able to shrink a bright star down to a very small point no matter how perfectly you focus the telescope. The brighter the star, the larger it will be in your image. The movement of air in the atmosphere spreads out the light from the star and causes the stars to twinkle (the technical term is scintillation). If the air is very steady, you will get smaller star images ; when the air is turbulent, you will get larger "bloated" star images. In later sections, you'll learn about Full Width at Half Maximum (FWHM), a tool that measures the width of the star image. FWHM isn't as complex as it sounds. If you were to plot the brightness of a star as a curve, the FWHM is the width of the curve at half brightness.

TIP: To judge focus accurately, first estimate how much the star's size is being affected by turbulence. If there is a wide range of focuser movement that shows no improvement in focus, then turbulence is making star images larger than normal. Focus will be difficult to optimize (longer focal lengths are more prone to this). Likewise, if very small amounts of focuser movement show changes in star size, seeing is more stable and you will have small, tight star images. Such conditions lend themselves to capturing excellent images.

The border of the star image – When a star is out of focus, there will appear to be a small cloud or halo around it that indicates poor focus. You may need to adjust the image contrast to see this border area. Figure 2.2.3 shows two highly magnified images of a very bright star. Can you guess which star image is in focus, and which is not? It's easier doing it with real software, because the difference is very subtle. It takes time to develop an eye for the differences.

TIP: The method you use to adjust image contrast is similar in various camera control packages, but the names of the adjustments differ. These are just different names for the same things. For example, in CCDOPS and CCDSoft you adjust the settings for Back and Range; in Maxim/DL you adjust Minimum and Maximum. See chapter 8 for more information.

The left image in figure 2.2.3 is only a little bit out of focus. The edge of the star in the left image is just a little less distinct than in the right image. The trick here is that an out of focus star image does not have as hard an edge as an in-focus star image. If the star is too bright, it will be harder to see the difference. If the star is too dim, it will disappear entirely if you move outside of focus.

To see the visual difference between in focus and out of focus most clearly, zoom in on the star image to get a good look at the edges of the star. If they are soft, you are still outside of critical focus. If the edge of the star shows a sharper cut-off between the star and the background, you are very close to critical focus. This technique requires experience to do well, however. There is always some degree of fuzziness at the edge, and learning how much is just right takes trial and error. The difference between focused and not focused can be very subtle, and hard to distinguish for the unpracticed eye. That is why more sophisticated focusing techniques have been developed, such as Hartmann masks, diffraction focusing, using dim rather than bright stars for visual assessment, automated focusing, etc.

To complicate bright-star focusing further, two other conditions have symptoms that are similar to the out of focus star: turbulence and poor collimation. Turbulence in the atmosphere will scatter the star's light – the greater the turbulence, the greater the scatter. The results of turbulence are nearly identical to poor focus.

 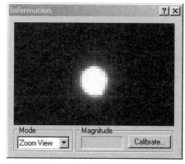

FIGURE 2.2.3. THE RIGHT IMAGE IS SLIGHTLY BETTER FOCUSED. IT SHOWS SUBTLE SIGNS OF A HARDER EDGE.

Poor collimation also scatters light, but in a characteristic fashion. In focus, poor collimation shows up as a stretching of the star image radially, with a gradual fading along the axis of the stretch (see figure 2.2.4). Slightly out of focus, the star image forms a ring with one side of the ring being brighter than the other. If the focus is too far out, you won't see this effect. When a scope that is poorly collimated is used for imaging, the image will never quite come to focus, no matter how hard you try. See chapter 3 for information about collimating telescopes.

FIGURE 2.2.4. COMA LOOKS SOMEWHAT DIFFERENT THAN A POORLY FOCUSED STAR. THE BRIGHT CORE OF THE STAR IS OFFSET FROM THE CENTER.

TIP: The amount of zoom you use to examine focus quality determines the technique to use. If you use a zoom factor of 300-400% or 3-4X), you should look for an overall appearance of crispness at the edge of the star. If you enlarge the image further, up to about 800% or 8X, you will be able to see individual pixels – you can actually count the number of pixels that the star occupies. However, the FWHM measurement, discussed later, is a much easier way to determine the width of the star. Best focus is typically achieved when the number of pixels spanned by the star is at a minimum. If you are counting pixels, and the number of pixels spanned by the star changes even when you are not making changes to the focuser, then you are dealing with turbulence.

Telescopes vary dramatically in how well they preserve the contrast of an object. If a telescope has poor contrast, that will adversely affect image quality and make it harder to detect best focus. The better the optics are, the better the telescope's contrast will be. Another important factor is light scattering. Internal baffles can reduce reflections inside the telescope, and a dew shield can reduce the amount of off-axis light entering the telescope, which will help reduce internal reflections. Smooth, well-finished optics also reduce scattering.

The Focusing Process

The typical focusing process starts with invoking the focus routine in your camera control software. For SBIG's CCDOPS program, the focus dialog (see figure 2.2.5) presents you with four options:

Exposure Time – The length of time to expose the CCD chip. Bright stars require short exposure times, typically on the order of a fraction of a second. The dimmer the star, the longer your exposure time must be to get enough of an image to evaluate focus. Exposures over 2 seconds average out turbulence effects.

Frame size – This is a somewhat deceptive term. What you are changing here is the bin mode. The idea is to use a coarse bin mode (2x2, 3x3) for rough focusing, and unbinned mode (1x1) for fine focusing. In CCDOPS, the coarsest bin mode is called "Dim" because it is useful for locating dim objects using short exposures. Once you get a good rough focus, switch to Planet mode.

FIGURE 2.2.5. THE FOCUS DIALOG FROM SBIG'S CCDOPS CAMERA CONTROL SOFTWARE.

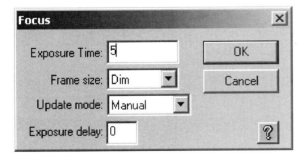

Update mode – Determines how the focus exposures occur: manually or automatically. Manual mode will take just one exposure, and then wait for you to click a button to take the next exposure. Automatic mode takes one exposure after another, with a delay between exposures if you enter a value into the Exposure Delay box. Once you gain experience with focusing, you can do most of your rough focusing in automatic mode. I prefer manual mode for final focusing because I like to study the image carefully to evaluate focus.

Exposure delay – How long to wait between exposures in automatic mode. Enter the number of seconds to delay. Once you are comfortable with focusing procedures, you can use the Automatic update mode, and set an exposure delay of around 3-10 seconds. During the delay, you can evaluate focus quality and adjust focus position.

Other camera control programs offer similar settings that allow you to manage the focusing process. Maxim/DL offers additional focusing options that are covered throughout this chapter. CCDSoft offers an excellent set of focus tools, as well as an automated focusing tool called @Focus which is covered in detail later in this chapter. Version 3 of MaxIm DL also includes automatic focusing tools. See the book web site for information.

Whatever software you use, the focusing routine is fairly standardized:

- Determine the appropriate exposure time by imaging a star field.
- Take an exposure (or initiate automatic exposures).
- Adjust the focus position, and take another exposure. Does it make the focus better or worse? This step determines the direction to move the focuser to improve focus.
- Repeatedly take exposures and adjust the focus position to improve focus.
- Continue until you have moved past the point of best focus, taking note of the appearance of the star at best focus.
- Move back to the point of best focus, and verify focus quality.

> **TIP:** The only way to know if you really have reached best focus is to go past it and observe a decline in focus quality. Otherwise, there may be a better focus point than the one you currently have – you just won't know it! By continuing until the focus gets worse, you can make sure you reach the best focus position.

On some nights, the seeing will be poor, and there will be a zone where focus won't get any better. Stars will be larger than on other nights. Your best strategy is to position the focuser in the middle of that zone. On other nights, you will have a very small range of positions where focus is best, and you will get stars that are small dots of light. Those are the nights to stay up all night imaging!

Rough focusing should be done using the binning features of your CCD camera. Start with a full frame. Binning combines multiple pixels, and results in faster download times. (See chapter 1 for details on binning modes.) Binning results in less time to download the data from the camera because there are fewer pixels to download. If you bin 2x2, for example, you can download an entire frame in one-fourth the time, since each virtual pixel is now made up of four actual pixels.

1x1 binning – This is really no binning at all, but you will often see this phrase used anyway. It simply means that the camera has been used at its highest resolution: one pixel in the camera equals one pixel in the image. Final focusing should be done at 1x1 binning; higher levels of binning can mask focus errors. However, if you don't normally use 1x1 binning because your focal length is very long, then focusing at 1x1 will just give you a larger fuzzy star, and may not help. In that case, use whatever bin mode you use for imaging for focusing as well. See figure 2.2.6 for an example of a 1x1 binned image taken with an ST-8E camera.

2x2 binning – Pixels are binned in groups of four, two pixels on a side. If 2x2 binning is the largest available bin mode, you can use it for rough focusing. Figure 2.2.7 shows an example of 1x1 and 2x2 binned images with the 2x2 image (left) enlarged to match the 1x1 image (right). The 2x2 image has much less resolution. The area of the 2x2 image is just one-quarter of the 1x1 binned image (see figure 2.2.8). This is why imagers typically use the smallest bin mode on any given night, limited only by the seeing conditions and focal length.

FIGURE 2.2.6. AN IMAGE OF M15 UNBINNED (1X1).

3x3 binning – Not all cameras offer 3x3 binning. It's the fastest way to do rough focusing. With many telescopes, but especially SCTs (Schmidt-Cassegrains), you may start with a star image that is dramatically out of focus. Binning 3x3 lets you use the entire chip for rough focusing with fast download times.

Other special-purpose binning modes are sometimes available. They are used for special purposes such as spectroscopy, and can be ignored for focusing.

Larger bin modes are more sensitive for the shorter exposures used in focusing. You can also use larger bin modes to quickly see if the object of interest is within the field of view. This allows you to see what you are pointing at using shorter exposures than would be possible without binning. This works extremely well for bright objects like clusters, but it is also surprisingly effective with galaxies and nebulae. I am always amazed when I can clearly see dim objects in my focus images, even though they are only 5 to 10 seconds long.

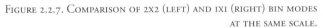

FIGURE 2.2.7. COMPARISON OF 2X2 (LEFT) AND 1X1 (RIGHT) BIN MODES AT THE SAME SCALE.

FIGURE 2.2.8. A 2X2-BINNED IMAGE OF M15, SAME SCALE AS FIGURE 2.2.6.

Moving Primary Mirror Issues

Most Schmidt-Cassegrain telescopes, and many scopes of similar design such as Maksutov-Cassegrains, focus by moving their primary mirror. Most such systems do not support the mirror rigidly. The mirror will shift when slewing and tracking, and when reversing focus direction. The shifting will alter the focus position, and/or cause the field of view to move to a different part of the sky.

When the primary mirror is moved to adjust focus, it moves both along the telescope axis, as intended, and it also moves laterally and it may change its tilt. This can cause image shift and slight astigmatism. The problem is most noticeable when you change the direction of focus travel. The amount varies from one scope to the next, but it is often annoyingly large.

When changing focus direction, not only does the mirror shift laterally, but it may also alter the focus in large jumps. This makes it challenging to get accurate focus with a moving primary mirror, and accounts for the large market in add-on focusers. It takes some practice to get good at focusing with the moving primary. One approach is to go past best focus, then go back through focus, and approach focus from the original direction and be very careful not to go past it. Otherwise, if you try to return to focus by reversing direction, your object may be moved too much. You typically wind up repeating the focus procedure until you get to best focus without going past it. When the seeing is marginal, that adds to the frustration of this maneuver. That is why alternative focusers, covered later in this chapter, are often used on scopes with moving primary mirrors.

If you do mount an alternate focuser on a scope with a moving primary, you may want to lock the primary down. Several web sites offer different methods for locking the mirror, and the method you choose will depend on the make and model of your telescope. It is easiest on many Meade SCTs, because you can simply put the locking bolt that was used for shipping back into the scope. With other models, you will probably need to drill into the back of the scope and add your own locking bolt(s). This is a non-trivial procedure, but it can make the scope a much better one for CCD imaging. Make sure you put a soft tip of some kind on the lockdown bolt to prevent damaging the mirror, and never apply excessive pressure to any of the bolts.

The moving primary rides on a hollow tube. A layer of grease keeps the mirror from sliding around too much, but it does not keep the mirror perfectly still. When you move the telescope to point at a new object, the mirror may shift a little as the weighting changes. This makes it more challenging to use a finder scope, digital setting circles, or a goto mount with one of these scopes. The longer your focal length, the more of an issue this will be.

When you flip the telescope across the meridian (that is the line directly over head running from north to south), the mirror may shift by a larger amount because the weighting has changed by 180 degrees.

This is not to say you cannot use a telescope with a moving primary for CCD imaging. Many of the CCD images out there have been taken with such telescopes. The issue is important but not as nasty as it sounds. The end result of having a moving primary mirror is that a certain percentage of your shots will be ruined by mirror movement, but it's typically no more than 20% of the time. If you can't live with that (and your results could be better or worse than that average), you can lock down the mirror, and use an alternative focuser such as the JMI NGF-S.

The Zen of Focusing

It might seem from the discussion so far that focusing is too complex to deal with easily. In some ways, this is true. Focusing involves a lot of steps and varaiables, any one of which can get in your way on a given night of imaging.

On the other hand, focusing is the key to getting good CCD images. If you can master focusing, you have gone a long way toward your goal of obtaining great images.

Many first-time CCD imagers bring a set of assumptions to the job of focusing. A typical assumption is to compare CCD focusing to regular camera focusing. I'm not referring here to astrophotography with film. I'm talking abou using a typical everyday camera, whether it be film or digital. Focusing such cameras is either automatic, or involves a simple process of turning a focusing ring while observing some obvious feature of the image to identify best focus.

In other words, the assumption for regular cameras is that focusing is simple.

Focusing *can* be simple when you are doing CCD imaging, but it won't necessarily *be* simple. Consider two situations that involve both ends of the focusing spectrum. Both are drawn from my own experiences with focusing.

The first situation involves a worst-case secenario. I had just bought a 4" refractor and an ST-5C CCD camera. These were my first tools for astrophotography, and I brought a full set of useless asusmptions with me to the process. I didn't have anyone around to let me know this, however, so I proceeded to try to use the equipment as though it were a giant camera.

There was an unending stream of frustrations. The focus knob seemed incapable of making the small adjustments I needed for accurate focus. The camera took forever to download an image. The images looked terrible, and no amount of adjustments would improve them. The object would move off of the camera's CCD chip before I could even find focus! I was ready to pitch the entire collection into the nearest waste basket.

Now if I had only known a few things, I wouldn't have been so frustrated. There were a few simple truths that would have saved me much frustration:

- Rack and pinion focusers, commonly found on refractors, are not designed for ultra-fine focusing. A motorized focuser is a great asset when working with a refractor. These come in t he form of motors that drive the refractor's own focuser, or add-on Crayford-style focusers with motors. Either approach gives you much finder control over focus position, and takes the hassle out of focusing a refractor. Even oversized focus knobs are a big help.
- Most camera control software has a feature that will speed up your focusing session. Instead of downloading the entire image every time, you can download a small portion of the image, called a subframe. The subframe downloads in a fraction of the time needed for a full frame. This streamlines the focusing process, but you have to know it exists to look for it.
- Unlike a conventional photograph, a CCD image starts out as a mess. You have to take steps to reduce the noise. These steps are unlike anything required with a conventonal camera. These include esoteric things like dark frames, bias frames, and flat-field frames. These mysterious frames can make the difference between garbage and beauty, and they are well worth learning about.
- It is very important to have a good polar alignment if you are going to image. This is true even if you are taking very short exposures, such as of the moon or planets. Not only does a good polar alignment help you take longer exposures, it keeps objects on the CCD chip during the time it takes you to focus. Of all the frustrations I expierienced in my first attempts at imaging, the failure to polar align was my silliest. When I finally started to take the time to get good polar alignment, I wished that I had learned to do this sooner.

The bottom line is that you probably have some assumptons of your own. You don't know in advance how they might trip you up as you learn how to make images with a CCD camera. Keep an eye out for these assumptions any time you start to feel really frustrated. The problem might well lie with your hardware or software, but it also might lie with your assumptions about how things should work.

Sometimes, when you enter a new field like CCD imaging, your assumptions are going to get turned on their heads. When this happens, take a deep breath, and ask yourself if there isn't a completely different approach available to solve the problem. Learning to see with new eyes is more than using the CCD camera to expand your vision. It's also the process of finding creative solutions to the various problems that crop up, especially in the early part of your CCD career.

Section 3: Software-Assisted Focusing

If you have followed the instructions on focusing by eye in the preceding section, you have probably noticed that when you get very close to focus, changes in focuser position have less and less visual impact on the star you are using to achieve focus. However, being out of focus even a little will lend a soft look to your overall images. Bright stars don't show focus as clearly as dim ones do, at least not to the eye.

Short of taking five-minute full exposures to learn how good your focus really is, there is a better way to check your focus. Most camera control software includes special features that make it easier (if not always easy) to locate the best possible focus position.

The types of focusing aids you will find in camera control software include:

- Numeric readout showing the brightness of a star
- Graphic representation of the brightness of a star
- Numeric readout showing the width of the star's image

The two most commonly used software aids that include these features are Brightest Pixel, and Full Width at Half Maximum (commonly abbreviated as FWHM). CCDSoft version 5 has introduced a combination of these (and a few other) factors, called a Sharpness value. I'll have more to say about that later.

In addition, I have some detailed advice on using the dimmer stars in combination with the bright ones to achieve an exceptionally fine focus.

Brightest Pixel Focusing

The concept behind brightest pixel is simple: the better the star's light is focused, the tighter and smaller the image will be. More photons are hitting a smaller area on the CCD chip. As a result, the pixels at the center of the star's image get brighter as you get closer to focus.

Most camera control software will show you the value of the brightest pixel during focusing. In theory, you can simply watch this number, and the point at which the number is greatest represents the best focus.

That's the theory. The practical application of brightest pixel is a different story. Stars twinkle, and that twinkling is the bane of the CCD astrophotographer. Not only does it spread out the light, making stars fatter and less point-like, it also causes random changes to the brightness of a star during any given exposure. Longer exposures can help with this, but at the cost of sharpness. Longer exposures help even out the variations, but they also spread out the star images making focus more challenging in a different way.

You can still get good results with brightest pixel focusing if you keep a few things in mind:

- The random changes in the star's brightest pixel value are greatest when the star is at best focus. If the value of the brightest pixel jumps from 34,000 to 43,000 from one exposure to the next, you are likely close to best focus.

- Combine brightest pixel with visual observation. If you have two focus positions that show rapid variations in brightest pixel, use the techniques described in the section on visual focusing to help you figure out which one is the closest to best focus.

- Take more than one exposure at a given focus position, and average the results. For example, if three exposures at focus position #1 yield brightest pixel values of 1,300, 1,400, and 1,450, that's an average value of 1383. If focus position #2 yields values of 1250, 1475, and 1495, that's an average value of 1406, so position #2 is the better focus position.

- Make sure that your brightest star doesn't saturate the chip during the exposure. If the value creeps up toward saturation for your camera, shorten your exposure. Saturation is often not the maximum possible value; see chapter 6 for information on calculating your saturation level.

- Brightest pixel works with almost any star. You can focus using dim stars if you need to; the brightest pixel technique works whether the values are a few hundred, or tens of thousands. A very dim star, however, will not work as well due to the fluctuations from atmospheric turbulence.

Figure 2.3.1 shows brightest pixel readings from focusing session using the CCDOPS camera control program from SBIG. "Peak" refers to the brightest pixel. "Planet" refers to the focusing mode in use; this is described in the section "Subframe Focusing" below. To be certain that you are comparing the same star, use the X and Y coordinates in this dialog to confirm the position of the brightest pixel. The value of 8026 (left) was the first reading, and moving the focuser in a very small amount, and taking a second exposure, resulted in a value of 9847. This indicates that focus improved in the second exposure.

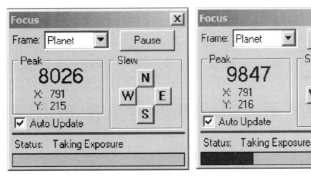

FIGURE 2.3.1. THESE DIALOGS SHOW THE NUMERIC VALUES FOR THE BRIGHTEST PIXEL IN AN IMAGE DURING TWO DIFFERENT EXPOSURES.

The most effective approach to use with the brightest pixel method is to use it in combination with other methods. The seeing may play havoc with your numbers, but if you are also visually observing the star, and perhaps using other focusing tools available in your particular camera control program, you will get better results than by any one method alone.

Figure 2.3.2 shows how MaxIm DL reports the quality of focus on the Inspect tab. The brightest pixel information is in the box labeled "Max" at bottom center. There is some additional information here as well, such as a graph and FWHM (Full Width at Half Maximum) in the X and Y directions, (covered in the next section). The graph provides very useful feedback about the state of focus. Note that when the brightest pixel is a higher value, the peak of the graph is very sharp (right), and when the focus is not perfect, the peak is a little rounded (left). The better the seeing, the more likely you are to get that nice sharp peak.

TIP: As focus improves, the brightest pixel value is increasing. Depending on your exposure time, the star may get so bright that it saturates (reaches maximum value). As you start to get close to saturation, the peak brightness will become spread out and give you a false indication of poor focus. In addition, the peak may spread out in a line due to blooming, which distorts the FWHM values in one direction. It will also flatten the top of the graph. If the brightest pixel exceeds 80% of the maximum allowable value, stop the focusing routine, shorten your exposure time (try cutting it in half), and resume. You can perform your analysis of focus quality with low brightness levels, but working at about 10-50% of saturation provides less noise and better results.

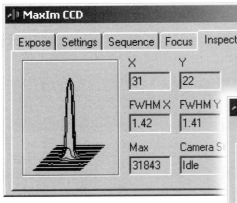
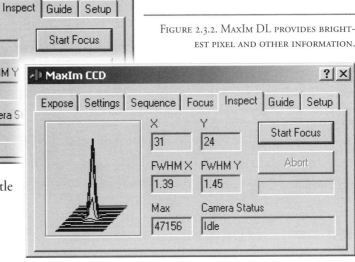

FIGURE 2.3.2. MaxIm DL provides brightest pixel and other information.

FWHM Focusing

The image of a star has a typical brightness profile (figure 2.3.3). The curve shows an idealized picture of the brightness level of the pixels across a star image. This is a bell shaped curve, with brighter values at the top. A bell curve has most of the values falling near a central value (called the centroid of the star image). There is no definite edge to such a distribution, so it can be hard to measure the actual width of the star image.

Fortunately, there is a way to characterize the width of the star image, called Full Width at Half Maximum (FWHM). To find the FWHM, you take the highest value, divide it in half, and measure the distance across the curve at that point. The line AB in figure 2.3.3 shows the FWHM for the idealized curve.

Most camera control software includes some way to measure FWHM. In CCDOPs, you can use the Display | Crosshair menu choice, and then pass the cursor over a star to get the FWHM. The numbers for the ver-

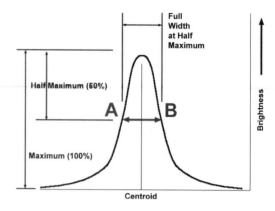

FIGURE 2.3.3. THE BRIGHTNESS PROFILE OF A TYPICAL STAR NEAR BEST FOCUS.

tical and horizontal FWHM are shown next to the heading "Seeing." In MaxIm DL 2.x, the Focus routine shows you two FWHM values (see figure 2.3.2). One is calculated from a vertical slice through the star image (FWHM Y), and the other is a horizontal slice (FWHM X). These values often differ by at least a small amount; if blooming occurs, they will differ by a larger amount. In version 3, MaxIm DL provides a more accurate single FWHM value.

With Mira, change the cursor to a cross hair (right click on the cursor and choose cross-hair). Position the cursor over a non-saturated star, and use the Measure | FWHM menu item. This displays several measurements at one time in a small window, as shown in figure 2.3.4. You get values for FWHM, peak value, and background value displayed at the same time. It is useful to compare the brightness of your peak pixel against the back-

FIGURE 2.3.4. DETERMINING FWHM IN MIRA AP.

SECTION 3: SOFTWARE-ASSISTED FOCUSING

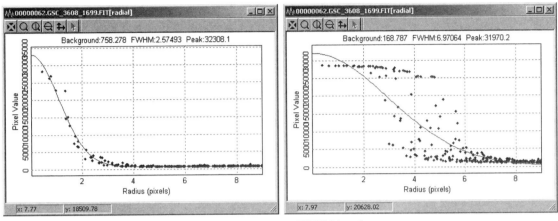

FIGURE 2.3.5. A NON-SATURATED STAR REPORTS ACCURATE FWHM (LEFT), BUT A SATURATED STAR HAS A NON-NORMAL SPREAD OF VALUES AND REPORTS AN INACCURATE FWHM.

ground. On a moonlit night, you might have a 1000-count brightness for brightest pixel, but if your background level is 800, then you aren't working with a very bright star.

The MaxIm DL method for viewing FWHM is ideal, while the CCDOPs and Mira methods are slow because you have to restart the focusing process to measure FWHM. CCDSoft v5 doesn't report the FWHM during focusing, but it does have a similar measurement called the Sharpness value that is covered later in this chapter. In CCDOPS and Mira, I rarely use FWHM because of the inconvenience. I use FWHM on a routine basis with MaxIm DL and CCDSoft v5.

Avoiding saturation when measuring FWHM is critical to success. Mira provides a radial plot of a star's brightness from center to edge (Plot | Radial Profile menu item). Figure 2.3.5 shows two examples of radial plots for stars from the image in figure 2.3.4. The plot on the left shows a non-saturated star. The FWHM is measured accurately at 2.57 pixels. In the plot on the right, the data points have a flattened, very broad top. This results in a cruve that is wider than it should be. These are classic indications that the star has saturated. The plot on the left is for the star with the cross hairs in figure 2.3.4. The plot on the right is for the bright star straight below the cross-hairs.

Table 2.1 shows some statistical information taken from a series of images during an actual focusing session. The focus position for successive images varies by a very small amount – the smallest manual change in focus I could make, as a matter of fact. A new focus image was taken after each change in focus position (see samples in figure 2.3.6), and the following values were obtained for two different stars:

Average value for star #1 – This number reflects the average of the values in a circle 9 pixels in diameter.

Max Brightness for star #1 – The brightest pixel value in the star image.

FWHM for star #1 – Taken by averaging the FWHM X and FWHM Y values in MaxIm DL.

FWHM for star #2 – Same as above, for a second star.

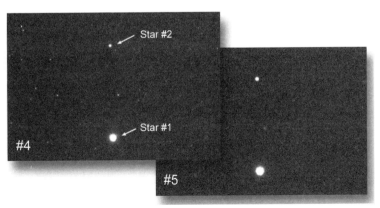

FIGURE 2.3.6. THESE ARE TWO OF THE IMAGES USED TO CONSTRUCT TABLE 2-1.

THE NEW CCD ASTRONOMY   51

Table 2.1: Brightness and FWHM

	Star 1		FWHM	
Image #	Average	Max Brightness	Star 1	Star 2
1	2222	49220	2.283	2.124
2	2037	49694	2.219	1.884
3	2208	50180	2.229	1.826
4	2275	50421	2.269	1.759
5	2124	49307	2.213	2.024
6	2423	48358	2.496	2.465
7	2510	34565	3.028	3.024

The brightest pixel value is for image #4. The lowest FWHM value for star #1 is in image #5, while the lowest FWHM value for star #2 is at image #4. In the original data, there was a slight amount of blooming in star #1, which was revealed by a significant difference in the FWHM for X and Y. A large difference indicates blooming and/or saturation. Because of the blooming on star #1, use star #2 to determine best focus.

Figure 2.3.6 shows the images referred to as #4 and #5 in table 2.1. Visually, image #4 looks slightly sharper, further confirming it as the right focus position. It is easy to compare images when you have brightest pixel and FWHM data handy, but it's a lot harder in real time to be sure visually. Having additional sources of data about the sharpness of an image will help you reach focus faster and with greater confidence.

Figure 2.3.7 shows a graphic representation of the data from table 2.1. The FWHM values have been multiplied by 10,000 in an Excel spreadsheet to get all values on roughly the same scale; this does not affect the relative values in each series. Note that the brightest pixel and lowest FWHM coincide on image #4. Note also that the values don't shift much right around best focus, so it takes careful work to measure the point of best focus. This is yet another reason why combining methods will work best to determine critical focus.

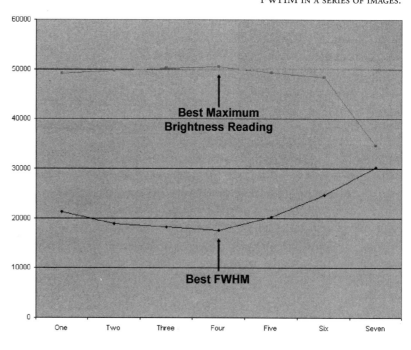

FIGURE 2.3.7. A CHART SHOWING THE VALUES FOR BRIGHTEST PIXEL AND FWHM IN A SERIES OF IMAGES.

Subframe Focusing

Depending on the size of your CCD chip, it can take a long time to download a full frame of information. Since final focusing is always done at 1x1 binning, downloading times are at their maximum. Most camera control programs allow you to select a sub-frame for focusing so that fewer pixels are downloaded for each exposure. This greatly speeds up the focusing process. If your polar alignment is good enough to keep stars stationary for a few minutes, you can use a very small focus frame and get focus feedback practically in real time.

In CCDOPS, this technique is called Planet mode, but you use it for much more than just planets. In CCDSoft and MaxIm DL, you can select a portion of the chip at any time by clicking and dragging. This makes focusing fast and easy.

To start Planet mode in CCDOPS, select Planet as the frame size when you enter focus mode. For convenience, you can set Update Mode to Auto, and enter a delay time between exposures. Auto update tells the camera to repeatedly take exposures during the focusing session. The delay gives you a pause in which to adjust focus position. For your first focusing sessions, you may wish to leave auto update turned off so you can work at your own pace.

Figure 2.3.8 shows what the CCDOPS Planet mode looks like in action. Two star images are readily visible, and they are very far out of focus. The Focus dialog is at upper right, and it remains visible throughout the focusing session. It shows that Planet mode is active. When you first enter Planet mode, CCDOPs will take a full-frame exposure of the length you requested at 1x1 binning, and display the image as in figure 2.3.8. It then pauses, and you drag out a rectangle to show the area you want to use for focusing. You must take the full-frame exposure at 1x1; you cannot image at 3x3, drag out the rectangle, and then jump to 1x1. Other packages, such as MaxIm DL and CCD-Soft v5, do allow you to switch fluidly from one bin mode to another. CCDSoft is slightly better in this regard.

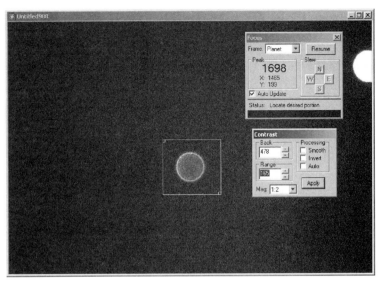

FIGURE 2.3.8. CCDOPS PROVIDES A PLANET MODE THAT ALLOWS FOR FAST, EFFICIENT FOCUSING.

TIP: Depending on the accuracy of your polar alignment, you may need to allow more room around the star you will be using for focusing. Unless the mount is extremely well aligned to the celestial pole, the star will drift during the focusing session. If the box is too small to accommodate the drift, the star will move out of the subframe and you will have to restart focusing.

Once you have set the rectangle to the size and location you want, click the Resume button at the upper right of the Focus dialog. The camera will take the first exposure. If you chose automatic update, you can adjust the focus position during the delay. To change the delay in CCDOPS, restart the focusing session.

The best overall procedure for CCDOPS focusing:

1. Use Dim mode to get a rough focus. Focus visually, and get the smallest, best-focused star image possible. Dim mode uses a 3x3 or 2x2 bin mode, so will probably still be a little off.

2. Switch to Planet mode for final focusing. Use the visual cues outlined in the first part of this chapter and the brightest pixel method to confirm when you have best focus.

3. Continue focusing until you have definitely passed the point of best focus, then back up to it again.

In figure 2.3.8, the stars are way out of focus, so I'm not really following my own advice! The Peak value in the Focus dialog is only 1698. However, there's no reason not to use Planet mode or a subframe for all of your focusing, and if your mount is well aligned to the celestial pole, and doesn't drift much, you'll be able to use Planet mode comfortably – the star won't drift out of view.

Focus Analysis

Figure 2.3.9 shows the brightest pixel values for a series of 24 images during a focusing session. This sequence is fairly typical of how things go when you are using a camera for the first time – it's hard to know where focus position is going to be. That's why the brightness values at far left are so low -- the star was very far out of focus. The frames numbered 1 through 5 are described a little further on.

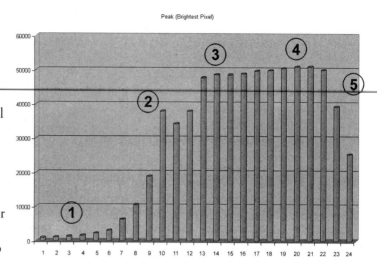

FIGURE 2.3.9. BRIGHTEST PIXEL VALUES FOR A SERIES OF 24 EXPOSURES DURING A FOCUSING SESSION. SEE TEXT FOR A DESCRIPTION OF THE FIVE KEY POINTS IN THE FOCUS PROCESS

The point of focus for a CCD camera will usually be different than the point of focus for your eyepieces. There are parfocal eyepieces that will come to focus at the same place as a CCD camera, such as the Software Bisque IFocus. You can make your own eyepieces parfocal using a ring described later in this chapter.

If you can point reliably at a star, such as with a goto mount, it will be easier to put objects on the chip. A bright star (mag 3 or brighter) will be visible even if you are extremely far out of focus. If your finder is very well aligned, it can also help you put a star on the chip. But if you have a telescope with an internal focuser, such as an SCT, an eyepiece that is parfocal with your camera can make it easier to get close to correct focus. If you can measure the focus position, as on a refractor, you can easily return to the CCD focus position.

TIP: If you have an SCT, you can count the turns of the focus knob required to reach focus with the camera. If you have a refractor, consider cutting a piece of plastic that is the same length as the focuser extension to make it easier to find the right starting position in the dark. Anything you can do to quickly find the general area of the correct focus position for CCD will help you speed up your focusing session.

It's worth pointing out a few characteristics of the curve in figure 2.3.9; the numbered positions in the figure correspond to the points that follow:

1. If you start very far out of focus, the value of the brightest pixel will initially change very slowly, then speed up as you get closer to focus. This can make it hard to determine the correct direction initially. The out-of-focus star image will shrink noticeably as you improve focus.

2. Don't' be shocked, surprised, or give up just because you suddenly get a lower value during the focusing session. Many things can cause a lower value – a cloud may have temporarily moved over the star, or the wind may have smeared the star out a little more than usual. Don't panic; if the star doesn't look focused, it isn't!

3. Once you get close to best focus, start making smaller and smaller changes to focus position. Sneak up on the best focus with small steps. Keep going until you are sure you have best focus.

4. The highest value is the best candidate for critical focus. If your focuser has digital readout, you can make a note of your best focus and return to it. Check the brightest pixel value to be sure you are at the correct position. If your focuser has backlash (free play when reversing direction), the numbers

will probably not match exactly. You may have to spend some time estimating how much movement gets used up in backlash in order to use the numeric readout to return to best focus. Develop a feel for how the focuser behaves when reversing direction.

5. Once you are past best focus, values will drop off rapidly. When you are first learning how to focus, it is a good idea to go too far. Way too far is OK. You want to develop your sense of where best focus is, so don't hesitate to go back and forth, back and forth, until you get a feel for where it is. Invest a little time now to master the best focus position, and reap dividends forever after. Amaze your friends with your ability to bring complex equipment to a complete and safe focus.

TIP: I've mentioned this elsewhere, but it bears repeating right here: if you aren't sure if you are at best focus, take 2, 3 or more exposures to see how the brightest pixel values changes. Average them, on paper or in your head. If the seeing is poor, you will see large variations in the brightest pixel values, and there will be a wide range of positions where best focus might be. Try to judge where the middle of that range is, and set your focus at that point (interpolation). If the seeing is very good, you will have much better control over the situation, and brightest pixel values will be more consistent and useful in finding the exact spot of critical focus. Oddly enough, when the seeing is good, the brightest pixel values will be so high that they may actually fluctuate more than they will with poor seeing. However, when the seeing is good, this fluctuation will be most pronounced right around critical focus, so you can use this to help you determine where the best focus position is.

Figure 2.3.10 shows eight of the images I used to create the chart in figure 2.3.9. Each is labeled with the value of its brightest pixel, and the image number is from the sequence of 24 images charted in figure 2.3.9. I pulled out these eight images because each of them shows something useful about the focusing process.

Image 5 – This image is taken early in the focusing sequence. You can clearly see the diffraction rings typical of a good scope when a star image is out of focus and at high magnification. The brightest pixel value is far lower than what it will eventually be when the image is in focus, just 4% of its final value.

Image 7 – This image is in somewhat better focus. There is a second star appearing – as you improve focus, dimmer stars become bright enough to stand out against the background. The background level appears to be different, but that's not the case. The camera control software has adjusted image contrast automatically. You can turn off this feature using the Auto checkbox in the Contrast window, and if you are using the visual

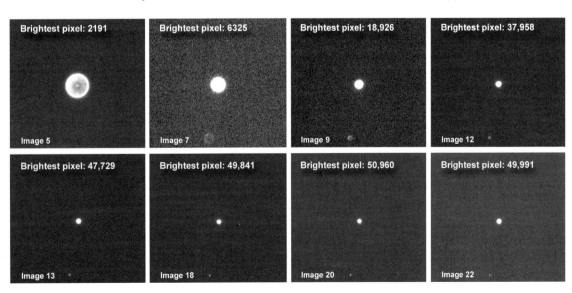

Figure 2.3.10. Eight images from the focusing series shown in figure 2.3.9. See text for details.

appearance of the star for focusing, turning off automatic adjustments can help you track what's happening more consistently.

Image 9 – The star image is smaller and the brightest pixel value is now much higher, about 35% of its final value. The dimmer star is giving you clear feedback that the image is not in focus, because you can still make out its diffraction rings. **It is ideal to have a mixture of dim and bright stars in the focus window, because each type of star provides different kinds of clues about the quality of focus.**

Image 12 – In this image, the brighter star is now quite small, and actually appears to be very close to focus – but it is not! The dimmer star gives better feedback here: although it is now very small, it still has not become very bright. The visual clues are becoming more subtle, but the brightest pixel value is still only about 75% of its final value. Despite how much better things look, we have a long way to go yet!

Image 13 – In this image, focus is clearly better. Two very dim stars can be seen just below the text "Brightest pixel." You may or may not be able to see them printed in the book – the printing process can loose subtle details. One of the new stars is below the letter "B," by about the height of that letter; and the other is below the colon following the "l," at about the same vertical position. The dim star at the bottom is a bit sharper, with a hint of a bright point within the tiny cloud of light. The bright star is just ever so slightly smaller; you would need to view the image at 400x or even 800x and count pixels to see this, however.

Image 18 – This image looks like it is very, very close to focus – and it is, but it is not quite there yet. The brightest pixel value is now 98% of its best possible value, and you might be tempted to stop here because the image looks very good. Even though we are very close to focus, we are not at focus! The dim star at the bottom is still a little cloudy, but has a very clear bright center. The two very dim stars at the top are a little bit clearer in this image, but still quite dim. This is why I recommend going past focus before you settle on what the best focus is like. You have to see how good you can get the focus on any given night to know when to stop, and the only way to be sure is to go past the best focus position.

Image 20 – This image shows critical focus. There are two important changes from image 18: the brightest pixel value is now consistently above 50,000; and the dim star at the bottom is a perfectly clear dot, with no fuzziness, no cloudiness at all that would indicate any amount of out-of-focus. The two very dim stars are also a bit more visible. See figures 2.3.12 and 2.3.13 for magnified views of the dim stars in this image.

Image 22 – In this image, we have gone a bit past perfect focus. The dim star is now slightly fuzzy, and the two very dim stars near the top are just barely visible.

> **TIP:** This example uses the SBIG camera control software, CCDOPS, to illustrate how to analyze a star image's brightness for best focus. You could just as easily use a program like MaxIm DL, and use both brightness and FWHM data to determine best focus. FWHM will show the same variations, and the same overall pattern as CCDOPS. However, having both terms available to cross-check focus gives you a better shot at finding the critical focus position.

Figure 2.3.11 shows extreme blow-ups of the dim star at the bottom of these images. The left side of figure 2.3.11 shows a magnified view of image #18. The star is so small that it *looks* like it is fully illuminating just a single pixel. However, upon close examination, you can see that it partially illuminates the adjoining pixels. Other surrounding pixels have a small amount of illumination, and the star is at least partially illumi-

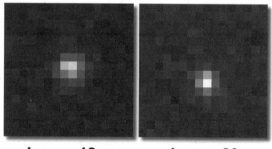

FIGURE 2.3.11. THE DIM STAR AT LEFT IS SLIGHTLY OUT OF FOCUS; THE BRIGHTEST PIXEL ISN'T MUCH BRIGHTER THAN SURROUNDING PIXELS. THE STAR AT RIGHT IS VERY WELL FOCUSED; THE BRIGHTEST PIXEL IS MUCH BRIGHTER THAN THE SURROUNDING PIXELS.

nating a box of pixels that is five pixels wide, and four pixels high.

Now look at the star in image #20 (right of figure 2.3.11). The central pixel is much brighter than the surrounding pixels. And the box which the star is illuminating is now only 3x3 pixels. All of this points to better focus in image 20. Now you know what I mean by subtle differences!

Figure 2.3.12 shows more evidence. These are the very dim stars at the top of the images. In image 18, the brightest pixel isn't much different from the surrounding pixels. In image 20, the brightest pixel is much brighter than the surrounding pixels. It really stands out, demonstrating the value of dim stars when focusing.

You won't always have dim stars that match your pixel size this closely – it will vary with the focal length of your telescope. A long focal length or poor seeing will smear the dim stars and make them less useful in assessing focus quality. But when these one-pixel stars are available, they can help you reach critical focus on nights when the seeing is truly excellent, when critical focus is so important.

TIP: In poor or average seeing conditions, the air will not be steady enough to bring the dim stars to such perfect points. But when the seeing is better than average, you can achieve a very exact focus by combining information about the brightest pixel in a bright star, the appearance of a dimmer star, and the visibility of any very dim stars. And whatever the seeing conditions, dim stars offer you yet another way to analyze the quality of your focus position.

Focuser Issues

The focuser built into your telescope is an important factor in your ability to get to perfect focus. As you get close to focus, you need to make very, very small adjustments to focus position. The shorter the focal length of your telescope, the smaller the moves you will need to make.

Your ability to move your focuser in very small increments will vary from telescope to telescope. Some telescopes, such as refractors, may have fairly gross focus movement, while others, such as the 9.25" Celestron SCT, have geared focusing or digital readout. If you can't get the degree of fine focusing you feel you need, there are aftermarket focusers that can be a good solution. There are two-stage focusers (coarse and fine); motorized focusers (some work great, others don't help much because they have so much backlash); and DRO (digital readout) focusers. See the last section of this chapter for information about these alternative focusing mechanisms.

If all of this focusing information seems like overload, take heart! There are even better ways to focus than what you've seen so far. Nonetheless, focusing with nothing but your eyes and the camera software is a worthwhile skill. As you get good at focusing, you may well return to your roots and focus quickly and easily with nothing more than you've seen so far. You develop a sense for where best focus is over time, and the need for clever aids goes away after a while. As I've said before, and will say again many times, the time you spend focusing is time well spent. And since a variety of factors can conspire to change the point of focus through the course of the evening, the ability to focus quickly and effectively will make it easier to bite the bullet and refocus often.

FIGURE 2.3.12. IN FOCUS (IMAGE 20, BOTTOM), DIM STARS ARE BRIGHTER.

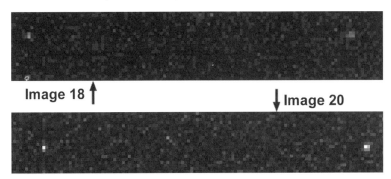

Section 4: Aids to Focusing

The techniques covered so far -- visual focusing, and focusing using numeric data from the software -- can work and work well. But a variety of variables, including everything from the optics to the seeing conditions, can conspire at times to make focusing still seem like a chore. There are additional tricks available to give you more options for focusing. One of the more interesting options is to partially block the light path, which creates diffraction patterns in the image. You can then use those patterns to give you a clearer indication of when you are at critical focus.

As always, you can also combine methods to better identify the best focus position.

We will explore two techniques in this section: a cross made out of masking tape, and various types of masks. There are other hardware focusing aids available (many of which you can make at home from readily available materials), and I'll give you a rundown on those at the end of this section.

Focusing with Diffraction Spikes

The spider that supports the secondary mirror in a Newtonian and many other kinds of telescopes serves as a built-in focusing aid. The vanes of the spider create diffraction spikes that sharpen noticeably at optimal focus. If your telescope doesn't have vanes, you can temporarily add various items to the front of your telescope to create diffraction spikes. Refractors don't have a secondary at all, and Schmidt-Cassegrains have a corrector plate instead of a spider. Both types are often used for CCD imaging, so I have a few suggestions for creating "artificial spiders" to create diffraction spikes on such telescopes.

When you are done focusing, you remove whatever you've added and image in the normal manner.

One very practical method probably seems like the ultimate in low-tech: slap some masking tape across your telescope's dew shield. It probably even sounds ridiculous. But it not only works, it works very well.

Figure 2.4.1 shows how simple it is to set up your telescope with the MTFA (Masking Tape Focusing Aid). Just apply the tape to the end of the dew shield in a cross pattern. You don't even have to be super accurate about making everything equal; the technique works quite well with a slap-dash application of the tape. I used one-inch tape, but anything from 1/4" to an inch and a half should work fine. For those who are squeamish about applying anything sticky to their telescope tube, Paul Walsh, a Seattle area CCD imager, suggests building a cardboard lens cap with the tape affixed to the cap. As shown in figure 2.4.1, the tape will work best if it is at 45 degrees to the camera's "up" ori-

FIGURE 2.4.1. POSITION THE MASKING TAPE AT 45 DEGREES FROM THE CAMERA'S "UP" ORIENTATION.

SECTION 4: AIDS TO FOCUSING

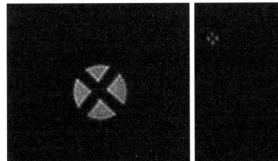

FIGURE 2.4.2. LEFT: STAR WELL OUT OF FOCUS, SHOWING SHADOW OF MASKING TAPE. CENTER: CLOSER TO FOCUS, SHADOW IS LESS PROMINENT. RIGHT: CLOSE TO FOCUS, SHADOW DISAPPEARS.

entation. The puts the diffraction spikes from the tape into the image diagonally. Running the spikes diagonally through the image pixels makes it easier to judge the exact width of the spikes. If the spikes line up too closely with the columns and rows of the CCD chip, it becomes harder to judge the exact width of the spikes.

The tape blocks some of the light, and changes the diffraction pattern of the scope. Figure 2.4.2 shows three images at various stages of focus. At left, the star is far out of focus, and the shadow of the tape is obvious. The middle picture shows a point closer to focus, and the shadow is only clear in a dim star at upper left. At right, the star is close to focus and the shadow has disappeared. Very bright stars work best.

As you get close to focus, you may or may not see the desired diffraction spikes. If auto contrast is on, you may need to manually adjust contrast to get a clear view of the spikes. The effects from the tape are fairly subtle, and you may also have to increase your exposure time to see them. If you are using a binned image to do rough focus (which I recommend highly), you will need to switch to the highest available resolution mode in order to do final focus. Figure 2.4.3 shows what you should expect to see near focus in unbinned (1x1) mode, with a long enough exposure, and properly adjusted contrast. Note that outside of focus the diffraction spikes aren't simply thick, as they

appear in 3x3 binned images; they are actually made up of two completely separate lines.

When the star is near focus, you will see spikes extending out from the star along the lines where the tape's shadow once was. The brighter and thinner the diffraction spikes are, the better your focus. The actual brightness and thinness will depend on your focal ratio and the physical characteristics of your tape or other material.

In CCDOPS, use Planet mode for final focusing. With other programs, take a binned image, select the subframe for final focusing, and then switch to 1x1 (unbinned) mode. Whatever software you use, make

FIGURE 2.4.3. THE APPEARANCE OF A STAR VERY CLOSE TO FOCUS IN BINNING MODE 3X3 (ST-8E CAMERA).

THE NEW CCD ASTRONOMY

sure you select an area around the star that is large enough to show the diffraction spikes at their full length. As you improve focus the spikes will get thinner and longer.

Figure 2.4.4 shows the process of selecting the star for focusing in Planet mode. When the lines merge, and are as narrow as you can get them, you are at critical focus.

TIP: Make sure you apply the tape to your refractor so that it makes a 45-degree angle with the camera's CCD chip. This will orient the spikes diagonally, which makes it easier to determine if the spikes are as thin as possible. If the spikes are lined up with the rows and columns of the pixels, you won't be able to analyze them as clearly.

Figure 2.4.5 shows the spikes as they would appear at best focus. Note that there is a single line for each spike. The lines are not perfectly thin lines; they still have some width. They are a little thicker close the to star, and they thicken slightly as they get more distant from the star. Both of these effects are normal. Gauge the thickness of the spikes at their thinnest part.

Figure 2.4.5 also reveals a few other interesting details that tell us that we are in focus. There is a small amount of blooming occurring (the small black line near the bottom of the star image). The exposure is just a touch too long to be perfect, but since this image was taken during a real-world imaging session, I didn't let a little blooming make me start over -- the spikes are the important part of the image, and a little blooming does no harm. In fact, to get bright spikes, you may have to use a long enough exposure to cause some blooming. Don't sweat it -- it does no harm. The important thing is to get spikes that are bright enough to be easy to evaluate.

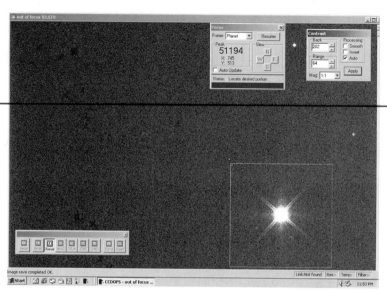

FIGURE 2.4.4. SELECT THE AREA AROUND THE STAR YOU ARE USING FOR FOCUSING, ALLOWING ENOUGH ROOM FOR THE DIFFRACTION SPIKES.

FIGURE 2.4.5. THE APPEARANCE OF THE DIFFRACTION SPIKES AROUND A WELL-FOCUSED STAR. NOTE POINT-LIKE DIM STARS AS WELL.

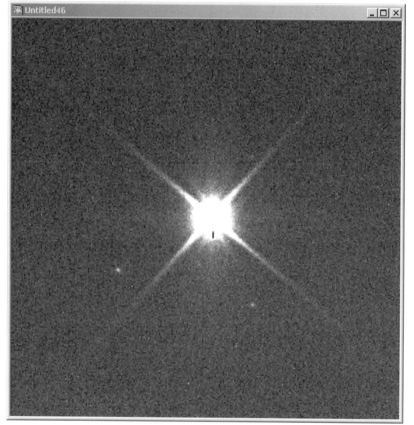

On the other hand, if they get too bright, you may not be able to judge their thickness accurately.

> **TIP:** You may need to adjust the contrast of the image in order to see the spikes clearly. Play around with the Back and Range settings in CCDOPS, or the Min/Max settings in MaxIm DL, so that the spikes show up clearly. Don't make then too contrasty, however, or you won't be able to judge them properly. If they ar still to faint, use a longer exposure.

Also visible in figure 2.4.5 are two dim stars. As with the examples of dim stars shown earlier, these stars strongly illuminate a single pixel. I examined images just inside and outside of this focus position, and found that the two dim stars no longer had the bright core that indicates best focus. As I often recommend, I was able to use several techniques to cross-check the point of best focus. I also could have used brightest pixel or FWHM, or both, to further confirm best focus. Even with the tape, these other methods still work.

Some telescopes use a 3- or 4-vane spider to support a secondary mirror. The spider can act as a built-in focusing aid. The spider vanes are thin, so they are not as obvious in out-of-focus images as the masking tape is, but they can be useful for focusing on very bright stars as you get close to critical focus. Figure 2.4.6 shows an image of the Bubble Nebula taken with a Takahashi Mewlon 210, a Dall-Kirkham design Cassegrain. It has a 4-vane spider, and you can clearly see the vanes in this in-focus image.

Focusing with a Mask

The principle behind the mask is very similar to that behind the masking tape: put an obstruction in the light path, and observe the changes in the diffraction pattern that occur as you move closer to focus. However, where the masking tape blocks a small portion of the aperture, and allows most of the light to go through, the mask blocks most of the light, and allows only a small amount to go through.

Figure 2.4.7 shows the pattern for one kind of mask. The dimensions shown are for an 8" mask, but in reality you need not be terribly careful about the size or placement of the two holes. You can even use three holes if you like. A piece of cardboard from a box, poster board, or other sources of cardboard can be cut and taped or wedged into the front of the scope. You can't have enough masking tape handy when CCD imaging!

Figure 2.4.8 shows two other patterns that I made and tested. I cut the masks out of cardboard, and tried them out on an Astro-Physics Traveler, a 4" APO refractor.

Figure 2.4.9 shows one of the masks I made. As things turned out, this funny little one with the two oddly-pointing triangles turned out to be the most useful. I made the masks from thin cardboard purchased at a local office-supply store. Any cardboard should work fine -- shirt cardboard, boxes, etc. It just has to be firm enough to hold its shape when you put it on the tele-

FIGURE 2.4.6. THE SPIDER VANES ON SOME TELESCOPES MAKE A GOOD TOOL FOR JUDGING FOCUS QUALITY.

FIGURE 2.4.7. THIS SIMPLE MASK IS INTENDED TO HELP YOU FOCUS MORE ACCURATELY.

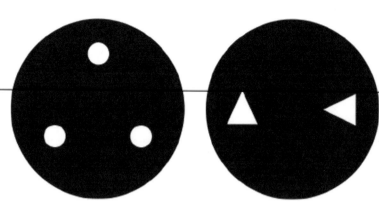

FIGURE 2.4.8. Two of the alternative masks I tested. The triangle mask worked best.

scope. I've even made masks from paper plates. Some paper plates, in fact, make a perfect force fit inside the rim of popular 8" SCTs.

The little tabs you see in figure 2.4.9 are designed to support the mask on the front end of your telescope. I found that they also work to wedge the mask inside of the dew shield on a slightly larger telescope when they are spread out a bit. No tape is needed unless there is a fair bit of wind. Even with the scope nearly horizontal (figure 2.4.10), the mask will stay in place. The trick is to put two of the tabs in a "V" at the top of the scope. If you try to hang it by only one tab at the top, it will slide right off unless the scope is nearly vertical.

Figure 2.4.11 shows the results of using three different types of masks to focus on a star. Each vertical column shows images taken through one type of mask. Each column stops at the point where I last found the mask useful approaching focus. The three holes start to crowd each other, and I didn't like using them. The two-hole mask holds up a little better, but when the two holes get close together, it's impossible to tell how close they are, or when they are actually co-incident. The two-triangle mask neatly solves this because it has diffraction spikes that line up at perfect focus.

One thing is immediately obvious from figure 2.4.11: the shape of the mask strongly affects what you will see when you are focusing. If there are two holes, you will see two bright spots. If there are three holes…you get the idea, I'm sure!

FIGURE 2.4.9. A simple focusing mask. The little tabs hold the mask on the front of the scope.

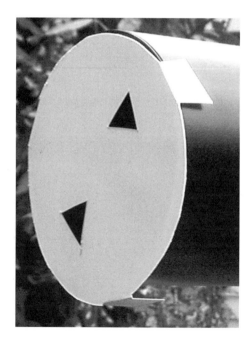

FIGURE 2.4.10. A mask in action.

SECTION 4: AIDS TO FOCUSING

The bottom row of images in figure 2.4.11 shows the appearance of the star when it is well outside of best focus. Everything is well spread out, and it's easy to tell that you are out of focus. The next row up shows the results after moving the focuser closer to best focus. The images are now smaller and closer together. This is a trend: as you get closer to focus, the separate images shrink and merge. The idea is to get everything to merge into a single image, indicating that you are at best focus.

In practice, this was often hard to do. The third row up from the bottom shows the problem: as the images get closer together, they merge with one another and it becomes hard to tell when they are exactly merged. In fact, I found it easier to focus visually than to try to tell when merging occurred.

The mask with the two triangles was an exception, however. The sharp corners of the triangles create diffraction effects, which are slightly visible in the second row, and much more visible close to focus in the third row. The triangle method was the most useful, and it was the only one that I could count on for telling when I was close to focus, as shown in the top row. There is only one image in the fourth row because there was only one technique that actually worked effectively close to focus.

Figure 2.4.12 shows the full extent of the diffraction spikes using the traingle mask. There are twelve of them, two for each point on each triangle. As with tape and spiders, you may need to increase your exposure time and/or adjust contrast to see the spikes clearly. The spikes help determine where the center of each triangle is, even when you can no longer see the triangles. Each pair of diffraction spikes define a line, and when all the lines appear to meet at the same point, you are at best focus.

FIGURE 2.4.11. RESULTS OF USING THE THREE MASKS. LEFT: THREE ROUND HOLES. MIDDLE: TWO ROUND HOLES. RIGHT: TWO TRIANGLES. OUT OF FOCUS AT THE BOTTOM; IMPROVING FOCUS AS YOU MOVE UPWARD.

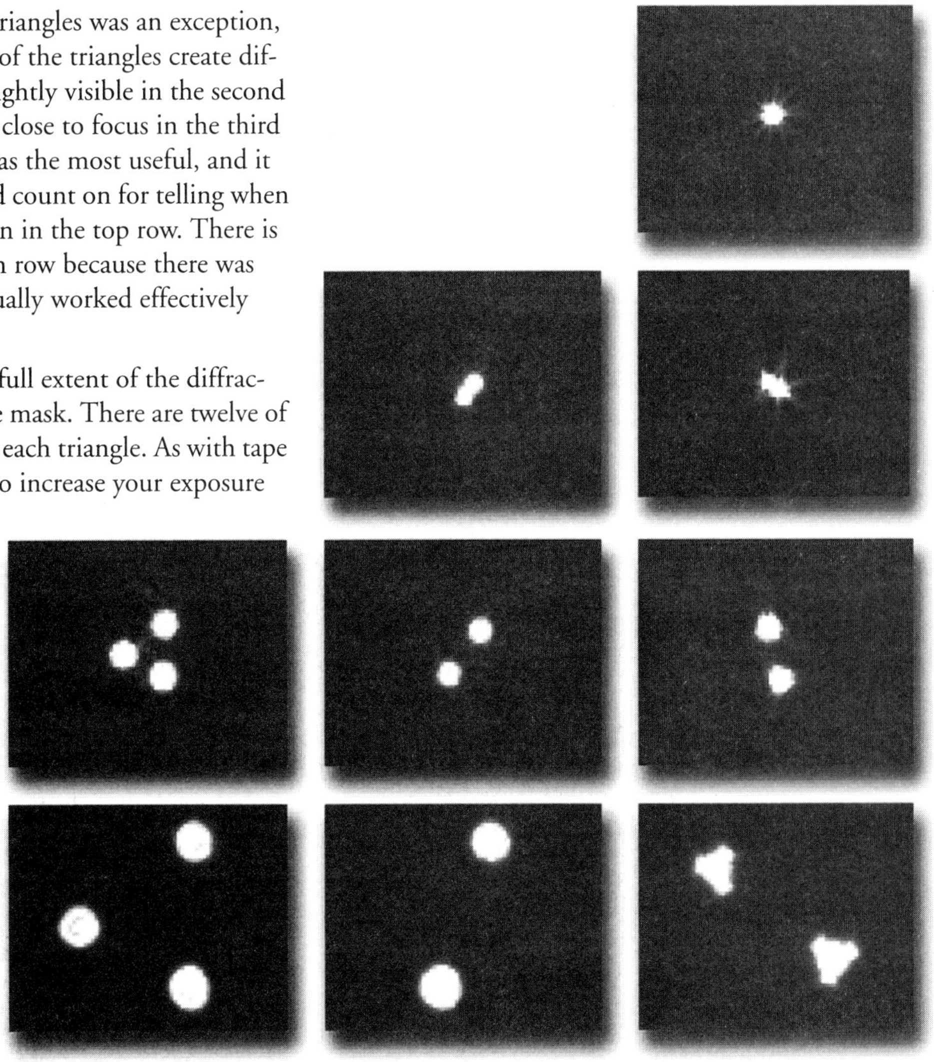

THE NEW CCD ASTRONOMY

Figure 2.4.12. The mask with triangles creates diffraction spikes, which makes it easier to tell when the two images are merged. The spikes will all point to the same place when the images overlap.

This is an interesting area for experimentation. As long as you don't do any damage to your front optic, just about anything handy might be the perfect answer to perfect focus. Different items will provide different diffraction patterns, and if you can find one that works especially well for you, that's what really counts.

In addition, the diffraction spikes get thinner as you get close to focus, so they function in two ways at once. I was able to use brightest pixel and FWHM as well while merging the two images, to provide additional data on when I had achieved best focus. Magnifying the image also helps.

Making the masks is extremely simple; they don't have to be very complex or precise. In fact, the "circles" don't have to be circles -- and the triangles work best in any case. I made all three masks in less than ten minutes using a carpenter's knife and light-weight cardboard.

Other Focusing Aids

There are a number of other objects and masks that you could put at the front of your telescope to help you figure out when you are at best focus. There are too many possibilities to cover them all here. Some work better than others; almost all of them are somewhere between cheap and free. For example, you can tape two small dowels in parallel across the front of your dew shield, creating a pair of lines in the out-of-focus image. As you approach focus, the lines start to merge. This technique is similar to the masking tape approach in ease of use, though the tape is more portable; you can carry a small roll in your pocket. A cross made of dowels also works.

Section 5: Alternative Focusing

The focusers built into many telescopes can be a challenge to use for the critical focusing required for good CCD images. Really good focusing requires the ability to make extremely small changes to focus position. Visual focusing is less critical, and most focusers are designed with visual focusing in mind. For example, most refractors have fairly coarse focusing mechanisms -- even the best ones. These focusers work extremely well for visual use, since the eye can accommodate small variances from perfect focus. When you try to use the typical refractor focuser for CCD imaging, it can get frustrating because it's very hard to make the small changes in position you need to make.

This is true of even very expensive telescopes. They have focusers that work well for visual observing, but imaging demands more precise focusing. It was challenging to focus Takahashi and Astro-Physics refractors for CCD imaging until I found a good alternative.

> **TIP:** Various vendors offer larger focusing knobs for refractors. The larger knob allows you to make finer adjustments to focus, and if you will be focusing a refractor manually the larger knobs are more accurate for focusing.

Schmidt-Cassegrain telescopes have focusers with a finer range of motion, but they can be troublesome in their own way. SCTs focus by moving the primary mirror, and the mirror slides on a tube that allows it to shift slightly when you reverse direction. This can be a real pain when you are doing fine focusing on a single star in the small window. The star may jump completely out of the window if you reverse direction. Since the whole point of focusing is to go just past the optimal position, and then reverse, this qualifies as a Major Pain. You will find suggestions for dealing with moving primary mirrors later in this chapter.

One approach to focusing problems is to motorize your focuser. This provides some advantage in most cases, but the real question is whether or not it is enough of an advantage. I've discovered that it depends on the type of motorized focuser you use. Simply motorizing an existing focuser may not be good enough for CCD imaging. Sometimes, what you really need to do is add a complete new focuser to your scope. You can replace the one you have, or add a supplemental focuser. In some cases, you use the existing focuser for rough focusing, and the supplemental focuser for fine focusing.

Don't forget that the faster your focal ratio, the shorter the critical focus zone is.

I just can't overstate how much easier life can be with the right focuser. Trying to focus a CCD camera, even when using the techniques in this chapter, can be frustrating if you can't control your focuser adequately. If it seems like I'm beating you over the head on this point, there's a reason for it. At some point in every CCD imager's career, the urge to upgrade the focuser strikes. When it hits, you need good options, and I've outlined them for you here.

The JMI NGF-S Focuser

The NGF-S focuser from Jim's Mobile (JMI) is one of the best compromise between cost and performance among the motorized focusers available. You can spend less, and you can spend more, but you get an excellent value with the NGF-S. If you have critical focusing needs, look to a more precise unit such as the Optec focuser described below.

The really nifty thing about the NGF-S is that, while it is designed for SCTs, you can also use it in refractors and many other telescopes, often without making any changes at all to the existing focuser. As long as you have enough focus travel, the NGF-S can be an excellent solution. You can attach it to other types of telescopes. It comes with an adapter that inserts into any 2" focuser. The adapter presents an SCT thread, to which you attach the NGF-S.

The fact that the NGF-S mounts on an SCT thread is an obvious clue clue to the origins of this focuser. It was designed to be mounted in place of the visual back on many of the popular Schmidt-Cassegrains. It has a short focus travel, and was intended for fine focus only. It moves about half an inch to adjust focus.

One problem with using the 2" adapter is that many focusers don't grip it as securely as you would like. It's important to keep the camera square to the optical axis, and you may get flexure if you use the supplied adapter. A better alternative is to buy or make a threaded adapter that will attach the NGF-S directly to your focuser tube. American Takahashi dealers sell an adapter called the Feldstein, which has Tak refractor threads on one end, and SCT threads on the other. This allows you to attach the NGF-S to any of the 4" or larger Takahashi refractors easily. For other scopes, you may need to contact a local machine shop to have them make an appropriate adapter for your scope.

The NGF-S consists of a Crayford-style focuser (see figure 2.5.1) with a motor and hand control. The unit sells for about $270 (US dollars) as of the time of writing. As pictured, the unit is shown with a DRO hand controller, which provides numeric position information and is a more costly option. A non-motorized version of the NGF-S is available, but the real beauty of this unit is the motor, which provides a very fine level of control.

To make use of the NGF-S, you will need to have a little over 2" of focus travel available. For example, if you normally move your focuser tube out 3.5" from the innermost position to bring your CCD camera to focus, then you have plenty of room for the NGF-S. SCTs in the 8", 9.25" 10", 11", 12", 14" and 16" sizes will all have enough focus travel for the NGF-S. An adapter is available from JMI to fit the 3" visual back of the larger scopes. The unit comes with an adapter for the 2" visual back found on small to intermediate SCTs. A 2" to SCT-thread adapter also ships with the NGF-S, which allows you to use it on any telescope with a 2" focuser as described above, but make sure that you use bungee cords or something similar to secure the focuser and camera. This will reduce flexure, and reduce the risk of the entire assembly falling out of your focuser! You can order a 1.25" to SCT-thread adapter separately from JMI if the supplied 2" adapter does not meet your needs. However, I recommend a 2" connection whenever possible; it is more stable.

Any motorized focuser is an improvement, but not all motorized focusers are enough of an improvement to be worthwhile for imaging. The motorized focuser must be capable of making extremely small adjustments to focus position. If you are using a refractor for imaging, the motorized focusers that move your rack-and-pinion focuser may not have the level of fine control you need for best results. If you are using an SCT, the motorized focusers that slip over the hand focus control may have too much looseness to be reliable.

JMI makes motorized focusers for various telescopes, but the NGF-S is the most versatile for CCD. With the DRO (Digital Read Out) option, you can maintain extremely fine control over your CCD focusing. The NGF-S also works reasonably well with some automated focusing routines, such as @Focus in CCD-Soft version 5. It's not quite as good for this as the Optec TCF-S or RoboFocus, but those focusers are in a higher-performance class.

Note: An add-on focuser like the NGF-S will not work in all situations. Celestron's Fastar, for example, requires a motirzed focuser attached to the SCT's focus knob. Newtonians often do not have enough back focus for an NGF-S. Try a RoboFocus or a replacement motorized focuser.

The Optec TCF-S (Temperature Compensating Focuser)

For those who desire the ultimate in focusing accuracy and repeatability, the TCF-S focuser from Optec fits the bill. Figure 2.5.2 shows the TCF-S, temperature probe, and hand controller. The unit also comes with a power supply, and cables to connect the hand controller to the focuser, and the computer to the hand controller.

FIGURE 2.5.1. THE NGF-S FROM JMI INCLUDES A MOTOR, ENCODERS TO READ FOCUSER POSITION, AND A HAND CONTROLLER.

The TCF has several features that make it very useful for CCD imaging:

- The unit will move, even during an exposure, to compensate for temperature-induced focus shift.
- The unit has repeatable focus position, using digital encoders built into the focuser body.
- The unit has extremely small steps for obtaining ultra-accurate focus position: just 0.0008", or 0.0203mm.

FIGURE 2.5.2. THE TCF-S, FROM OPTEC, IS NOT JUST A DIGITAL MOTORIZED FOCUSER: IT ALSO INCLUDES A TEMPERATURE PROBE AND WILL ACCURATELY ADJUST FOCUS POSITION WITH CHANGING TEMPERATURE.

Note: The TCF-S requires about 3" of back focus. This is fine on systems with generous backfocus, but if you have a scope with fast focal ratios or minimal back focus, consider the RoboFocus instead.

To use the focuser, attach the temperature probe to the outside of the telescope tube. A supplied piece of foam insulates the probe, so it measures only the temperature of the telescope, not the nighttime air.

You then put the unit in Learn mode, and push a button to cause the unit to note the current focus position and the current temperature. As the temperature changes, you manually adjust focus. When the temperature has changed at least 5 degrees, you push the same button, and the unit calculates the amount of movement required to compensate for temperature changes. It will test the temperature every half second, and adjust the position of the focuser as needed.

Most of the time, I use the TCF-S without the temperature feature. The accuracy of the unit for automated focusing is the most important feature.

The TCF focuser has extremely high precision, and it also can be computer controlled. I have included a simple program for controlling the TCF-S on the book web site:

http://www.newastro.com/newastro/book_new/samples/tcf_control.zip

For more information about the TCF-S:

http://www.optecinc.com/astronomy/products/tcf.html

RoboFocus

The RoboFocus, from Technical Innovations, offers similar features to the Optec TCF-S in a completely different kind of package. The TCF-S is a self-contained focuser. The Fobofocus is a motor and electronics that turns many existing focusers into remoe-control digital focusers.

RoboFocus thus takes up no back focus, and is suitable for an installation where there is limited back focus available. Since it uses your existing focuser, the quality of focus depends heavily on the quality of your existing focuser. If you have a high-quality focuser, then Robo-Focus will add digital focusing, automatic focusing remote focus, and other features to your existing hardware. Contact Technical Innovations to find out if your specific focuser is available for a RoboFocus installation.

Both CCDSoft v5 and MaxIm DL 3 support the RoboFocus for remote and automatic focusing. For more information, visit the Technical Innovations web site:

http://www.homedome.com

Automatic Focusing

If you choose to use a motorized focuser, you can use some of the emerging automated focus techniques. CCDSoft version 5, for example, includes an automated focusing tool called @Focus. It moves your motorized focuser in small increments, assessing focus as it makes changes. It calculates the optimal focus position based on the sharpness of the image, and then moves the focuser to that position. If you do not have a focuser that attaches directly to the serial port of your computer, you will need TheSky, CCDSoft, and a mount with an output for controlling focus. TheSky supports mounts with focuser control, and CCDSoft can use TheSky to control the focuser through the mount. Mounts that will work for this include many that support the LX200 communications protocol, the Paramount, and Astro-Physics GTO mounts.

CCDSoft supports two types of focusers: those that use encoders and have exact repeatability, like the Optec, and standard motorized focusers, such as the NGF-S. MaxIm DL 3 will support automatic focus. It will only support focusers with exact repeatability.

Not all motorized focusers will work with automatic focusing. This is similar to choosing a mount for CCD imaging: you want a unit with low backlash and the high accuracy of motion. If the focuser has backlash or couples loosely to your existing focuser, it won't make precise movements and thus won't work as well with automated focusing. And if the focuser can't move precise distances, the automated focusing software can't move the focuser reliably.

The list of focusers that specifically support automated focusing is fairly short. Optec and RoboFocus have models available, and JMI should have serial support available for the NGF-S. Prices for such focusers range from a few hundred dollars to nearly a thousand dollars. The book web site will be updated with information about additional automated focusing tools, both hardware and software, as they become available. In particular, I'll have a full review of the Optec focuser with MaxIm DL when version 3 of MaxIm ships.

Using @Focus

@Focus is an automatic focusing tool that refines focus after you do a rough focus. @Focus uses Shaprness to adjust focus so that the CCD detector is in the critical focus zone. The Sharpness parameter is graphed on the Focus Tools tab during the @Focus run. Sharpness is calculated from a variety of properties in the image, including image contrast, FWHM, and more. By combining measurements, the Sharpness value becomes less susceptible to the noise that plagues brightest-pixel focusing methods.

@Focus requires a focuser than can be controlled by TheSky, or one that is directly connected to your computer and has a CCDSoft-compatible focus driver installed.

You can use the entire CCD frame with @Focus, but sub-frame focusing is much faster and is reliable. As long as you start with a decent rough focus, a subframe will give good results. The steps for using @Focus are:

1. Click the Settings button on the Setup tab of the Camera Control panel. Set the parameters to match your equipment's capabilities.
2. Obtain a good rough focus. Move six large step sizes away from focus (see below to determine step sizes).
3. Click the @Focus button on the Focus Tools tab. Select whether to start by moving in or out. (If you are uncertain about direction, pick one at random and let @Focus figure it out. @Focus will take longer if the initial direction is not toward focus.)
4. Run @Focus, and verify that it finds the best focus position.

Note: If you are using TheSky to control the focuser, some parameters must be set in TheSky, such as the large and small step sizes for moving the focuser.

@Focus automates the focusing process, but it needs an accurate focuser and appropriate settings to be successful. The following sections examine how various conditions affect @Focus performance.

Mechanical quality in the focuser

Best case: The focuser should have virtually no backlash, and be able to be positioned with an accuracy of one half or less the size of the critical focus zone. (Backlash is free play in the focuser mechanism.)

Potential issue: Poor mechanical quality in the focuser results in an inability to find focus accurately. Typical mechanical quality issues include:

- Backlash (looseness) in the focuser, which results in varying movement when reversing direction
- Varying voltage levels in the mount or the focuser, which results in variable movement of the focuser (a problem with LX200 mounts)
- Non-orthogonality (focuser does not hold camera square to the optical path), which results in image distortions that make it harder to define the best focus position

Position feedback from the focuser

Best case: If the focuser reports its actual physical position to CCDSoft, this allows CCDSoft to position the focuser with a very high degree of accuracy. The TCF-S focuser, for example, matches numeric values to specific focuser positions. Zero always means that the focuser is fully in; 3,500 is exactly halfway; and 7,000 always means the focuser is fully out.

Potential issues: Focusers that do not report their position do not allow CCDSoft to absolutely position the focuser. These include the NGF-S, MotoFocus, etc. Some focusers, such as the NGF-S with DRO, report a number but the number does not necessarily map to a physical focus position. This type of focuser doesn't have absolute position, but if backlash is minimal, accurate focus is possible. Also, if the focuser reaches the end of travel and the motor continues to turn, the numbers change but focus position does not. Avoid the end of travel condition to avoid errors.

There can also be delays within the device that communicates with the focuser. Some mounts, such as the LX200, introduce variable delays while processing focus commands. Because the length of the delay cannot be predicted, the effects on @Focus are variable and do not allow @Focus to reliably position the focuser. In such cases, you may need to tweak focus manually after the @Focus run.

No movement in the optical system

Best case: Nothing in the telescope/camera system moves other than the focuser controlled by CCDSoft.

Potential issues: If you have a moving primary mirror, even a small movement of the mirror will change focus, and probably the field of view as well. A mount that tracks poorly or that is not well aligned to the pole can also change the field of view. Such changes confound the data collected by @Focus, and that may make it impossible to locate the critical focus zone. Locking down the mirror in an SCT and adding an external focuser helps. Focusers that move the primary may or may not have sufficient accuracy to work well with @Focus. If any part of the focuser has backlash, or if the field of view changes too much, it becomes very difficult for @Focus to move the focuser reliably.

@Focus parameter settings

Best case: @Focus must be tuned to the focal length of your telescope, the focal ratio, the speed of your focuser, and other factors in order to perform effectively.

Potential issues: If the large step size is too small, @Focus will not see enough difference from one focus position to the next, and may judge focus prematurely. If the large step size is too large, @Focus may miss focus because it won't generate enough data to know where best focus is. See the "Set @Focus Parameters" section for additional information about step sizes.

Start close to good focus

Best case: @Focus works best if it takes 6 large steps away from rough focus when starting out. This allows it to discover the direction toward focus unambiguously, and results in the fastest focus. The best way to do this is to first get close to focus (you don't have to be fussy about it). Then move 6 large step sizes away from focus, and start @Focus.

Potential issues: If you start too far from focus, the change in sharpness will be too small to provide useful feedback, and @Focus may not work properly because it is starved for data. If you start too close to focus, @Focus will not discern direction until it has moved past focus, and it may mistake passing through focus for noise and get the direction wrong. @Focus is best at achieving that last bit of focus, and if you use it correctly, that's where it will excel.

Very good to excellent seeing

Best case: Steady seeing with minimal changes in apparent focus quality allow @Focus to converge upon focus quickly and easily

Potential issues: If the seeing is poor, the noise level in the focusing data makes it difficult to measure focus. A high noise level makes it more difficult for @Focus to determine focus. The averaging parameter allows @Focus to accommodate nights with poor seeing by averaging images and thus averaging the noise. @Focus will take longer to find focus when you use averaging. @Focus will often work under surprisingly poor seeing conditions, but I would recommend that you monitor focus quality when the seeing is average or worse.

Dark skies

Best case: Dark skies without serious light pollution, moon glow, or skyglow.

Potential issues: Any kind of sky illumination can decrease the signal to noise ratio in your images. The ultimate impact of a bright sky depends on whether you are using a light pollution filter, how bright the objects in the image are, and other factors. A bright sky by itself is usually not enough to confound focusing, but it can team up with other issues and make those problems harder to deal with.

Focuser types

@Focus can either communicate directly with a motorized focuser, or use TheSky's focus control via mounts that support focusers (e.g., LX200, Paramount, etc.). There are two types of motorized focusers available:

- Focusers that have positional feedback. The Optec TCF-S is an example of this type of focuser. These focusers can be positioned with repeatable accuracy. They are the best type of focuser for critical focusing because when @Focus says, "go to a specific position for best focus," the focuser goes to exactly that position.

- Focusers that do not have positional feedback. Typically, the focuser is pulsed for a brief period of time by the mount. These focusers can be positioned with variable success, depending on how accurate the mount's pulse length is; how long it takes the mount to send the pulse (latency); how much backlash the focuser has; and other electrical and mechanical factors. When @Focus tries to go to a specific position for best focus, the focuser may or may not be able to go to that position.

Based on my tests, @Focus is extremely reliable with focusers that provide positional feedback. Other types of focusers, and various combinations of focusers and mounts, provide varying degrees of success with @Focus. The current list of compatible focusers, in roughly the order of quality of focus provided, includes:

Best: Optec TCF-S. This is the serial-port model; the TCF version cannot link to a computer. Contact Optec to get a TCF converted to a TCF-S.

Very Good: JMI NGF-S attached to a Paramount GT-1100 or GT-1100S mount.

Good: JMI NGF-S attached to an Astro Physics GTO series mount. Small delays may occur in making position changes; verify quality of final focus.

Fair: JMI NGF-S attached to a Meade LX200. Key problems include delays in carrying out focus position changes, and variable voltage applied to the focuser motor. Both problems result in random errors in positioning. You should use the methods described later in this chapter to evaluate the quality of focus before imaging. If you cannot achieve focus with a non-positional focuser, bypass the LX200 and use a focuser than can be controlled by your computer's serial port. Or use @Focus to get close, and then manually refine focus.

Variable: RoboFocus attached to your computer's serial port. Since the RoboFocus controls the existing focuser on your telescope, the results you get with a RoboFocus depend heavily on the quality of your telescope's focuser. If your focuser is capable of small movements with little or no backlash, the RoboFocus will work very well.

Variable: Adding a motor to an existing focuser. Results vary over a very wide range, from not acceptable version of Motofocus to the high reliablility of units such as the RoboFocus. The higher the quality of your existing focuser, the more likely you are to be satisfied with motorizing it.

Keys to Successful Automated Focusing

There are some things you can do to get the best possible results with @Focus. The underlying principles apply to any autofocus software, but the details of implementation will vary.

First and foremost, not all motorized focusers are accurate enough to work with automated focusing routines. Small, repeatable movement is the key. The more accurately a focuser can make extremely small movements in both directions, the more likely it will deliver excellent focus. Likewise, if your focuser makes large movements, or has excessive backlash (and it doesn't take much to rate excessive in this particular application), results will be more variable.

Here are some suggestions for getting the most out of automated focusing:

- For the most consistent and accurate focusing, use a focuser that provides accurate, absolute position feedback to @Focus (TCF-Sm RoboFocus, etc.).
- For a high degree of consistency and accurate focusing, use a focuser with an absolute minimum of backlash. The greater the backlash (free play or looseness), the less likely the focuser is to achieve accurate, repeatable focus. @Focus attempts to correct for backlash problems, but the less backlash, the better the results. If your focuser has too much backlash or other problems, @Focus should get you close, and you can manually refine focus following the @Focus run.
- Set the correct small and large step sizes for focuser movement. The large step size must be large enough for @Focus to be able to detect changes in focus, and yet small enough so that @Focus does not completely move through focus in less than three steps. See the detailed information on step sizes in the section "Set @Focus Parameters" below.
- A single star usually leads to successful focus, but if that doesn't work, using several stars will be more effective. @Focus can also focus on extended objects, such as the moon, planets, galaxies, etc.

The ability to focus on non-stellar objects may vary with the capabilities of your focuser and mount, however.

If you experience ongoing problems with automated focusing, you should also look at the physical aspects of your system to determine if there is something besides the software contributing to the problem. Among the things to look for:

- If your telescope uses a moving primary mirror for focus, lock down the primary mirror and use a motorized external focuser. Motorized focusers that move the primary mirror are less likely to be successful with @Focus. Techniques for locking down the primary mirror vary with the make and model of telescope, and are well documented on the web.
- Make sure that focus can be reached within the travel of the motorized focuser controlled by @Focus. Rough focus should be done with any add-on focuser in the middle of its range of travel.
- Your mount must be well aligned to the celestial pole. Software Bisque's Tpoint software, sold separately, can quantify polar alignment and is ideal for imaging. TPoint is often thought of in connection with observatory telescopes, but it works just as well in the field. It takes 30 to 60 minutes to get an extremely accurate polar alignment with TPoint. If objects drift in and out of the field of view during focus, this will affect the sharpness values and could lead to spurious results. Minor drift will not have significant impact, but objects moving into and out of the field of view will cause problems.

Start Close to Focus

Automated focusing software is designed to handle the hardest part of focusing: getting into the critical focus zone. Before you use @Focus, you need to get the focus position near the critical focus zone. Typically, the recommended starting position will show many stars in your image as hollow circles or doughnuts. See below for information about determining the large and small step sizes.

FIGURE 2.5.3. A GOOD EXAMPLE OF A STARTING POINT FOR @FOCUS.

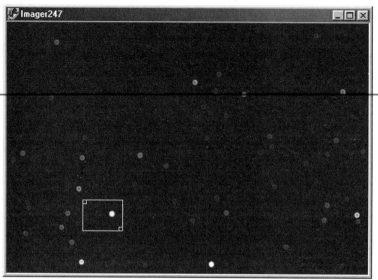

FIGURE 2.5.4. SELECTING A SUBFRAME.

Note: @Focus can find focus even if you do not start at the preferred starting point. It will take longer while @Focus determines which direction improves focus.

Figure 2.5.3 shows an example of a good starting point for an @Focus run. The image was taken with an SBIG ST-8E camera, binned 3x3. Binning provides faster downloads, and it's an effective way to speed up the rough focusing process. However, for best focus, I recommend 1x1 binning in a subframe when using @Focus or other automated focusing software.

TIP: Once you know your large step size (see "Set @Focus Parameters" below), you can determine the starting point for an @Focus run quickly. The ideal starting point for a fast focus is 6 large steps away from the critical focus zone.

@Focus works well with a small subframe (small white box in figure 2.5.4), even if there is only one star in the subframe. As you reach best focus, you will often see a few dim stars pop out, providing visual confirmation of success (see figure 2.5.5). A subframe downloads much faster than a full frame, and will speed up the focusing process.

In the left side of figure 2.5.5, there are two faint doughnuts to the left of the bright star. The right side of figure 2.5.5 shows the appearance at best focus. Not only are the two dim doughnuts now resolved into stars, but several other dim stars are visible as well. The left image in figure 2.5.5 is typical of a good starting point for @Focus. The right image is a good example of the kind of focus accuracy possible using @Focus.

Seeing conditions will have an impact on your ability to use faint stars for evaluating focus quality. This is especially true at longer focal lengths (greater than about 1500mm). However, @Focus will still be able to find the best possible focus for the seeing conditions.

Setting @Focus Parameters

If you have a focuser that attaches to your serial port, there are two places where you set @Focus parameters: on the Setup tab of the Camera Control panel, and in the @Focus dialog when you run @Focus.

Parameters specific to your focuser are found on the Setup tab of the Camera Control panel. Click on the Settings button to open the Settings dialog for your focuser. Figure 2.5.6 shows samples for two serial-port focusers, the Optec TCF-S (left) and the RoboFocus (right). Different focusers support different parameters, so the appearance of the dialog will be different for other focusers.

The following parameters are in the Setup dialog (or in TheSky). Although they can be changed at any time, once you find optimal settings there is usually not

FIGURE 2.5.5. DIM STARS ARE BLOBS WHEN OUT OF FOCUS (LEFT), BUT RESOLVE TO SHARPER DOTS AT FOCUS (RIGHT).

SECTION 5: ALTERNATIVE FOCUSING

much reason to change them unless you change telescope or camera. Not all focusers will have all of these parameters.

Large step size - This defines how far @Focus will move the focuser at each step during a focusing run. Setting the large step size requires some analysis of the behavior of your telescope/mount/focuser system, and is covered in detail below. If you set TheSky as your focuser type, you must change the large and small step sizes using TheSky. For the LX200 and compatible telescopes, use the Telescope | Options | Initialize menu selection, then click on the Focus Settings button to set your large step size. For the Paramount use Telescope | Options | More Settings menu, then change focus step sizes.

Small step size - This is a smaller step size that is used to position the focuser once @Focus has determined the best focus position. @Focus will use the Large step size to move the focuser close to the calculated best position, and then use the Small step size to get as close as possible. If you know the size of your critical focusing zone and the step size of your focuser, you should set a small step size that is about one quarter of the size of the critical focus zone or smaller. Otherwise, a good starting value for the small step size is about 1/10th to 1/25th of the large step size. Use smaller step sizes for faster focal ratios. See "Large step size" above if you are setting small step size in TheSky.

Backlash - If your focuser has backlash, the last movement of the focuser during the @Focus run will typically fall short of best focus (see inset in figure 2.5.10). This is caused by backlash, which prevents @Focus from moving the focuser in a repeatable manner. To determine a value for backlash, note the position of the focuser at the end of the focusing run. Manually move the focuser to best focus using the large step size, and note the position. Subtract the smaller number from the larger, and then use the result as your backlash setting. Enter the value for backlash in the focuser setup, and run focus again. If you undershoot/overshoot best focus, adjust the backlash accordingly. Continue until you reach focus reliably. If you still cannot reach reliable focus, you may have multiple sources of backlash, variable backlash, too large of a small step size or other limitations.

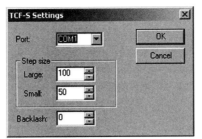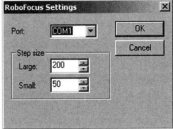

FIGURE 2.5.6. SETTING FOCUSER PARAMETERS.

The key parameter for success with @Focus is setting the large and small step sizes correctly. On many nights, you can successfully use @Focus with a range of large step sizes. However, finding and setting the optimal step sizes will speed up @Focus and give you more consistent results. See "Setting Step Sizes" below.

Additional parameters are included in the @Focus Setting dialog, which opens each time you click on the @Focus button on the Focus Tools tab (see figure 2.5.7). The parameters have a significant effect on the nature of your focusing run, so check out the explanations below before you try your first run. Depending on the focuser you are controlling, some parameters may not be available.

The following parameters are in the @Focus dialog, and can be changed each time you run @Focus.

Samples - This is the number of images @Focus will use to achieve best focus position. The available range is 10 to 50. For most situations, a value of 10

FIGURE 2.5.7. SETTING @FOCUS PARAMETERS.

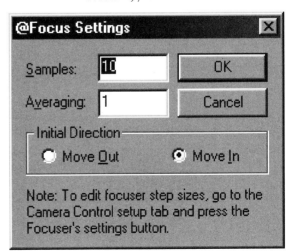

works well. The large step size recommendation is based on a sample setting of 10. Larger sample sizes are available for applications that require a smaller step size. @Focus requires at least 3 samples on each side of the critical focus zone to detect the rise and fall of the Sharpness parameter, and calculate the optimal focus position. For critical applications, you can increase the sample number, but you will need to reduce the large step size and the averaging parameter at the same time.

Averaging - This is the number of images per sample. Use a value of 1 under good seeing conditions, and values of 2 or 3 when seeing is poor. When seeing is creating serious problems, averaging 5 or even 10 images will smooth out the Sharpness curve significantly. It will take longer for @Focus to reach focus because of the multiple images needed for averaging. You can also use averaging when the seeing is good to reduce the overall noise level in the focusing data. Experiment to determine the optimal setting for your system. Generally, the smallest setting that regularly achieves excellent focus is the right value to use. I recommend starting with the default of 1. If that doesn't consistently deliver good focus on a given night, try a larger number. The impact of the Averaging parameter also depends on your exposure time. Exposures of several seconds duration will also tend to smooth out the effects of seeing. I usually use exposures of 3-5 seconds and get consistently good results without averaging.

Initial direction – This is the direction that @Focus moves the focuser when it begins the focusing run. If you know the direction to move for better focus, click the appropriate radio button. If you don't know the direction to move, let @Focus figure it out. @Focus always tries to determine the direction in which focus lies, even if you give it a starting direction. If it is moving in the wrong direction, @Focus will recognize this and reverse itself.

Setting Step Sizes

Figure 2.5.8 shows an optimal @Focus graph of Sharpness during a focusing run. The graph appears on the Focus Tools tab during an @Focus run. Note several features of this graph:

- There is a point where the Sharpness value begins to increase rapidly (low shoulder).
- There is a very small zone where the Sharpness has a peak value (high shoulder).

TIP: The peak of the curve corresponds to the critical focus zone. For most telescopes, the critical focus zone is so small that it will not show up as a separate plateau on the curve.

The area between the low shoulder and the high shoulder is what I call the active focus zone. This is the area where the Sharpness value changes rapidly with changes in focus position. Outside of this area (to the left of the low shoulder), the Sharpness value doesn't change much even with fairly large changes in focuser position. This is why I describe @Focus as a tool that will do your final focusing. Don't rely on @Focus to make large changes in position. The ideal session works like this:

1. Verify that AutoDark is turned on (Take Image tab of the Camera Control Panel).
2. Use the large step size Move Focus button to get close to focus. Judge the quality of focus the same way that @Focus does: the highest Sharpness value.

FIGURE 2.5.8. AN IDEAL GRAPH OF THE CHANGES IN SHARPNESS DURING AN @FOCUS FOCUSING RUN.

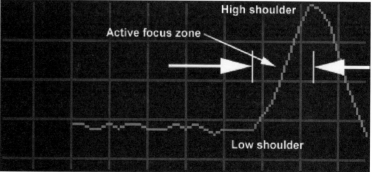

3. Move in or out of focus by 6 large step sizes. Use the Move Focus button to change focus. Your large step size must be small enough that you are still in the active focus zone, but large enough to move you close to the low shoulder. Adjust exposure time if necessary to make sure that the brightest pixel value is greater than 1000.
4. Click the @Focus button, tell @Focus which direction to move, and click OK to begin.

If your focuser is good enough to use with @Focus, the above routine will work consistently well. Don't start outside the active focus zone because the changes in the Sharpness value will be too small to be reliable.

The hardest part of using @Focus is setting step sizes. Here is a practical approach you can use:

1. Get as close to focus as you can manually. Use the Sharpness value to guide you. You don't need perfect focus; just get a decent rough focus.
2. Adjust the duration of your exposure to get a peak brightness value in the range of 20,000 to 25,000.
3. Note the current focuser position. For example, it might be 5240.
4. Move the focuser in one direction, in or out of focus, until the peak brightness value is 1,500 to 2,000.
5. Note the new focuser position. For example, it might 4750.

Calculate the difference between the two focuser positions (490 in this example). Divide by 6 to get the large step size (80). Depending on your focal ratio, the small step size should be 1/10th to 1/25th of the large step size. For fast focal ratios, something close to 1/25th is necessary because of the short critical focus zone. For slow focal ratios, you can use a larger small step size. For my f/5 refractor, I use a large step of 60 and a small step of 3. The same focuser on a C11 at f/10 works best with a large step of 250 and a small step of 20.

A Sample @Focus Run

The Sharpness graph clears automatically when you start a new @Focus run. During the run, the Current and Highest Sharpness values appear below the graph, and each Sharpness value is plotted so that you can see the progress of the focusing run clearly. If the Sharpness line is moving up, and the out-of-focus doughnuts are getting smaller, focus is improving. Figure 2.5.9 shows a sequence of graphs from an actual focusing run, with the corresponding Sharpness values.

The initial Sharpness value is always 1.00. The second Sharpness value is the ratio of the second reading to the first. If you are beyond the active focus zone, you might see very small changes in the Sharpness value -- 0.98, 1.02, 1.01, etc. This is just natural fluctuation

FIGURE 2.5.9. EIGHT STEPS IN A TYPICAL @FOCUS RUN.

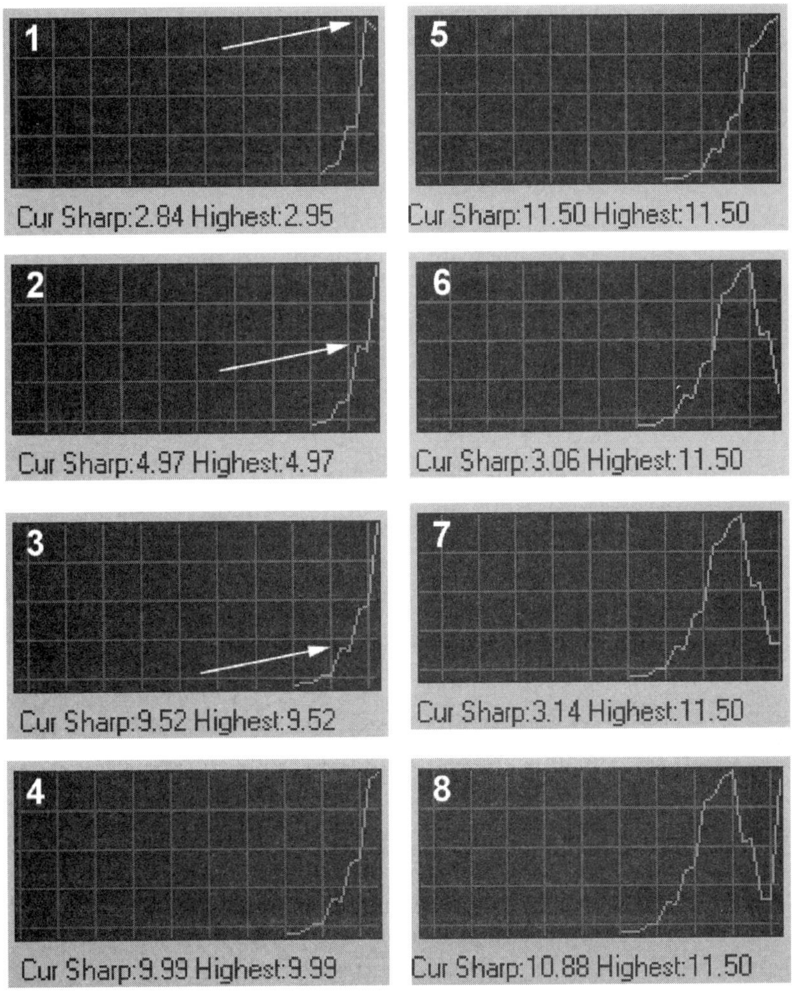

due to seeing conditions. When you reach the low shoulder, the Sharpness value starts to rise quickly. The top of the graph is always the highest Sharpness value so far -- the graph is self-scaling so that it can adapt to whatever values occur during the run.

The three arrows in figure 2.5.9 show the same data point, a value of 2.95 for Sharpness. As increasingly larger values come in, the graph automatically scales itself.

Note in step 8 that the final focuser position does not result in a Sharpness value that is exactly the same as the highest value. If you are using a focuser with absolute positioning, such as the Optec TCF-S, such a small difference is usually due to seeing fluctuations. You can test focus by taking additional exposures and noting the resulting Sharpness values. Ideally, you should expect @Focus to set a focus position that gives you a Sharpness value that is close to the highest sharpness value recorded during the run. Changes in the environment -- varying sky brightness; the presence of thin, high clouds; and other factors can interfere with this. When in doubt, try another @Focus run. If the final focus position is consistently short of best focus, you are probably dealing with backlash.

@Focus needs at least six samples in the active zone to determine best focus (three on one side, three on the other). It also needs 4 to 5 samples before reaching focus to verify direction. Starting 6 large steps from focus guarantees you will meet these conditions. @Focus then calculates the optimal focus position, and moves to it using a combination of large and small steps.

Note that the main graph of Sharpness in figure 2.5.10 is very similar to the curve shown in figure 2.5.8, but it also includes the final move to the best focus position. The inset shows what happens if your focuser has too much backlash. Backlash will eat up some of the focuser travel, and @Focus will not reach focus when it reverses direction. @Focus is doing what

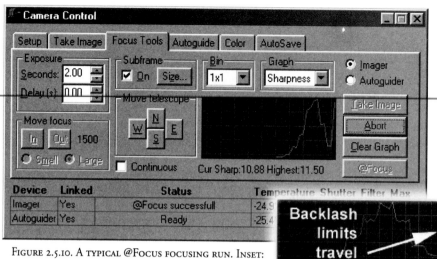

FIGURE 2.5.10. A TYPICAL @FOCUS FOCUSING RUN. INSET: BACKLASH CAN PREVENT REACHING FOCUS.

I recommend for focusing: go past the point of best focus, so that you know exactly where it is, and then reverse direction to return to it. @Focus has the advantage in that it can calculate the optimal focus position mathematically. A focuser with zero backlash allows @Focus to move right to the best focus position.

If you have backlash, you can measure it and enter the result as the Backlash parameter described above (if available for your focuser). If the Sharpness graph looks like the inset in figure 2.5.10, make a note of the focuser position, then continue moving the focuser with the Large step button until you get close to optimal focus. Look for the focuser position where the Current Sharp value is close to the Highest Sharp value. The backlash value is the difference between this final focuser position and the one at the end of the @Focus run.

Some focusers have so much backlash that you will not be able to get to focus reliably. If that's the case, you can manually refine focus using Sharpness values to guide you, or upgrade to a more precise focuser.

The Sharpness curve in figure 2.5.10 is typical of a successful @Focus run. Note that the focusing session started close to the low shoulder, reached a peak, and then came back nearly to the low shoulder on the other side. This profile is characteristic of a successful run. The exact proportions will vary with your starting point, large step size, focal ratio, etc.

SECTION 5: ALTERNATIVE FOCUSING

If you started at the appropriate distance from rough focus, @Focus has enough data to determine the best focus position. It then moves the focuser to the calculated position and reports "@Focus successful" in the Status section at the bottom of the Camera Control panel.

If you know the step size of your motorized focuser, you can use the equation for calculating the size of the close focus zone to calculate the initial large step size for your system. For example, if you are using the Optec TCF-S focuser on an f/5 telescope, you know that the critical focus zone is 55 microns. You also know that the step size of the TCF-S is 20 microns.

> **TIP:** If you do not know the step size of your focuser, you can determine the large step size using the procedures outlined earlier. If you have access to a dial indicator, you can measure the step size of your motorized focuser directly. You will need an indicator that can measure movement as small as one thousandth of an inch. Measure movement when controlling the focuser using the smallest step size available.

A good starting point for your large step size is somewhere in the range of 20 times the size of your close focus zone. For example, an initial large step size for an f/5 scope with the TCF-S would be (20 * 55) = 1,100 microns. At 20 microns per step, this yields about 55 TCF-S steps to make up a large step. This is very close to the 60 steps I obtained using the methods described earlier.

Figure 2.5.11 shows a very different example of a successful focusing run. In this example, the run started further away from focus, and the large step size was smaller. The net result is that @Focus did not get all the way down to the second low shoulder, but it still had enough data to accurately find the correct focus position. Note also that this run was accomplished with binning set to 3x3 (ST-8E camera), and the exposure time was ten seconds because no bright stars were in the field of view.

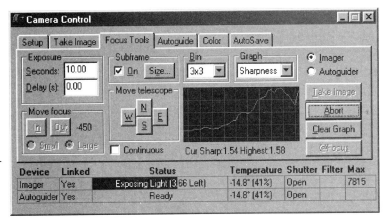

FIGURE 2.5.11. ANOTHER EXAMPLE OF A SUCCESSFUL @FOCUS RUN.

Verify @Focus Results

If your focuser has absolute positioning, you will rarely if ever have poor focus with @Focus. If you have a focuser controlled through TheSky, or if your focuser has excessive backlash, positions are not as repeatable. You can either repeat the @Focus run looking for a better ending Sharpness value, or tweak the focus position using the In and Out buttons to get the highest Sharpness value.

You can also verify the results of an @Focus run by examining the final image visually. Use the various techniques described earlier in this chapter to evaluate focus quality. @Focus will even work with most of the masks in place on your scope.

THE NEW CCD ASTRONOMY

Section 6: Other Focusing Aids

There are other tools out there that can help you master the CCD focusing process. These include rings to make one or more eyepieces parfocal with the camera, and devices that allow you to have both an eyepiece and a camera inserted into the telescope at the same time.

Rings for Parfocal Eyepieces

Figure 2.6.1 shows rings that can be attached to the barrel of an eyepiece to make it parfocal with your camera. These rings are normally sold to allow observers to have all of their eyepieces come to focus at the same point of focuser travel, but they are useful for CCD imaging as well.

The rings prevent the eyepiece from going all the way into the eyepiece holder. This allows you to make an eyepiece come to focus at the same focus position as your CCD camera. The ring has a setscrew in it, and you tighten the setscrew to position the ring. This controls how far the eyepiece goes into your focuser.

Use these rings to make one or more of your eyepieces parfocal with your CCD camera. The simplest way to do this is to hunt and peck your way to focus with the camera in the focuser, and then insert an eyepiece with one of these rings on it. Move the ring until it causes the eyepiece to come to perfect focus, and then tighten down the setscrews in the ring.

From then on, you can insert your parfocal eyepiece, bring an image to focus, and then insert your camera for fine focusing. Software Bisque sells parfocal eyepieces for SBIG cameras (IFocus), but the ring method lets you use your existing eyepieces.

Parfocal eyepieces have their greatest value for telescopes that do not have a visible focusing tube. On a refractor, you can easily measure the correct focusing position for your CCD camera. To return to that focus point, just pull out your ruler and back out the focuser to the appropriate distance.

On an SCT, however, the focusing position isn't visible. A parfocal eyepiece can make it easier to get close to the CCD camera's focus position. The best solution for an SCT, or any scope that moves its primary mirror to achieve focus, is to lock down the mirror and install an alternative focuser such as those described earlier in this chapter.

You can take two basic approaches to choosing an eyepiece to make parfocal. You can choose an eyepiece that has the same approximate field of view as your camera, and use it both for focusing and to frame your subject. Or you can choose a wide-field eyepiece, and use it to assist you in centering objects before you put the camera in. The latter approach is most useful when you don't have highly accurate goto or digital setting circles. You can put the object into the field of view of the eyepiece, center it, and then insert your CCD camera.

Flip Mirrors and Off-Axis Guiding

Flip mirrors provide yet another approach to solving the problem of finding focus. With a flip mirror you can have your cake and eat it too. The flip mirror allows you to have both a CCD camera and an eyepiece attached to the telescope at the same time. Using a mirror, either the eyepiece or the CCD camera receives the light from the telescope. This differs from an off-axis

FIGURE 2.6.1. PARFOCALIZING RINGS [COURTESY GARY'S ASTRO FABRICATING, GARY WOLANSKI]

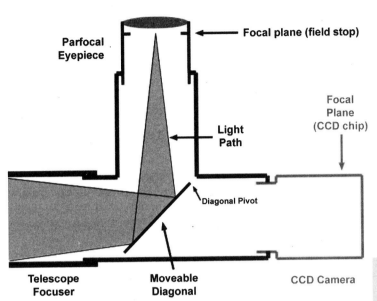

FIGURE 2.6.2. A FLIP MIRROR CAN DIRECT LIGHT TO EITHER AN EYEPIECE OR A CCD CAMERA. AS SHOWN HERE, THE MOVEABLE DIAGONAL IS POSITIONED TO REFLECT LIGHT UPWARD TO AN EYEPIECE.

of the term "off-axis." The mirror reflects to an eyepiece or autoguider, so that both the camera and the eyepiece/autoguider receive light from the telescope simultaneously. Neither interferes with the other since they "see" different parts of the light beam.

With the flip mirror, the image you see is identical to what the camera sees, but you have to switch from one to the other. A flip mirror is useful for focusing and framing, but not for guiding. With an off-axis guider, both devices get a portion of the incoming light at the same time, so you can guide manually or with an autoguider during an exposure.

TIP: To use a flip mirror or an off-axis guider, you will need to have enough focus travel on your telescope to accommodate the rather large space required for the unit. Measure the longest light path in the unit to determine whether it will be suitable for your scope. Don't forget to include the length of any adapters or accessories that might also be needed for proper operation of the unit.

guider, which uses a small pick-off mirror to direct a small portion of the incoming light to an eyepiece or autoguider.

A hypothetical flip mirror is shown in outline form in figure 2.6.2. It has a moveable diagonal mirror. When the mirror is in one position (figure 2.6.2), it reflects light into an eyepiece.

When the mirror is flipped, as shown in figure 2.6.3, it moves out of the light path, and the light travels directly to the camera. The eyepiece and camera can be adjusted so that both come to focus. This makes the eyepiece parfocal with the camera. To use the flip mirror, you frame and focus using the eyepiece, and then flip the mirror out of the way. The light goes to the CCD camera for final focusing and imaging.

An off-axis guider is similar to a flip mirror, but it has a small mirror that "picks off" an unneeded portion of the light beam (see figure 2.6.4). That is, the mirror is in a portion of the incoming light that does not cast a shadow on the CCD chip. This is the source

FIGURE 2.6.3. THE FLIP MIRROR IS SET UP FOR IMAGING. THE MOVEABLE MIRROR IS FLIPPED INTO THE UP POSITION, ALLOWING LIGHT TO REACH THE CAMERA.

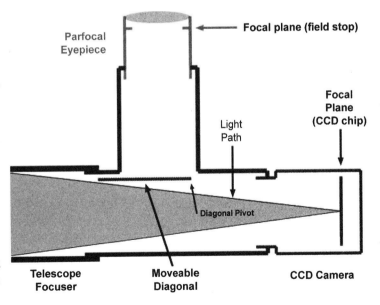

The autoguider you use with an off-axis guider can be any CCD camera that can output guide corrections. The imaging camera is placed in one position, and the guiding camera is placed in the other position. The entire assembly must be very rigid and stiff for this to work; if there is any flexure, the two cameras could become misaligned and spoil the guiding.

Although flip mirrors and off-axis guiders can be useful, they are sometimes a challenge to use. I personally prefer doing without, but many imagers get excellent results with both types of devices. You can swap camera and eyepiece instead of using a flip mirror, for example. The alternative to an off-axis guider is a separate guidescope, which presents its own problems unless you can make sure that both sets of optics are rigidly mounted and will stay stable with respect to each other.

This is why cameras that self-guide are so popular. They remove a lot of time and trouble (and often expense) required for more complex arrangements.

However, the experienced imager gains some flexibility from the use of a separate guidescope. It is easier to find guide stars, and you can use high-end stand-along guiders such as the STV.

FIGURE 2.6.4. THE OFF-AXIS GUIDER DIRECTS PORTIONS OF THE LIGHT PATH IN TWO DIFFERENT DIRECTIONS.

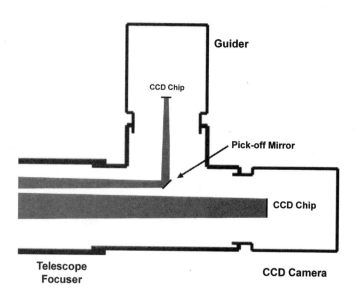

3 *Practical Imaging*

CCD imaging requires the utmost in precision.

The optical system in your telescope must be in the best possible alignment. Your mount must be tuned to provide the best possible pointing, tracking, and guiding. The goal is to optimize every element of the system so you can get the best possible images.

Section 1: Setting Up Your Telescope and Mount

Alignment of the optical system is called collimation. Some telescopes are more difficult to collimate than others. Refractors, for example, are most often collimated on an optical bench. Fortunately, they hold collimation exceptionally well, and may never need recollimation. Many reflectors, such as Newtonians and most Cassegrains, are easier to collimate, and usually can be collimated in the field. Many reflectors require frequent collimation, though some will hold their collimation fairly well.

This section provides instructions for collimating scopes in the Cassegrain family of telescopes. They are the most commonly used types of reflecting telescopes for CCD imaging. The two most common are the Schmidt-Cassegrain and the Maksutov-Cassegrain.

The other essential ingredient for imaging a well-tuned mount. The key issue is backlash. You need to know how much you have, you need to reduce it to a practical minimum, and you need to know how effectively you have compensated for what remains.

Collimation: First, Last, and Always

Collimation is simply the act of aligning the optical elements of your telescope. Not all telescopes can be user-collimated, but those that can should be collimated often. If you own a Newtonian, a Schmidt-Cassegrain, or any other type of telescope which provides some means of field collimation, you will always get better results if you take the time to carefully collimate the optics. Collimation is work, but you gain so much from good collimation that it is always worthwhile to make sure your collimation is correctly set.

Figure 3.1.1 shows a Schmidt-Cassegrain with an exaggerated secondary mis-collimation. The essence of the problem with poor collimation is that one or both of the optical components is titled with respect to the optical axis. This causes the focal plane to tilt. Since the focal plane is most likely curved rather than flat, the result is a stretching out of the star images or other optical aberrations.

Figure 3.1.2 shows a properly aligned system. The mirrors are parallel and at right angles to the optical axis. The focal plane is now lined up with the CCD chip, and stars focus to points. Even if the focal plane is still slightly curved, the CCD chip is usually small enough that this is not an issue.

Figure 3.1.3 shows why this is true. The upper example shows the coverage of medium-format film with respect to a typical curved focal plane. The curve is slight, but the film is large enough to be at a relatively large distance from the focal plane at its outer edges. In practice, there are two solutions. One is to curve the film to match the focal plane, as is done in a Schmidt camera. The other is to put a corrector lens between the secondary and the film to eliminate the curvature and provide a flat field of view. Curvature is a natural artifact of the Cassegrain and many other designs. Different types of telescopes have different amounts of field curvature. Astrographs are telescopes designed with an absolute minimum of field curvature.

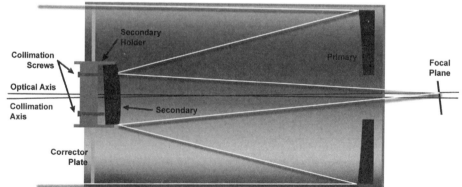

FIGURE 3.1.1. IF THE SECONDARY MIRROR ISN'T ALIGNED TO THE OPTICAL AXIS, THE FOCAL PLANE BECOMES TILTED.

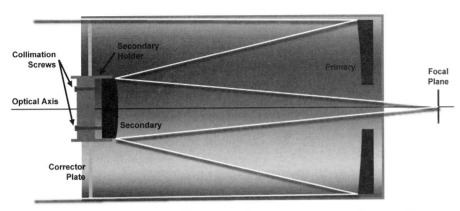

FIGURE 3.1.2. A PROPERLY COLLIMATED SCHMIDT-CASSEGRAIN.

collimation. The exact appearance will vary. Figure 2.2.4 in chapter 2 shows one example of aberrations from poor collimation. Stars will bloat and extend, and you won't be able to resolve fine details. Contrast is also reduced with poor collimation.

Telescopes with a fast focal ratio are among the most likely to be affected by collimation. A fast Newtonian, such as an f/4.3, will show lack of collimation very easily.

Certain telescope designs, however, are especially sensitive to collimation errors irrespective of focal ratio. The f/11 Dall-Kirkham design of the Takahashi Mewlons, for example, will show the slightest error in collimation all too clearly. These designs are optimized for a sharp but small field of view. The small size of most CCD chips works well with such a design.

The middle example in figure 3.1.3 shows why most CCD chips do not show adverse effects from a curved focal plane. The chip is often so small that the amount of curvature across its surface is minimal. In fact, unless the chip is very large, it is likely that the distance from the focal plane will be less than the critical focus zone, so that every part of the CCD chip will be in focus when you examine the image. In effect, the CCD chip is using the best part of the focal plane.

The lower example in figure 3.1.3 shows what happens when the focal plane becomes titled with respect to the CCD chip. The focal plane is no longer in close contact across all of the chip, and aberrations result wherever the distance between the chip and the focal plane is greater than the size of the critical focus zone. This is why collimation is so important to getting crisp images.

The symptoms of poor collimation are as bad as being significantly out of focus. Star elongation is the most typical result of poor

FIGURE 3.1.3. TOP: A CURVED FOCAL PLANE CAUSES POOR FOCUS ON FILM. MIDDLE: A SMALL CCD CHIP IS NOT AS AFFECTED BY A CURVED FOCAL PLANE. BOTTOM: A TILTED FOCAL PLANE CAUSES PROBLEMS WITH FOCUS.

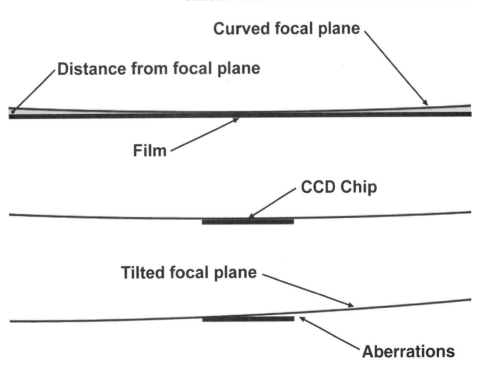

Although refractors hold their collimation very well, you will occasionally see a collimation problem in a refractor. All it takes is a misalignment of the optics to create collimation problems in any telescope.

Collimating a Cassegrain's Secondary

Collimation seems like such a simple thing, but there are plenty of subtle touches you can learn to make collimation easier. The techniques described below cover collimation of a Dall-Kirkham Cassegrain (specifically, a Takahashi Mewlon 210), but they apply to many other types of telescopes in the Cassegrain family, such as Schmidt-Cassegrain, Ritchey-Chretien, classical Cassegrain, etc. Collimation is important on any telescope, but it is particularly important on many Cassegrain designs because even a little mis-collimation causes problems. Collimation of Newtonians is well documented in many places, and is therefore not covered here.

There are several signs that indicate a need for collimation:

- Elongation of star images. This is a trailing of star images away from some common point in the image frame (not necessarily the center). If stars aren't pinpoint sharp no matter how well you focus, you may be dealing with a collimation problem. Examine the stars by zooming in to see if they are oblong rather than round.
- Mildly out-of-focus star images with one side of the diffraction rings brighter than the other side. The focal plane is tilted, and one side of the star images is brighter because it is receiving more illumination.
- Far out-of-focus star images show a secondary shadow that isn't at the center of the out-of-focus image. This indicates serious miscollimation. Minor collimation errors won't be visible when you are very far out of focus.

The best way to collimate is on a star at night. You can also create an artificial star to collimate with during the daytime. To be effective, an artificial star has to be as close as possible to a point source, and be located at an appropriate distance from your telescope. The longer your focal length, the greater the distance must be. The close-focus point of your telescope will also constrain the distance to an artificial star. If you have access to a very large interior space, such as a warehouse, a flashlight suspended above a small metal ball makes a good artificial star. The hardest thing about indoor collimation is finding a space big enough. Outdoor collimation is limited by air currents that disrupt the star image and make it hard to see just how well collimated the telescope is.

Stars at night are ideal, but you will need reasonably steady seeing to achieve good collimation. The better the seeing, the better you'll be able to judge collimation. To collimate, you will use the diffraction rings around a slightly out-of-focus star. If the air is turbulent, the diffraction rings will be tossed around and you won't be able to see them clearly.

You will need the following to perform a secondary collimation:

- Two or three eyepieces that offer a range of magnification from about 200x to 600x.
- An Allen wrench or screwdriver appropriate to the screws that you will use to set collimation. These are the screws on the secondary mirror holder. See figures 1 and 2 for the typical location for accessing these screws.
- A flashlight.
- Patience!

Collimation Guidelines

You should always put the eyepiece directly into the visual back of the telescope for collimation. Never use a diagonal. You want the straightest possible light path for collimation. A diagonal could (and usually does) introduce alignment error that will throw off the collimation.

You should make one adjustment at a time. An adjustment is usually a combination of loosening one screw, and tightening two others. On some scopes, the screws may be spring-loaded in which case you can adjust one screw at a time. On many scopes, the screws all need to be tight at the same time to steady the secondary, and if this is the case, a single adjustment involves three screws. Otherwise, you will think the scope is collimated, and when you lock down the screws, it will no longer be collimated.

An exception to this would be when doing final tweaking. You will find that you can make very small adjustments by loosening one of the screws. Tightening just one screw can lead to over-tightening, which could distort the secondary in some arrangements. Most of the time the secondary rides on a platform and can't be distorted, but this cannot be relied upon blindly.

Moving the collimation screws is like trying to tie your shoes by looking in a mirror. Everything is backwards and the familiar suddenly becomes unfamiliar. Making one change at a time helps you learn the system.

Always (repeat: always!) re-center the star after every adjustment. Collimation pertains only to the exact center, the exact optical axis. Adjusting for a star that isn't on the optical axis leads to miscollimation.

And most important of all: be patient! It may take you a half hour or an hour to collimate the first time. Once you understand how it all works, you can finish a good collimation in minutes.

Setting Up for Collimation

To start collimation, point your telescope at a moderately bright star. "Moderately bright" will vary based on seeing, the aperture and focal ratio of your scope, the eyepiece you are using, etc. So there is no hard and fast rule. The key, however, is to choose a star that is bright enough to give you diffraction rings just outside of focus, yet not so bright that the rings are thick or too bright.

The object of collimation is to make adjustments until the diffraction rings of a slightly out-of-focus star are as perfectly concentric as you can make them. Start with an eyepiece that gives you about 200x, and center the star in the field of view. You *must* have the star in the center of the field of view to collimate. When an adjustment to a collimation screw moves the star out of the center of the field of view, move the mount to put the star back at the center. Defocus the image slightly. You want to see a few diffraction rings, not a broad doughnut of light.

Figure 3.1.4 shows some of the things you might see while collimating. You need a slightly out-of-focus star image to work with during collimation. The image at far left of figure 3.1.4 shows what you don't want: an almost solid doughnut of light. This star image is too far out of focus to be useful for collimation. The problem is that the inner and outer edges of the doughnut are fairly far apart, and it is difficult to judge when they are precisely concentric. On the other hand, if the doughnut is obviously not concentric, then you know you are very far out of collimation, and there is serious work to do.

The image in the center of figure 3.1.4 shows what a slightly out of focus star will look like in a slightly miscollimated scope. The diffraction rings are not concentric. They are pinched or bunched up in one direction. You may also see some flaring or fuzziness on the side away from the pinching, or the rings may look oval instead of circular. These are all typical symptoms of a minor miscollimation. If the collimation is very poor, you may not even see the diffraction rings very clearly because they are so stretched out. In such a case, you need to make larger adjustments.

The image at far right is what you can expect to see when you've got collimation exactly right. The seeing conditions may blur or fragment the diffraction rings, but the rings can only be concentric when collimation is right.

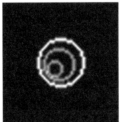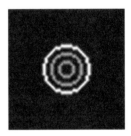

FIGURE 3.1.4. EXAMPLES OF WHAT YOU MIGHT SEE THROUGH THE EYEPIECE WHILE COLLIMATING. SEE TEXT FOR DETAILS.

Secrets of Collimation

There is a very simple rule you can follow that will make collimation a pleasure rather than a chore. I have watched people (including myself, once upon a time) begin collimation by making random changes to the collimation screws, and try to learn which screw controls which direction. Granted, after 20 minutes or so, you will be an expert on which screw moves collimation in which direction.

When you start with the out-of-focus star in the center of the field of view, one side of the star will bulge out like the middle image in figure 4. What you want to do is find one collimation adjustment which, when loosened or tightened, will move the star in the direction of that bulge. You don't want or need to do anything that doesn't accomplish this simple goal.

So loosen one collimation screw a small amount. "Small amount" usually means about 1/8th of a turn, or something close to that. During this test phase, do not tighten the other two screws before checking your results. Observe whether or not the adjustment has moved the out-of-focus star image in the direction the bulge is pointing. If the answer is no, make a note of the direction of movement (on paper if necessary) and re-tighten the screw so that the out-of-focus star image is again centered. Adjust the pointing of your scope if the re-centering is not exact. Then try a different collimation screw. Repeat until you find one screw that moves the image as close as possible to the desired direction. Tighten the other two screws, and then re-center the out-of-focus image. You have now made your first collimation adjustment.

Evaluating the Collimation Adjustment

Examine what has happened to the out-of-focus star image. You should see an improvement in collimation (unless you made too large of an adjustment, in which case the bulge will be pointing in the other direction). The bulge should be smaller, and the pinch on the other side of the star should be reduced. This evaluation must be made *after* you re-center the star!

Note whether the bulge points in a new direction. This will affect which screw to use for the next adjustment. If collimation looks perfect or very close to it, change to a higher power eyepiece and continue until perfection is achieved, or whatever the seeing will allow. It is only when you get to around a 600X eyepiece that you will get the kind of collimation that will knock your socks off while viewing planetary detail on a still night. Unfortunately, the seeing is often not good enough for that level of collimation.

When you have gotten good collimation while slightly out of focus, you can improve it further by collimating in focus. It takes really steady seeing and a high-power eyepiece to collimate in focus. The principles are the same, but you are working with the very faint diffraction rings around the in-focus star instead of the out-of-focus rings. The in-focus rings are harder to see, and require a high power eyepiece, a bright star, and superb seeing conditions. The goal is to make the diffraction rings around the Airy disk as concentric and evenly bright as possible. I have never been able to do this with less than a 600X eyepiece in the telescope, and only on extremely steady nights.

Free Play (Backlash) Adjustments

Visual astronomers sometimes take their mount for granted when they track and slew across the sky. Imagers are less likely to get complacent about their mounts. Tracking the movement of the stars accurately enough to take long exposures is closer to a miracle than not. The accuracy required is phenomenal, on the order of a couple of arcseconds. Given that there are 1.3 million arcseconds in a circle, following a star with that accuracy for minutes at a time is a tough job.

If the word backlash isn't in your vocabulary yet, it will be soon. Backlash is the looseness in the mount's gears. Some backlash is necessary so that the gears are free to turn. Without at least some small amount of backlash, even the finest mounts would seize up with friction. The amount of backlash is part of what separates the capable mounts from the also-rans.

No matter how eager you are to start imaging, you will almost certainly get better images if you take some time to understand the level of backlash in your mount, and then do a few things to bring it under control. The ability to track the stars is based on a mount's ability to react immediately to any errors in tracking. Excessive backlash can prevent that immediate response, resulting in flaws in long exposures. Knowing your backlash, taming it with backlash compensation, and then keeping it under control will give you better images.

Backlash is best dealt with by prevention rather than attempting cures. Knowing your backlash means understanding the fundamental behavior of your mount. The steps to dealing with backlash are:

- Find out how much backlash you have.
- Reduce backlash to the lowest practical point by tuning your mount.

- Compensate for whatever backlash remains. Some mounts are able to run their motors at a higher speed for a very brief period of time to take up backlash. This is called backlash compensation.

Once you understand how much backlash you have, and have done what you can to reduce it, you are ready to start imaging with much greater confidence.

You'll learn the details of evaluating and tuning your mount in later chapters.

Your mount is a key element in the imaging process. It's impossible to overstate how important a well-tuned mount is. In addition to the tips you'll find in this book, you should scan the Internet for web site that offer tips and tricks specific to your brand and model of mount. These can be invaluable in getting the most out of your mount.

Measure Your Backlash

You can use your camera control software to measure your backlash. To measure the current physical backlash, turn off any backlash compensation, set your mount to move at guiding speed, and then follow these steps. To make it easy to evaluate your results, insert the camera so that it is square to the mount's axes and with the top of the CCD chip oriented toward north.

The following procedure assumes that you have a very good polar alignment, and that the camera is set up, cooled, and ready to image. You can measure backlash with most camera control programs. The autoguiding features of such programs are the most convenient because they usually provide a means to move the mount at guide speed manually. During the procedure, if you wind up reversing direction other than as directed, start over to make sure that you measure backlash accurately.

1. Pick an axis and a direction, and move the scope at guide speed in that direction long enough to get past any backlash. For example, if you are measuring backlash in Dec, it will be the Y direction (up and down when the camera is set with North at the top of the frame). Move +Y for a long enough time to remove any possibility of remaining backlash. This could be 10 seconds; it could be a minute if you have a lot of backlash. (If necessary, take an image to verify that you have gotten past the backlash and are moving the mount.)

2. Take a 5-10 second image using the guide chip (or imaging chip if using a one-chip camera). For best results, make sure you have a bright star that is noticeably brighter than the other stars so you can find it on subsequent images. If you don't have a bright enough star on the chip, continue moving in the +Y direction until you find one. Take an image and save it as your reference image.

3. Now pick a time interval for a move. It should be long enough to move your chosen star about 10 pixels or more, but not so long as to cause the star you chose in step 2 to move off of the chip.

4. Move in the -Y direction for the chosen time interval. Take an image. Measure the amount that the star has moved. If the star has NOT moved, or moves less than 10 pixels, your move time was not long enough to take up the backlash. Start over from step 1, and use a longer move time. If you cannot find a time long enough to move the star in step 3, then your backlash is extreme and you should take steps to reduce it before starting over.

5. Move in the +Y direction for the same time interval. Take another image. If you have a very small amount of backlash, the star will return almost exactly to the starting place in the reference image. If it does not return to the starting point, you have backlash, and you have just measured it in pixels. To convert to arc seconds, determine your image scale in arcseconds per pixel, and multiply the number of pixels by the image scale.

You can now adjust backlash compensation as needed. After setting the compensation, measure backlash again to see how accurately the compensation is set. If your second +Y move goes too far, reduce the amount of compensation. If the second +Y move winds up short, increase backlash compensation. Repeat from Step 1 each time you change the backlash compensation until you are satisfied. Be careful not to overdo compensation; always leave at least a bit of backlash in the system. Too much compensation will have a worse effect on your images that too little.

Physical Adjustments

Compensation isn't the only way to deal with backlash. You can also remove excessive backlash by tuning your mount. Different mounts require different amounts of backlash to operate properly, and not all mounts provide a simple way to adjust backlash. Check your documentation, or contact the manufacturer, to find out what the proper amount of backlash is for your mount and to learn the method for adjusting it.

Use the following generic procedures for evaluating and adjusting most mounts used in astrophotography:

- You can check for gross backlash or other looseness by attempting to move the mount in the RA and Dec axes manually while the mount is set up with telescope and counterweights. Don't force things! Just a gentle to and fro motion will tell you if there is a large amount of backlash present. It is possible to have too much backlash for your setup and not be able to feel a thing, however, so this is just a check for really large amounts of backlash.

- Check the amount of endplay in the worm gears. Endplay, if present, will create some non-intuitive behaviors during guiding. If there is endplay, the mount will start moving in the opposite direction briefly before reversing and moving in the expected direction. It may also move in the opposing axis as the mount takes on loading after reversals. These kinds of behaviors are deadly for guiding, since an attempted correction in a given direction results in movement in the opposite direction. This leads to another, larger guide adjustment, which causes further movement in the wrong direction. Adjustment of endplay typically involves snugging some kind of retaining ring or nut on one end of the worm. Consult your mount's documentation or contact the manufacturer to determine what to adjust if you have any doubts. Don't over tighten, or you will create binding that could be very bad for the health of your gears! Finding the sweet spot on gears and bearings is something of an art form; when in doubt, try to find someone with some experience in this area.

- Check the amount of backlash in RA and Dec. This is usually due to loose mesh between the worm and worm gear. Adjust if necessary. Some mounts do not provide this adjustment, and may not be suitable for imaging. There will also be some backlash in the gear train between motor and worm, but this is not usually adjustable. As with any gears that mesh, some backlash is required here. The goal overall is a minimum of backlash in both axes, without being too tight. If any of the gears are too tightly meshed to their neighbors, the motors will strain to move the mount, or the gears may even bind and prevent movement. The amount of necessary backlash will vary with the quality of the mount and the torque of the motors. Higher quality mounts and high-torque motors can work properly with a tighter mesh. Backlash is adjusted by varying the distance between the worm and the worm gear. The way you do this varies from mount to mount. Some mounts make it easy to adjust, while others hide this adjustment and require you to tear the mount half apart. When in doubt, contact the manufacturer to learn how it is done. It's also important to get the worm square to the driven gear; having it off at an angle can result in guiding problems or uneven wear.

- Set backlash compensation for RA and Dec if available. This is usually found on the mount's hand controller. This is a trial and error process to get the right settings; measure as outlined above if you want reliable settings. The idea is to add compensation sufficient to eliminate any pauses when switching directions in RA or Dec, but not so much as to make the mount jump in the new direction. I have found that visual testing is not, repeat *not* sufficient for setting backlash compensation. I have generally found that visual adjustments tend to result in overcompensation. If you measure with your camera using guide speed, you will get much more accurate backlash compensation. It takes a significant chunk of time, but it's not something you have to do often. If your mount doesn't have hardware backlash compensation, many camera control programs offer software compensation.

Following these adjustments, your mount is tuned and you are familiar with its behavior. You can now calibrate the camera control software to the mount. For more information, see the section on "Mount Calibration" in chapter 5.

Section 2: Signal versus Noise

Noise is always present in the CCD imaging process. Noise lurks around every corner, ready to foil your efforts. Learning to control, limit, and reduce noise is an important key to successful imaging.

Where does noise come from? You can find it in the CCD chip itself, in the camera, in the heat and light that find their way into the camera, in the process of reading the data from the chip, in image processing, even in the quantum nature of light itself. Virtually everything you do with a CCD camera has the ability to introduce some noise.

The good news is that many of the sources of noise can be dealt with effectively. Most CCD cameras are very sensitive to heat energy, and cooling the camera greatly reduces noise from this source. A camera like the SBIG ST-7E has 50% less noise for every 6 degrees Celsius you cool it. Since the camera comes with about 35+ degrees of cooling capacity, the noise level when cooled is just 1/64th of what it would be without cooling.

Still, there is always some residual noise. The trick is to get as much signal as possible to overwhelm the noise. The best CCD images always have a high signal to noise ratio. This means that there is a lot of signal (the image) and very little noise. The camera designers have done a lot to reduce noise, but there are several things you can do to get the highest possible signal to noise ratio:

- Image under dark skies
- Take long exposures
- Use as much cooling as possible
- Combine multiple images

The best CCD images will always come from taking multiple, long exposures at dark sites with a well-cooled CCD chip. But you don't have to do it all to get decent results. For example, if you are imaging from your back yard, light pollution may be the norm. If that's the case, then you can lean more heavily on the other techniques: cool the camera as much as possible, or select a camera that has a very high degree of cooling such as the MaxCams from FLI. You could also benefit by taking long exposures and combining them.

Similarly, if you are unable to take long exposures because you don't have a guider, you can use dark skies, extra cooling (perhaps the SBIG secondary cooling package) and combining images to get a better signal to noise ratio. You can get decent images of deep sky objects even from the city if you are willing to take the long, multiple exposures required. They won't be as deep as images taken from a rural site, and there's no way around that. Light pollution doesn't stop you from imaging; it just makes you work harder.

Signal to Noise Ratio

The signal to noise ratio is simply the ratio of the signal in your image to the noise in your image. If the signal is 1000, and the noise is 50, then the signal-to-noise ratio is 20 (1000/50). The signal to noise ratio is often abbreviated as S/N.

> **TIP:** Technically, S/N is measured in decibels, which is abbreviated dB. The formula for expressing S/N in decibels is 20 times the log of the S/N ratio. This is convenient for engineering types, but the details are beyond the scope of this book. S/N will be measured as a simple ratio throughout the book.

The noise in an image is the uncertainty in the brightness level. This point is often misunderstood, so it's worth a moment's examination to be clear about it. Noise is more of a mathematical concept than a simple intuitive concept, so it's easy to latch onto analogies that don't quite fit. A common misperception is that signal to noise can be measured by comparing the brightness of the image background with the brightest details in the image. It's not that simple.

To measure noise, you must repeat a measurement many times and analyze it statistically. Figure 3.2.1 shows an example that should help you get a handle on what noise is and why it's a problem. You are looking at highly enlarged rows of pixels from two images of the same part of the sky. I selected the pixels from the same area in each image. That area was one where the brightness level should be the same for all of the pixels in each row. The top row shows a lot of variation in brightness, while the bottom row shows much less variation. Since variation is a measure of noise, the top row of pixels is noisier.

Figure 3.2.2 shows enlargements of the two images from which the rows of pixels were taken. The image on the left is the noisier of the two. The variations in brightness due to noise create a grainy appearance. The image on the right has much less grain, and is therefore the less noisy of the two.

Fortunately, you don't need to measure noise to take steps to reduce it. Dark skies, long exposures, a cold chip, and combining images each can work to improve the quality of your images. Each technique contributes to noise reduction. If you are lucky, you have control over all of these factors, allowing you to create images with superb signal to noise ratios.

FIGURE 3.2.1. COMPARING NOISE IN HIGHLY MAGNIFIED ROWS OF PIXELS. TOP: NOISY. BOTTOM: LESS NOISY.

Imaging under dark skies improves S/N because there is very little background illumination to mask the signal from distant sources. Big signal and little noise deliver high S/N.

Long exposures improve S/N because signal always increases faster than noise. Time is on your side, and it delivers higher S/N.

Combining improves the signal to noise ratio because signal increases faster than noise when you combine images, too. This means you could take shorter images and still get excellent results. But you will get even better results by taking the longest possible exposures and combing them. There is some camera noise in each individual exposure. Long exposures will be less noisy than many short ones.

Cooling lowers noise by reducing the thermal energy in the camera. Heat generates more stray photons than cold, so a cold CCD chip is struck by fewer unwanted photons.

Taking one very long exposure will always deliver a little better S/N than combining images. But a number of factors limit the maximum exposure time:

- Non-antiblooming cameras allow stars to bloom. Longer exposures have a greater potential for blooming.

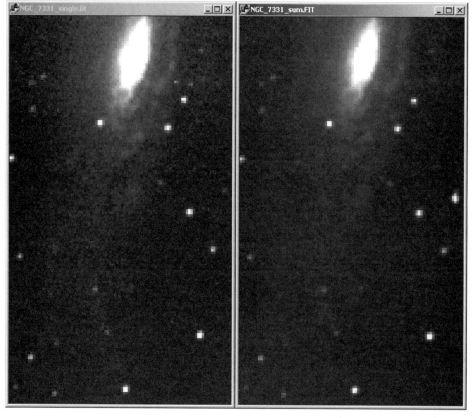

FIGURE 3.2.2. COMPARING A BLOW-UP OF A NOISY IMAGE (LEFT) AND A CLEAN IMAGE (RIGHT).

FIGURE 3.2.3. A SINGLE IMAGE OF THE CONE NEBULA.

- Environmental conditions can limit exposures times. Sky glow, for example, can create excessive background levels in long exposures, resulting in a poor signal to noise ratio.
- The risk of hazards increases with longer exposures. Hazards include satellite tracks, meteors, cosmic rays, etc. Even a bump on the mount is a hazard. The longer your exposure, the greater the likelihood of a problem.

For example, if a 30-minute exposure of the Cone Nebula results in excessive background levels from light pollution, you could take three 10-minute exposures, or six 5-minute exposures, or any combination of exposures that suits your conditions. Combining these exposures would get you close to the signal to noise ratio of a single 30-minute image. The longer your indivudual exposures, the closer you get. The decision about how much to risk by taking long exposures is up to you. Whenever possible I like to take 30-minute exposures with an ABG camera, but if there are problaems from any source, I'll take 10-minute exposures.

Note: You cannot make copies of an image and then combine them to reduce noise. The images must be taken separately and then combined in order to reduce noise. The noise in one image tends to cancel some of the noise in the other images. No cancellation can occur if the images are identical.

Reducing Noise

The risk of hazards influences the choice of exposure time. I like to always take at least three images so that I can use them to reduce noise from hazard and other sources. Combining images often results in a better signal to noise ratio for a given amount of exposure time. During a single 30-minute exposure, you would probably wind up with at least a few cosmic-ray hits, and perhaps a satellite track. If you took six 5-minute exposures, you could use median combining to reduce both the cosmic ray hits and the satellite track to insignificance. You wouldn't simply add the images instead of median combining them because you would need to

FIGURE 3.2.4. A SUM OF THREE SEPARATE IMAGES PROVIDES A BETTER SIGNAL TO NOISE RATIO.

throw out all of the hazard images. Summing is not effective at removing that type of noise.

In other words, I don't simply take one 30-minute image instead of three 10-minute images or five 6-minute images. If I'm taking one 30-minute image, I'm in for at least two more in order to control noise. In fact, I like to get at least four or five images because that allows me to choose how to combine them based on the results I get. With that number of images, I can either toss out the bad images and sum, or use median combine, depending on what provides the best result.

The relative advantages and disadvantages of summing and median combining are discussed in detail later.

Because of the reduced risk of hazards when using multiple exposures, you can take large numbers of short images with little fear of disaster. You can discard any that are ruined by hazards. And if you want the ultimate in S/N, increase your total exposure time beyond the longest exposure you might consider taking. For example, you could take a series of images adding up to 45 minutes, and combine them to get a better signal to noise ratio than a single 30-minute exposure would give you.

On the other hand, I usually take single images of 30 minutes. Yes, there is a risk of hazards, but they don't occur all that often. And the greatest hazard is sometimes the camera operator. My own mistakes have cost me more imaging time than any other hazard.

Figure 3.2.3 shows a single image of the Cone Nebula. The exposure duration was 3 minutes, which was as long as it was possible to go without serious blooming using an SBIG ST-8E camera binned 2x2. The image was adjusted and tuned to display as much nebulosity as possible. Note that the dim areas of the nebula are grainy. Graininess is a sure indication of noise.

Figure 3.2.4 is also an image of the Cone Nebula, but this time three exposures of three minutes each have been summed together using CCDSoft. The dim areas are less grainy than in the single image (see figure

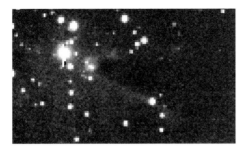 

FIGURE 3.2.5. DETAIL SHOWING LOWER NOISE FOR COMBINED IMAGE (RIGHT).

3.2.5 for enlarged detail). This lack of graininess is characteristic of good signal to noise ratio. Summing images improves S/N; signal increases faster than noise.

Figure 3.2.5 shows details of two images. Both are noisy, but the image on the right, a combination of three exposures, has less noise and better contrast.

Figure 3.2.6 shows a small detail from four different images of the Cone Nebula. The two images on the left are single images. The two images on the right are combined images. The bright patch at lower left is a good area to examine for comparing the noise levels.

Both of the single images clearly show noise (graininess). The top left image has had a simple histogram adjustment (setting black and white points). The lower image has had a histogram stretch to show more dim details. These images are noisy.

The combined images show less noise. The top image is an average of three images. Each pixel is the average of the values of that pixel in three images. The bottom image is a sum of the same three images. Each pixel is the sum of the pixel values in all three images.

The averaged image at top right has less noise (grain) than either of the single images. There is still some noise, but the overall appearance is smoother.

The summed image at lower right also has low noise. The grain is less than in the single images.

The advantages and disadvantages of different image combining methods are discussed in detail in the "Combining Images" section later in this chapter.

Note: Before you can combine images, the images must be aligned. Camera control software typically provides image alignmen tools. For information about aligning and combining images, see chapter 6.

FIGURE 3.2.6. COMPARING THE NOISE LEVEL (GRAININESS) OF DIFFERENT IMAGES.

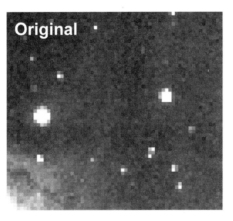 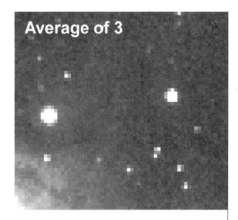
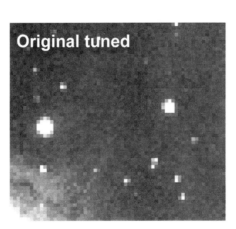 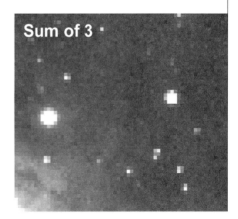

Section 3: Imaging the Sun, Moon, and Planets

Bright objects like the sun, moon, and planets require short exposures. This means that issues such as the tracking of the mount, guiding corrections, and other complexities are eliminated or at least reduced in importance.

This simplification makes taking images of the brighter solar system objects easier, but there are still a few gotchas to watch out for.

With most astronomical objects, the problem is collecting enough photons to get a good image. The sun and the moon, however, are so bright that you need to reduce the amount of light striking the CCD chip. Many cameras do not have sufficiently short exposures to image the moon, and all cameras require a solar filter (and sometimes more) to image the sun. Planets do not require filters to attenuate the light; magnification with a Barlow usually solves that problem by increasing the focal ratio. Focal ratio controls exposure time. If an f/5 imaging system gives you overexposed images, an f/10 system or slower will tame the excessive light. You also get a larger image with an increased focal ratio.

Figure 3.3.1 shows what you can expect with a small aperture (5" in this case) and an appropriate solar filter. If you have superb seeing conditions, you can magnify using a Barlow and record even more detail on the Sun, Moon, and planets.

Filters are mostly used on the sun and moon. Long focal ratios are used most often on planets, but also on the sun and moon when the seeing conditions are good enough. While magnification on the sun and moon is optional, planets are tiny, and magnification is almost essential. Planets will leave a very small image on the CCD chip without some kind of magnification. The shorter your focal length, the more likely you are to need supplemental magnification in the form of a Barlow or eyepiece projection.

Choosing and Using Filters

Filters, including both moon and solar filters, vary widely in quality and suitability for CCD imaging. For solar imaging with film cameras, special solar filters exist that pass more light than visual filters. For CCD solar imaging, exactly the opposite is needed: the light must be reduced dramatically, frequently even more than for visual observing. The idea is to reduce the incoming light to a fairly extreme degree so that the CCD chip won't saturate. This is critically important with non-antiblooming chips. You can mask off part of the aperture if necessary, or add additional filtering to reduce the incoming light for solar imaging.

Many CCD cameras also require filtering for lunar imaging. The moon is much less bright than the sun, but still bright enough to overwhelm many CCD cameras.

Figure 3.3.2 shows one version of what to expect if you have too much light coming in when you are using a non-antiblooming

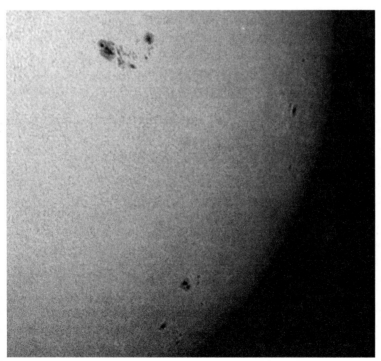

FIGURE 3.3.1. AN IMAGE OF THE SUN SHOWING SUNSPOTS AND FACULAE.

FIGURE 3.3.2. SATURATION OF PART OF THE IMAGE RESULTS IN EXCESSIVE BLOOMING.

CCD chip. In this example, the moon's image has saturated some but not all of the pixels. The vertical streaks are evidence of the electrical charge leaking from one pixel to the next in a blooming cascade that affects a large section of the image. The same thing can happen with solar images; you might even wind up with a completely saturated image (see figure 3.3.3) even if you are using a solar filter. With an anti-blooming camera, you will also lose important details if your exposure is too long, or if your filtering isn't strong enough.

If you are using an anti-blooming camera, and you do manage to saturate it, you will see something similar to what an NABG camera would show. The non-antiblooming camera will saturate and bloom far faster, however.

Solar Imaging

Of the various white-light (full spectrum) solar filters I have used, one stands out as the best for both visual and CCD imaging: the Baader Planetarium solar film. The images and visual observations are sharper than what I have gotten with other filters.

Whatever solar filter you use, a single layer of a solar filter will not be enough for many cameras, even with the shortest available exposure. Cameras with ultra-short exposures, such as the ST-237, can take exposures down to a millisecond. Such cameras will work with small apertures and medium to slow focal ratios, such as 60-100mm f/8 refractors. For larger telescopes, an aperture mask will reduce the effective aperture and therefore increase the focal ratio. For example, an 8" f/10 SCT masked so it has a 80mm aperture will have a focal ratio of 2000/80, or f/25.

You can make a simple aperture mask for a Schmidt-Cassegrain out of cardboard. Simply cut a cardboard mask as large as the front of the scope, and then cut a circular hole in it that has a diameter equal

FIGURE 3.3.3. EVEN AN ABG CAMERA CAN GET TOO MUCH LIGHT.

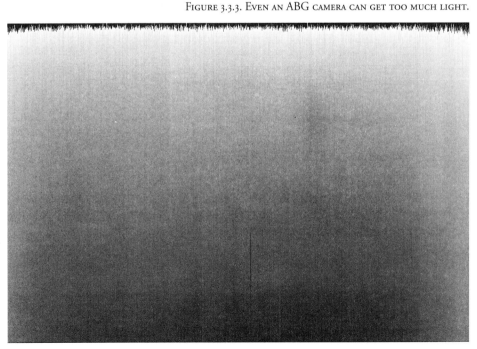

to the distance between the outer edge of the secondary mirror and the edge of the corrector plate. This gives you the largest possible unobstructed aperture. If necessary, you can make an even smaller aperture to get to a focal ratio that will give you a workable exposure time.

TIP: Many paper plates have an outer diameter that exactly matches the inner diameter of the ridge at the front of Celestron SCTs. Cut the circular hole in the plate, and then carefully wedge it into the front of the scope. You can use the circular hole to easily grab the plate and remove it.

Cameras that can't take ultra-short exposures require more extreme measures. One option is a second filter that will further reduce the incoming light. An ST-7E camera, for example, can only take exposures as short as 0.11 second, which is much longer than the millisecond exposures of the ST-237. A second layer of Baader film will cut the light, though for some telescopes this may require longer exposures than optimal. If the exposures get too long, the turbulence that results from solar heating may blur your images.

Additional filtering options for such cameras include neutral density moon filters (often used for visual observation of the moon) and polarizing filters. Both would be used in addition to a conventional solar filter. The two-piece type of polarizing filter is especially useful because you can twist one of the two filters to adjust the amount of darkening that occurs. Unfortunately, the additional optical surfaces may reduce the sharpness and contrast of your images. The better the quality of your filters, the less likely this is to be true.

The bottom line is that a camera with an ultra-fast shutter will give you the best options for white-light solar imaging.

Another approach is to use a non-white-light filter, such as a hydrogen-alpha filter. Such filters pass a narrow band of light, allowing you to use longer exposures. These are typically two-part filters. One is called an energy-rejection filter, and its job is to filter out most of the light coming from the sun. The second filter is a narrow-band filter. It passes a very narrow wavelength of light, as small as a fraction of a nanometer in wavelength. The filter's bandpass is selected to match the wavelength of light emitted by specific elements. The hydrogen-alpha filter is the most commonly used. It passes light at a wavelength

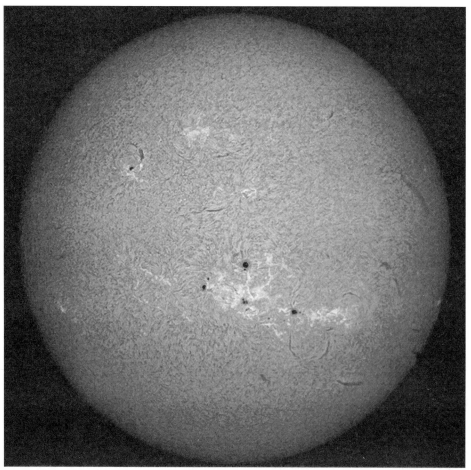

FIGURE 3.3.4. TAKEN BY ADRIAN CATTERALL USING AN ASP90 CORONADO SOLAR FILTER ON A TAKAHASHI SKY90 REFRACTOR.

emitted by hydrogen atoms at an electron energy level that is useful for analyzing solar surface activity. Similar filters are available for other narrow bands, such as those associated with specific electron energy levels of calcium, oxygen, and sulfur atoms.

These narrow-band filters are much more costly than white-light solar filters. You can buy high-quality white-light filters for under $100 for a small scope. Small narrow-band filters, with bandwidths about a nanometer wide, can be found in the $800-900 range. These replace what used to be called prominence filters, which did a reasonable but not stunning job of displaying the prominences at the edge of the solar disk. The first example of the new narrow-band economy filters is the Solar Max from Coronado. The Solar Max has a very small 40mm aperture, and this accounts for its low cost. The small aperture makes it well suited for imaging, since CCD cameras won't be bothered by the limited light-gathering power of such a filter.

Much more costly large-aperture, ultra-narrow-band filters, typically in the range of 0.5 to 1.5 nanometer wavelengths, cost from $2,500 to $10,000 and more. They provide truly stunning views of the solar surface, however, showing incredible detail. Figure 3.3.4 shows an example of an image taken with a Coronado ultra-narrow-band Hydrogen-alpha filter.

Whatever type of filter you use, sharpening will almost always reveal additional detail. Raw images of extended objects often look blurry, but various sharpening technique will reveal hidden details. Unsharp masking is an effective method for solar images.

Figure 3.3.5 shows an image of the sun taken during the recent Solar maximum. Both sunspots and faculae are clearly visible in the top half of the image, which has been sharpened with an moderate unsharp mask. The lower portion is unsharpened, and shows less contrast and fewer details. The inset on the right side of the iamge shows sharpened and unsharpened portions of the image. It clearly shows how sharpening reveals additional detail in sunspots.

You can also experiment with deconvolution of solar images, such as Lucy-Richardson and Maximum Entropy. Since there are no stars in the image from which to generate a point spread function (PSF), experiment with different sizes of Gaussian PSFs. Astroart is a good choice for deconvolution.

FIGURE 3.3.5. SHARPENING USING UNSHARP MASKING REVEALS ADDITIONAL DETAIL IN SOLAR (AND PLANETARY/LUNAR) IMAGES.

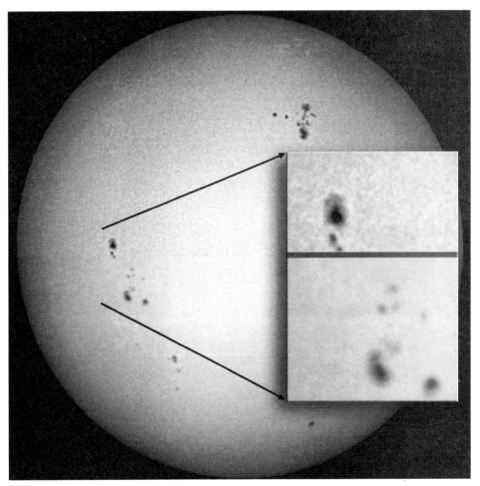

Lunar Imaging

The moon is not nearly as bright as the sun, but it still presents a challenge because of its brightness. Cameras with ultra-fast minimum exposures, such as the ST-237 and STV, are ideal for lunar imaging. The STV even has a built-in neutral-density filter that will attenuate the moon's light with no fuss or bother.

For other types of cameras, a filter is required to reduce the amount of light. A simple neutral density filter will get the job done. Optical quality is the number one issue. The type of filter used for visual observing (the so-called Moon Filter) will work, but make sure you purchase one that has superior optical quality. There are some cheap lunar filters out there that will destroy detail in your images. Polarizing filters also work to reduce moonlight enough to get a good image. The faster your focal ratio, the more likely it is that you'll need filtering.

If you have a color filter wheel, you can use one of the color filters to cut the light. This will be enough for some setups, while others will require additional filtering to cut the light adequately.

Figure 3.3.6 shows the hazards of too much illumination. The bright areas at the right of the image are completely washed out. If the over-illumination is severe, you will also see blooming as seen in figure

FIGURE 3.3.6. A MOON IMAGE THAT WASN'T FILTERED ADEQUATELY

3.3.2. Figure 3.3.7 shows the histogram for the moon image in figure 3.3.6. Note that the right-hand side of the curve ends abruptly (A), and that there is a very large peak (B) at that edge of the curve. This peak is made up of all those white pixels, and detail is lost in those areas. There is no one exposure that will work for the moon; it depends on the sensitivity and capabilities of your camera and the focal ratio of your telescope. If you see washout like the example in figure 3.3.6, shorten your exposure or add filtering.

Figure 3.3.8 shows a proper lunar exposure. Note that even the very bright areas show clear detail. Tycho, at bottom right, is now clearly visible, and the rays can be traced for their entire length. The two bright areas at top right, and numerous other small, bright impact features, show lots of detail.

Lunar images have a large range of brightness values. There are so many, in fact, that if you show all of them the image won't have good contrast. It is challenging to set the brightness and contrast of a lunar image without giving up some details.

Figure 3.3.9 shows the histogram for figure 3.3.8. Note that there is a much more balanced distribution of brightness levels. The lack of an abrupt peak at the right edge tells us that there are no details lost to overexposure. The full range of bright values remains in the image. The trick is to compress such a large range of

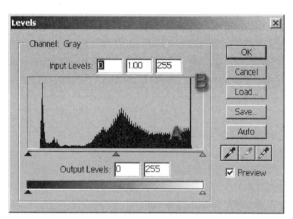

FIGURE 3.3.7. THE HISTOGRAM FOR THE IMAGE IN FIGURE 3.3.6.

FIGURE 3.3.8. A PROPERLY EXPOSED IMAGE OF THE MOON.

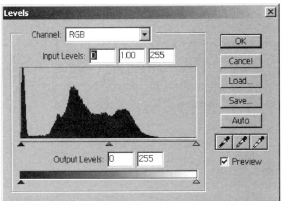

FIGURE 3.3.9. THE HISTOGRAM FOR THE PROPERLY EXPOSED MOON IMAGE.

FIGURE 3.3.10. USING AN S-SHAPED HISTOGRAM ADJUSTMENT.

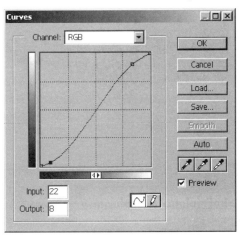

values into the small range that the eye can actually distinguish.

You can adjust the contrast of the image to emphasize fine detail using a specific type of histogram curve. Figure 3.3.10 shows the shape of the curve, using Photoshop's Curves dialog as an example. The dip in the curve at lower left darkens dim details. The top right portion of the curve increases the brightness of the already bright areas. The net effect of these changes is to compress the subtle details in the shadows and highlights. This makes more brightness values available for the middle range of brightness values, where most of the detail lives. The net effect of applying this curve is shown in figure 3.3.11. Overall contrast has improved, and many details that were too subtle in the original are now clear.

Figure 3.3.12 shows the resulting histogram; it shows several changes from figure 3.3.9. The spike at far left is gone; this is a result of darkening the dim areas of the image. The right-hand side of the curve has moved a little further toward the right edge. The overall shape of the curve is the same, but it is more stretched out, so more detail is visible.

Lunar images almost always benefit from some sharpening. Unsharp masking is the best method to use because it gives you a high degree of control and is less

FIGURE 3.3.11. AN IMAGE OF THE MOON AFTER CONTRAST ADJUSTMENTS.

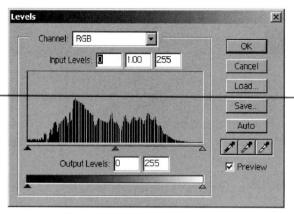

FIGURE 3.3.12. THE HISTOGRAM FOR FIGURE 3.3.11.

likely to create false detail when used with reasonable restraint. The amount of sharpening is limited by the seeing conditions. If you have poor seeing, a little sharpening will be all you can do without introducing artifacts of the sharpening process. If the seeing is very good, you can typically sharpen more and get truly awesome results. To get good results from sharpening, you should have good brightness levels. If your exposures are too short, sharpening will be less effective. A good exposure is one whose histogram looks like figure 3.3.12. There is no sharp spike at the right side, so there isn't any overexposure. Instead, the data peaks dip neatly down at the right edge, indicating that there is adequate exposure. If there is a long blank area on the right, then your exposure may be too short to be sharpened effectively.

Figure 3.3.13 shows a moon image with the left half sharpened, and the right half untouched. The seeing was above average and the image had a long enough exposure for effective sharpening.

You might want to try imaging the portion of the moon illuminated by Earthshine. Figure 3.3.14 shows an example. Careful exposure choice and processing are necessary to get good results. You'll need a long enough exposure to get detail in the darker portion of the moon, but if you go too long the brightest portion of the moon will bloom (when using a non-antiblooming camera). The bright portion of figure 3.3.14 is bloomed, but I used an extreme histogram adjustment to mask it. This was a case of making the best of a difficult situation.

Figure 3.3.15 shows the first step in bringing out the Earthshine details: adjusting the histogram. Note how different this histogram is from the other moon images. There is a clump of dim pixels (the background and the earthshine portion of the moon) and a clump of bright pixels (the brightly-illuminated portion of the moon). You won't be able to show detail in the bright and dim areas simultaneously, so something has to give. Since you want to see the earthshine portion, lower the white point dramatically as shown by the cluster of triangles at the left under the histogram. Photoshop was used for this example, but any histogram tool will work just as well.

FIGURE 3.3.13. LUNAR IMAGES OFTEN NEED AT LEAST A LITTLE SHARPENING TO LOOK THEIR BEST. THE LEFT HALF OF THIS IMAGE IS SHARPENED, AND THE RIGHT HALF IS NOT.

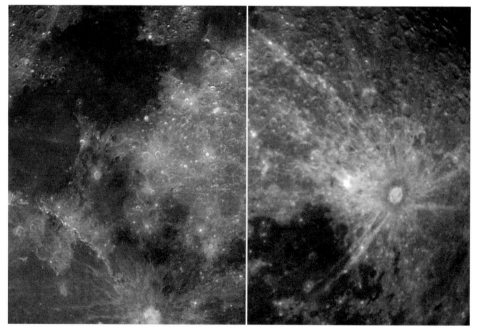

Planetary Imaging

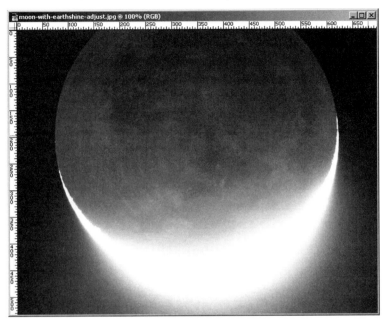

FIGURE 3.3.12. IMAGING THE MOON IN EARTHSHINE.

Planets are bright, but they do not present the same kinds of problems you encounter with the moon and the sun. The image scale at prime focus of most telescopes is quite small. Most of the time, the best way to reduce brightness is to increase your focal ratio by using a Barlow or eyepiece projection.

For example, if you are using an f/10 Schmidt-Cassegrain for imaging, you can use a 2X Barlow to increase the focal ratio to f/20. You could use eyepiece projection to achieve the same result. But Barlows are much simpler to use, and I prefer them for that reason alone. As long as you use a quality Barlow, you will get excellent sharpness. Eyepiece projection involves additional equipment. While it isn't as simple to set up and use, it offers more flexibility in the amount of magnification.

Which method you use depends on your patience and interests. Barlows are the simplest way to start out. If you plan to use a digital or video camera for planetary imaging, you must use eyepiece projection because most digital cameras and video cameras have a lens attached, and they will not work with a Barlow. They require an eyepiece to project an image into the camera's lens.

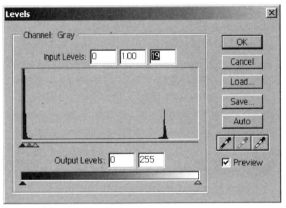

FIGURE 3.3.12. LOWERING THE WHITE POINT TO SHOW EARTHSHINE.

Working with a Barlow

Figure 3.3.16 shows two examples of images of Jupiter. The image on the left was taken at prime focus, while the image at right was taken using a Barlow lens well ahead of the camera, which yielded significant magnification. The images were taken with two different telescopes, but most of the size difference is due to using a Barlow.

TIP: To show widely separated portions of the histogram use techniques such as Layers and Masks. Chapter 9 contains a section that describes using Layers and Masks with Nebulae, but you could also use it for difficult images like figure 3.3.14.

Once you have lowered the white point to show the dim Earthshine-illuminated portion of the moon, you can use non-linear histogram adjustments to improve the contrast. The S-curve shown back in figure 3.3.10 will work here as well.

You could use a Barlow for any kind of CCD imaging, not just for planets. Keep in mind that using a Barlow requires longer (sometimes much longer) exposures. Exposure time is based on focal ratio, and a Barlow increases your focal ratio. Planets are bright enough that this is a benefit, not a hindrance. But imaging deep-sky objects with a Barlow could lead to several hours of exposures with small-aperture telescopes. If the long exposure times don't discourage you, it's a great way to get a larger image with the equipment you already have.

Seeing will limit how much magnification you can use effectively. Seeing affects magnification when imaging just as it does during visual observing. As you increase magnification, you reach a point where the atmospheric turbulence creates so much fuzziness that you see no additional detail. This isn't as much of a problem when imaging as when observing visually because you can reduce the size of the image in software later to get a clearer result.

Digital and video cameras also make good tools for imaging planets. Video cameras will require some kind of capture card for your computer. Most digital cameras allow direct download to your computer. In both cases, you can combine and edit the images just as you would CCD images.

Moving Targets

Planetary images need the highest possible resolution. Atmospheric turbulence is the main obstacle to successful planetary imaging. No matter how short your exposure is it takes some finite amount of time to capture the image of a planet. During that time, if the seeing is average or worse, the planet's image is likely to jump

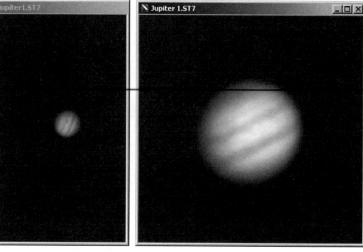

FIGURE 3.3.12. A BARLOW INCREASES THE IMAGE SCALE OF YOUR PLANETARY IMAGES.

around enough to make for a poor image. If you take a lot of images, you can usually get a few good images out of the bunch even on a poor night.

The Jupiter images in figure 3.3.16 are the result of above-average seeing conditions. They demonstrate that excellent planetary imaging requires superb seeing, not just above average. Figure 3.3.17 shows two images of Jupiter taken a few minutes apart. The left image was taken during a moment of especially good seeing. The right image was taken during a moment of especially bad seeing.

Poor seeing will not only lead to smearing and blurring of planetary images. It can also lead to geometric distortions. For example, half of the planet might appear smaller than the other half due to varying atmospheric refraction. A portion of the planet might appear displaced laterally. Or the planet might appear pinched in the middle, or flattened. Such geometrically distorted images might appear crisp, but the distortion renders them less useful. They are especially troublesome for combining because the planet's features won't line up from one image to the next.

Speaking of exposure times, the shorter the better. A short exposure time will reduce the risk of all types of seeing-induced prob-

FIGURE 3.3.13. CHANGES IN SEEING LEAD TO GOOD (LEFT) AND POOR (RIGHT) IMAGES EVEN ON THE SAME NIGHT.

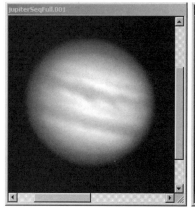
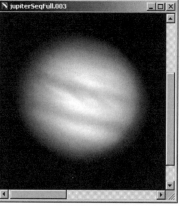

lems. If the exposure time is too short, the image will be noisy and you will have to combine many more images to get a good result.

The optimal strategy for planetary imaging is:

1. Take a large number of images
2. Select the sharpest and least distorted
3. Combine them using sum or median combine.
4. Sharpen the combined image.

Combining even five or six images makes a big difference. If you want the best possible planetary images, take dozens to hundreds of images to make sure you get enough good ones.

Figure 3.3.18 shows two images of Jupiter. The left image is a single image. It suffers from several major dust shadows near the top. They look like planetary features unless you look closely. The image on the right is a median combine of a six images. Because the planet shifted slightly from one exposure to the next, the dust shadows cancel out in the median combine. There was some geometric distortion among the images, however, so the edge of the planet is slightly fuzzy.

Color imaging of planets can also be very rewarding. Jupiter represents a special case because of its rapid

FIGURE 3.3.12. COMBINING MULTIPLE IMAGES REMOVES NOISE. NOTE THE ABSENCE OF THE DARK BLOTCHES AT THE TOP OF THE RIGHT-SIDE IMAGE.

rotation. You must make sure to get your red, green, and blue images taken very quickly -- no more than 10 minutes from start to finish for the entire image sequence. Otherwise, rotation will become apparent when the colors do not line up properly. Figure 3.3.19 shows a combination of rotation and geometric distortion due to seeing. When the images are combined, color fringing messes up the result.

If you take all of your images in a short enough time, and the seeing is good enough, the combination will be much more effective. Figure 3.3.20 shows a color combination that is much more accurate.

FIGURE 3.3.13. IMAGES OF JUPITER TAKEN TOO FAR APART IN TIME RESULT IN ODD COLOR PATTERNS.

FIGURE 3.3.14. A BALANCED COLOR IMAGE OF JUPITER.

 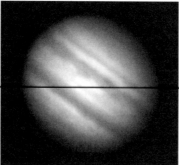 

FIGURE 3.3.21. FROM LEFT TO RIGHT: NO SHARPENING, SOME SHARPENING, EXTRA SHARPENING.

Shadow transits of Jupiter's moons require faster imaging because the shadow moves quickly across the surface. A time lag will create color fringing on the shadow.

Figure 3.3.21 shows a sequence of color images that demonstrates the benefits of sharpening. The left image is the raw color image. The middle image has been sharpened to bring out additional details. The right image has been sharpened more heavily to emphasize the maximum amount of detail.

If you go too far with sharpening, however, you will bring out details that aren't there (artifacts). The trick to sharpening is to find the point where you gain the maximum benefit without falling over into false details. Figure 3.3.22 shows a tragic case of too much sharpening. An image with poor resolution can't be sharpened effectively, and can wind up looking just as bad as an over sharpened image like figure 3.3.22.

FIGURE 3.3.22. SHRINKING AN IMAGE (RIGHT) CAN IMPROVE THE OVERALL APPEARANCE AND HIDE FLAWS.

Unsharp masking is my favorite sharpening technique for planets. It generally gives you the best results, but if your signal to noise is excellent, and the seeing conditions are unusually steady, you can sometimes get slightly better results with deconvolution (especially Lucy-Richardson). You will need to manually set the

FIGURE 3.3.21. TOO MUCH SHARPENING CREATES ARTIFACTS.

size of the point spread function since no stars will be visible in the image. Experiment with values from 0.7 to 1.5 pixels for the PSF. The worse the seeing, the larger the PSF should be.

If you can't sharpen the image as much as you would like, or if the image simply lacks good resolution no matter how you try to process it, you can sometimes salvage the situation by reducing the image size. Figure 3.3.23 shows an image of Saturn full size (left). The image has been sharpened, and it looks grainy. The right-hand image has been reduced. The grain is less noticeable, and the image looks sharper as a result. This effect is similar to switching to a lower powered eyepiece when observing visually.

In addition to shrinking the image, you can use products like Visual Infinity's Grain Surgery to reduce the grain of an image. This is a Photoshop plug-in that is often used by film photographers, but it is also very useful for cleaning up noise and sharpening artifacts in astro images. Figure 3.3.24 shows the Grain Surgery dialog in beta form. It allows you to adjust various parameters (including amount of sharpening) to avoid blurriness in the finished image.

Figure 3.3.25 shows a half-and-half image of Saturn. The upper half has been de-grained using GRAIN SURGERY, and the lower half shows the after-effects of sharpening. GRAIN SURGERY is very effective at making sharpened images more presentable. It removes the artifacts of sharpening without making the image look fuzzy. GRAIN SURGERY has its own sharpening routine built in, but I usually prefer using unsharp masking first, and then cleaning up with GRAIN SURGERY.

Using Eyepiece Projection

Eyepiece projection delivers similar results to using a Barlow, but you have greater control over the image scale. If you elect to go with eyepiece projection, I recommend getting a unit that will allow you to vary the spacing between the camera and the eyepiece. This allows you to finely tune the amount of magnification.

Figure 3.3.26 shows the TeleVue eyepiece projection unit, with a Takahashi eyepiece. Most eyepiece projection units can be used with a variety of eyepiece types, but most require a 1.25" eyepiece, and the eyepiece must not be too long so it will fit inside the unit.

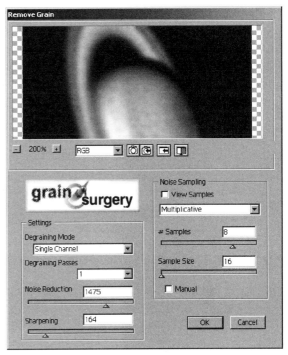

FIGURE 3.3.21. USING GRAIN SURGERY TO REMOVE GRAIN.

FIGURE 3.3.22. THE UPPER HALF OF THE IMAGE SHOWS THE RESULTS OF USING GRAIN SURGERY.

FIGURE 3.3.23. THE PARTS OF A TELEVUE EYEPIECE PROJECTION UNIT.

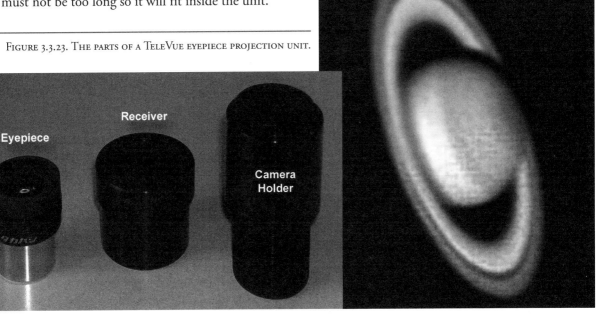

If the eyepiece has a rubber eyecup, chances are you will get better results by removing it. The rubber eyecup can interfere with the movement of the parts of the eyepiece projection unit, or it may prevent the eyepiece from fitting inside the unit. Also pay attention to any rubber armor or padding on the side of the eyepiece barrel. The clearances inside the average eyepiece projection unit can be tight, and the last thing you want to have to do is bang out an eyepiece with a hammer because it got jammed!

The receiver of the TeleVue unit holds the eyepiece. Some eyepiece projection units have receivers that will hold the eyepiece securely, such as the Takahashi TCA-4. Others, like the TeleVue unit, do not hold the eyepiece securely until the unit is completely assembled. In the case of the TeleVue unit, the camera holder screws into the receiver and effectively clamps down on the eyepiece. Thus the TeleVue unit will not allow you to conveniently vary the eyepiece to camera distance. Figure 3.3.27 shows the TeleVue unit assembled, with the 1.25" barrel of the eyepiece extending out from the unit at far left, and an ST-8E camera attached to the camera holder portion at right.

You can use a wide variety of eyepieces and eyepiece types in an eyepiece projection unit. Plossls and Orthoscopics are the most commonly used types because they project a relatively good image to the camera. Some eyepiece manufacturers make eyepieces that are specifically intended for projection, and such eyepieces have the flattest fields and will provide the best results. With the small chips in most CCD cameras, however, this is often not a major concern. The larger your CCD chip, the more thought you should give to obtaining one of the special projection eyepieces.

You may need to use extension tubes between the telescope and the eyepiece projection unit in order to come to focus. The same is true with Barlows. Both techniques can make major changes to the focus point, and you may need to experiment to find the correct focus point. If this is a problem, consider using one the Power-

Mates from TeleVue. They have a unique design, and they have less impact on the focus position.

When possible, such as with the Takahashi TCA-4, remove the camera holder and focus the eyepiece manually. The exact focus point when used with a camera will vary with the distance between the camera and the eyepiece, but visual focusing will get you very close.

If you want to use a digital camera or a video camera for imaging bright objects, you will need to use a technique called afocal projection. This is the same setup as for eyepiece projection, but it is called afocal because the camera also has a lens. When you are doing simple eyepiece projection, there is no lens on the CCD camera. Lensless cameras are the easiest and most flexible to use for imaging because they don't require projection. However, very few video cameras and digital cameras come without lenses. For video cameras without lenses, check out the catalog from SuperCircuits.

FIGURE 3.3.24. THE EYEPIECE PROJECTION UNIT ASSEMBLED AND ATTACHED TO AN ST-8E CCD CAMERA.

Section 4: Imaging the Deep Sky

FIGURE 3.4.1. An example of a deep-sky image, a 10-minute exposure of the Lagoon nebula.

To take an image of a distant cluster, galaxy, or nebula, you need time. Quality time. The kind of quality and time that allows your mount to point accurately, to track accurately during the entire exposure, and that allows your CCD camera to collect enough photons to make an image that will make you happy. Figure 3.4.1 shows the ultimate goal when imaging deep-sky objects: low noise and excellent details in both dim and bright areas of the image.

This sounds hard to do. The reality is that it's very hard to do. You can start with short exposures to catch your first exciting glimpses of deep-sky objects, but you need good equipment and good technique to do exceptional deep sky imaging. For advanced details on processing deep-sky images, see chapters 8 and 9.

Taking Longer Exposures

To take long exposures, your mount must track accurately. To get accurate tracking:

- The mount must be solidly built and have well made gears. There is no substitute for quality. Price ranges from the Vixen GP-DX at the low end to the Paramount GT-1100ME at the high end.

- The mount must be adjusted for minimal backlash. Figure 3.4.2 shows one possible result of excessive backlash: guiding errors. See earlier sectons of this chapter and chapter 4 for details on adjusting and tuning your mount, and see chapter 5 for details on autoguiding.

- The mount must be accurately aligned to the celestial pole. The more accurately aligned your mount is, the less need there is for guiding corrections.

FIGURE 3.4.2. BACKLASH IN A MOUNT CAN LEAD TO PROBLEMS WHEN GUIDING DURING AN IMAGE.

FIGURE 3.4.3. POOR POLAR ALIGNMENT CAUSES STARS TO ELONGATE DURING UNGUIDED EXPOSURES.

If these three criteria are met, good images become routine. Your mount is by far the most critical link in the imaging chain. If you can make these three things happen consistently, you'll get the best possible results.

Why is polar alignment essential? Several things happen when your mount isn't polar aligned:

- The mount doesn't track accurately. This results in stars becoming elongated (see figure 3.4.3). The greater the misalignment, the faster the elongation occurs. When well-aligned to the pole, stars are round in unguided exposures, subject to the periodic and random error of the mount at longer focal lengths.
- Guiding corrections must be made more frequently if your mount isn't well aligned. A guiding correction is a movement, and unnecessary movement should be avoided. The greater the random and periodic error of your mount, the more critical this is. Figure 3.4.2 shows one type of problem that results when guiding ruins an image.
- Field rotation occurs when your mount isn't aligned to the pole. This is true even if you are using an autoguider. Field rotation can be a small effect, just a few pixels, but if you are combining images for color you will have to de-rotate the images to align them. If you want to do single images of 10, 20, 30 minutes or more, field rotation will limit your exposure length. Figure 3.4.4 shows blow-ups of misalignment due to field rotation. Two images are being compared in MaxIm

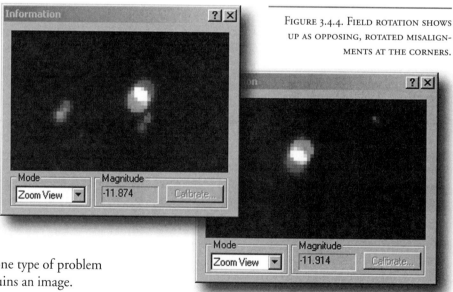

FIGURE 3.4.4. FIELD ROTATION SHOWS UP AS OPPOSING, ROTATED MISALIGNMENTS AT THE CORNERS.

DL using the Zoom mode of the Information window. One image shows up in magenta, the other in green. The upper-left example is from the upper left corner of the image. The offset shows the green star lower than the magenta one. The lower-right example is from the lower right corner of the same image. Because of field rotation, the green star is higher than the magenta one. This flip-flop in opposite corners is typical of misalignment due to field rotation.

FIGURE 3.4.5. A WIDEFIELD IMAGE OF M101.

- If you have a goto mount, a poor polar alignment will reduce goto accuracy. It is very frustrating to go to an object and not see it on the chip. You have to spend time hunting for the object, since it is nearby rather than right in the middle of your chip. Investing 10 to 30 minutes for a really good polar alignment means you can find objects much more easily. The time you spend on polar alignment is an investment. It pays dividends all night long.

Long exposures are highly desirable for imaging deep-sky objects. The deeper you want to image, and the slower your focal ratio, the more critical polar alignment becomes. Both galaxies and nebulae have lots of very dim detail, and long exposures will help you get those details clearly in your images. Shorter exposures contain more noise, while long exposures will reveal subtle details.

Unguided Imaging

Because of the longer exposures required for deep-sky imaging, unguided images are more challenging. If your budget or your personal preferences point you toward unguided imaging, then a fast focal ratio and a short focal length can make your life easier. A 10" Schmidt-Cassegrain at f/10 will be very difficult to use unguided, while a 4" f/5 refractor or an f/2 Fastar will be much easier to use unguided.

Short focal lengths provide wide fields of view. Many celestial objects are small, but wide-field views of such objects can often be pleasing nonetheless. And there are many objects, like the Lagoon Nebula in figure 3.4.1, that are ideally suited to wide fields of view. Figure 3.4.5 shows an example of a wide field image using a short-focal length refractor (Takahashi FSQ-106, 530mm focal length) and a CCD camera with a relatively large chip (SBIG's ST-8E). APO refractors are an excellent choice for wide-field imaging because they have superb contrast and sharpness. These qualities enable you to get excellent detail despite the small image scale.

The Celestron Schmidt-Cassegrains equipped with a Fastar are also a good choice for wide-field imaging. The C8 model offers an f/1.95 focal ratio and an ultra-short 400mm focal length. The fast focal ratio delivers very short exposure times, which is perfect for making unguided exposures. Put the Fastar on a good mount, such as a Vixen GP-DX, for best results unguided.

If you want to do unguided imaging at longer focal lengths, you'll need a superb mount with excellent tracking and very low periodic and random errors. The high-end example of such mounts is the Paramount from Software Bisque, which has periodic error under 5 arc seconds and virtually no random error.

For any given mount, you will find a maximum focal length that will work for unguided imaging. As an example, with a Vixen GP-DX, you should expect to get up to 2 minute unguided exposures at short focal lengths (500-700mm). If you increase the focal length to 1000mm, you will find that 1 minute unguided exposures are more typical. Going beyond 1000mm, the length of your exposures with most mounts drops to a point where unguided exposures are no longer practical. This is due in part to the heavier weight that is typical of many scopes with longer focal lengths.

The bottom line is that the better the quality of the mount, the longer you can go unguided, and the longer the focal length you can use unguided. Guiding can be expensive because it requires a second CCD camera or a CCD camera with a built in guider. The ability to do long unguided exposures can help you get more for your budget.

Stacking (Combining) Images

Your mount and/or camera may limit the length of unguided exposures you can take. If you are limited to taking 1 minute unguided exposures, for example, you won't be able to go as deep as you might like for galaxy and nebula images. If your camera saturates from skyglow after 2 minutes, that will limit how deep you can image. The trick is to take multiple images and combine them. Figure 3.4.6 shows a single image on the left, and a combined image on the right. Notice how much deeper the combined image goes, even though all of the individual images look just like the left-hand image. Notice also that the combined image is less noisy.

Combining images isn't quite as effective at going deep as taking a single long exposure, but it comes very close, and is a great way to cope with limitations of your mount and/or camera.

Dealing with Light Pollution

Light pollution can limit your ability to "go deep" and image distant galaxies and nebulae. Long exposures and combining will help overcome this problem, and you can also use light-pollution filters to cut out some of the pollution. Dark skies will always be the best solution, but if you must frequently image from a light polluted location, you can take some steps to improve your images.

Think in terms of long total exposure. Don't hesitate to take 30 to 60 minutes of exposures, either single or combined. For example, to get good detail in the Trifid Nebula from a suburban location, try taking at least 30 minutes of exposures. In general, the longer your individual exposures, the better. But if your mount or blooming limits your exposure length, simply increase the number of exposures. For example, you might take 30 to 50 1-minute images instead of 3 ten-minute images. You won't get quite the detail of the longer exposures, and your noise levels may rise a bit, but you will get surprisingly good results with this many-image approach.

FIGURE 3.4.6. M101 WITH A SINGLE IMAGE (LEFT) AND WITH FOUR IMAGES COMBINED (RIGHT).

The best light pollution filter I've used is the Hutech Light Suppression Filter. It's available to fit a wide variety of thread sizes, including standard 1.25" and 2" filter threads. It won't remove all light pollution, but it will improve your images by removing a good portion of it. The Hutech filter is especially good at removing light pollution from mercury-vapor light sources. Sodium-vapor and broad-spectrum light pollution will still be a problem, but every little bit of light pollution reduction helps. You will need to increase your exposure times due to light loss, but you will still get better results with a filter.

If you are using an IR blocking filter already, then the Hutech LPS filter will require about 10-15% longer exposures. If you are not using an IR blocking filter, the LPS filter will require you to approximately double your exposure times. The IR blocking will be an advantage with refractors because refractors don't focus IR as well as visible light. IR blocking will be a disadvantage for other types of telescope that do not have chromatic focus shift. You will lengthen your exposure time without as much benefit. The reduction in light pollution effects, however, still makes the filter worthwhile.

Light pollution typically creates gradients in your images. Figure 3.4.7 shows an example of a gradient near M42, the Great Nebula in Orion. The left-hand side shows how badly a gradient from light pollution can affect an image. The right side of figure 3.4.7 shows how the image can be improved by removing the gradient. Gradient removal is challenging but very worthwhile; see chapter 6 for details. A light pollution filter will reduce light pollution gradients, and that will simplify your image processing.

FIGURE 3.4.7. AN IMAGE OF M42 WITH A SEVERE LIGHT POLLUTION GRADIENT (LEFT) AND WITH THE GRADIENT REMOVED (RIGHT).

Section 5: Fun Science with a CCD Camera

A CCD camera doesn't just take images; it collects data. With the proper tools, you can use that data as the basis for some scientific investigation. Although there are a wide variety of tools out there, I have found that CCDSoft version 5 offers the best combination of usability and functionality, so this section will use CCDSoft to show you the kinds of things you can do with your data. CCDSoft works in conjunction with TheSky to perform research functions, so you will need both products to follow along.

This section covers minor planet searching and supernova hunting, as well as the old standbys of astrometry and photometry. The material here is based on the CCDSoft documentation (which I wrote), with the kind permission of Software Bisque. I've added new material here, and most of the step-by-step instructions can be found in the CCDSoft documentation.

You'll learn about:

- Generating astrometric data from your images
- Searching for minor planets and supernovae
- Generating light curves for minor planets and variable stars

CDDSoft uses a source extraction tool called SExtractor, which incorporates advanced algorithms that scan your image for objects. SExtractor is very effective at finding source: stars and galaxies in CCD images.

Principles of Astrometry and Photometry

Astrometry allows you to determine the coordinates of the sources in your images. Photometry allows you to measure the magnitude (brightness) of the sources in your images. A source is any object in an image (star, galaxy, etc.), and the process of finding them is called "source extraction."

You can use astrometric data to determine the location of a suspected minor planet or comet, or to help you identify a dim galaxy by its RA and Dec coordinates. You can also use astrometry to measure the separation or position angle of any two objects, a task common when working with double stars. Astrometry gives you a precise street map that allows you to perform any kind of position-dependent task. And if you or CCDSoft finds something that could be a minor planet, you can even use CCDSoft to prepare a submission to the Minor Planet Center.

Photometry measures the brightness of stars. This gives you accurate information about the magnitude of a suspected supernova, for example. You can also create a light curve from the brightness data in multiple images. You could use the light curve to determine the rotation period of a minor planet. You can also use photometric data to create a light curve for an eclipsing variable star, or create light curves for longer-term phenomena such as a novae or supernovae.

In the past, it has taken a lot of detailed work to turn images into hard, useful data. With CCDSoft version 5, you can perform many steps automatically. "Automatic" doesn't mean trivial, however. To get the best use out of the tools in CCDSoft, you may need to spend some time learning the science behind the tools.

CCDSoft performs astrometry by identifying stars and then passing the image to TheSky. TheSky performs an Image Link, a feature that matches the center of the image to a specific RA and Dec. (The image already contains RA and Dec information, but it may not be exactly accurate, depending on the polar alignment accuracy of the mount at the time the image was taken.) TheSky passes information about the stars in the image back to CCDSoft. This includes the IDs (e.g., GSC 5554:910), the equatorial coordinates (RA and Dec), and the magnitudes of the stars.

CCDSoft uses this information about the stars to create what is called an astrometric solution. This is also sometimes called a plate solution. It includes a list of the stars in the image, as well as an assessment of how accurately the star positions in the image match the star positions in the databases used by TheSky.

> **TIP:** Image files must be saved to disk using the FITS format, so make sure that CCDSoft's AutoSave is on and set to this format. The data reduction and research tools are designed to use certain features of the FITS format.

Using Astrometry and Photometry

You can use CCDSoft to perform different kinds of research. These include minor planet searches, supernova searches, and light curves.

Minor Planet Search - CCDSoft can identify moving objects in a series of images. The images should have a time delay between them sufficient to show motion. The actual time interval depends on the apparent motion of the minor planet and the focal length of your telescope. Intervals of 15 to 45 minutes are commonly used. Shorter intervals work best for fast-moving minor planets and longer focal lengths (greater than 2500mm).

Supernova Search - A supernova search is simpler than a minor planet search because you are looking for the presence or absence of the supernova in a fixed location. Instead of taking a series of images with a short delay, your best strategy is to take images of the same area over a long period of time, such as nightly or weekly images. You can then compare the current image to your own reference image to see if a supernova has appeared. There are many other checks to perform in order to determine if you have a supernova, such as taking another image to confirm that the suspected supernova isn't just a cosmic ray hit on the CCD detector.

Minor Planet Light Curve - CCDSoft can measure the magnitude of both moving and stationary objects. To improve accuracy, CCDSoft uses three objects to generate the light curve: two reference stars whose magnitudes do not vary (one is used to check the validity of the other), and the star or minor planet you wish to analyze.

Figure 3.5.1 shows a minor planet light curve created with CCDSoft. There were a total of 55 images of minor planet 7505 1997 AM2, taken during a three hour imaging session. The curve at the top shows the variations in brightness between the two reference stars, which defines the noise level since their brightness does not vary. The curve with crosses shows the magnitude calculated for the minor planet. The rotational period of the minor planet, about 2.5 hours, is easily seen in the plotted light curve.

Variable Star Light Curve - You can also take a series of images of a variable star, store them in a folder, and CCDSoft will analyze the images and produce a light curve. The procedure is nearly identical to that for a minor planet, except the object being analyzed is not moving. You can also create light curves for any object that changes in brightness, including supernovae and comets, as long as the same two reference stars are present in all of the images. If the object moves too fast, you can create the light curve in sections and combine and graph the data manually using a spreadsheet.

A Minor Planet Search using TheSky

Minor planet searches involve taking at least three images with a delay between them. Three images are needed to decrease the likelihood of false identifications such as cosmic ray hits. A very efficient technique for minor planet searching is to take a series of images at different locations, and then repeat the series two more times. This gives you three images of each portion of the sky.

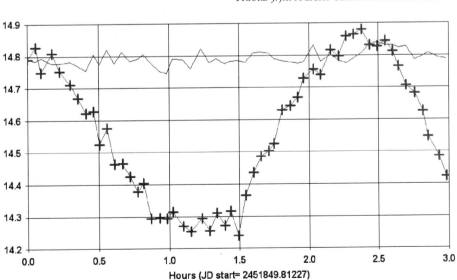

FIGURE 3.5.1. A LIGHT CURVE FOR A MINOR PLANET.

Figure 3.5.2 shows how you can use TheSky's mosaic feature to scan for moving targets such as minor planets. The sample mosaic covers an area of approximately .75 x .75 degrees. There are sixteen image areas in the mosaic. If it takes two minutes to take each image and download it, that is a total of 32 minutes to cover the entire area. You could then take a second and third set of images over the course of about an hour and a half, and then analyze each trio of images with CCDSoft to search for minor planets.

You can also perform this type of search manually. If you know the image scale of your camera/telescope combination, you can calculate your field of view by multiplying the pixel dimensions of the chip by the image scale. Divide by 60 to get the field in arcminutes. For example, if your image scale is 3.51 arcseconds per pixel, and your camera has a chip that is 765x510 pixels, then your field of view is (3.51 * 510)/60 by (3.51 * 765)/60 or 30 arcminutes by 44 arcminutes. The formula for calculating image scale is shown in the next section.

RA	Dec
11h 05m 38s	+15°02'18"
11h 02m 54s	+15°02'18"
11h 00m 10s	+15°02'18"
11h 05m 38s	+14°35'58"
11h 02m 54s	+14°35'58"
11h 00m 10s	+14°35'58"

You can then take a series of images whose centers are offset by slightly less than those dimensions, which will give you an overlapping series of images. For example, using the above FOV, you could set up a series of images with the following coordinates to search for minor planets in a small section of Leo. The points allow for a 10% overlap, about 3-4 arcminutes:

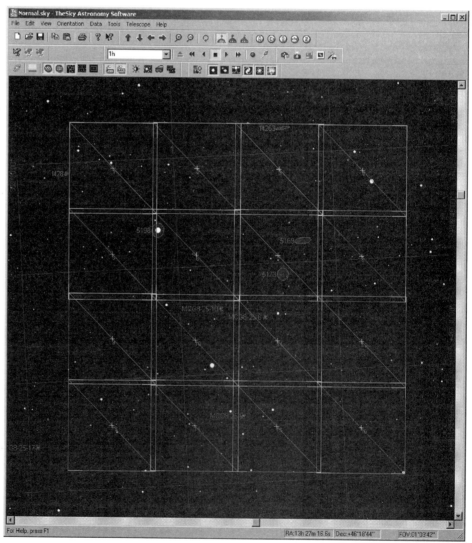

FIGURE 3.5.2. USING THESKY'S MOSAIC FEATURE TO SCAN FOR MINOR PLANETS.

AutoAstrometry for a Single Image

Figure 3.5.3 shows the AutoAstrometry dialog. If you used CCDSoft version 5 with TheSky also present, the image coordinates are stored in the image header. They will appear in the "Image equatorial coordinates" box. You can also click the "Get Previously Entered Coordinates" button to recall the last set of coordinates. If you took the image with another program, you can enter the image coordinates using the format shown in figure 3.5.3.

Specify the image scale in arcseconds per pixel. You can calculate the approximate image scale for your camera/telescope combination using the following formula:

$$\left(\frac{pixel\_size}{focal\_length}\right) * 206$$

where pixel_size is expressed in microns, and focal_length is expressed in millimeters. For example, to find the arc seconds per pixel for a 10" f/10 SCT using an f/5 focal reducer and an ST-8E CCD camera:

$$\frac{9}{(25.4*10*5)} * 206 = 1.46$$

The focal length is calculated using 25.4 (mm in an inch), 10 (aperture in inches), and 5 (focal ratio with reducer). If the ST-8E is binned 2 x 2, the calculation then uses 18 microns as the pixel size, with the result:

$$\frac{18}{(25.4*10*5)} * 206 = 2.92$$

Get the pixel size of your CCD chip in microns in the camera documentation or web site. The image scale you get from this calculation is only approximate, but it will be close enough to allow the AutoAstrometry routines to perform an Image Link with TheSky.

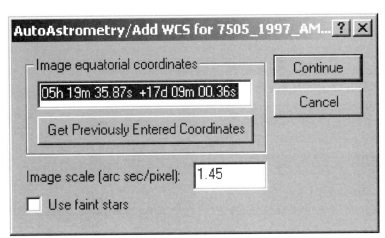

FIGURE 3.5.3. ADDING ASTROMETRY AND WORLD COORDINATE SYSTEM (WCS) INFORMATION TO AN IMAGE.

If AutoAstrometry is successful, you will see a list of the stars in the image (see figure 3.5.4), including:

- Whether the star is used in the astrometric solution. If the star's position doesn't match the catalog, it is not used. The default is within 1.5 arcseconds.
- The star's catalog ID. For USNO stars, the RA and Dec serve as identification.
- The error factor for the star catalog used, in arc seconds.
- The star's coordinates.
- The X and Y coordinate of the star in the image.
- The residual error for the star, in arcseconds. Residual error is the amount by which the position of the star varies from the catalog position.

FIGURE 3.5.4. AN EXAMPLE OF AN ASTROMETRIC SOLUTION.

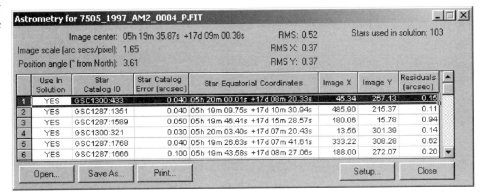

The numbers at the top of the Astrometry dialog (figure 3.5.4) tell you how good the astrometric solution is. The calculated image center is shown at top left. It may vary from the image center originally stored in the image header if the mount was not perfectly polar aligned and positioned at the time the image was taken. Also shown are:

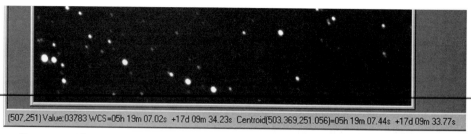

FIGURE 3.5.5. AFTER AN ASTROMETRIC SOLUTION IS AVAILABLE, ADDITIONAL INFORMATION APPEARS IN THE PROGRAM STATUS BAR.

- Image scale in arc seconds per pixel
- Position angle (the amount in degrees by which the vertical axis of the image varies from celestial north)
- Overall RMS (root mean square) error of the solution. Values under 0.50 indicate a very good solution.
- RMS for the X and Y axes. Values under 0.35 indicate a very good solution.
- The number of stars used in the solution. The leftmost column "Used in Solution," tells you which specific stars were used in the solution. By default, stars with a residual error of 1.5 arc seconds or less are used in the solution.

CCDSoft automatically adds the stars used in the astrometric solution to the image header.

When astrometric data is stored in the image header following AutoAstrometry, you will also see additional information about the current cursor position in the CCDSoft status bar at the bottom of the program window (see figure 3.5.5). From left to right, the information shown includes:

- Cursor coordinates (X, Y in pixels)
- Brightness value of the current pixel
- WCS (World Coordinate System) RA and Dec of the current cursor position
- Centroid of the current cursor position (X, Y in thousandths of a pixel)
- RA and Dec of the centroid to hundredths of an arc second

Troubleshooting AutoAstrometry

If you are having problems with Auto-Astrometry, verify that the image coordinates (RA and Dec) are correct, and that the image scale is within +/-0.25 arcseconds of the actual value. You can test coordinates and image scale by manually pasting the image into TheSky and performing an image link using the Link Wizard menu option. The Link Wizard can tolerate a large error in RA and Dec, and can work without an image scale. When the link is successful, you can obtain RA, Dec, and image scale from the Object Information window.

CCDSoft relies on TheSky to find all stars on the image that match the corresponding stars in the stellar catalogs present. These catalogs include the Hipparcos and Tycho catalogs, the Guide Star Catalog and the optional US Naval Observatory catalog (on a single CD-ROM of 54 million or 11 CD-ROM set containing 526 million stars). Although the Hipparcos and Tycho catalogs are preferred because their coordinates and magnitudes are very accurate, most of the time these relatively bright stars are saturated on the CCD image and therefore cannot be used for astrometry.

Astrometry may also fail if there are not enough stars in the image, or if the image contrast is adjusted so that the background is too bright.

AutoAstrometry for Multiple Images in a Folder

It would be tedious to add astrometry to each individual image one at a time. To make this process faster and easier, CCDSoft also allows you to operate on all of the images in one or more folders. These folder-based tools are located on a submenu of the Research menu, at Research | Analyze Folder of Images. The available tools are:

Pre-analyze - add astrometric and/or WCS coordinates to one or more folders of images, and/or create an inventory of objects in the images.

Minor Planet Search - Scan three or more images (or multiple groups of three or more images) for evidence of minor planets. Optionally, generates Minor Planet Center observations reports.

Supernova Search - Scan a folder of images for evidence of supernovae

Minor Planet Light Curves - Analyze a folder of images, and construct a light curve for a moving object

Variable Star Light Curves - Analyze a folder of images, and construct a light curve for a stationary object.

The first step for both searches and light curves is always to pre-analyze the image files. The following section explains why this is necessary, and how you perform a pre-analysis.

Pre-analyze Images

In order to make use of the data contained in your images, CCDSoft must pre-analyze the images to generate astrometric information. For example, when building a light curve for a variable star, the brightness of selected stars in the image is used to build the light curve. In order to track a known minor planet as it moves from image to image, CCDSoft needs the expected positions of the minor planet.

The most efficient way to pre-analyze images is to use AutoSave to put them all into a folder as you are capturing the images. Otherwise, you can copy them to a folder. If you have only a few images, you can also add astrometric information one file at a time as described above.

Click on the Research | Analyze Folder of Images | Pre-analyze menu item, which displays the Data Analysis panel with the Pre-analyze tab active (see figure 3.5.6). There are several options on the left side, and an Image List on the right side. The options on the left are applied to the images on the right.

A typical pre-analysis run consists of these steps:

1. Use the Folders button to select one or more folders containing the images you want to pre-analyze.
2. (Optional) Click the Open button to verify that the images are the ones you want.
3. Set options (checkboxes and image scale).
4. Click Start to begin pre-analysis.

There are four checkboxes and a text box on the Pre-analyze panel:

AutoAstrometry/Add WCS - Uses the Image Link feature of TheSky to generate an astrometric solution.

Generate inventory of celestial objects - Causes a .SRC file (list of objects in the image) to be created and written to disk, with the same name as the image file.

Overwrite existing WCS solution - Causes any existing .SRC file to be overwritten. Out-of-date .SRC files are always overwritten. An .SRC file is out-of-date when it is older than the image file it belongs to.

Image Scale (arc secs/pixel) - Sets the image scale for all images.

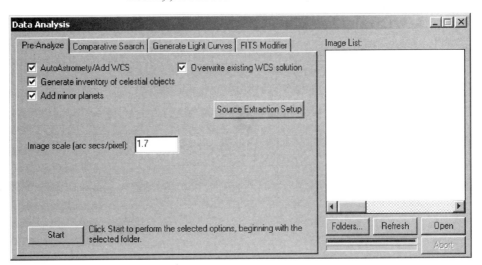

FIGURE 3.5.6. THE DATA ANALYSIS PANEL, WITH THE PRE-ANALYZE TAB ACTIVE.

In addition, there are five buttons on the Pre-analyze tab:

Start - Begins the pre-analyze process.

Folders - Opens a dialog that allows you to choose one or more folders for pre-analysis. The folder(s) should contain only images relevant to the current analysis task.

Refresh - Rescans the current folder(s).

Open - Opens the currently highlighted image file. If no image file is highlighted, the first image is opened.

Abort - Stops pre-analysis.

A typical pre-analysis run starts with a click on the Folders button to select the folder(s) where the images are located. You can add as many folders as you need to. For example, if you stored minor planet images in separate folders for each of three passes over the same area of sky, you can add all three folders.

The objects in the folder(s) are listed in the Image List by object or by RA and Dec, but not by folders. Images with similar coordinates are grouped together. Figure 3.5.7 shows two examples of the Image List. The example on the left shows what you see when there is only one object/location, in this case a minor planet. The example on the right shows what you see if there are multiple objects/locations present in the folder, such as when you have a series of images of different positions in the sky.

Click on the plus icon to the left of the object name or RA/Dec coordinates to see which images are present for that object (see figure 3.5.8. Scroll to the right to see the full path and filename.

The key to working with the Image List is knowing that it is object- and location-based, not folder-based. As you add folders, CCD-Soft sorts out the objects/locations present, and groups the images based on the object/location. If the location of a group of images is within five arc-minutes of each other, CCDSoft interprets that as a single location.

When you have the settings you want, click the Start button to perform the pre-analysis. The progress bars to the left of the Abort button will give you feedback on how many of the images have been processed so far. The top bar indicates progress on the current image; the bottom bar indicates overall progress. In addition, the lines of text below the "Image scale" entry will count off files as they are processed. Figure 3.5.9 shows a view of the Data Analysis dialog during a pre-analysis run of 55 images. Nineteen of the images have been processed so far.

Once pre-analysis is complete, you can move on to analyze your data for minor planets and supernovae, or create light curves.

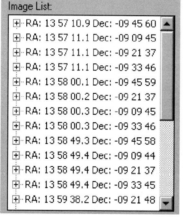

FIGURE 3.5.7. ABOVE: AFTER CHOOSING ONE OR MORE FOLDERS OF IMAGES, THE IMAGES ARE ARRANGED BY OBJECT NAME (OR BY RA AND DEC IF NO OBJECT NAME IS PRESENT).

FIGURE 3.5.8. BELOW: EXPANDING AN OBJECT BY CLICKING ON THE PLUS ICON DISPLAYS THE IMAGES PRESENT FOR THAT OBJECT.

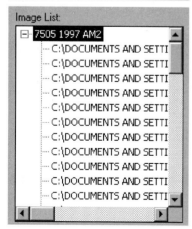
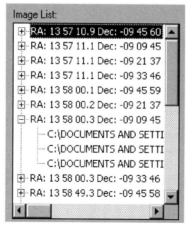

Searching for Minor Planets and Supernovae

The minor planet search routine takes as input three or more images of the same area of the sky. This constitutes one set. You can have more than one set in the image list; they will be scanned one set at a time. Each set can cover a different area of the sky. In the example that follows, there are three sets of three images. The search routine is extremely sensitive, and can locate minor planets that might not be readily visible to the eye. That also means that blooming, ghosts, hot pixels (or the remnants of cold pixels in a dark frame), and other things can show up as false positives. The minor planet search is looking for three items in a roughly linear relationship, which eliminates many false positives.

The time interval between the images within a set should be long enough to show the motion of any minor planet that might be present. Most minor planets move approximately 0.25 to 1.0 arc seconds each minute. A good separation between subsequent images would be about 5-10 pixels, but you can configure the search routine to recognize any number of pixels. Three pixels is the practical minimum for recognition. Anything smaller, and you will start identifying all kinds of noise as candidates.

If your image scale is 2 arc seconds per pixel, then a minor planet must move through 10 arc seconds to show a 5-pixel movement, or 20 arc seconds to show a 10-pixel movement. The length of the exposure required to do this depends on the rate of movement of the minor planet across the sky. A slow, distant minor planet might move at about .25 arc seconds per minute. That would require a 40-minute delay between exposures to yield a 5-pixel change in position. A faster, nearby minor planet might move at 1.25 arc seconds per minute. That would require only a 7-minute delay. In other words, the longer the time interval between exposures, the greater the likelihood of finding slow-moving, distant minor planets. If your aperture is small, or your light pollution is at suburban or worse levels, you won't be able to image dim minor planets, so you could choose to focus your efforts on nearby minor planets using shorter intervals. If you are using a large aperture instrument, you can make good use of longer intervals.

You can also use the difference in movement rates to focus your attention on different types of minor planets. If you are looking for NEOs (Near Earth Objects), then short intervals will work well. If you are looking for very distant comets or minor planets, an hour or more between images would be reasonable.

You can use the following formula to come up with a reasonable range of times between exposures of the same area of the sky: The estimated shortest useful delay in minutes (based on movement of 1.0 arc seconds per minute):

$$image\_scale * number\_pixels$$

Estimated longest useful delay in minutes (based on movement of 0.25 arc seconds per minute):

$$\frac{image\_scale * number\_pixels}{0.25}$$

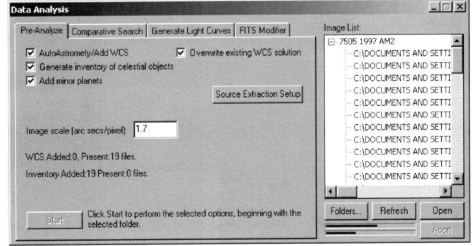

FIGURE 3.5.9. THE PROGRESS BARS AND TEXT AT CENTER LEFT INDICATE HOW MANY IMAGES HAVE BEEN PROCESSED SO FAR.

For example, if you are imaging with an ST-7E on a 10" SCT at f/10, your image scale is 0.74 arc seconds per pixel (see the formula earlier in this section). If you want to see a 5-pixel movement, and you have dark skies that justify using a long delay that will show dim/distant minor planets, then:

$$\frac{0.74 * 5}{0.25} = 14.8$$

A delay of at least 15 minutes between exposures of the same area of the sky would give you a good shot at being able to find any minor planet that might be lurking in that area.

Exposure Duration

Various exposures can be workable when searching for minor planets and supernovae. Exposures of 5-10 minutes are a good starting point. The following guidelines will help you choose an appropriate exposure for your equipment and interests:

- Undiscovered minor planets within reach of typical amateur instruments tend to have magnitudes in the range of 16-20. Supernovae often peak around mag 14-15, but this varies with distance. Your exposure time should be long enough to show this magnitude. Use AutoAstrometry to determine how deep your system can image successfully in a given amount of time.

- Avoid blooming of stars. Blooming will interfere with astrometry, or could cover up a potential find. Blooming will not necessarily be fatal, since you can still blink-compare the images manually should AutoAstrometry fail.

- The sensitivity of your camera will make a difference in how long you expose. If you are imaging with an ST-8E, which has relatively small pixels, you will need long exposures and/or binning to get deep enough. If you are using an FLI Dream Machine, which has 80% QE (quantum efficiency) and 24-micron pixels, you might find yourself imaging for less than a minute to avoid blooming.

- Make sure your exposure is long enough to overcome any skyglow.

- The ability of your mount to track accurately may limit your exposure times. If you are performing an automated search without autoguiding, then the accuracy of your polar alignment and your mount's periodic and random errors will set an upper limit on exposure duration.

- The focal ratio of your telescope determines exposure duration. Slower focal ratios (e.g. f/8, f/10) require longer exposures; faster focal ratios (e.g., f/2, f/5) require shorter exposures.

Search Parameters

The following example shows you how to search for moving objects (e.g., minor planets, but the search will work with comets as well). The steps for a supernova search are very similar; the only difference is that CCDSoft is not following a moving target.

Click on the Comparative Search tab (see figure 3.5.10) to start searching. The difference between the Minor planets and Supernovas buttons is that the Minor Planets radio button searches for moving objects.

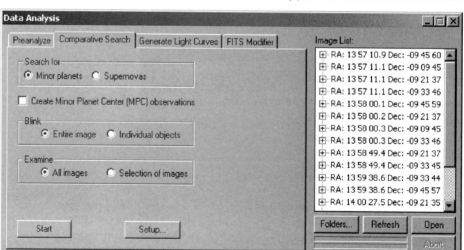

FIGURE 3.5.10. THE COMPARATIVE SEARCH TAB.

If you want to create a Minor Planet Center observation report for any suspected minor planets, check the "Create Minor Planet Center (MPC) Observation." I suggest that, for your initial searches, you leave this unchecked. If you find anything that looks like a good candidate, you can then check this button and run through the search again to generate the observation report. See the CCDSoft manual for details.

The "Blink" setting determines how you view the suspected finds. If you click on "Entire Image," you will see the entire image in the Blink Comparator. If you click on "Individual Objects," you will see each object in each image, and the area around the object will be enlarged so you can see it more clearly.

The comparative search uses the SExtractor (source extraction) features built into CCDSoft. The default settings for SExtractor will usually work well, but you can modify them if you are not detecting minor planets effectively. To access the SExtractor settings, click the Setup button at lower center of the Comparative Search tab. Figure 3.5.11 shows the large number of settings available (with their default values). The settings you enter here apply to both minor planet (moving object) and supernova searches.

FIGURE 3.5.11. SETTING PARAMETERS FOR MINOR PLANET AND SUPERNOVA SEARCHES.

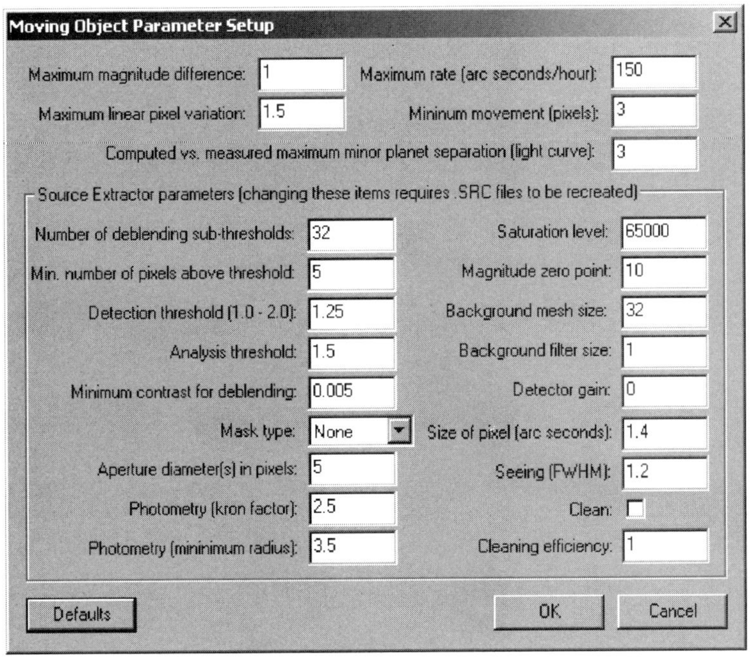

This is a huge list of parameters, and they are complex enough that it's hard to know what to change, and what not to change in order to alter the way that the search routines work. The CCDSoft documentation contains a short description of each of the parameters, but there are really just a few that are the most effective at altering the way that the search operates.

Maximum magnitude difference - Determines how much the magnitude of a minor planet can vary from image to image, and still be viewed as the same object. If clouds or other conditions cause the magnitude between images to vary, you can use this number to make the search routine more forgiving of such changes. The downside is that you may increase the number of false matches.

Maximum rate - Determines how sensitive the search routine is to minor planet movement from image to image. The number refers to the maximum movement rate in pixels per hour. If you want to focus on NEOs, use larger values because they move faster. Smaller numbers will reduce the number of false positives.

Maximum linear pixel variation - Determines how far out of a line a minor planet can be and still be viewed as the same object. The first two images define a line, and this parameter tells CCDSoft how far off of that line to look in the third image to find the minor planet. Under most circumstances, you should leave this parameter alone because if the objects are not in a line, then they aren't a minor planet. The default value of 1.5 is very strict, allowing only a very small deviation from a line.

Minimum movement - Determines how much movement determines a minor planet candidate. The default value of 3 pixels is about as low as you can go without creating a lot of spurious detections. If you know that your images are spaced far enough apart in time to generate larger movements, you can use a higher number to cut down even further on spurious detections.

Detection threshold - Determines how aggressively SExtractor should search for objects in your images. Larger numbers mean that fewer objects will be found. If searches are taking too long, or if the presence of galaxies or gradients is messing up the results, a higher number can reduce the sensitivity.

Saturation level - The maximum brightness count for your camera. This is defined as the full-well capacity divided by the gain. For an ST-9E, this would be 180,000/2.8, which is approximately 64,000. It is good for SExtractor to know your saturation level. Stars that are brighter than that level are not reliable for astrometry. If you are using an antiblooming camera (not recommended!), I suggest you set this value to 50% of actual saturation. Antiblooming cameras start to bleed off electrons at about 50% of saturation, and values above that threshold are not reliable for photometry.

Filter size - Defaults to 1. If you have a noisy image, or an image with processing artifacts, try using a larger number. That will help filter out false matches, but it will also decrease overall sensitivity. Think of this as an emergency adjustment only.

Detector gain - Get this number from the specs for your camera, or contact the manufacturer. This helps SExtractor scale several parameters accurately.

Size of pixel - This is your image scale. See earlier sections for the formula for image scale.

Performing a Search

When you click the Start button, CCDSoft will display a group of three or more images for each location in the Image List using the Blink Comparator (see figure 3.5.12). Possible minor planets are outlined in a box with a dashed edge. In figure 3.5.12, there are three such boxes, numbered for convenient reference in the illustration. As the blink comparator rotates from one image to the next, look for a minor planet moving within the confines of each box. If there are no minor planet candidates, there will not be any boxes.

If you cannot see any objects in the boxes, right click on the image and click "Histogram" to display the Histogram tool (shown at lower right in figure 3.5.12). Use this tool to lower the white point. Click approximately in the area of the asterisk shown in figure 3.5.12 to lower the white point. You may need to do this more than one time to see extremely dim objects.

Figure 3.5.13 shows the effect of lowering the white point. This reveals more dim objects, including a minor planet candidate shown by the arrow in box #3. The planet can be seen to move in the three images thanks to the Blink Comparator.

In figure 3.5.13, the white point has been lowered from 1467 to 861. Boxes #1 and #2 are false positives, caused by the bright area streaming away from the bright star at the left of both boxes.

FIGURE 3.5.12. USING THE BLINK COMPARATOR AND HISTOGRAM TOOLS.

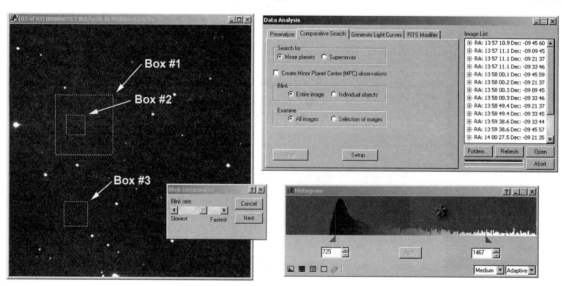

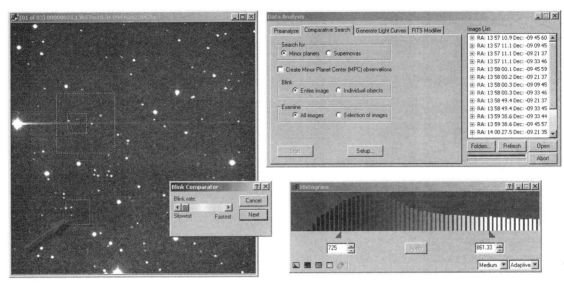

FIGURE 3.5.13. MINOR PLANETS ARE OFTEN EASIER TO SEE WHEN THE WHITE POINT IS LOWERED.

You can change the rate of the Blink Comparator by moving its scroll bar left (slower) or right (faster).

If you clicked on "Individual Objects" on the Comparative Search tab, you will see only the area around an object instead of the entire image. Figure 3.5.14 shows what this magnified view looks like. The image contains just the immediate area around the suspected minor planet (or supernova). I have added an arrow that points to the suspected minor planet; the arrow is not part of the CCDSoft interface. (In actual use, the blink comparator would show the minor planet moving as the images blink, so no arrow is necessary.)

The Blink Comparator makes it very easy to identify candidate objects. When you are performing a Detailed Search, the appearance of the Blink Comparator dialog changes, as shown in figure 3.5.15. The Blink Comparator asks you a question: Is this a minor planet? If you click Yes, a Minor Planet Center observation report is created in the CCDSoft installation folder, and the next object appears for your examination. If you click No, the Blink Comparator will present the next object without writing the report. If

FIGURE 3.5.14. THE MAGNIFIED VIEW ZOOMS IN ON A SUSPECTED MINOR PLANET (ARROW).

FIGURE 3.5.15. THE BLINK COMPARATOR ASKS YOU IF THERE IS A MINOR PLANET CANDIDATE PRESENT.

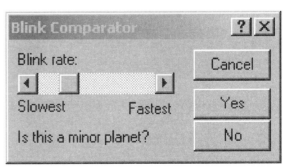

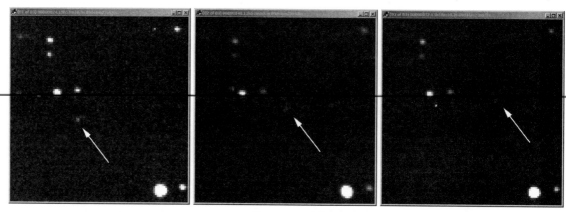

FIGURE 3.5.16. A MINOR PLANET CHANGES POSITION BETWEEN IMAGES.

there is no next object, the Blink Comparator returns you to the Comparative Search tab.

Figure 3.5.16 shows a typical series of three images that include a minor planet. The minor planet is the only object that moves from one image to the next. The arrows point to the minor planet's position in each image. Using the Blink Comparator with the magnified view, the movement of the minor planet is immediately obvious. If you don't see a minor planet, use the Histogram tool as described earlier to lower the white point. You can access the Histogram tool by right clicking on the image and choosing Histogram from the popup menu that appears.

Supernova Searches

Supernova searches are very similar to minor planet searches. The primary difference is in the strategy that CCDSoft uses to find the desired object in each image. For minor planet searches, the key indicator is movement. If an object occurs in three positions in a straight line, it is a candidate minor planet. This is why at least three images are always needed to locate a minor planet.

To search for supernovae, select Supernova instead of Minor planets in the "Search for" section of the Comparative Search tab. Otherwise, proceed as described for minor planets above.

For supernova searches, the key indicator is a star-like object in the vicinity of a galaxy-like object. The source extraction capability within CCDSoft is able to differentiate between these two types of objects with reasonable effectiveness. If the galaxy is very large relative to your frame size, detection will be less effective. The trained human eye is still the most effective search tool when examining blinking images.

Generally speaking, automatic detection of galaxies and therefore of supernova candidates is a tougher challenge than detection of minor planets. Detection of galaxies is subtler, and may require considerable effort to find SExtractor settings that match your local equipment and conditions. The better the quality of your images, of course, the easier it will be to identify objects.

It takes a considerable amount of time and effort to learn how to adjust SExtractor parameters effectively. As you learn how to manipulate the parameters, you gain an enormous amount of flexibility in your searching. You can also use the blink comparator alone for supernova searching if you have only a few images. But for large volumes of data, it's worth the time to optimize the SExtractor parameters for your equipment and local sky conditions.

Creating Light Curves

A light curve graphs changes in brightness in a celestial object. You can create light curves for moving objects, such as minor planets and comets, and for stationary objects such as supernovae and variable stars.

The following example of creating a light curve uses a sequence of images of a minor planet. A minor planet presents a moving target, but the differences in processing for moving and stationary targets are minimal. Figure 3.5.17 shows one of the 55 images taken over almost three hours of minor planet 7505 1997 AM2.

There is some trailing in the image, but SExtractor is amazingly forgiving of this kind of problem, and will find objects in these kinds of images. However, the trailing will lead to less accurate magnitude readings from image to image.

Figure 3.5.18 shows the location of minor planet 7505 1997 AM2 on November 1, 2000 at 6:29am local time (the time that the image in figure 3.5.17 was taken). The chart was created in CCDSoft using the Research | Comparison | Star Chart menu item. I changed the Labels settings in TheSky to show the magnitude of stars and the ID of extended minor planets. To change Label settings, use the View | Labels | Setup menu item, then click the Extended tab, choose the type of object, and then set the options for that type of object. For star magnitudes, set a magnitude limit just dimmer than the minor

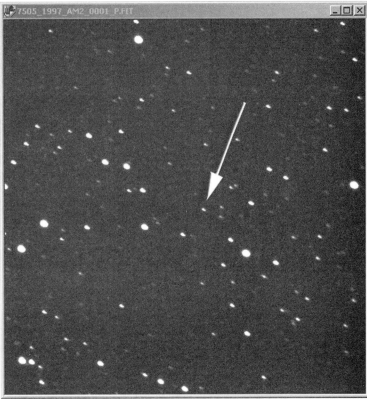

FIGURE 3.5.17. ABOVE: ONE OF A SERIES OF IMAGES TAKEN OF A MINOR PLANET.

FIGURE 3.5.18. BELOW: A STAR CHART CORRESPONDING TO THE POSITION OF A MINOR PLANET.

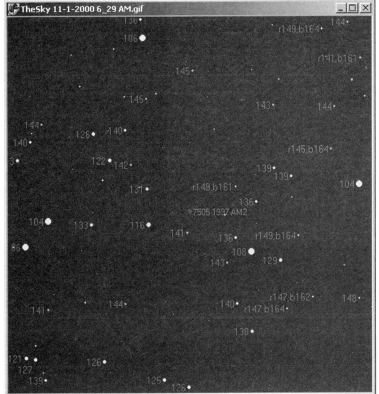

planet. Otherwise, your star chart may be too crowded with data to be useful.

The CCDSoft documentation contains an example of a light curve. You can download a sample set of data for the light curve and perform the operations yourself. The following summary will give you an idea of how you generate a light curve in CCDSoft.

You start by selecting the folder with the images (55 in this example), and running a pre-analysis. CCDSoft and TheSky will work together to identify any known minor planets in the field of view, and enter the name automatically in the Object Name text box, as shown in figure 3.5.19.

Make sure that the "Moving (minor planet)" radio button is selected. For other types of objects select the "Stationary (variable star)" radio button.

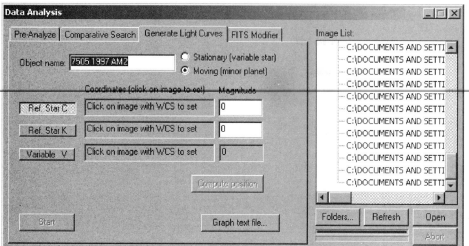

FIGURE 3.5.19. SETTING UP TO GENERATE A LIGHT CURVE

then click in the image to select the C, K and V objects. CCDSoft will get the coordinates of the object from TheSky and show the location of the minor planet with a yellow circle.

CCDSoft will automatically fill in the coordinates for the V object when it is a minor planet. Make sure you pick objects that are visible in all images. Processing stops if an image doesn't have one of the objects in it.

Figure 3.5.20 shows a sample image with good V, C and K objects selected. The reference stars are bright enough to provide good accuracy. The brightest non-saturated stars are your best bet for reference stars.

The light curve will be generated using the brightness of three objects in the images, using a technique called differential photometry. The three objects are:

Reference Star C - A comparison star that does not vary in brightness.

Reference Star K - A check star that does not vary in brightness.

Variable V - The object that varies in brightness.

The difference in brightness between the comparison star and the variable object is used to generate the light curve. The difference in brightness between the comparison star and the check star is also plotted. This shows the variations in brightness from image to image, and allows you to judge the quality of the curve, or to normalize the curve in a spreadsheet.

Note: To see actual magnitude values in the final light curve, you can manually enter the magnitude of the C and K stars in the boxes labeled "Magnitude." The magnitude values will only be as accurate as the figures you enter here, however. If possible, choose stars whose magnitude is known very accurately.

Be sure the "Moving (Minor planet)" radio button is active. Highlight an image in the Image List, click the Open button, and

FIGURE 3.5.20. THE V, C, AND K OBJECTS ARE MARKED.

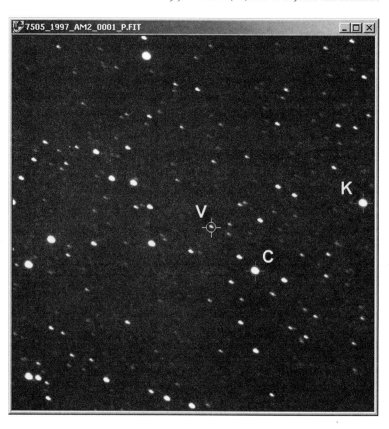

If there are any images with excessive star trails or other problems, an error message will pop up to let you know. You can remove the problem files, and generate the light curve with just the good images. Figure 3.5.21 shows a sample light curve for a minor planet.

The blue line (top) is the difference in the magnitudes of the check star and the comparison star (K-C). The red line (with crosses) is the difference in magnitude between the variable object and the comparison star (V-C). The magnitude scale is at left, and time is at the bottom. The start time is shown in Julian Date format (2451849.81227). To re-open a Light Curve window, click on the "Graph text file" button on the Generate Light Curves tab, and navigate to the text file.

If the light curve has major irregularities, the problem usually is with one or more images in the set. Typical problems with individual images include:

Poor focus - If focus isn't accurate, then the magnitude measurements will be lower for that image.

Bumping of the mount during the exposure - This creates trailing stars, and also spreads the light energy out so that accurate readings are less likely.

Excessive periodic or random tracking error - Same result as above: if stars trail, accuracy is reduced.

Excessive guide corrections - Same result as above.

Failure of guide corrections - Same result as above.

Poor polar alignment - Same result as above.

Loss of guide star - Could result in long streaks or in zigzag star trails as the camera control software hunts for the guide star.

CCDSoft can tolerate errors and still locate objects on the images, but there may be blips or inconsistencies in the light curve from problem images. The images just before the second peak in the light curve in figure 3.5.21 show these kinds of problems, but they are small enough that the data isn't unduly compromised.

You can also save the light curve data and import it into a spreadsheet for additional analysis, such as normalizing the curve based on the variations in the relative brightness of the C and K reference stars. The CCDSoft documentation contains a detailed example of this.

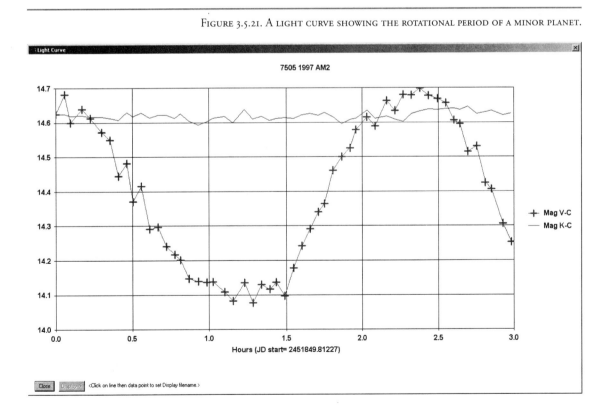

FIGURE 3.5.21. A LIGHT CURVE SHOWING THE ROTATIONAL PERIOD OF A MINOR PLANET.

4   *The Hardware Explained*

PART TWO: TAKING GREAT IMAGES

Given the high price of CCD cameras, it might sound a little odd to say you should spend more money on your mount than on your camera.

But investing a significant amount of your budget into a mount is usually the best strategy. Without a good mount, it won't matter how good your optics are or how fancy your CCD camera is. You can't take great pictures if you can't follow the stars.

Section 1: Start with a Solid Mount

You don't have to spend a ton of money in order to get a good mount, but you do need to spend a fair amount. There is a direct relationship between the focal length of your telescope and the cost of your mount. Short focal lengths (under 700mm) can track and guide fairly easily. By the time you get up to the focal length of the ubiquitous f/10 8" SCT (2000mm), accurate guiding is quite a challenge. Beyond 2000mm, guiding becomes even more challenging.

This relationship between the mount, the scope, and the camera is often something CCD imagers learn about after they have bought their equipment. It's never too late to deal with the issue, however. For example, many telescopes can be fitted with a focal reducer to shorten the focal length for CCD imaging.

If you already own a mount, the first step is to qualify the mount for CCD imaging. No matter how much or how little you paid, no matter how much weight your mount is rated to carry, the place to start is with a test of your mount. You need to know how well it actually performs for CCD imaging. You might find that an otherwise humble mount does OK, or that a high-end mount needs some adjustments to perform at its best. By understanding your mount's capabilities, you know what to do and what to expect with regard to CCD imaging.

I've seen heavy-duty mounts that couldn't track well enough to image for 10 seconds. I've seen lightweight mounts that could track accurately for several minutes at a time. In many cases, a mount can be tuned or upgraded, and you can get decent results. In some cases, a mount may not be unable to track accurately enough for CCD imaging. It is good to know where you stand up front.

This advice runs counter to the most natural desire to get out there and image. You could wait to qualify your mount if and when you run into problems. But by taking some time to test your mount, you'll know what it can do, what it needs in the way of a tune-up or fix, and whether you need to upgrade to accomplish your CCD goals.

Types of Mounts

There are many types of mounts out there, but some form of equatorial mount is by far the most common for CCD imaging. A suitable equatorial mount has an axis that lines up with the earth's rotation, and a motor that tracks the motion of the stars. That motor, as well as the bearings inside the mount and the electronics, are responsible for the accuracy (or inaccuracy) of the mount.

In visual use, minor tracking errors are not noticeable. Many mounts are made primarily or only for visual use, and these mounts are the ones that are likely to cause you the most grief with CCD imaging. Some mounts are intended for both visual and photographic use, but may have different limits for visual and photographic use. For example, the venerable G-11 mount from Losmandy is often rated as able to carry 60 pounds of equipment. That may be true for visual use, but most imagers who have used this mount will tell you that photographic loads should be on the order of 30-35 pounds at most.

German Equatorial Mounts

Most amateur astro images, CCD and film, are taken using some form of the German Equatorial mount. This type of mount is relatively light, flexible as to what kind of telescope you put on it, and if done right can be made to track exceptionally well. Needless to say, not every equatorial mount is done right from a CCD imager's perspective.

Figure 4.1.1 shows a few of the German Equatorial Mounts (GEMs) I have owned myself, ranging from the ultra-light Takahashi Sky Patrol II at upper right to the NJP-160 at bottom right. I have enjoyed the flexibility of these mounts because I have imaged through a wide variety of telescopes. You can mount almost any kind of telescope on a GEM.

The GEM has two axes of rotation. One axis is aligned with the celestial pole, and turning the mount around this axis tracks the stars. This movement in

SECTION 1: START WITH A SOLID MOUNT

FIGURE 4.1.1. A VARIETY OF EQUATORIAL MOUNTS. FROM LEFT, CLOCKWISE: ASTRO-PHYSICS 400 QMD, TAKAHASHI SKY PATROL II, TAKAHASHI NJP, AND AN ASTRO-PHYSICS 600E.

Right Ascension (RA) is motor-driven. Gears connect the motors to a worm, which drives a gear at sidereal rate. The sidereal rate is the rate at which the stars move across the sky. The second axis, called the Declination (Dec) axis, moves at right angles to the RA axis. The combination of the two axes can point a telescope to any area of the sky.

Figure 4.1.2 shows the worm and worm gear from an AP 600E mount. The worm is on the left. Note the gear at one end (top in the picture), which is driven by a gear train between the motor and the worm. You can see the worm end of the gear train at the lower right of the right-hand image. The upper left of the right-hand image shows the gear that the worm drives, which turns one of the mount's axes (in this case, the Declination axis).

The key to success with a GEM is choosing one with high-quality gears. If the gears are not made with exquisite precision, there will be large periodic and random errors as the mount tracks. There is always some level of periodic error no matter how perfect the gears are, of course, but a quality mount will keep it to a minimum. The best mounts made, such as the Astro-Physics 1200 series and the Paramount 1100 series, have periodic error down around 2-5 arc seconds and almost no random error. Once periodic and random errors get up around 20 arc seconds, it becomes hard to use such a mount for imaging, even at short focal lengths.

Periodic error results from anything that is out of round. Random error results from lack of smoothness on the gears and bearings, or contamination such as dust, packing materials from shipping, etc. I've had several expensive mounts that arrived with crushed packing materials in the gears, and that is a lousy way to start your experience with a mount! In one case, a trip to the local gas station and some high-pressure air cleaned out the crushed Styrofoam. High-end mounts are heavy, and most come with a tough plastic wrap

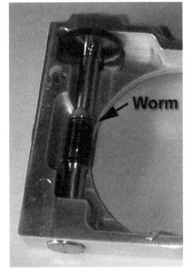

FIGURE 4.1.2. AN EXAMPLE OF A WORM AND WORM GEAR.

around them, with many layers of bubble wrap for shipping. You can tell a lot about the quality of the mount from the quality of the packaging it comes in.

A GEM, especially a small one, can be awkward to point near the meridian. The mount must be flipped to cross the meridian, which means that you can't take a long exposure very far across the meridian. Some designs will allow you to cross the meridian for an hour or so, but if you aren't careful the scope or camera might strike the pier when you go too far. The other hassle with flipping the mount has to do with accuracy. High-end GEMs can flip and maintain accuracy reasonably well, but low-end and mid-level GEMs may not handle the flip very accurately.

Unless the mount is large enough to allow the camera to clear the mount, the camera can also strike the pier or tripod when approaching the zenith. A large mount allows you to mount the scope well clear of the RA axis centerline, and makes it easier to shoot overhead. Smaller mounts require the load to be carried closer to the center of the axis, so you are closer to the pier when pointing straight up.

Among the various makers of equatorial mounts, several stand out as being at the top of the heap: Astro-Physics, Software Bisque, and Takahashi are my personal favorites. As a general rule, there's no free lunch. You get what you pay for most of the time.

Astro-Physics

Astro-Physics has a long-standing reputation for quality in both their telescopes and mounts. I have owned several AP 400 and AP 600 mounts over the last few years, and I have had occasional use of AP 900 and 1200 models. The non-goto versions are a superb buy on the used market, though they do command a fairly high price. The AP 400 is the over-achiever of all time, and in my opinion is the best deal in the AP line-up. It's small, light enough to pack into a Pelican case for airline travel, and will carry a larger load effectively. The non-goto handles a C9.25" SCT surprisingly well, for example. I've imaged with this combination effectively, though careful balance is critical to success.

The AP 600 non-goto is also a good mount. It's heavier and larger than the 400, but only slightly more capable of carrying heavier loads. As long as you don't overload it, it's superb for imaging. If you have to choose between the 400 and the 600 in GOTO, I like the 400 GOTO better because it can handle nearly as much as the 600, yet is *much* lighter and easier to transport.

The AP portable piers are rock solid. The wooden tripod for the 400 and 600 mounts is very good for a tripod, but I recommend using it only when you can plant the feet solidly into medium-hard ground. On concrete or other hard surfaces, the tripod can slip too easily and destroy your hard-won polar alignment. If you really need a tripod, however, it's one of the best out there and competent for imaging if you take care to keep the legs from sliding. The tips of the legs are very sharp, and don't offer good purchase unless you can sink them into the ground. Once sunk, they don't move. If the ground is too soft, however, they may sink further as they are very slim and sharp.

The AP 900 mount is significantly beefier than the 600 series, and is an excellent mount for imaging. It's heavy enough to be a good observatory mount, but it's just light enough to be reasonable for transport, too. The AP 1200 is a real heavy-duty mount, and a joy to use. Both the 900 and the 1200 goto have enough capacity to carry large loads easily, but the 1200 doesn't break a sweat even with a 100-pound load.

As with almost all AP products, the waiting list is long. The AP 400 has been produced in larger numbers lately, and the wait may not be too bad. Of the entire line, the AP 400 is something special because of its combination of light weight and carrying capacity. The AP 1200 is something special because of its carrying capacity. It breaks down into two pieces, and for the hardy imager, it could be the ultimate "portable" mount if you can handle the weight.

All of the AP mounts are well-made, with smooth, low-level periodic error and excellent response to guide corrections. There were some issues with the first hand controllers, but ongoing revisions to the firmware in the controllers have dealt with most of them at this point. The most important feature to look for in a used AP goto mount is probably the ability to react properly to an overloaded motor condition. Such units have a light on the "brains" of the goto unit that changes color when the motors stall, and can recover from the stall without a loss of alignment. If the used mount you are looking at doesn't have this installed already, it's definitely worth the time and trouble to send the goto unit in for the upgrade.

Software Bisque

Software Bisque makes just one mount as of the time of writing, but it's a real dandy: the Paramount 1100ME. Designed primarily for observatory use, the Paramount is superb in every respect. It has extremely high pointing accuracy, incredibly low periodic error, and the ability to make minute corrections flawlessly. These are the things you dream of in a mount for imaging. The Paramount is pricey at $8500, and heavy as well, but the feature set is the best available.

The ME is the third version of the Paramount, following the 1100 and the 1100s. Each generation has seen significant improvement of an already impressive design. The ME features a lighter head weight, as well as an increase in carrying capacity. It is worth serious consideration for both observatory and semi-portable use. An important advance in this mount will be a feature dubbed Pro-Track™, which adjusts the tracking rate to compensate not only for atmospheric refraction, but also tube flexure and other measureable deflections. Early tests showed an ability to take 10 minute unguided exposures at very long focal lengths, so this is a major feature. If these mounts turn out to be as good as they appear on the drawing board, they could well turn out to be the number one mount available. The existing model, the 1100s, is already an outstanding performer and is my personal choice in a mount.

Losmandy

The Losmandy G-11 has been the entry-level mount of choice for imagers on a budget for some years. The GM-8 is not stiff enough to be a good imaging mount, but some folks have had good success by putting the G-11 tripod and saddle on the GM-8. If you are going to take that approach, you might as well just go all the way and get the heavier G-11 head, too. The price difference between the GM-8 and the G-11 is not large.

The G-11 has larger periodic error than most of the other mounts mentioned here, but you can usually guide out the error satisfactorily if the focal length of your imaging scope is under 2000mm. There is some variability in the periodic and random errors in tracking from mount to mount, and some web sites have sprung up with suggestions on how to improve tracking accuracy. The G-11 Tuning page is an excellent reference:

http://www.tfh-berlin.de/~goerlich/cg11tune.html

Unguided exposures with the G-11 are often problematic due to tracking errors. I recommend using a guider with the G-11.

The G-11 is a good choice if you are on a tight budget, but expect to put in some time tuning the mount and learning about its behavior under load. The G-11 is often spoken of as capable of carrying 60 pound loads, but for imaging, somewhere around 30-35 pounds is more realistic.

Now available in a goto version called Gemini, the G-11 is excellent for visual observing, and the servomotors of the Gemini may better for imaging than the stepper motors used in the non-goto version. The physical mount is identical in goto and non-goto versions.

Takahashi

Takahashi makes some of the best mounts available anywhere in the world. The EM-10 is a classic portable mount; the NJP-160 is an awesome photographic platform for scopes up to about 60-70 pounds. Goto is now available for the Takashi mounts directly from Takahashi. The slewing speed is slower than for many other goto systems. Until other software packages such as TheSky support the new system, you must use the planetarium and mount-control software that comes with the mount instead of your current software. Tak mounts are fairly expensive -- usually the most expensive in their class -- but the combination of superbly smooth and accurate gears along with great design and manufacturing make them excellent mounts for imaging. Match your load and imaging requirements to the right mount, and you can be sure to get good results.

The Takahashi mounts have the added advantage of being readily available from Tak dealers. The AP mounts, by contrast, can take years to obtain. If you want a great mount for imaging, and can afford the Takahashi prices, you aren't likely to be disappointed with any of these mounts. Even the little Sky Patrol II (top right in figure 4.1.1) is the best in its class.

William Yang

A new mount came out in 2001 from William Yang Optics, the GT-ONE. It is available in standard and heavy-duty (HD) models. The gears in these goto mounts are extremely well made, yet the price is exceptionally affordable for such quality (about $3000). All reports point to this being a great bargain for imaging. The first version uses the SkySensor 2000 PC which is very full-featured for visual use, but you should expect to take some time to get familiar with the features you'll be using when you image as they are mostly buried in the menu structure. I highly recommend reading the full documentation for the SkySensor, as otherwise you may be unaware of what it can actually do for you.

The SkySensor is oriented toward the visual observer, but the features you need are available once you know where to find them. The tracking of the GT-ONE is exceptional. The GT-ONE has less than 10 arcseconds of periodic error. The unit I tested had a periodic error of 7 arcseconds. The GT-ONE has very smooth tracking, with very low random tracking errors, again a pleasant surprise for this price range.

Other Manufacturers

Photo-quality mounts are available in small quantities from various other manufacturers, but I have not personally tried them out or formed an opinion. I have heard good reports from a number of owners of Mountain MI-250 mounts, and they are in growing use as astrophotography platforms. The MI-250 is now available in a goto configuration.

Based on my conversations with owners of various mounts, the general rule is that you get what you pay for. Each mount may have one or two things it does better than others, and as long as you are careful to match your requirements to what's special about a boutique mount, you can expect good results.

Dob with Equatorial Platform

The better equatorial mounts start around $3,000-3,500, and move rapidly to $5,000-8,000 if you want to put a lot of weight on the mount. You would have to spend about $7,000 for a 4" APO refractor and a high-quality mount, and about $20,000 for a high-end 10" scope and a mount to support it. You can do some wonderful imaging with these setups, but for $7,000 aperture is limited, and $20,000 is not going to work for most of us.

The Equatorial platform provides an alternative way to image with larger apertures. Just buy yourself a great Dob, such as a StarMaster, and put it on an equatorial platform. A superb 10" Dob would run you about $2,000, and a superb equatorial platform would be another $2,000. The end result is the ability to image with a larger aperture for a fraction of the cost of the more traditional approach using a GEM.

So what is an equatorial platform? Figure 4.1.3 shows a high-end aluminum equatorial platform made by Tom Osypowski. Tom's web site is at:

http://www201.pair.com/resource/astro.html/regular/products/eq_platforms/

FIGURE 4.1.3. A HIGH-END EQUATORIAL PLATFORM MADE BY TOM OSYPOWSKI.

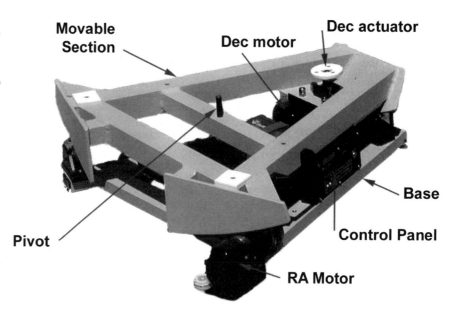

The lower half has motors and roller bearings, and the upper half has curved surfaces that ride on the rollers and track the sky. Your Dob sits on top of the upper half, and rides on Teflon pads. In other words, the equatorial platform replaces your ground board with a device that tracks the stars accurately. You can still point your Dob in the usual fashion while it's on the platform.

The features shown in figure 4.1.3 include:

FIGURE 4.1.4. THE EQUATORIAL PLATFORM ROTATES TO TRACK THE STARS.

Pivot - Engages the rocker box of the Dob, and provides a center of rotation in azimuth. Think of the EQ platform as a giant ground board, and you won't be far wrong.

Moveable Section - This is the top half. The RA motor drives it to track the stars.

Dec motor - Powers the Dec actuator, and allows guiding correction in Dec. It is a limitation of the design that you can't get effective guide corrections to the low northeast and northwest, but you will seldom image at such low elevations.

Dec actuator - Lifts and lowers the Dob's rocker box to provide Dec guiding corrections. Not all EQ platforms have this feature, but it's important to have for imaging.

Base - The stationary bottom half of the platform.

Control Panel - You'll find the on/off switch, a button for lunar/sidereal tracking rate selection, guider input, etc. here.

RA Motor - Drives the platform to track the stars.

Figure 4.1.4 shows the two extremes of movement of an equatorial platform. Most platforms will take from 1 to 1.5 hours to make their full swing. You manually reset the platform at or near the end of its swing. This limits you to 1 to 1.5 hours of imaging in one go, but that's more than enough time for CCD imagers.

Fork Mounts

Fork mounts are found most commonly on Schmidt-Cassegrain telescopes from Meade and Celestron. These mounts are real bargains when you consider that they include a mount and a telescope for just a few thousand dollars. When you compare that price to the cost of high-end mounts, which cost the same or more but don't include a telescope, it's evident that something has to give. The fork mounts are a good value, but they will not have the same performance as a stand-alone quality mount. You will find a little more periodic error, a little more random error, and a little less of an imager-centric approach to mount design on the commonly available fork mounts (see figure 4.1.5).

These are not problems inherent to the fork design; there are excellent high-end fork mounts out there. Some of the largest telescopes in the world sit on fork mounts, for example. The issue has to do with the amount of quality you can get at a given price point. The more you pay, the more you get. The less you pay, the harder you work to get good images.

Most SCT fork mounts support the telescope at two points. Meade's LX200 line and the Celestron higher-end lines do an acceptable job of keeping the telescope steady for imaging. There is some variability in tracking ability from one mount to the next, however. The low-end fork mounts available from these

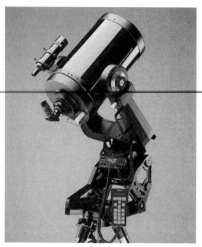

FIGURE 4.1.5. TWO EXAMPLES OF COMMONLY AVAILABLE FORK MOUNTS: LEFT, LX200; RIGHT, CELESTAR 8 DELUXE.

manufacturers, on the other hand, may hold the telescope well enough for visual use, but the tracking is usually inadequate for serious imaging. This is typically due to high periodic and random errors.

Another issue with the fork mount is swing-through. If you mount your CCD camera at the back of the scope, you may not be able to pass it through the base of the forks. Be sure to check the distance available between the back of the telescope and the base of the forks to see if your camera will fit. If not, you will not be able to image a portion of the sky because the scope will not have complete freedom of travel.

German equatorial mounts have issues with pointing, too, but you can usually cover all areas of the sky. A German equatorial can't go too far past the meridian without striking the pier, while a fork can cross the meridian effortlessly. Conversely, a fork mount may have an area it can't point to at all if the camera won't fit between the scope and the fork base, while an equatorial mount can usually point anywhere in the sky.

High-end fork mounts do not suffer from these problems, and there are several available. But at the high end, the German Equatorial reigns as the most commonly available option because of its flexibility. Fork mounts by their nature must be designed with a specific size of telescope in mind. If you start with a fork mount and make a change in telescope, you will likely be shopping for a new mount as well.

Despite their limitations, many successful images have been and will be taken with the Meade and Celestron fork mounts. As with any budget mount, careful attention to the mount's behaviors will teach you a lot about how to keep it under control when imaging. If you have trouble with a fork mount, you can bring it under control (most of the time) by using shorter guide exposures, or by turning off guide corrections in one direction in Dec. This will limit the amount and frequency of corrections, and turning off one direction of Dec corrections will avoid reversals of direction that can bring backlash into play. For example, you can turn off one direction in Software Bisque's CCDSoft by simply entering a zero for one Dec direction in the Calibration Results dialog (See chapter 5, "Autoguiding in Action" section). You must do this only after you perform a calibration. If you turn off the wrong direction, as evidenced by uncontrolled drifting in Dec, zero the other direction instead.

Another issue specifically with the LX200 is the rate of guide corrections. The ideal guide correction is less than the sidereal rate, but the LX200 uses 2x the sidereal rate for guide corrections. This magnifies any issues with guide corrections, and my recommendation is to use short guide corrections if guiding proves to be erratic.

If that still doesn't tame the mount, try imaging with a focal reducer. Shortening the focal length reduces the demands on the mount. A camera with larger pixels will also place fewer demands on the mount.

I also suggest you consult the MAPUG archives for many tips and tricks related to imaging with an LX200:

http://www.mapug.com/AstroDesigns/MAPUG/ArhvList.htm

Section 2: Selecting a Telescope

The selection of telescope and CCD camera are intimately related. Whether you already own a telescope or camera, or are buying both at the same time, it's important to see how well the two fit together. The goal is to pick a telescope and camera that are well-matched to your location, budget, and aspirations.

This is not to say that you can't mix and match. You can make almost any combination work. But it pays to consider how different telescopes and cameras match up well with each other.

Focal Length Issues

Focal length is the Great Separator. The biggest differences in ease of use, and the type of objects you can image, are controlled by the focal length of your scope.

Shorter focal lengths make almost everything easy. The scopes are usually short, or light, or both. Long focal lengths make almost everything much harder. The scopes are usually long, or heavy, or both. A camera with larger pixels makes things easier, but the weight and length still remain. And the temptation is always to use small pixels for higher resolution.

If you want to just have some fun with CCD, get a scope with a short focal length. How short is short? The boundary is fuzzy, but I lump everything from 400 to 800mm as short. Under 400mm is the realm of very short focal lengths, often achieved with camera lenses.

Short focal lengths provide a wide field of view. Long focal lengths give you a narrower field of view. Wide fields are less demanding of mount and operator. Narrow fields, on the other hand, are very demanding.

The longer your focal length with a given camera, the more dependent you are on seeing conditions. The shortest focal lengths make seeing irrelevant; you can image on any clear night.

The following discussion assumes a camer with 10-micron pixels for all cases. Smaller pixels will give shorter focal lengths more resolution, and larger pixels will give longer focal lengths less dependence on seeing conditions, but weight and length of long focal length scopes remains an issue to deal with.

<400mm - This is the realm of two kinds of hardware: camera lenses, and extremely fast/short telescopes. Fields of view are extremely wide, measured in degrees not arcminutes. This is an extremely forgiving range. The fields are so wide that they lead to low resolution. There are a limited number of single objects that require such large fields of view, and some benefit from specialized accessories such as Hydrogen-alpha filters. Milky Way images can be breathtaking. These fields of view are so wide that even poor seeing conditions are not likely to disturb your imaging. Unguided exposures are a good approach for ultra-short focal lengths.

400-800mm - This is a great range for beginning CCD imagers. The demands on your mount range from very light at 400mm, to moderate at 800mm. The Celestron Fastar with SBIG ST-237 camera comes in at a 400mm focal length. Many small refractors (mostly 3" to 4") fit in the 500mm to 800mm range. Smaller Newtonian reflectors also fit into this range of focal lengths. The fields of view are wide, perfectly suited to many nebulae. You can get nice detail in many of the larger, brighter galaxies. The larger open clusters are also often well suited to this focal length. The fields of view at the short end are wide enough to make seeing a rare issue. At 800mm, seeing will sometimes make imaging impossible, but most of the time you are fine as long as seeing is average or better. Unguided exposures are reasonable, but that approach gets more demanding as you approach 800mm.

800-1500mm - Beyond 800mm in focal length, imaging becomes a little more technical. The demands on the mount are high enough that any old mount will not do; you need to have a mount that is specifically suitable for imaging purposes. You will find a lot of 5" and 6" refractors fitting into this range at the high-quality end, but an 8" Schmidt-Cassegrain with an f/5 or f/6.3 focal reducer also fits in this range. There are always some larger Newtonians to pick from in this range as well. Scopes in this focal length are excellent at galaxies that go beyond the Messier catalog, globular clusters, and tight shots of interesting detail in larger nebulae. Unguided exposures are challenging.

1500-2000mm - This zone is a gray area. An f/10 SCT fits in at the high end, and many large, fast Newtonians fit in at the low end. But the technical requirements, while higher than for telescopes under 1500mm, are still reasonable and if you are serious about CCD imaging, you can start out in this zone successfully, but expect a steeper learning curve. A good mount is fundamental to success. This range of focal lengths really opens up the galaxy imaging options. You can also image details in nebulae, and faint globulars really come to life. In this range of focal lengths seeing is almost always a factor. Average seeing will affect the appearance of your image unless you use a large-pixel camera to reduce resolution. Unguided exposures require the utmost in precision mounts.

>2000mm - This is the zone of "serious imaging." You probably want to cut your teeth on shorter focal lengths before diving in here, but experienced film imagers and patient newcomers can succeed. Above all else, you must have a superb mount to image in this range, with nearly perfect tracking, very low backlash, and the ability to carry the weight of larger scopes. If your mount is marginal, you may be able to compensate using extremely short guide exposures. The list of potential targets is endless. Galaxy imaging is wide open, and you can really zoom in on details of larger objects. Seeing is a dominant factor. If the seeing isn't above average, you wont' be able to image at these focal lengths without large pixels or binning. There are some locations, like the east slope of the Rocky Mountains, where the seeing is seldom good enough for these focal lengths. Planetary imagers enjoying steady Florida skies, on the other hand, can image at focal length of 7000mm. If you want to buy a telescope with a focal length longer than 2000mm, take the time to get to know your local seeing conditions first.

You can use reducers, reducer-flatteners, Barlows, eyepiece projection, and other techniques to alter your native focal ratio. There are always some trade-offs involved in changing your focal ratio, however, so having a scope that works natively at the focal length you prefer is often the best choice.

So how do you choose a focal length range that fits your interests? It's hard to do until you experience both wide-field and high-magnification imaging.

Table 4.1: Focal Length Effects

Length	Seeing	FOV	Aperture	Targets	Flexibility	Cost
< 400mm	Who cares?	Wide, wide, wide	2-4"	Big targets and multiple targets	Minimal; wide only	Wide range, mostly camera lenses
400-800	Rarely	Wide	3-6"	Big targets, small targets with lots of empty space around them	Good; add Barlow to increase focal length	Wide range, with refractors very high
800-1500	A consideration	Medium	5-10"	More targets, tighter framing	Good if focal reducers available	Wide range, but many good low-cost options
1500-2000	Always matters	Getting narrow	7-14"	Many more targets	Good if focal reducers available	Medium to high with exceptions
>2000	A dominant factor	Can be extremely narrow	8-16" and larger	Huge numbers of targets	Minimal; specialized for deep sky	High, with rare exceptions

FIGURE 4.2.1. A WIDEFIELD IMAGE TAKEN WITH A SHORT-FOCAL-LENGTH SCOPE IS RELATIVELY IMMUNE TO SEEING.

The most challenging and difficult choices lead to narrow fields of view and high magnifications. See table 4.1 for a summary of the conditaions at various focal lengths. In practice, you either choose an image scale and then take images appropriate to that scale, or you choose a scale appropriate to the target you want to image. If you have one scope and one camera, you can use focal reducers, Barlows, and eyepiece projection to alter your focal length and therefore your field of view. However, most telescopes perform best at their native focal length, and there is usually a trade-off involved in changing the focal length.

For example, a focal reducer often creates a hot spot in the center of the image. Extreme focal reduction may introduce optical problems away from the center of the image, such as elongation of stars. Barlows and eyepiece projection, on the other hand, increase the focal ratio and require substantially longer exposures.

Longer focal lengths also make you more vulnerable to seeing conditions. A short focal length is relatively immune to seeing problems, while poor seeing can make it impossible to image at long focal lengths. Figure 4.2.1 shows a wide-field image taken with a Takahashi FSQ-106 4" f/5 refractor with an ST-8E camera. The image was taken on a windy night, yet M42 is magnificent in spite of the poor seeing. The image was taken at a focal length of 530mm. Such short focal lengths are a pleasure to use and typically provides wide fields of view.

Short focal lengths aren't useful for small objects, however. Figure 4.2.2 shows the Ring Nebula imaged with the FSQ-106 and an ST-8E. The short focal length (540mm) provides a wide field of view that fails to show details. The main strength of shorter focal lengths is their sweeping field of view, but there is nothing interesting in this view besides the Ring. Of

FIGURE 4.2.2. A WIDEFIELD OF VIEW IS UNAFFECTED BY SEEING CONDITIONS, BUT SHOWS LITTLE DETAIL ON SMALL OBJECTS.

There is no one right image scale. It all depends on what you want to image, and on what equipment you have. Figure 4.2.4 shows three commonly imaged subjects, M42/NGC1977, M27, and M57. All are at the same image scale, taken with the same telescope and camera (FSQ-106, ST-8E). The M42 image is almost exactly one degree from top to bottom. The FOV is well suited to show the M42/NGC1977 complex. The wide field of view sacrifices fine detail to show how everything fits together.

The Dumbbell image at this scale is small, but there is still reasonable detail in the image. The Ring Nebula, on the other hand, is just too small with this image scale to show any detail.

course, it *is* fun to see the Ring Nebula in context. And if you have a wide-field setup and want to image the Ring, you can certainly do that. But to get details, you need a longer focal length.

Figure 4.2.3 was taken at a focal length of 2350mm, using a Celestron 9.25" SCT and an ST-7E. The seeing was actually better than for figure 4.2.2 or the M42 image, but not good enough for imaging at this focal length. In addition, you can see a slight left-to-right smearing of the star images. The mount was not quite up to the requirements of such a long focal length. In order to image at long focal lengths, you need good seeing and high-quality equipment.

FIGURE 4.2.3. IMAGING AT LONGER FOCAL LENGTHS PUTS YOU AT THE MERCY OF THE SEEING CONDITIONS AND THE LIMITATIONS OF YOUR MOUNT. UNLIKE THIS EXAMPLE, IT CAN ALSO REVEAL EXTRAORDINARY DETAIL UNDER IDEAL CONDITIONS.

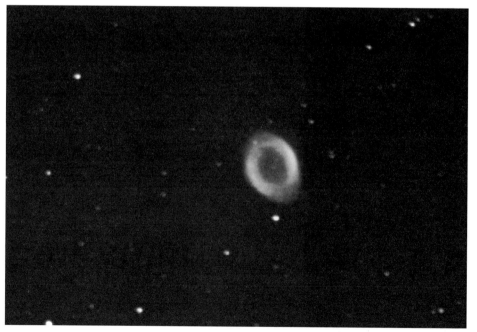

FIGURE 4.2.4. ALL OF THE OBJECTS ABOVE ARE SHOWN AT THE SAME SCALE. DIFFERENT OBJECTS SHOW UP BEST AT DIFFERENT MAGNIFICATIONS.

When you are observing visually, you can adjust the size of the object by changing eyepieces. You could do the same thing by changing CCD cameras, but that's a little on the expensive side. Most CCD imagers take a practical approach to image scale: find objects that are a good match for your scope and camera and stick to imaging those. You can also change the camera and/or scope periodically to image different sizes of objects. You can also use a focal reducer, Barlow, or eyepiece projection to alter the magnification. But increasing magnification increases the focal ratio, requiring significantly longer exposures.

An imager's first impulse is usually to buy equipment that allows you to zoom in and get gloriously detailed images. That means a large telescope, an expensive mount, and lots and lots of patience on those rare, steady nights. Maybe ten percent of starting CCD imagers have the time, money, and ambition to get started with a long focal length telescope and all that it implies. Be honest with yourself about whether or not you are in that ten percent. You will enjoy CCD imaging all that much more if you start out in the right range for your situation. You should have a good feel for where you sit by the time you sift through the information in this chapter.

When I was starting out, I did not take my own advice. I started out imaging with a 4" refractor, which I immediately sold (after just a few nights of imaging) and bought an 8" SCT. I spent the next several months filling the starry nights with complaints about the complexities of CCD imaging. I managed to take some good images, but there were 20 bad ones for every good one. If I weren't as stubborn as I am, I'm sure I would have given up CCD imaging entirely.

So imagine my surprise when, quite without thinking about the consequences, I acquired another 4" refractor and set to imaging with it. It was like being let out of prison. Suddenly, I could image on any clear night because turbulence doesn't matter nearly as much at 400-800mm focal lengths. I could place objects on the CCD chip with unbelievable ease. I was having so much fun it was hard to believe I ever tried to do it any other way.

When I returned to working at longer focal lengths, it was with a renewed appreciation for the challenge. In addition, I could apply the fundamental skills I learned at shorter focal lengths; I was no longer trying to learn everything at once. I reserve long focal length work for those nights when the seeing is something special. The rest of the time, I get out the short focal length hardware and have some fun.

If it sounds like I'm saying, "Lean toward shorter focal lengths," you are reading me right. Venture into longer focal lengths either when you have established your skills, or when you are clear about the challenge of longer focal lengths and find it exciting.

Within each range of focal lengths, there will be a variety of telescope types available for you to choose from: refractors, SCTs, Newtonians, Maksutov-Cassegrains and others. The next section will help relate telescope, focal length, and CCD camera to each other.

Telescope types for CCD imaging

Almost any telescope will work with a CCD camera. The primary reason for a telescope *not* to work for CCD would be an inability to reach focus with the CCD camera. A film or CCD camera sits further up in the focuser than an eyepiece does. The most common type of telescope to suffer from this problem is a Newtonian, especially those with fast focal ratios (f/5 and faster). Newtonians are optimized for visual use, and the focus point is very close to the tube in order to keep the size of the secondary mirror small. You can adapt such a telescope by putting in a larger secondary, and/or by moving the primary closer to the secondary. In many cases, simply using a coma corrector, such as the TeleVue Paracorr, will move the focus point out far enough to be successful.

Otherwise, the field of available telescopes is wide open. Different types of scopes have different advantages and disadvantages for CCD imaging.

Small APO refractors

Small APO refractors are a wonderful entry point for the beginning imager. They offer very good optical quality, and with a good mount you can take long exposures that will reveal a lot of detail. You will be imaging a wide field of view, so those details will be small but quite clear because of the excellent optical quality (see figure 4.2.5). The demands on your mount will be minimal, and you can concen-

Figure 4.2.5. A small APO can take amazingly crisp widefield images.

FIGURE 4.2.6. AN APO REFRACTOR PROVIDES BETTER OVERALL SHARPNESS.

trate on developing general CCD skills without having to worry too much about mount issues. The image quality of small APO refractors is extremely high, and you can take amazingly sharp images with these small instruments. They are also very portable, so you can take them to dark-sky sites with less hassle.

At some future time, if you want a bigger challenge or more detail in your images, you can go for a larger aperture and/or a longer focal length telescope. But small APOs are pure fun to use, so don't count on ever selling one if you get one!

Achromatic (non-APO) refractors

Small non-APO refractors have problems with bringing all colors of light to sharp focus at the same place. If you intend to use a non-APO refractor, plan on buying a yellow or green filter to image through. This removes the extremes of red and blue light, and allows you to get a finer focus. Figure 4.2.6 shows a comparison of an image taken with an achromat (left) and an APO (right). The APO image is sharper, and has

better overall contrast, too. You can come closer to APO quality if you choose a quality achromat (sometimes called a "semi-APO"), and use it with a yellow or green filter.

For color imaging with achromats, you must refocus when you change to another color filter. Otherwise, the variations in focus from one color to the next will result in poor focus for one or more of the filtered images. Figure 4.2.7 shows images taken through red, green, and blue filters without refocusing. Focus was done with the green filter. Note that the red is slightly out of focus, and the blue is way out of focus. This is with a semi-APO refractor; a true low-cost achromat would be worse in red and much worse in blue.

Figure 4.2.7 also shows why you can get better results in monochrome by using a yellow or green filter. Such filters block the unfocused red and blue light. Color shots taken without refocusing, as shown in figure 4.2.8, clearly reveal the contributions of the unfocused colors. There is a violet halo around the brightest stars, the result of the semi-APO not bringing the blue light to focus as well as the green and red. This is much

FIGURE 4.2.7. A SEMI-APO DOESN'T BRING ALL COLORS OF LIGHT TO THE SAME FOCUS.

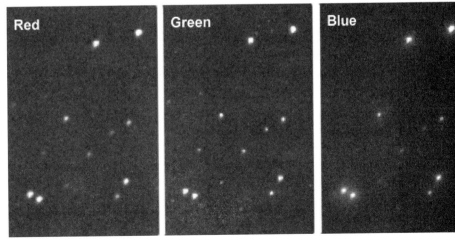

more obvious in color than in black and white. You can correct this in Photoshop using the Color Range tool to select the halo, and then use Image | Adjust | Levels to darken it. You can also desaturate the halo using Image | Adjust | Hue & Saturation.

Rich-Field Imaging

Many small APO refractors provide a wide field of view, called a rich field. Figure 4.2.9 shows an image taken with an ST-8E and a Takahashi FSQ-106. It is 1 by 1.5 degrees. When you consider that this camera on a Celestron 11" SCT would provide a 10 by 15 arcminute field of view, 1x1.5 degrees is a wide field indeed. That's a difference in area of 3600%.

If you are interested in rich field imaging, look for a relatively fast focal ratio that will give you a short focal length (f/6 or better for small refractors; f/4.5 or better for other small scopes; and f/3 or better for medium-sized scopes). Another good rich-field imaging setup is a Fastar-equipped Celestron SCT and an SBIG ST-237 camera.

FIGURE 4.2.8. VIOLET HALOS RESULT WHEN IMAGING WITH AN ACHROMAT.

FIGURE 4.2.9. A RICH-FIELD VIEW OF THE VIRGO CLUSTER, SHOWING MARKARIAN'S CHAIN OF GALAXIES.

Large APO Refractors

Refractors 4" and smaller are small, and refractors 5" and larger are large. For one thing, there is a big price break between these two sizes. For example, the Takahashi FS-102 typically costs a little over $2300, while the Takahashi FS-128 runs more than double that, around $5300. D the difference in performance justify making the move from a small refractor to a large one, especially the move from 4" to 5"?

It's not just the price that jumps from 4" to 5". There *is* a big jump in performance, especially visually. A really good 4" refractor will show good planetary detail, but a 5" APO refractor gets into "Wow!" territory on those really steady nights. The jump between 5" and 6" isn't as dramatic, but there is still a significant increase in available detail because the 6" will support even higher powers on those steady nights.

The difference for imaging comes part in details and partly in image scale. An f/5 4" refractor like the Takahashi FSQ-106 has a focal length of 530mm, which provides a wide field of view (60x90 arcminutes with an ST-8E). An f/8 4" zooms in to 39x58 arcminutes. A 5" at f/8, on the other hand, zooms in more to give you 30x45 arcminutes. You can also find 5" refractors at f/6 and 6" refractors at f/7, both of which provide a wide field of view and superb detail at the same time. Many serious film and CCD imagers really love their 5" and 6" fast refractors for this reason -- it's a sweet spot in terms of field of view, magnification, and aperture. Cost, however, is very high.

For example, the image of IC443 in figure 4.2.10 was shot with a Takahashi FCT-150 and an SBIG ST-8E camera. The contrast and excellent color are a tribute to the incredible optics of this incredibly expensive refractor, as well as the skill of the imager, Robert Gendler.

There are two price classes when it comes to large refractors: doublets and triplets. The doublets, such as the Takahashi FS-128 and FS-152, are less costly because they have one less piece of expensive glass. Triplets provide the ultimate in color correction for today's available refractors, but they cost substantially more than their doublet cousins. Triplets also tend to be better corrected in other ways, and to offer better sharpness overall. You can do superb color work with a doublet, but pay attention to the corners of the image to make sure your color images line up the way they should. Even a triplet, however, may require a field flattener with today's large CCD chips such as found in the ST-8E and ST-10E. Images taken with doublets also tend to be slightly less accurate geometrically. This doesn't mean much at all for single images, but if you are going to create mosaics, an accurate geometric projection is highly desirable. If you can afford a large triplet refractor, you will enjoy some of the best imaging available.

FIGURE 4.2.10. AN IMAGE OF IC443 TAKEN WITH A LARGE APO REFRACTOR, THE TAKAHASHI FCT-150.

Reflecting Telescopes

While premium refractors occupy a large place in the hearts of CCD imagers, the fact remains that there are many other good options available. In fact, contrary to popular assumptions, other telescope designs can actually offer better sharpness than a refractor. High-end refractors have no obstruction, of course, but designs such as the Newtonian often have smaller spot sizes that lead to tighter star images. And the cost per inch of aperture is incredibly high for large refractors, so when it comes to larger apertures, reflecting telescopes of one design or another are a very good option.

The single biggest reason why some imagers shy away from a reflecting telescope is the need to collimate. Mirrors move (more so with some scopes than others), and when they do collimation is the only way to get things back into alignment. Most premium reflectors hold collimation quite well, however. The toughest part of collimation is learning how to do it. Once you get the hang of it, it's not nearly the challenge it is made out to be. Given the huge cost difference between a great Newtonian and a great APO refractor, collimation might not seem so bad. Whatever the type of telescope, great optics can always deliver great images.

Newtonian Telescopes

Newtonian telescopes are probably the sleeper when it comes to CCD imaging. Now you can't just take any Newtonian and expect superlative results. As with any telescope, quality optics are needed for quality results. There are a ton of Newtonian telescopes out there that have mediocre optics, and I wouldn't recommend that you rush out and pick up a $100 special. But if you get a high quality Newtonian, such as from Excelsior Optics, you can expect superior results. You can even use many high-quality Dobs for imaging if you put them on an equatorial platform. The image in figure 4.2.11 was taken with a 12.5" StarMaster EL on an equatorial platform from Tom Osypowski. See the first section of this chapter for information.

You can even do casual imaging with a goto Dobsonian. Figure 4.2.12 shows my image of M82 taken with a 16" StarMaster goto. Figure 4.2.12 is a sum of 14 images taken with an ST-9E, each of which was a mere 5 seconds long. The fast focal ratio of the Dob allows short exposures. With longer exposures, field rotation would occur, extending the stars into streaks instead of dots. You can't do this kind of imaging with any scope on an alt-az mounting. You'll want to use as much aperture as possible to get the most magnification, and the fastest possible focal ratio to benefit from those short exposures. You won't get incredible results with such short exposures, but you can have some CCD fun.

FIGURE 4.2.11. A NEWTONIAN IMAGE SHOWS VERY SMALL STARS AND EXCELLENT DETAIL.

TIP: The 20-micron pixels of the SBIG ST-9E match up well with most larger Dobs. If the focal length is in the range of 1500-2000mm, you will find that a Paracorr and an ST-9E camera will deliver excellent results with short exposures. You'll get a lot out of those 5-10 second exposures!

FIGURE 4.2.12. EVEN A DOB MOUNT CAN DELIVER DECENT IMAGES IF THE CAMERA IS SENSITIVE ENOUGH TO ALLOW VERY SHORT EXPOSURES.

Maksutov-Newtonians

The Mak-Newt, as the Maksutov-Newtonian is commonly known, is making a name for itself with CCD imagers. The design offers a very small secondary mirror, and a high proportion of Mak-Newts have good optics. Unlike Newtonians, which flood the market, Mak-Newts remain a specialty item and must earn their place in your heart by their quality. The results with Mak-Newt scopes can be really exceptional, rivaling APO refractors in contrast and often providing small spot sizes.

Figure 4.2.13 shows an image of the Sculptor galaxy taken with a 6" f/6 Mak-Newt by Rockett Crawford. Note the exceptionally fine detail. This ability to provide sharp detail makes the Mak-Newt a real bargain in the CCD imaging world.

Large-aperture Dobs are great for imaging planets. A digital camera, STV, ST-237, or a one-shot color camera can work well if the focal ratio is fast.

An important point with Newtonians is that you need to be careful about quality. Also pay close attention to available backfocus. Make sure that you can bring your camera to focus. A simple test is to make an eyepiece parfocal with the camera, or to know the difference in focus position between some eyepiece and your camera. You can quickly test a parfocal eyepiece, and if you know the difference in focus position, you can focus with the eyepiece and then check to see if you can move the distance required to come to focus with the camera.

Newtonians also have a somewhat curved field, which can cause elongated stars away from the center of the field. The faster the focal ratio, the more likely this is to be a problem. The TeleVue and Lumicon coma correctors can reduce or even eliminate the problem, depending on your chip size. The smaller the chip, the less of a problem this is. Some high-end Newts have their own correctors available, though often at a high cost.

As with all Newtonian designs, check the available back focus of a Mak-Newt. Make sure that you have enough room to move the focuser to reach focus. Many 5" Mak-Newts, for example, do not have enough back focus to accommodate color filter wheels. The ST-237 can be a good choice for such small Mak-Newts because it has an internal color filter wheel that uses no back focus. The 5" Mak-Newts tend to have the most serious problems with adequate focus travel for CCD imaging, and as the aperture increases the

FIGURE 4.2.13. SCULPTOR GALAXY IMAGE TAKEN WITH A MAK-NEWT.

problems are less frequent. A 7" or 8" Mak-Newt makes a really good scope for deep sky and planetary imaging.

Takahashi Epsilons

The Takahashi Epsilons look like simple Newtonians, but they are outfitted to make them exceptionally useful for imaging. They qualify as astrographs, special purpose telescopes intended primarily for imaging. The Epsilons have extremely heavy-duty focusers, for example, and Takahashi makes correctors for these scopes that provide an unusually large flat field for a Newtonian.

The downside, if there is one, is that this line of exceptional telescopes was designed with film in mind. You may need to order or have made special adapters to use the camera and filter wheel of your choice.

The image of the Veil in figure 4.2.14 by Wil Milan shows what you can expect from the Epsilon series of telescopes. Details are extremely sharp, and colors are very rich due to the excellent contrast. The one thing I don't care for in the Epsilon line is that the diffraction spikes tend to be short and stubby. But this is a matter of personal taste. Many imagers find this of no consequence, so judge for yourself. If the wide field, very fast focal ratio, and top quality of the Epsilon scopes intrigues you, visit Wil Milan's web site to see what's possible with these scopes:

http://www.airdigital.com/astrophoto.html

Schmidt Cassegrains (SCT)

Schmidt-Cassegrains, often referred to simply as SCTs, are among the most commonly used telescopes. I have found than many folks get started in imaging because they own an SCT and want to see what kind of astrophotography they can do with it.

I would divide the SCT universe into three distinct categories, based on the mount that accompanies the scope:

- SCTs that are sitting on low-end, visual-only mounts, such as the Celestar 8 or an inexpensive equatorial. If the SCT has above-average optical quality, consider moving the optical tube to a higher-end mount, or a complete upgrade.

- SCTs that are sitting on mounts that are reasonably competent for CCD imaging, but that require attention to function well at long focal lengths. This includes mounts such as the LX200 and the Celestron Ultima. Although frustrating at times, you can make this class of SCT work with some care and effort. One option is to get a focal reducer so you can work at a shorter focal length.

- SCTs that are sitting on conventional equatorial mounts and that are already suitable for CCD imaging.

If you are in doubt as to the capabilities of your SCT's mount, you can either try with what you have, or upgrade. Even the best SCT fork mounts are looser and less capable than mounts such as the GT-ONE, EM-10, AP 400

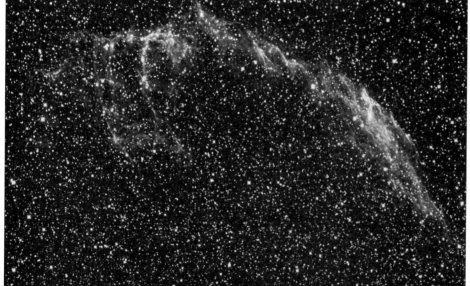

FIGURE 4.2.14. AN IMAGE OF THE VEIL NEBULA TAKEN WITH A TAKAHASHI EPSILON 210.

GTO, etc. But if budget is a concern, and you already have an SCT, you can use the many web resources for Meade and Celestron telescopes to learn the ins and outs of using this equipment for imaging.

Most SCTs use a moving primary mirror for focusing. This is convenient for visual use, and it is also a factor in keeping the price of the scopes at a reasonable level. But a moving primary mirror is not as precise as other types of focusers, and can lead to problems when imaging. Focus tends to shift, with the worst case being a focus shift during an exposure. You also cannot return to an exact focus point because of the limited accuracy of this type of focuser. Most imagers who use an SCT buy an external focuser to improve accuracy and repeatability. You can also lock down the primary mirror by various means to avoid focus shift.

Another issue to consider is that there are quality variations from one SCT to the next, even in a single manufacturer's line. I've used a handful of SCTs, ranging in size from 8" to 14", and there were substantial differences in optical quality. If you suspect that the quality of your optics isn't as good as it should be, talk to the manufacturer about rectifying the problem. Most manufacturers will deal with problems if you are persistent.

The optical tube assemblies that are good can be really good. Figure 4.2.15 shows an image of the galaxy pair NGC 3190, taken with a C11 by Rocket Crawford. Note that the image is very clean, with excellent details.

The weakness of SCT scopes for imaging is that they lack really good contrast, due to the secondary obstruction and a lack of highly smoothed optics. You can see the lack of contrast when comparing the visual view with a refractor or high-end Newtonian. When imaging you can partially compensate for the lack of contrast using processing to bring up contrast. But the best CCD images will always also have the best contrast, so scope designs with better contrast will have an advantage. To get good contrast, you need excellent optical quality and a reasonably small obstruction. High optical quality is the most important thing to look for, as it can overcome a large obstruction more readily than a small obstruction can overcome poor optics.

Maksutov-Cassegrains

There are some outstanding Mak-Cass (Maksutov-Cassegrains) scopes out there. The Questar is legendary, and Meade's 7" model is often regarded as the choice of the LX200 line. The curves on the glass required for a Mak-Cass are easier to make accurately because they are less complex than for most other compound telescope designs. When extra care is lavished on these optics, the results can be outstanding.

Most Mak-Cass designs have a fairly slow focal ratio. The Questar 3.5" has a focal ratio of f/14.4, which means you will need to take very long exposures. The Meade Mak-Cass has an f/15 focal ratio. F/10 and f/12 models are available from Internet Telescope Exchange, as well as an f/6 model designed specifically for imaging.

Figure 4.2.16 shows a sample of four images taken with a 6" f/12 Mak-Cassegrain by Matthias Pfersdorff and Katharina Noee on the island of La Palma, Spain, and in Karlsruhe, Germany. Figure 4.2.17 shows an image of the moon taken with a TEC 10" f/20 Mak-Cass taken by Eric Roel.

FIGURE 4.2.15. AN IMAGE TAKEN WITH A C11 SHOWS HOW GOOD THE IMAGES FROM SCTS CAN BE.

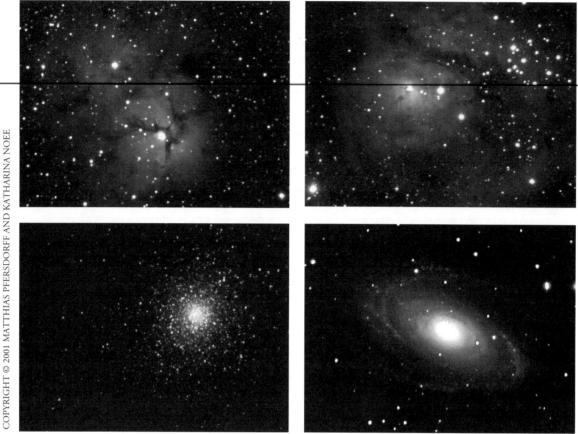

FIGURE 4.2.12. CLOCKWISE FROM TOP LEFT: TRIFID, LAGOON, M5, AND M81 (MAKSUTOV-CASSEGRAIN).

FIGURE 4.2.13. LUNAR IMAGE WITH TEC 10" F/20 MAX-CASS

Takahashi Mewlon (Dall-Kirkham Cassegrain)

The Takahashi Mewlons are superb instruments with excellent sharpness and contrast. They look like classical Cassegrains, but they use the Dall-Kirkham variation on that design. Excellent collimation is required to take advantage of all that sharpness. These scopes are not especially hard to collimate, fortunately, and they also hold collimation extremely well.

The image of M51 in figure 4.2.18 was taken with one of the smallest Mewlons, the 180, yet it shows some extremely fine detail. Image courtesy of B. Alex Pettit, Jr.

The Mewlon 180 and 210 use movement of the primary mirror for focusing. Although the mirror shift is not nearly as high as typically found on Meade and Celestron SCTs, there is a small amount of shift. You can avoid this by mounting a Crayford-style focuser on the rear of the scope, such as the JMI NGF-S. Locking

FIGURE 4.2.12. AN IMAGE OF M51 TAKEN WITH A MEWLON 180.

An RC without a field flattener or corrector generally works fine with the small size of most CCD chips. RCs intended for film use, such as the Takahashi BRC-250, often include an integrated flattener.

Figure 4.2.19 shows an image of M101 taken with an Optical Guidance Systems RC by Robert Gendler.

The downside of the RC is high cost, about $1000 per inch of aperture. What you get for the extra money is a faster focal ratio for shorter exposures and wider fields of view, without sacrificing image quality.

As with any Cassegrain, collimation is very important. With the faster focal ratios, collimation is critical, in fact. A quality RC, however, should hold collimation well.

down the mirror is usually not required because the focus shift is much smaller than on most SCTs. You will have to get a custom adapter made up to do this, however. The 250 and larger Mewlons use a motorized secondary for focusing, so there is no mirror shift and you can focus remotely right out of the box. All of the Mewlons provide superb sharpness and contrast, and are excellent choices for deep-space imaging with their long focal lengths. You don't need the optional field flattener for CCD imaging unless you have a camera with a very large chip. A focal reducer can be useful when you want to work at a slightly faster focal ratio for shorter exposures.

Ritchey-Chretien

The Ritchey-Chretien (RC) is another variation on the Cassegrain design. The RC uses hyperbolic (and therefore expensive) optics to deliver very sharp, high-quality views at a more reasonable focal ratio than is typical for Cassegrains. Cassegrain have focal ratios of f/11 to f/20, but many RCs have focal ratios faster than f/10. With a focal reducer, you can image at f/6 with some RCs.

FIGURE 4.2.13. A BEAUTIFUL IMAGE OF M101 TAKEN WITH A RITCHEY-CHRETIEN.

Section 3: Choosing Camera and Software

It may seem strange to address camera selection as the third choice, but your camera should be well matched to your telescope, and you want an adequate mount for both. I recommend that you choose the mount first, the telescope second, and then your camera.

If you already have a telescope and mount, read the earlier sections so you can learn how well your existing setup will work for CCD. It's not uncommon to upgrade, or at least tune up, your mount and/or telescope in order to be successful with CCD imaging.

Although there are many different cameras available, they fall into two major categories: antiblooming (ABG, short for anti-blooming gate) and non-antiblooming (NABG). The former are simpler to use, but the latter are more sensitive and more accurate when measuring (e.g., astrometry). The choice between the two isn't a simple one, so get ready for a flood of data. Fortunately, it's also not a life and death choice; you can take good images with both cameras. If, after you've read all about it, you still can't make up your mind, you probably should get an antiblooming camera. On the other hand, if astrometry and photometry are your goal, then the non-antiblooming NABG camera will be required.

The Blooming Facts

Figure 4.3.1 shows a number of examples of bloomed stars, from minor to major. The example at far left shows what happens when a very bright star is in your image. Minor blooming can be fixed by hand. Medium-sized blooming takes much longer to fix, and a major bloom can take hours to clean up by hand. All booms cover up some of the image, so unless the bloom

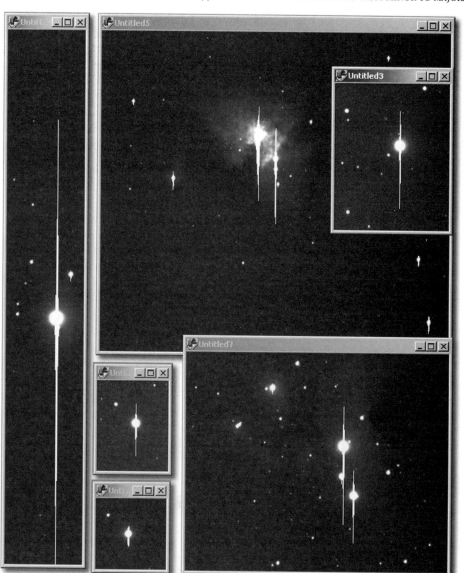

FIGURE 4.3.1. EXAMPLES OF BLOOMED STARS FROM MINOR TO MAJOR.

FIGURE 4.3.2. THE BLOOMING OVERWHELMS THE NEBULOSITY.

ABG versus NABG: Some Theory

I used an NABG ST-8E for more than a year, and I took thousands of images with it. The camera did a great job, but I spent a lot of time cleaning up blooming. Certain objects were very challenging because of bright stars near dim nebulosity. Examples include the Cone Nebula, the Pleiades, the Flaming Star, etc. With an antiblooming camera, it would be a simple matter to image these objects.

The problem with these subjects was that I had to take very short exposures to limit blooming. As a result, I wasn't getting much (if any) nebulosity. With longer exposures, I began to pick up more nebulosity, but the blooming was excessive and took too much time to clean up. And then there was the problem of the lost data in the area of the blooming.

I then had a chance to use an ABG ST-8E for several months, courtesy of SBIG. I found that I really enjoyed not having to deal with blooming. It was easy to set up the camera for any subject, and take really long exposures without worrying about how

covers an empty area of sky, you are not going to be able to replace that data.

The image at bottom right of figure 4.3.1 has two very large blooms and one or two very minor blooms. You can barely see it, but this is actually an image of NGC 1977, the Running Man. Figure 4.3.2 shows the image with the histogram adjusted to reveal the nebulosity. The blooms are so severe they overwhelm the nebula.

As a test, I cleaned up the blooms manually. It took about a half hour, and figure 4.3.3 shows the result. There is no remaining evidence of the blooms, so it is possible to clean them up effectively. I used the Rubber Stamp tool in Photoshop to copy nearby bits of the nebula over the blooms. When the blooms are large, it takes a fair amount of effort to do the cleanup. You can never exactly reproduce the nebulosity covered by a bloom, so the image isn't as accurate after fixing the bloom. An antiblooming camera would be preferable for such situations.

FIGURE 4.3.3. THE BLOOMING HAS BEEN REMOVED FROM THE IMAGE.

big the blooms would be. This is why I say if you have any doubt about which to choose, the ABG camera is most likely the one that will meet your needs.

Here's a summary of what I've discovered about ABG and NABG cameras:

- NABG cameras are more sensitive than ABG. ABG chips require about a 30% longer exposure that their NABG equivalents.
- NABG cameras have a bigger full-well depth, but the smaller well depth of the ABG cameras doesn't really come into play because the excess charge bleeds away, allowing you to image longer to obtain dim details.
- NABG cameras have a linear response to light. This means that a star that is twice as bright shows up twice as bright when you image with an NABG camera. ABG cameras avoid blooming by bleeding off electrons after the pixel gets about half full, so any pixel that is 50% or more full is not going to deliver an accurate brightness level for astrometry or photometry. If you are doing astrometry and photometry, that pretty much makes the ABG/NABG decision for you. Get the NABG camera because of its ability to measure brightness accurately over a wide range.
- I used my NAGB camera mostly with short focal lengths. This resulted in blooming on nearly every image. Longer focal lengths were easier in this regard, as the narrower field of view contains fewer bright stars on each image.
- Minor blooming can be fixed in a few minutes by hand using Photoshop or any image editor with a Rubber Stamp or Cloning tool. Chapters 8 and 9 include information about fixing bloomed stars.

More severe blooming represents lost data. This can be fixed up by hand if you are careful and patient, but the data covered by the blooming cannot be recovered. You can take a mixture of short and long exposures to preserve that data, but in that case you might just as well take a largeer number of short exposures and avoid serious blooming in the first place.

- One way to deal with blooming that doesn't require much hand editing is to rotate your camera about 5-10 degrees between exposures (or a full 90 degrees for square chips), and then to use a median combine. It takes about 5 images to get a good result. The median combine cancels out the majority of the blooming since the blooming does not overlap. If you combine multiple images with blooming, they can actually reinforce one another and make the blooming worse.

From the foregoing, you might assume that the default choice is to get an antiblooming camera. In my opinion, the ABG cameras are in fact the safe choice. Unless you feel that the NABG makes the most sense for you, the ABG camera will be easier to use.

Before I actually tried an ABG camera, I had always thought that the additional sensitivity of the NABG was a slam-dunk argument in its favor. But the situation is more complex than that. The bottom line is that either camera works really well, and it's hard to imagine being really unhappy with either as long as it fits your basic needs. The greater well depth of the NABG chip has just not been an issue; it's only the bright stars that reach those levels on 99.9% of my images. The non-linear response of the ABG is also not an issue for taking great-looking pictures, but if you are doing astrometry/photometry it would be the deciding factor.

FIGURE 4.3.4. LONG EXPOSURES (LEFT) ARE BETTER THAN SHORT ONES (RIGHT).

 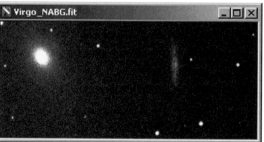

FIGURE 4.3.5. LONGER EXPOSURES ARE BETTER (LEFT), BUT YOU MAY NEED AN ANTIBLOOMING CAMERA TO TAKE THEM.

Choosing between ABG and NABG is like choosing between an APO refractor and a high-end Newt with superb optics. There isn't a perfect choice; you are always making some trade-offs. Study the differences to determine which better meets your needs. If you can't make a firm decision, the ABG chip is the safer choice. And some cameras only come with one type of chip or the other, and that may make the choice for you.

Exposure Times

With an ABG camera, you can image faint nebulosity close to bright stars by taking long exposures. The sensitivity of the ABG is noticeably less than for the NABG, but the ability to "go long" without blooming enables you to take much longer exposures. This is why the ABG camera makes a good default choice: you need to take 30% longer exposures, but longer exposures are what ABG cameras are best at.

So I don't just add 30% for my ABG exposure times. With the NABG chip, my exposure times are limited by blooming, and are in the range of one to ten minutes. One, three, and five minutes are my most commonly used exposure times. I use 1 minute for objects with bright stars, 3 minutes for objects with medium-bright stars, and 5 minutes for objects without any bright stars in the frame. Occasionally I find an object that has only very dim stars nearby, and I can image for as much as 10 minutes. If the object has extremely bright stars involved, I simply do not try to image it with an NABG camera. I take from 3 to 25 images and stack them to improve signal to noise ratio and avoid grain.

With the ABG chip I take 10-, 20-, 30-minute, or even longer exposures. There is nothing to be concerned about with long exposures; the results are great.

When I use an ABG camera I wind up with really deep images with good detail in the dim portions of the images. Figure 4.3.4 shows a comparison of ABG and NABG images of the same area of the Virgo galaxy cluster. The right image was taken with an NABG camera, and it was limited to a maximum of 5 minutes because of blooming. The left image was taken with an

ABG camera, and is 30 minutes long as there was no risk of blooming.

The difference between the two images is immediately obvious. The right image is very noisy. The core of the small edge-on galaxy at center right is approximately 500 units brighter than the background. In the left image, the core is 2,000 units brighter than the background. This results in a much smoother-looking image, with excellent contrast, detail and clarity. Overall, the image on the left is smoother and more satisfying. It has an excellent signal to noise ratio.

You can take multiple short exposures and add them together to approximate the quality of the longer exposure. But longer exposures are still slightly better. You can even combine long exposures as well as short ones. The left image in figure 4.3.5 combines three 30-minute ABG exposures. The right image combines three 5-minute NABG exposures.

A single 30-minute exposure with an ABG camera (left in figure 4.3.6) is also better than 3 5-minute exposures with an NABG camera (right), but the difference is not as dramatic. The bottom line: if you want to enjoy the benefits of owning an ABG camera, take lots of long exposures.

The end result of taking long exposures is better detail, better signal, and less noise. However, I am able to take fewer images of fewer objects on a given night because I spend more time taking exposures for a particular object. The good news is that I enjoy the appearance of the long-exposure images much more.

The biggest difference when I started imaging with an ABG version of the ST-8E was that I could suddenly image as long as I wanted to. Granted, I had to image 30% longer anyway. (The anti-blooming gate that bleeds away excess electrons covers 30% of the pixel area, and exposures must be longer to compensate). But the lack of blooming motivated me to try longer exposures. Ten minutes is now nothing; I started imaging 20, 30, even 60 minutes at a time. With blooming no longer forcing an upper limit, I was free to do really long exposures easily. I was using the same techniques I

FIGURE 4.3.6. A SINGLE 30-MINUTE ABG EXPOSURE HAS BETTER DETAIL THAN THREE 5-MINUTE NABG EXPOSURES.

FIGURE 4.3.7. AN EXAMPLE OF AN IMAGE WITH MINOR BLOOMING.

The ability to take long single exposures fits my needs, but that might not be true for everyone. And there are times when I find the 30-minute exposures frustrating, not least when I make a mistake. It's not fun to find out that you've just taken three 30-minute exposures of the Virgo cluster binned 3x3 instead of 1x1! Instead of wasting 5 minutes as in the past, I've blown a half hour. Anti-blooming will make you pay attention, that's for sure!

Practical Issues in the ABG/NABG Battle

Figure 4.3.7 shows a typical example of an image with blooming. You don't generally expose long enough to get the dramatic blooms in figure 4.3.1. The galaxy in the image (M74) isn't affected, so you could either hand-edit the blooms to

had used with the NABG -- stack multiple images that were as long as possible -- but "as long as possible" was no longer limited by blooming.

The downside is that where I used to spend an hour collecting L, R, G, and B image sets, I now spend that much time just on the luminance. I get deeper images with better signal, so there is a benefit as well as a cost. The ABG chip changes the way I approach imaging. I capture much more light for any given image, which results in a dramatic improvement of image quality. I could ultimately accomplish this with an NAGB camera, but I would have to take *many* more exposures, involving additional download time, and that pretty much erases the 30% difference in exposure time right there.

FIGURE 4.3.8. AN IMAGE WITH BLOOMING RIGHT IN THE MIDDLE OF THE NEBULOSITY.

remove them (see the Galaxy section in chapter 9), or crop the image to show just the galaxy. If this image were taken with an ABG camera, there wouldn't be any blooms, of course.

Blooming isn't such a problem when it occurs outside the object of interest, but it gets to be more annoying when it's right in the middle of your subject. Figure 4.3.8 shows an image of the Cone Nebula taken with an NABG camera (an ST-8E binned 2x2). This is much more challenging to clean up, and you run the risk of changing the appearance of the subject. After all, the portion of the image hidden by the bloom is lost data.

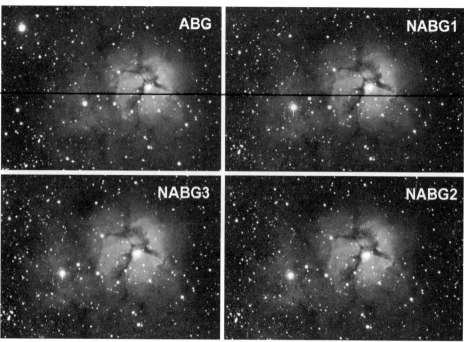

FIGURE 4.3.9. COMPARING THE RESULTS OF INDIVIDUAL ABG AND NABG EXPOSURES.

Figure 4.3.9 shows a comparison of individual exposures of the Trifid Nebula taken with both ABG and NABG cameras. The ABG exposure (upper left) is 30 minutes long. As noted previously, the reduced sensitivity of the ABG camera becomes a moot issue because most of the time you will *want* to take a much longer exposure because of the great results you get.

The NABG1 exposure is 120 seconds long. Several of the bright stars in the nebula have started to bloom, obscuring details in those areas. A longer exposure would suffer from even worse blooming. The NABG2 exposure is 60 seconds, short enough that there is a small amount of blooming. This image is very noisy and has a lot of grain. The NABG3 exposure is a sum of five 2-minute images, and is less noisy.

Figure 4.3.10 shows details from the four images, blown up 4 times actual size. Note that the ABG image (top left) and the summed NABG images (NABG3, at bottom left) are similar in quality. Both have low noise, although the ABG image is overall less noisy and has better contrast. Due to differences in seeing, the NABG3 image has slightly brighter dim stars, while the

FIGURE 4.3.10. BLOW-UPS OF THE FOUR IMAGES SHOWN IN FIGURE 4.3.9.

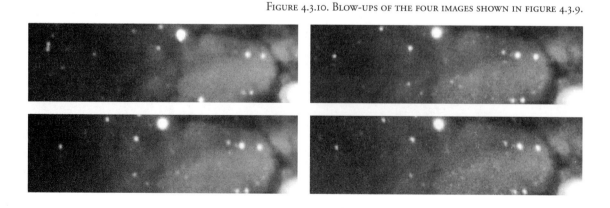

ABG image has more low-contrast details visible due to the length of the exposure and the lower noise.

The shorter individual NABG exposures at right in figure 4.3.10 are clearly noisier. That is, they show more graininess. Thus, your basic strategy with an NABG camera is almost always to take multiple images with short exposures, and combine them to reduce the noise level.

Which Should You Choose?

The bottom line is that you can take excellent images with either type of camera. How you take those images will be different. The ABG allows you to take images of any length as long as the skyglow doesn't saturate the chip. The NABG camera requires you to take shorter exposures, but if you take more of them, you will be successful in getting good details and low noise.

The ABG camera has the advantage when imaging dim objects that have exceptionally bright stars in the field of view. The Pleiades is a prime example of this type of object. The nebulosity is impossible to image effectively with an NABG camera because of the severe blooming that occurs long before you can record the nebulosity. Not many objects are this extreme (the Flaming Star Nebula and M42 have bright stars, but not as troublesome as the Pleiades), so you can use an NABG camera to image many, but not all, objects.

There are four deciding factors when it comes to choosing between the ABG and NABG versions of a camera:

- Focal ratio of your telescope. ABG cameras are a trade-off. They are less sensitive, but they are capable of long exposures without blooming. If you have a faster focal ratio, then the longer exposures required with an ABG camera are less of an issue.
- Light pollution levels at your imaging site. You need long exposures to fight off the effects of light pollution. An ABG camera is more likely to be capable of long exposures in more situations.
- The ability of your mount to track and guide effectively. If your mount can handle long exposures -- 10, 20, 30 minutes or more -- then the longer exposures required with an ABG camera are less of an issue.
- The availability of an ABG chip for the camera. Many cameras only come in an NABG version. The availability of an ABG chip is determined by the chip manufacturer (e.g., Kodak).

An ABG camera gives you something, and it takes something away. It gives you the ability to take long exposures without worrying about blooming. It takes away some sensitivity, so you also need to take longer exposures. So if you expose for 10 minutes with an NABG (non-antiblooming) camera, then you'll need to expose for 13 minutes with an ABG camera. In practice, you actually are more likely to wind up taking multiple images to cover that 10-minute exposure. Consider a typical set of NABG and ABG exposures with an ST-8E camera. The NABG camera is likely to require three separate exposures, which include 10 minutes of exposure time and 3 minutes of download time, for a total of 13 minutes. The ABG camera will require one exposure and one download, for a total of 14 minutes. When you look at it this way, the additional time doesn't matter nearly as much.

You could adjust focal ratio when matching camera and telescope to balance the ABG/NABG exposure times better. For example, for an ST-7E camera, you would get about the same exposure times if you use an NABG camera on a 5" f/8 (1000mm focal length) or an ABG camera on a 5" f/5.6 (700mm focal length). Not that you couldn't still use the 1000mm scope with an ABG camera. Your exposures would need to be about a third longer. Note the difference in focal length; this could also affect your decision.

An NABG camera also gives you something and takes something away. It gives you better sensitivity, and it also gives you a linear response that lends itself to measuring the light output of stars and other objects. It takes away your ability to take arbitrarily long single exposures.

This means that an NABG camera is not just desirable for photometry and astrometry; it's required. An ABG camera is ideally suited to taking "pretty pictures." You don't need to worry nearly as much about your exposure duration with an ABG camera -- longer is almost always better. My own typical exposures with an NABG camera are in the range of 1-5 minutes, depending on when blooming becomes objectionable. My typical exposures with the ABG camera usually start at 10 minutes. This is a big difference. As long as your focal ratio, light pollution, and mount capabilities don't make the longer exposures a problem, ABG makes a lot of sense for pure imaging.

The bottom line is that an ABG camera is easy to use because you have fewer things to worry about when taking an image. It's a fun camera to use. An NABG camera is much more precise as a measuring tool, and you can cope with many (but not all) blooming problems by taking shorter exposures and stacking them.

Of course, you can also stack your long ABG exposures and stay ahead of the NABG camera.

Telescope Focal Ratio as a Factor

As your focal ratio increases, exposure times get longer. The exposure times for an ABG camera are proportionally longer. An NABG camera makes more sense for long focal ratios because there is less likelihood of blooming with the smaller field of view.

Adding one-third to a two-minute exposure with an f/5 scope is trivial. The result is an exposure of two 2 minutes and 40 seconds. But if you are spending an hour collecting data for color images with an NABG camera, then you are probably up to an hour and 20 minutes with an ABG camera. If you are imaging at f/10, then you are spending two hours to collect the same color data with an NABG camera, and almost three hours with an ABG camera. An f/15 Cassegrain would not be a good match for an ABG camera. The NABG camera is less likely to bloom when used with a long focal length, and a long focal length usually comes with that slow focal ratio. This happens because of the smaller field of view. There are less likely to be bright stars in a given field of view if it is small. A wide field of view is more likely to have one or more stars that will bloom, so it makes more sense to use an ABG camera for wide-field imaging.

Light Pollution as a Factor

With respect to light pollution, the issues are again similar. The greater your light pollution, the longer your exposures need to be to overcome the poor signal to noise ratio that results from light pollution. If you get an ABG camera, your exposures must be at least a third to a half longer. The longer your exposures need to be because of the light pollution, the greater the increase in exposure time to compensate.

If you are patient enough to take such long exposures, however, an ABG camera may actually offer your best solution for imaging with light pollution. An NABG camera will limit the length of your individual exposures because of blooming. Those bright stars cut right through the light pollution.

FIGURE 4.3.11. LONG EXPOSURES OVERCOME LIGHT POLLUTION EFFECTIVELY.

You can take long individual exposures with an ABG camera without fear of blooming. This is especially true if you use a light pollution filter to reduce the impact of light pollution. Figure 4.3.11 shows the result of imaging under suburban, light polluted skies using a Hutech LPS (light pollution suppression) filter with an ABG camera. The exposure is 60 minutes (two each 30 minutes summed) using an F/6 5" APO refractor. The image was taken with an ABG camera because of the bright stars in the field of view. Shorter exposures would not show such good detail and contrast, especially in the dim galaxy. The long exposures help overcome most of the light pollution that makes it through the filter. From a dark sky site, an exposure of 5-10 minutes would have been about as effective, however, so plan on spending a lot of exposure time if you want to overcome light pollution effectively.

Mount Capability as a Factor

Since ABG cameras respond favorably to long exposures, you will need a mount that can deliver those long exposures in order to get the best use out of an ABG camera. Of course, any camera will benefit from being used with a high-quality mount. But if you are using a mount that can deliver one or two minutes of exposure and that's all, then camera sensitivity is an important factor to consider. In such a situation, a very fast scope and the additional sensitivity of an NABG camera will keep exposure times within the capabilities of the mount.

On the other hand, if your mount is capable of very long exposures, you can choose an ABG camera and not be concerned about your ability to take the long exposures of which the camera is capable.

One-Shot Color Cameras

There is another type of camera out there that you may find interesting: the one-shot color camera. It does not require a filter wheel to take color images. Cameras that do use a filter wheel require at least three separate images to generate a color image. One-shot color cameras have tiny filters built into the chip surface, which direct light of different colors to different pixels. This allows recording of all of the color data in a single image.

The best of such cameras include chips that use all of the data from all of the pixels to create monochrome images. These cameras, such as the MX5-C and MX7-C from Starlight XPress, deliver decent resolution because all pixels deliver luminance data as well as color data.

However, I have found that some one-shot color cameras are less sensitive than competing cameras. To get good color, you will often need to take longer exposures with the one-shot camera than the sum of the three exposures required with a camera plus filter wheel. If the exposure is too short, you will get little or no color. And dim areas may simply not have the color you would like, if they have any color at all.

Cameras with a filter wheel also offer additional flexibility. With many such wheels, you can choose the filters you use, including non-standard, special-purpose filters such as OIII, Hydrogen-alpha, etc.

For some subjects, one-shot color makes good sense. One example is Jupiter. Jupiter rotates rapidly, so anything that helps you get color data quickly is a plus. Jupiter is also extremely bright, so a slightly longer exposure is not a problem. However, a digital camera can also be a good choice for planetary imaging.

At the time of writing, the software for Starlight XPress one-shot color cameras had some serious flaws that made it difficult to obtain accurate color from the cameras. The manufacturer was working to resolve these problems. You should check to see which software version you are getting when you buy a one-shot color camera to make sure you get a version that fixes the color-accuracy problem.

Cameras by Manufacturer

Now that you have figured out whether to get an ABG or NABG camera, I'll offer a little advice on the cameras offered by some of the leading CCD camera manufacturers.

Apogee

Apogee makes a large variety of CCD cameras. They sell not only to astronomers, but to microscopists and industrial imagers as well. Their AP line of cameras has been a mainstay of professional and advanced imagers for many years; their other camera lines are generally intended for non-astronomy applications.

Many Apogee cameras come without software, so plan on buying a software package to control the camera and perform image-processing tasks. If you are interested in supernova or minor planet hunting, the back-illuminated chips in several of the Apogee cameras are among the most sensitive available. These chips have had a portion of the chip shaved away to thin them, and allow light to pass through to the sensors. These chips have extremely high quantum efficiency. At the time of writing they come in 512x512 and 1024x1024 arrays of pixels that are 24 microns square. These large pixels require longer focal lengths for critical sampling, so they are not good for all telescopes. Generally speaking, they are best on telescopes with focal lengths of 2000mm or longer.

Apogee also makes cameras with many other chips. Compare on price and features with the other manufacturers at the time you plan to buy, as prices on both chips and cameras are constantly evolving.

At the time of writing, Apogee had started to market, but was not yet shipping, a line of cameras called LISAA. These are lower-cost cameras with less-expensive CCD chips in them, intended to appeal to a wider audience. The LISAA line includes some one-shot color cameras as well as conventional monochrome cameras, as well as a guider that can work with most film and CCD cameras. Check the book web site for information about this line of cameras once we have received some for testing. For more information:

http://www.apogee-ccd.com/products.html

Starlight XPress

Starlight XPress specializes in lower-cost cameras made with Sony CCD chips. Most of the other manufacturers are using CCD chips from Kodak, SITe, Thompson, and other manufacturers. The Sony chips are a different design, and the Starlight XPress cameras reflect this by offering some different features than other cameras do.

From what I have seen done with these cameras, they are capable instruments but not as good as the more expensive cameras from other manufacturers. The images tend to be a little noisier, in my opinion, but I have not had much chance to use the cameras since none were sent to me for evaluation and review. Judging from the images I have seen, however, these are capable cameras that can be a good choice for the imager on a budget. You should weigh the advantages of lower cost against the issues of noise and sensitivity.

The biggest problem I have had with the Starlight XPress cameras is that they come with limited software and documentation. Part of the price you pay for a lower dollar price is a need to spend more of your time figuring out how everything works. Starlight XPress recently took a very positive step forward by working closely with the providers of Astroart software to get their cameras supported by that software. It is easier to use Astroart to control the Starlight XPress cameras, removing a significant obstacle to recommending these cameras. Astroart is itself not the most well documented camera control program, but it is powerful and contains many useful tools. If you are willing to invest a little time learning how to use the hardware and software, you can save yourself some money.

Starlight XPress offers a guiding option for many of their cameras that uses an intriguing methodology. When guiding, the camera uses half of the pixel rows (every other row) to image, and half to guide. At the midpoint of the exposure, the camera switches halves, with the half that was guiding now imaging, and the half that was imaging now guiding. This doubles the exposure length (and the time required to take dark frames, since the dark frames must do the same switching), so take than into account when you make your buying decision. For more information:

http://www.starlight-xpress.co.uk/

SBIG

SBIG has a superb reputation in many areas, including design, manufacturing quality, camera control software, and technical support. This makes it very easy to recommend SBIG cameras. It's no accident that most of the images in the book were taken with SBIG cameras. They were fully cooperative throughout the writing of the book, and I have had tremendous success using a wide variety of their cameras. In all cases, in fact, whenever I have been dissatisfied with an image taken with an SBIG camera, it has been because I failed to do something right. As I learned more about proper technique for taking CCD images, my images with SBIG cameras got better and better. I realized that these cameras are first rate in every respect -- they never get in the way of taking a superb image. You can really grow into an SBIG camera over time; it will take you a while to reach the full potential of the camera.

The ST-237 is an ideal beginner camera. It is lightweight and compact, so it doesn't overwhelm your telescope and mount. Image quality is extremely high, and cost is reasonable for a 640x480 pixel chip. The small 6.8-micron pixels make it a good planetary camera as well. Best of all, you can buy an internal filter wheel for the ST-237 and take color images with it quite easily.

The ability of many SBIG cameras to self-guide is a real plus. Many models, including ST-7, ST-8, ST-9, and ST-10, include two chips in the camera. The larger chip is the imaging chip, and the smaller chip is a guiding chip. This allows the cameras to image and guide at the same time. All of the major camera control packages support this feature, making the SBIG cameras among the easiest to use for long, guided exposures.

The ST-7E camera is a great entry-level camera for someone with a bigger budget. The ST-7E (and most of the other SBIG cameras) can be matched up with the CFW-8 color filter wheel to do color imaging.

If you have a much more generous budget, the ST-8E and ST-10E cameras offer very large chips and small pixels. These cameras offer superb resolution and a large field of view -- a fantastic combination.

If you have a longer focal length (2000mm or longer), consider the ST-9E. It has large 20-micron pixels. The pixel array isn't that large -- 512x512 pixels -- but the field of view is nearly as large as that of the ST-8E and the cost is significantly lower.

SBIG is constantly developing new products and accessories. If you take the time to match an SBIG camera to your telescope and mount, as described later in this chapter, you really can't go wrong. I highly recommend their products, both for the technical quality and the level of support. For more information:

http://www.sbig.com/

Meade

Meade manufactures several CCD cameras. They are priced lower than the competition, but they do not include the same features, software, or technical support. The lower price is offset by the reductions in other areas. For more information:

http://www.meade.com/catalog/index.html

Finger Lakes (FLI)

At the time of writing, Finger Lakes Instruments was in the process of introducing a new line of budget-priced CCD cameras. These cameras are called MaxCams, and they offer some interesting features, such as more cooling than similarly priced cameras. Since cooling reduces noise, the MaxCams are an intriguing new option. Please check the web site for the book to learn more about these cameras when information becomes available.

FLI has been making research-grade CCD cameras for many years, and if the MaxCams are up to the same level of design and functionality, they will be a great addition to the field. For more information:

http://www.fli-cam.com/

Do-It-Yourself

If you are interested in building your own CCD camera, you can save a substantial amount of money. Probably the best online camera-building project is Genesis. Get more information about this project here:

http://www.genesis16.net/

Camera and Image-Processing Software

It would take about a thousand pages to do full justice to all of the image editing and camera control programs available. Most cameras come with at least some software, but most imagers wind up owning several software programs. Many of the camera control programs provide some degree of image editing as well, but for serious image editing you'll usually get the best results with a pure image editor. I like Photoshop, but Picture Window Pro and Paint Shop Pro are also well regarded by many imagers, and they are substantially less costly than Photoshop.

My own personal choices are:

Camera Control, including data reduction and astrometry: CCDSoft version 5.

Color Combining: Photoshop 6.0 or MaxIm DL.

Image Processing: Photoshop 6.0.

Supplementary Processing (deconvolution especially): CCDSharp or Astroart.

CCDSoft

The latest version (version 5) sets new standards for flexibility and ease of use in camera control, color image acquisition, data reduction, astrometry, and photometry. CCDSoft is best in class in all of these areas. You may have to learn some new things to take full advantage of its capabilities, but they are all things worth learning.

The data reduction capabilities are a good example of this. The previous best in class was MaxIm DL. MaxIm required you to prepare your darks, flats, and bias frames prior to actually applying them. For example, you had to do you combining before you could point MaxIm at your "calibration frames."

> **TIP:** MaxIm's use of "Image Calibration" is incorrect. The correct term is "image reduction." CCDSoft not only has better image reduction; it uses the correct terminology, too.

With CCDSoft, selecting, combining, and applying your dark, flat-field, and bias frames is done in one step. You create reduction groups, and CCDSoft then applies all frames in the group to the image or folder of images that you specify. This will work even better if you learn how to use and apply multiple frames. Once you understand the benefits of multiple frames (lower noise), it's clear why CCDSoft is the best in class.

CCDSoft is also the best program for image alignment. It's fast, automatic, and really accurate.

More information: **http://www.bisque.com**

Mira AP

Mira AP is technically very sophisticated software, and it will appeal to anyone who has a strong background in CCD imaging, mathematics, and/or statistics. The downside is that without such knowledge, you may not be very clear about how to use Mira to accomplish your goals. Many tools and steps are lightly documented, leaving you to guess far too often if you are new to the field of CCD imaging.

However, if you take the time to learn the ins and outs, and if you find others using Mira who can help you figure out what's what, you'll find that many of Mira's features are in fact rock solid. It also includes some unique and powerful image processing methods and tools that can be very handy.

However, overall the learning curve is quite steep, and for most users this is not the software package to start with. The more sophisticated your background is, the more likely you are to appreciate it and be able to use its raw power. Mira isn't often as convenient as the other tools, but it tends to be robust and accurate in what it does.

More information: **http://www.axres.com/**

MaxIm DL

MaxIm DL has been the popular choice of many imagers for the last few years. It was a breakthrough program, with author Doug George introducing a lot of features that were major improvements on what was previously available. So there is a lot of affection for MaxIm DL out there, and with good reason. For several years, MaxIm DL was the best in class in most categories.

The introduction of CCDSoft version 5 has taken some of the best in class away from Maxim DL. However, version 3 of MaxIm is due out in the latter half of 2001. It stands up well to the competition. Image reduction is more flexible, and image processing is more sophisticated and powerful.

MaxIm DL continues to be strong in many image-processing areas, and if you like Digital Development, MaxIm's implementation is very good. It allows you to quickly and effectively bring out dim details in your images. MaxIm DL also supports the widest variety of imaging hardware.

More information: **http://www.cyanogen.com**

Other Programs

Mira AP, CCDSoft version 5, and MaxIm DL are the Big Three of camera control. There are also some other useful programs out there that are worth a look. I haven't used all of them, but you can download free or trial versions of some of these programs to try them out.

Astroart - The documentation is slim, but the features are many. Astroart has some limitations with respect to color, but it has superb implementations of deconvolution and unsharp masking that alone make it worth having around as an extra program. Astroart also has an open architecture, with many third parties writing camera control and image processing routines for it that make it a heck of a deal. If your budget is a critical concern, Astroart is good enough to serve as your only camera control program at about a third the cost of the more full-featured and fully documented products such as CCDSoft. If your budget permits, add Astroart to your basic camera control program. A downloadable trial version is available at **http://www.msb-astroart.com/**

AstroPIX - Software for the CB245 Cookbook CCD camera. Information at **http://www.wvi.com/~rberry/astropix.htm.**

StellaImage3 - The latest version of one of the groundbreaking CCD image processing programs. A downloadable trial version is available. Developed in cooperation with the Japanese amateur astronomy community, StellaImage include data reduction; support for various file formats; scanner support for film imagers; RBG, ORGB, and WCMY color combining; 32-bit floating point operations; digital filters; deconvolution; vignetting removal; digital development; star sharpening, and many other functions.

SuperFix and MegaFix - SuperFix is a basic package intended for someone new to CCD imaging. It is very economical and includes the basic camera control and image processing functions. MegaFix is a more advanced, full-featured product, and is available for a special price to SuperFix owners. For more information, go to **http://members.aol.com/BJohns7764/BJCfix.htm.**

SSC Astronomy - This is a collection of shareware programs for the CCD imager, including DOSPVIEW, Specostropy, FTS Animator, and GPS Geodesy. For more information: **http://24.5.47.244/astrostf.html**

IRIS - A powerful freeware program, downloadable from **http://www.astrosurf.com/buil/us/iris/iris.htm.** The original is in French, so the English documentation tends to lag behind the latest version most of the time. IRIS uses a command-driven interface, but it is well regarded and very powerful. It is a great way to learn the fundamentals.

API4Win - This is technically a book, but it's a great book and it contains a CD with a wealth of image processing tools. If you have ever wanted to actually understand the how and why of a wide range of image processing operations, this is for you! It is a great complement to the book you hold in your hands. Info at **http://www.willbell.com/aip/index.htm.**

For pure image editing (no CCD support), Photoshop, Picture Window Pro and Paint Shop Pro are the packages most commonly used. Many other image editors have features that will be useful on CCD images. The most important features to have are tools for linear histogram adjustment (to set black point and white point); gamma adjustments; non-linear histogram tools; sharpening filters, especially unsharp masking; and smoothing tools, especially Gaussian blurring.

FIGURE 4.3.12. AN IMAGE OF M33 TAKEN WITH AN SBIG ST-8E CAMERA AND A TAKAHASHI FCT-150 REFRACTOR ON A SOFTWARE BISQUE PARAMOUNT 1100S.

Section 4: Matching Camera, Telescope, & Mount

Image scale is a key concept in the world of CCD imaging. Image scale describes how much sky each pixel "sees." Image scale varies with the size of the pixel and the focal length of your telescope. If you make pixels bigger, they see more sky. If you increase focal length, each pixel sees less sky. To measure image scale, you must know both the size of the camera pixels and the focal length of the telescope.

Image Scale Explained

Measuring the image scale helps you determine how well your scope and camera are matched to each other. If the pixels see too much sky, you may not get the resolution you would like but you do get short exposures and less dependence on seeing conditions. If the pixels see too little sky, you get more potential resolution but you also get longer exposures and a greater dependence on seeing conditions. The middle range of image scales are the most commonly used, but the extremes have their applications.

The amount of sky that each pixel "sees" is measured in arcseconds. An arcsecond is 1/60th of an arcminute, and an arcminute is 1/60th of a degree. A circle has 360 degrees, so an arcsecond is 1/1,296,000th of a circle.

The range of useful image scales varies from a fraction of an arcsecond per pixel to more than 40 arcseconds per pixel. Figure 4.4.1 shows one extreme. It is an image of the sky including M31, taken at 38 arcseconds per pixel using a 50mm camera lens and an ST-8E camera. Each pixel "sees" a square 38 by 38 arcseconds. An area that is one hundred pixels on a side covers approximately one square degree of sky. The entire image, which is 1530 x 1020 pixels, covers an area of sky that is 15.3 degrees wide and 10.2 degrees high. Longer camera lenses, such as 135mm or 200mm, would cover a smaller area of the sky. Shorter camera lenses, such as 28mm or 35mm, would cover an even larger area of the sky.

For both camera lenses and telescopes, a longer focal length means a narrower field of view. A shorter focal length means a wider field of view.

Figure 4.4.2 shows another extreme, an image of the Blue Snowball at 0.85 arcseconds per pixel. Each pixel "sees" an area of the sky that is 45 times smaller than in figure 4.4.1. The Blue Snowball image was taken with an ST-7E camera using a Mewlon 210 (8.4" f/11.5). The image covers an area that is 6 by 9 arcminutes. That's a very small area of sky. If you were to draw a box in figure 4.4.1 covering the same area, it would be only 9x14 pixels. For planetary imaging, I've used image scales as small as 0.25 arcseconds per pixel in order to obtain fine detail. A smaller value for the image scale means a greater level of magnification.

Figure 4.4.1. A widefield image of M31 taken at an image scale of 38 arcseconds per pixel.

FIGURE 4.4.2. A VERY NARROW-FIELD IMAGE (6x9 ARC MINUTES) TAKEN AT AN IMAGE SCALE OF 0.85 ARCSECONDS PER PIXEL.

At first glance, you might think that it would be great to have a really small image scale so that you get a lot of magnification. But several things get in the way:

- The seeing conditions on any given night limit the image scale. For most locations, an image scale of 2.5-3.5 arcseconds per pixel is going to be useful on most nights. Higher magnification at image scales of 1 to 2 will be possible on nights of exceptional seeing.
- The seeing conditions needed for high magnification (<1 arcsecond/pixel) are not only rare; they may never occur at some locations. Local hills, mountains, and cities can create turbulence.
- The capabilities of your mount may be a limiting factor in your quest for the ultimate image scale. A mount that tracks or guides poorly will not deliver sharp images at high magnification.
- The optical quality of your telescope determines how far you can push the image scale. Lack of sharpness and/or contrast will limit how much magnification you can use for CCD imaging.
- Many telescopes that deliver high magnification have slower focal ratios requiring longer exposures.

There have been long, intense debates about the role that image scale should play in camera selection for a given telescope. While it is possible to debate the advantages and disadvantages endlessly, I'd like to supply a some guidelines that will help you decide which image scale is right for you.

1. If you are just getting started, you can simplify the learning process by going with a medium to large image scale, around 2.5 to 3.5 arcseconds per pixel. This puts less demand on your mount, and allows you to image successfully in average seeing conditions. In fact, I would go so far as to say that, unless you have a specific reason for starting with higher magnification, you should start in the 2.5-3.5 arcseconds per pixel range.

2. If you have a modest mount, aim for a large image scale, in the 3-5 arcseconds per pixel range.

3. If you have a small refractor, or a fast focal ratio telescope, then smaller image scales are still readily achievable. You can use a Barlow to get more magnification at the cost of having to take longer exposures because of the increase in focal ratio.

4. If you want the highest possible resolution, then the best strategy is to aim for an image scale that will match your seeing conditions. The STV from SBIG can help measure seeing, or you can estimate it visually. For example, if you find that your seeing conditions are 3.1 arcseconds, then a camera that has an image scale half of that (1.55 arcseconds per pixel) will provide optimum resolution. Keep in mind that seeing conditions vary considerably at any given location.

5. If you have a strong desire to image at high magnification, don't let recommendation #1 above deter you. Yes, it's more of a challenge, but it's not impossible!

6. The ideal setup is to have a telescope/camera combination for large-scale imaging, and another for small-scale imaging. If you can't choose, then consider getting equipment that will give you both options. For example, you could use the same camera on a Fastar-equipped SCT at f/2 and on the back of the same SCT at f/6.3 or f/10.

The CCD Calculator

The trick with image scale is to choose a camera that is well matched to your telescope and mount. The combination needs to be suitable for your particular equipment and purposes. I have developed a Visual Basic application that will show you graphically and numerically how a given telescope and camera match up. Download the application from this web page:

http://www.newastro.com/newastro/camera_app.asp

The program is free to anyone who has purchased the book or who has a web subscription. If for some reason you cannot download the application, you can view a web page with similar features here:

http://www.wodaski.com/wodaski/pick_a_camera.htm

You can discover more than just the image scale of various camera-telescope combinations with the CCD calculator. It will help you visualize what the relationship is between these two and the sky. Figure 4.4.3 shows the overall appearance of the CCD Calculator program. The left window contains the program controls, and the right window shows the field of view projected onto an image of the sky. The top of the control window allows you to select a telescope and CCD camera from lists, or you can enter numeric information about a telescope and camera that are not in the lists.

The lower middle section contains options, and the bottom of the control window has drop-down lists that let you rotate quickly through your recent telescope and camera choices.

Telescope data entry:

- Pick a telescope from the list
- Modify aperture, focal length, and/or focal ratio as needed

or

- Enter aperture and focal ratio

Camera data entry:

- Pick a camera from the list

or

- Enter pixel size in microns (width and height required)
- Enter the array size in number of pixels, width and height
- Select a bin mode

As you make changes, the FOV data is updated automatically. You'll see the following information on the right side of the control window:

Chip size - This is the physical size of the CCD chip, in mm. The height is displayed first, then the width.

Image scale - Tells you how large a portion of the sky is covered by each pixel. Measured in arcseconds per pixel (averaged for non-square pixels).

Field of view - The area of sky covered by the entire CCD chip. Measured in arcminutes. The height is displayed first, then the width.

Chip size compared to 35mm film frame - the image of 35mm film has a black box on it. The black box is the physical size of the chip with respect to a standard 35mm film frame. If the black box is large enough, the chip size in mm is shown.

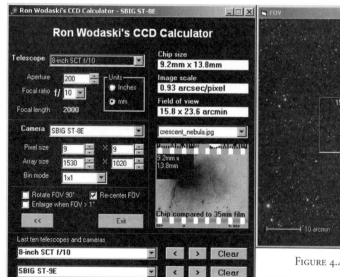
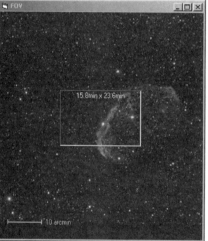

FIGURE 4.4.3. THE CCD CALCULATOR

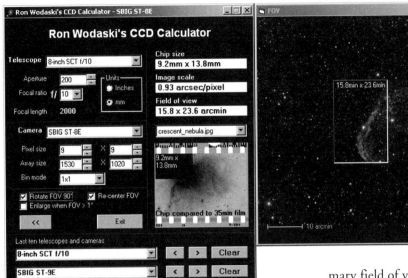

FIGURE 4.4.4. YOU CAN SWAP THE FOV INDICATOR'S ORIENTATION FROM HORIZONTAL TO VERTICAL.

The field of view is projected onto an image of the sky in the right-hand window. The default projection is an image of M51. All of the images are one degree square (60 minutes on a side). You can change the displayed image using the drop-down list just above the 35-mm negative image.

The field of view of the camera/telescope combination is drawn onto the sky to show you how much coverage you will get when imaging. Various objects of various sizes are available to illustrate how well the chip will cover some well-known objects when used with the selected telescope. You can change telescope, camera, or features of either to get quick visual feedback on how suitable the combination of camera and telescope is for your needs. You can also click and drag on the image to create a second, temporary field of view box. This second box will con-

tain text that tells you the size in arcminutes. This allows you to see how large a field of view would be required to image a given area of sky. Both the primary and secondary fields of view can be dragged around on the image to place them over a specific object in the image.

There are several options available at bottom left:

Rotate FOV 90° - Changes the orientation of the primary field of view indicator. By default the FOV indicator is wider than it is high. If you check this box, it becomes taller than it is wide. Figure 4.4.4 shows the rotated FOV indicator for an ST-8E camera.

Re-center FOV - When checked, the FOV indicator will jump to the center of the image whenever a recalculation occurs. It is checked by default.

Enlarge when FOV > 1° - When checked a field of view that is larger than one degree causes the FOV window to enlarge to show the extra space.

The "<<" button - Closes/opens the FOV window.

FIGURE 4.4.5. CHANGING THE FOCAL RATIO CHANGES THE IMAGE SCALE AND THE FOV.

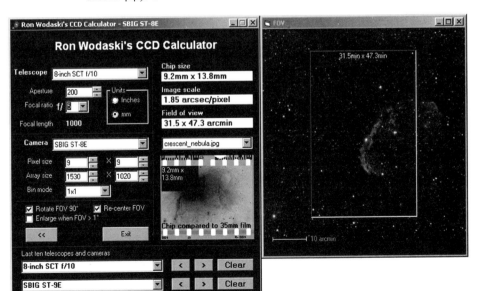

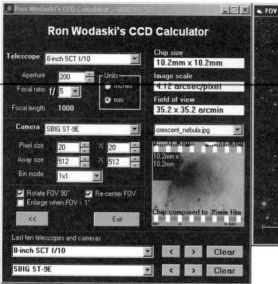
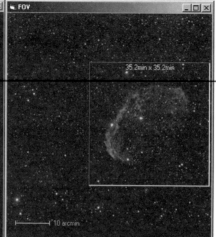

cover a larger field of view as the image scale changes. The chip size stays the same, but the image scale and field of view change.

Figure 4.4.6 shows the result of dragging the field of view indicator to see if the camera-telescope combination (an 8" SCT at f/10 and an ST-9E) can cover the Crescent Nebula adequately. To drag the FOV indicator, click on it and then drag it to the new location. This allows you to get a better idea of coverage for objects that are not in the center of the image in the FOV window.

Figure 4.4.7 shows the result of clicking and dragging on the image itself. A second field of view indicator appears, with text at the top telling you the size of the field of view in arcminutes. This is useful to learn what size field of view is required to image a given object (in this case, the Lagoon Nebula). To remove the secondary field of view indicator, click on the background image outside of the secondary FOV indicator.

FIGURE 4.4.6. MOVE THE FIELD OF VIEW INDICATOR BY CLICKING AND DRAGGING IT.

Last ten lists - There are two drop-down lists, one for the last 10 telescopes and one for the last 10 cameras. You can pick recent choices from these lists to make quick A/B comparisons.

The "<" and ">" buttons - rotate through the items in the "Latest ten telescopes and cameras" lists.

Clear button - removes all items from "Last ten" lists.

Scale bar - (at bottom of the FOV window) It shows the size of a 10-arcminute segment to give you an idea of relative sizes. Some functions of the program rescale the images, and this allows you to get a good idea of the magnitude of the change.

Changing a parameter, or choosing a different telescope or camera, changes the calculation results and the FOV window automatically. Figure 4.4.5 shows what happens when the focal ratio is changed from f/10 to f/5. The camera will

FIGURE 4.4.7. DRAGGING OUT A SECONDARY FIELD OF VIEW INDICATOR.

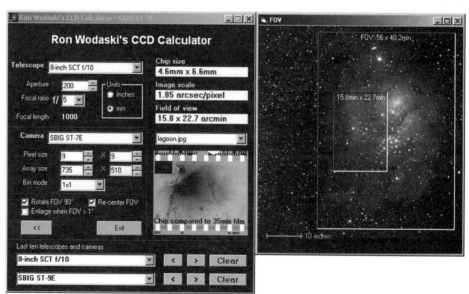

When "Enlarge when FOV > 1°" is checked, the FOV window grows automatically when the FOV is larger than one degree (see figure 4.4.8). For very short focal lengths and very large CCD chips, this may make the FOV window too large to fit on the screen. You can uncheck the checkbox and the FOV window will stay small, and the image will shrink. The percentage change shows up at the bottom of the FOV window.

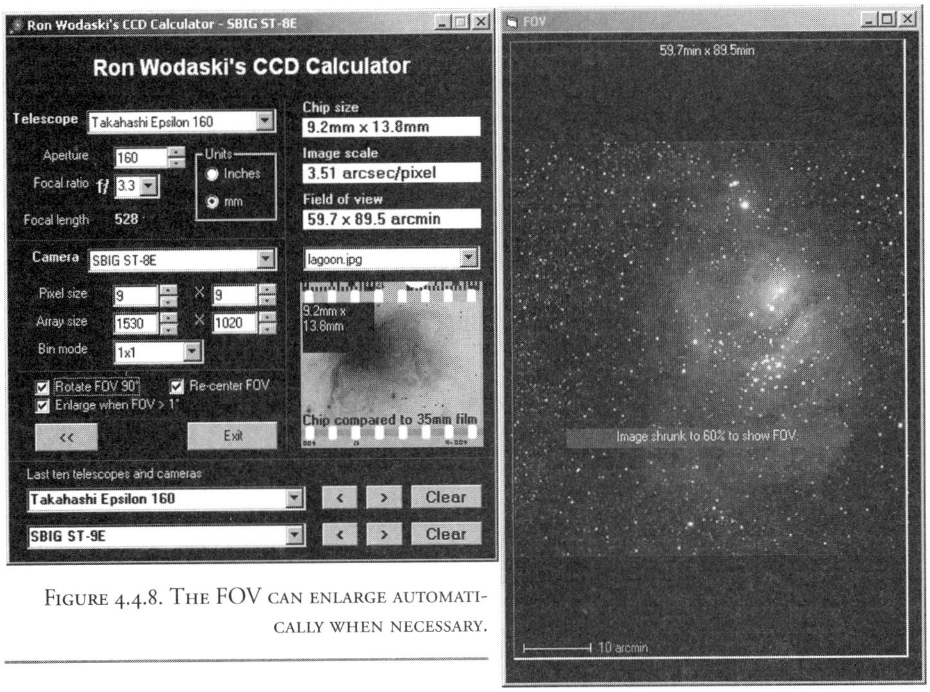

The data for telescopes and cameras is stored in a data file that comes with the program. The data file (camera_data.dat) describes the format for the data with comments. Comments start with a semi-colon. The sample below shows what the data file looks like.

You can add or remove cameras and telescopes. However, be very careful to follow the exact format shown. Do not include any spaces outside of the double quotes. The program won't accept them. Don't forget to add all of the data items for each telescope or camera; a missing data item will cause the program to fail when it tries to read the data file.

Check the web site for updates to the program. I will continue to add features and fix bugs. The program version can be found just under the image of the 35mm film negative.

FIGURE 4.4.8. THE FOV CAN ENLARGE AUTOMATICALLY WHEN NECESSARY.

Focal Ratio is King

For a given CCD camera, exposure length is determined not by the aperture of the telescope but by the focal ratio. That is, a 16" scope will require the same exposure time as an 8" scope if both scope have the same focal ratio and the same camera is used. The images will not be the same, however; the 16" scope will have a longer focal length, and thus more magnification. The 16" scope will show a larger image of the object (that is, it will have a smaller field of view).

Cameras behave the same way. A 50mm f/2 lens has the same exposure requirements as a 200mm f/2 lens. Let's look at what happens when you change aperture and focal ratio with the same CCD camera.

A well-known f/2 scope is a Fastar with an ST-237. So we'll use the ST-237 for all of the comparisons. I'm assuming roughly a 30% obstruction by diameter for all of the scopes. That's a little under for SCTs,

A SAMPLE OF THE CCD CALCULATOR DATA FORMAT.

```
; Telescopes
; type, name, aperture in mm, focal ratio
t,"Takahashi FSQ-106",106,5
t,"Takahashi FC-50",50,8
t,"Takahashi FC-60",60,8.33
; Cameras
; type, name, pixel width, pixel height, array width, array height, max bin
c,"SBIG ST-237",6.7,6.7,640,480,3
c,"SBIG ST-237A",6.7,6.7,640,480,3
c,"SBIG ST-7E",9,9,735,510,2
c,"SBIG ST-8E",9,9,1530,1020,3
```

and a little over for most Newtonians, but close enough for a valid comparison.

Scope #1: Fastar

- Aperture: 8"
- Area (approx): 50 - 18 = 32 square inches
- Focal length: 400mm
- Area of sky covered by chip: 30x40 arc minutes
- Area of sky covered by a pixel: 3.8 arcseconds

The image above shows the approximate field of view obtained with this scope and the ST-237 camera, with M51 as the subject of interest.

Now consider what happens with an f/5 version of the same scope using the same camera.

Scope #2: 8" SCT f/5

- Aperture: 8"
- Area (approx): 50 - 18 = 32 square inches
- Focal length: 1000mm
- Area of sky covered by chip: 12x16 arc minutes
- Area of sky covered by a pixel: 1.5 arcseconds

In this example, each pixel is covering less than half as much sky. Each pixel receives less than half as much light. Exposures will have to be more than twice as long to get the same quality. Note that the magnification is greater. Objects are more than twice the size on the chip as they were at f/2. M51 now nearly fills the field of view.

Now let's look at a scope that is also f/5, but has a 16-inch aperture.

Scope #3: 16" SCT f/5

- Aperture: 16"
- Area (approx): 200 - 72 = 128 square inches
- Focal length: 2000mm
- Area of sky covered by chip: 6x8 arc minutes
- Area of sky covered by a pixel: 0.75 arcseconds

Some interesting things jump out about this scope compared to #2:

- The focal length doubles, from 1000mm to 2000mm.
- The light gathering ability is 4x scope #2 (128 versus 32 square inches)
- The area of sky covered by the chip is one-quarter of #2

So we've increased the focal length by 2x, the aperture by 2x, and the light gathering ability by 4x. The net result is that four times the light is spread over four times the area, so there is no net change in the amount of light hitting the CCD chip. Figure 4.4.12 shows the difference in image scale. We have zoomed in quite a bit. So the change in aperture does cause a change, but not in the amount of light per pixel. The change is in the amount of sky per pixel.

What conclusions can we draw from this exercise?
- A faster focal ratio yields shorter exposure times.
- A slower focal ratio yields longer exposure times.
- Changing the aperture, but keep the focal ratio and the camera the same yields identical exposures and a different image scale.

That last point is interesting: if you increase the size of your telescope, but don't change to a camera with larger pixels, you gain *potential* resolution (you are limited by seeing). Objects will appear larger. If the magnification is too great, then the seeing will interfere with reaching full resolution. For example, if the seeing is 4 arcseconds and the image scale is .25 arcseconds, you are not going to get the advantages of that potential increase in resolution.

Now let's consider a fourth case: we keep the 16" scope, but we use a hypoethical camera that has pixels exactly twice the size of those on the ST-237.

Scope #4: 16" SCT f/5
(camera pixels 14.8 microns square)

- Aperture: 16"
- Area (approx): 200 - 72 = 128 square inches
- Focal length: 2000mm
- Area of sky covered by chip: 12x16 arc minutes
- Area of sky covered by a pixel: 1.5 arcseconds

Well, golly, the area of sky covered by a pixel is the same as for scope #2. So now we have a setup that images the same field of view as scope #2; that has the same image scale as scope #2; and that has four times as much light gathering ability. By increasing pixel size, we are able to use shorter exposures to get the same results. Or you can take an exposure just as long, and go deeper and get additional detail. Compare figure 4.4.13 to figure 4.4.11 -- the simulation shows the image with the larger scope and larger pixels going deeper.

The moral of this story: if you match the pixel size of your camera to your telescope, skills, and interests you will enjoy a happy balance of image scale, exposure length, and resolution.

For any given scope, what matters most is what you want to get out of it. Do you want a balanced setup that doesn't lean toward any extremes? Go for an average pixel size, one that will give you an image scale of 2.5 to 3.0 arcseconds/pixel. Do you want resolution (and do you have the skies to use it)? Go for smaller pixels that will give you an image scale under 2 arcseconds/pixel. Do you want the luxury of short exposures? Go for bigger pixels that will give you an image scale of 3.5 arcseconds/pixel or more.

The most convenient way to measure pixel size is in terms of the area of sky that the pixels cover; that's why you see so much emphasis on arcseconds per pixel. Image scale is just one criterion to use in selecting a camera for your scope. Planetary observers will take all they can get, down to .25 arcseconds per pixel, but you will need a very steady mount and you'll have to move to the steady skies of Florida if you expect to use a camera at that level of resolution! Or you can comfortably image at 3 or 4 arcseconds per pixel and just have a lot of fun. Somewhere around 3.5 arcseconds per pixel you wind up with a wide field of view and minimal precision required for good guiding. You can even image unguided at those image scales if your mount is good enough.

My advice to beginners is to shoot for a combination that gives you about 2.5-3.5 arcseconds per pixel. If you have good reasons for choosing differently, by all means do so -- it all works.

The Bottom Line

A good mount, CCD camera, and telescope are not cheap, so making a wise decision from the start feels especially important. If you buy quality components with a good reputation, and if you buy equipment that is also useful for visual observing, you are unlikely to suffer if you decide to sell down the road. Buying used has its hassles and risks, but the upside is that the price you pay is going to be close to the price you get when you sell an item if it remains in good condition.

Astromart (**astromart.com**) remains a good place to buy and sell used equipment.

If you are buying on a budget, here are my thoughts on what you can do in various price ranges. I mention quite a few products by name, but there are almost always numerous other products of equal or similar performance. I generally list products that I have used personally, or that have been used by folks whose opinion I trust. There are going to be many other products that are also worth your consideration, so don't dismiss a product just because I haven't mentioned it.

To get feedback on mounts, telescopes, and CCD cameras, join the following online resources where you can ask questions and get useful answers:

> **http://groups.yahoo.com/group/ccd-newastro**
> **http://groups.yahoo.com/group/telescopes**

Also check out the various product review sites; here are a few good ones to get you started:

> **http://www.weatherman.com/wxastrob1.htm**
> **http://www.scopereviews.com/**
> **http://www.cloudynights.com/**

Prices in the following sections do not include software or other accessories, just the mount, telescope, and camera. Prices are approximate since they go up and down over time.

As Cheap As Possible

If you want to get into CCD imaging economically, look into video cameras. You can use a video capture card to digitize images, and then combine and process the images like a CCD image. The downside is that video cameras can't do long exposures, so you are limited to lunar, planetary, and solor imaging. If you have some electronics experience, there are a growing number of web sites describing how to modify cheap webcams for astronomical imaging.

If you want the cheapest possible CCD setup, and you aren't building your camera from scratch, then a used, older-generation camera is the way to go. If you have a small refractor or any other scope with a short focal length, a used Cookbook camera, SBIG ST-5c, ST-237, Starlight XPress, or any other small-chip camera will be your least expensive point of entry. The longer the focal length of your telescope, the more challenging it will be to center objects on these small chips.

If you want the least expensive telescope, then your best choices at $500 and under would be:

- A small quality refractor, even a semi-APO. Get a yellow or green filter to remove chromatic focus problems; you'll get much sharper images that way.

- A used 8" SCT. Consider picking up an f/6.3 reducer to shorten the focal length a bit. That will make your learning curve more reasonable. Get a camera with the largest pixels possible. An old, used ST-6 or ST-6B would be a perfect way to learn with an SCT.

- Newtonians are an often-overlooked resource for inexpensive, quality telescopes for CCD imaging. Some Newtonians, of course, are terrible. But if you can locate a 6" or 8" Newtonian that is about f/6 or f/8, and if you can put it on a mount that will track reasonably accurately (the hard part for these often heavy scopes), you could get some outstanding results.

- Don't overlook 4" and 5" mirror-based scopes. If the optical quality is good, you'll learn that small aperture is not a hindrance to good imaging. Even an older Meade 90mm ETX OTA can work well if you can put it on a suitable mount. The long focal length will be a challenge, however, for many mounts and for beginning imagers.

There are probably many unique solutions out there if you are working on a budget. Think used, and you gain a lot of leverage for any budget. Just be careful to evaluate before you buy, and to walk away if the deal doesn't feel right. Think in terms of short focal lengths, and you'll find yourself with a more affordable setup that will be fun to learn with.

Under $3,000

It's a challenge to set up for good CCD imaging under $3000. The key is to include used equipment in your setup. For example, a used C8 that is Fastar compatible ($450-600), a Fastar lens assembly ($350 new), a Vixen GP-DX mount ($800-1200), and a low-end CCD camera bought with whatever is left in the budget. Starlight XPress has some low-cost entry-level cameras, and the SBIG ST-237 is one of the best entry-level cameras.

The new FLI MaxCams are aggressively priced. You can do even better if you can locate a used Kodak chip on your own. If you go without the Fastar lens assembly an ST-5C, or a used ST-7 will work, but be prepared for the challenge of imaging at long focal lengths.

The Fastar is f/1.95 and 400mm, and even a cheap mount will provide surprisingly good results. The Vixen mount, however, provide the best quality in their price range. To land a good deal on the scope, lens, and mount, figure on spending about $1200-1400 on the camera -- which is enough to buy even a new ST-237. The main advantage of the Fastar setup is that the short focal length puts very little strain on your mount, so you can economize on that key component.

The same setup will work well with a used small fluorite refractor (Takahashi FS-78, FC-60, or a TeleVue 85), or a new semi-APO achromat such as a Pronto or Megrez. The aperture is smaller, but that gives you wider fields of view. Shorter focal lengths (under 500mm) make the most out of your mount.

Under $5,000

If you have to fit under a $5000 ceiling, the big question is where to spend more money: mount, camera, or telescope. Based on my own experiences, I would spend more on the mount first, then on the telescope, then on the camera. So you might upgrade to a GT-ONE mount (full goto, great gears, average in convenience and suitability for imaging), but keep the scope and camera in the same spending levels as in the under $3000 category above (since you are spending the full $2000 additional money on the mount).

Other good mounts to consider include:

- Takahashi EM-10
- Takahashi P2Z (no Dec motor, but great tracking)
- Losmandy G-11 (risky due to variations in accuracy from mount to mount, and often requiring some TLC to get the most out of your mount).
- Used Astro-Physics 400 QMD (includes digital setting circles, and available for $2000-2500 used).

If you have a good-quality Dobsonian, consider getting a premium equatorial platform for it. Figure on about $2,200, and spend the rest on a camera.

Once you have a mount, you could upgrade telescope and camera choices as your budget allows. If you choose a mount with high quality and larger capacity, you can get a longer focal length scope if you wish to.

Some nice combinations for under $5000:

- Takahashi EM-10 mount, FS-102 telescope, and an ST-237. Get one of these used to fit under the $5000 budget. If you want all new, consider replacing the EM-10 with the P2Z, or the FS-102 with an FS-78 or a TeleVue 85 (OTA only version). The ST-5C camera will also work here, but the small chip size will have you thinking upgrade.
- Takahashi EM-10 mount, used C8 SCT, and an ST-7E camera. This pushes the budget limit, but it's a great setup. If you can go a bit beyond $5000, get a C9.25 instead. You'll have to work hard to make the mount carry that load, and you may lose some images if the wind kicks up or you just get unlucky. You'll have to work hard to put objects on the chip. But on a steady night, you'll have killer images. This goes against everything I say about short focal lengths and overloading the mount, but those are general guidelines. If you have the stubbornness to succeed, you'll succeed as long as the equipment is quality stuff.
- Here's an off-beat approach to the under-$5000 setup: a Vixen GP-DX with SkySensor 2000 PC; a C8 Fastar with Fastar lens, and an ST-237 camera with color filter wheel. All new, this setup will fit under the budget, will be easy to learn and operate, and can provide you with absolutely first-rate images. The compromise here is the smaller dynamic range of the ST-237, but the upside is complete color capability, goto-mount, and a wide field of view to make life easy. If you just want to have fun, hard to beat this setup. The SkySensor 2000 PC has way more features than you need for

imaging, so it gets in the way at times, but it tracks and guides capably. The GT-ONE from Wm. Yang Optics would be another, somewhat more expensive version of that choice. That way, you wouldn't have to buy another mount when you decide to upgrade camera or scope.

- If you can pick up a used EM-10 mount or equivalent, combine it with a 4" APO refractor and an ST-7E camera with the optional cooling package. You'll have a super setup that will give you a lot of imaging pleasure and superb, quality images.

The FLI MaxCams are less costly than the SBIG cameras, and could make a difference in what you can afford in CCD chip or camera features. Check the book web site for MaxCam information and images.

At the time of writing, the LISAA cameras from Apogee had not yet shipped. If these turn out to be good, their low cost will allow some interesting camera/scope combinations in the under $5000 price range. Check the book web site for the latest information.

Under $10,000

If you have up to $10,000 as your budget, your options open up considerably. You could use one of the following high-end items as the basis for a great system, combining modest equipment and high-end equipment with a focus on what's most important to you:

- A large-chip, high-resolution CCD camera, like the SBIG ST-8E/10E or an FLI MaxCam CM-10E. These cameras are good for wide-field, high-resolution imaging.
- A large-chip, high-sensitivity, medium resolution camera, such as an ST-9E or AP47p. These cameras are ideal for long focal length, deep-space imaging, minor planet and supernova searches, etc.
- A mount closer to the high end, such as the Takahashi EM-200 or NJP.
- An astrograph, such as a Takahashi FSQ-106 or Epsilon 210.
- A high-end scope with superb optical quality, such as a 5" APO refractor, a CCD-oriented Newtonian from Excelsior Optics, a Mak-Newt, a Mak-Cass from TEC, etc. The options are really too numerous to mention, and the differences between designs are likely to dominate your choices anyway.

The possibilities for combining quality components are numerous. Here are just a few to think about:

- Takahashi NJP mount, ST-7E camera, and a 4" APO refractor.
- Takahashi EM-10 mount, 4" APO refractor, and an FLI CM8-2 camera.
- Astro-Physics AP 400 GTO mount, 4" APO refractor, and camera of your choice with the $3500 left over. A used ST-8 would be sweet, as would an ST-7E or one of the FLI cameras.
- Mountain Instruments MI-250 or Takahashi EM-200, ST-7E camera, and a 6" Mak-Newt.

Over $10000

Once your budget gets over $10,000, some interesting choices start to open up. Among the most interesting possibilities are:

- Takahashi FCT-150, a 6" triplet refractor. One of the best telescopes for imaging. No longer in production, but available used now and then.
- Ritchey-Chretien telescopes. Cassegrain design, but with a shorter focal length and a very good spot size for sharp images. A 10" starts out around $12,000, and you can get up to 20".
- The premium mounts start with the Paramount from Software Bisque. At $8,500, it's everything you could want in a mount. The Astro-Physics 900 GTO and 1200 GTO are also excellent, but just a notch below the Paramount in features.
- Large-chip CCD cameras, including the ST-10E and ST-1001E from SBIG, the CM10-2 from FLI, or the really big chips in several of the Apogee cameras:
  - Apogee AP4, 2048x2048 (9-micron pixels)
  - Apogee AP9, 3072x2048 (9-micron pixels)
  - Apogee AP10, 2048x2048 (14-micron pixels)

The choices available when price is not the primary consideration are mind boggling. The best equipment blends both features and quality. CCD cameras feature huge chips and fast download times. And don't forget to budget for that home observatory.

Section 5: Imaging Options

There are a lot of things to plan for and control when you do CCD imaging. There are many things you can do to improve the convenience of taking images, or to make your setup as efficient as possible. This allows you to focus on the imaging process itself, and avoid the pitfalls of distraction.

There are many ways you can set up your mount, scope and camera. These range from setting everything up each time you image, to walking out to your home observatory. The options between these two extremes include running your equipment off batteries, or leaving equipment set up during a run of good weather. You can control your scope while standning next to it, or run cables into a warm house.

Figure 4.5.1 shows two setups I've used. The mount on the left is an AP-400 QMD. The mount on the right is a GP-DX with an AP Traveler. These pictures prove once and for all that your setup doesn't have to be pretty to take pretty pictures.

However you set up, there are many ways to get images into your computer. This section takes a look at the equipment you need and how to use it.

Setting up for Imaging

There are many ways to combine and set up scope, camera, and mount. CCD imaging ranges from taking a few images every now and then to gathering data every clear night. The type of setup you choose, and the way you use it, varies with the circumstances. Below you will find a variety of typical setups, with some advice on how to make the most out of each one.

Casual Imaging

You don't have to have an expensive setup to take CCD images. You can give up some of the conventional amenities, such as a high-end mount, and still accomplish some interesting results. Fast optics is one requirement for this, because that keeps exposure times short. The second is a goto mount that tracks reasonably well. Here's the part that may surprise you: the mount need not be an equatorial.

For example, goto Dobs are more common. The Dob is an alt-az (altitude/azimuth) mount, and thus subject to field rotation, but the fast focal ratio and large aperture of many Dobs allow you to use short exposures, down to 5-10 seconds. You must, of course, take a large number of images and combine them to get good results. Dark skies are a big help, but they are not as critical as you might think.

I recommend using a CCD camera light enough to avoid scope balancing problems. If you go heavier, be

FIGURE 4.5.1. THE MORNING AFTER.

prepared to find creative ways to balance the tube. Take short exposures to avoid field rotation. SBIG cameras with Track and Accumulate are a good choice, because you can stack many short exposures with less work. The results will never be as noise-free as you can get with a high-end equatorial mount, but you can have some fun with casual imaging.

Almost all CCD cameras require a computer. One exception is the STV, which has video output as well as enough memory to store 14 images. If you want the ultimate in imaging simplicity, the STV makes an OK CCD camera, is a superb guider, and has excellent video output. It is more sensitive than any other video camera because it integrates (takes long exposures). A conventional video camera can't image for more than 1/60th of a second because it is pushing out two frames 30 times a second. The STV doesn't output a video image until it is done integrating. This allows you to take a 10-minute image if the urge strikes. The small dynamic range of the STV makes it less capable than other CCD cameras, but if you need or like the video output, guiding capabilities, built-in tools and other features), then the STV is the simplest setup around. For serious Dob imaging, try an ST-9E or MX7-C.

But the STV is the exception. Most of the time you will be running power cords, either from an outlet or from a battery, to power your mount, camera, and computer, and data cables back the other way. The left image in figure 1 shows the nature of the problem.

The One Night Stand

If you are going to be setting up your CCD imaging equipment every night that you image, and then tearing it down when you are done, you want a setup that goes together easily. The following equipment will make your life as pleasant as possible:

- A folding or roll-up table. Set it up first so you can put equipment on it temporarily. Think about where in the sky you plan to image on a given night, and set up the table so that cables that run between table and scope will be well out of your way. The table is a great place for things like the water bucket for a water-cooled CCD camera; control box for STV; power strips (to avoid getting moisture in them); Allen wrenches, and so on.

- Cut up a large box and use it to keep dew off of your computer or other equipment on your work surface.

- A power strip. It provides a central location to plug everything in, not to mention enough outlets for everything you'll be using. And you can turn everything on and off from a central location.

- Wire ties. You can use them to keep all your cables tidy, as there is nothing quite so discouraging as tripping on a cable in the dark and bringing your setup crashing down to earth. They are also handy for storage of all those cables.

- See-through plastic bins for your small equipment and cables. They allow you to carry everything out easily and conveniently, and you can see what's in which bin. If dew occurs, bring the bins in the house and leave the tops open until everything dries out. For severe dew, or if you get caught in some rain, lay your cables loosely in the garage of family room until dry. The cables won't likely be damaged from moisture, but the connectors at the ends will slowly rust if water is allowed to stay in contact for extended periods of time.

- Have a tarp handy in case you need to cover up in bad weather. It should be large enough to reach nearly to the ground. If you are using a power strip, you can quickly disconnect power. If possible, stow the power strip well up under the tarp to keep your electronics dry. Buy a 10 or 20-foot length of thin bungee cord, and use it to wrap the tarp tightly around the scope (if you don't take it inside), mount, and pier/tripod. Tie it securely with two or more half hitches (a type of knot). Your cables will suffer a bit in an emergency, but the major equipment will be safer this way.

- Think about what can you leave together between setups. For example, could you leave most or all of your power cords plugged into the power strip? Could you leave the control cables connected to the mount and still carry it safely? Is the mount light enough to carry outside still on the tripod? Anything that reduces the number of trips between storage and setup location is a good thing, as long as you can do it safely.

- Consider a small outside storage shed just for your astronomy equipment. A small shed won't provide much protection against heat and cold, but it will keep your stuff dry and it will provide protection from direct sun. Whatever you use, it should be well ventilated so that heat and moisture do not accumulate to high levels for long periods of time.

- If you are going to be setting up frequently, you can help yourself out by choosing a mount that has a better than average polar alignment routine. All of the Takahashi mounts are excellent, and the NJP is the best of the best. The Takahashi polar scopes are designed to deliver outstanding accuracy. The NJP polar scope is better than anything else I've seen, even compared to the other Takahashi mounts. The Vixen GP-DX polar alignment procedure is a little complex but very accurate.

The AP GTO (goto) mounts allow you to polar align using the hand controller; this is very effective and you may not even need a polar scope with these mounts. Some digital setting circles can assist you in polar aligning, but tend not to be as simple or accurate as the others I've just listed. If you have goto, you can align relatively quickly and adequately for short exposures by following these steps:

1. If your mount requires that you orient it before slewing with it, do that before starting the polar alignment. It doesn't matter if you are a bit off in alignment at this point; all will be well eventually.

2. Do a rough polar alignment, either by centering Polaris in the bore (with or without a polar scope), or by sighting along the edge of the saddle when the Dec is at 90.

3. Point the scope at a bright star that is in the list of goto stars for your mount. You can also use your finder if it's got very fine cross hairs (e.g., Takahashi 7x50 illuminated) and is very well aligned. The star should have an RA that is at least 3 hours different from Polaris. Avoid stars that are 12 hours away; because of how alignment works, a star with a 12-hour difference is essentially on the opposite side from Polaris, and is as bad as a star with the same RA. A star with an RA difference of 6 hours is ideal, but anything from 3 to 9 hours, or 15 to 21 hours away, will work.

4. Tell the mount what star you are aligned on, using whatever menu feature is required (e.g, sync).

5. Use the mount to GOTO Polaris. Use the Alt and Az adjustments to center Polaris in the field of view.

6. Use the mount to GOTO the original star. Use the mount's hand controller to center the star, and then re-sync on the star.

7. Repeat going to Polaris and the bright star until your polar alignment is satisfactory. As you progress, both Polaris and the bright star should require smaller and smaller adjustments.

8. Test your alignment by using GOTO to put other stars or objects in the field of view. If the alignment isn't good enough for your purposes, use a star closer to the celestial equator for better accuracy.

You can polar align with your camera already attached to the scope. You can either use your finder if it's a good one, or use the camera to image Polaris and your bright star as you execute the above polar alignment routine. You need to determine where the center of the field of view (the optical axis) really is if you want precise alignment. It won't necessarily be at the center of the CCD chip. Flexure in your camera mounting makes it difficult to find the optical center.

The simplest way to find the optical center is to rotate the camera a little during an exposure. The stars will form arcs, and the center of the arcs is the center of rotation. This will be at the very least close to the optical axis, close enough for a successful polar alignment using this technique.

If you have digital setting circles, they may or may not provide enough accuracy to use this method for polar alignment. You can always give it a try to see what kind of results you get. But many digital setting circles provide only 3 or 6 arc minutes of accuracy. You need a better polar alignment than that (2 arc minutes would be excellent, 5 arc minutes would be adequate for shorter focal lengths) for really good imaging.

You may have a concrete pad near your home that looks like an appealing place to set up your equipment. This might be a bad idea for imaging, however. The concrete could be physically touching your home and able to transmit vibrations from refrigerators, hot tubs, people walking around, etc. A concrete pad will also transmit serious vibration as you walk on it. You are better off to put in a small, isolated concrete pad if you want to use one, or to cut a hole in the concrete, and pour a separate little pad that doesn't touch the larger pad.

Good-Weather Setup

A good-weather setup is one that you leave outside for as long as the weather stays dry. You may not image every night, but you are leaving all or most of your equipment set up during the day so it is ready to go the next night. A major advantage of this approach is that your polar alignment is only done one night. As long as you don't disturb your equipment, the polar alignment should be good to go on successive nights. This allows you to spend a little more time with the polar alignment on the first night if you want to do a really good job.

> **TIP:** With a non-permanent setup, you may find that heating and cooling during the day and night slowly alters your polar alignment. You can check your polar alignment each night to see if it is within acceptable limits for your equipment.

The main issue with a good-weather setup is how to protect your mount from late-night moisture (dew, fog), and how to prevent too much heating from the sun during the day. Your protection should also be good enough to handle a minor rain shower, but if it gets windy or a storm threatens, it's probably time to pull up stakes and bring stuff inside. If the weather will only be bad for a day, you might consider leaving your pier, and maybe the mount, standing. Bring cables, telescope, camera, etc. indoors for protection when bad weather threatens.

The simplest cover is the same one you would use for emergencies with a daily setup: a tarp and a long, thin bungee cord. The tarp should come nearly to the ground, and you can secure the bungee cord with two or three half hitches. Remember to tuck electrical cables well up above the ground and under the tarp. I wrap the cables loosely around the counterweight arm. The counterweights serve to hold the cables in place and prevent it from sliding off.

You can also get a Desert Storm cover, which is a giant bag that has a reflective surface to reduce solar heating. If you use just a tarp, a breeze can go a long way toward keeping things cool, even in 90-degree heat. Leave the tarp a little loose to allow some air circulation. You want to avoid trapping air; trapped air will heat up over time and create extremely hot conditions. When in doubt, take your stuff inside.

The next level of protection would be a shelter made from lightweight foam insulation. You can find such insulation with a shiny cover on one side, which will reflect the worst of the sun's heat. You can build a lightweight wood frame for your foam boards, or just glue them together. Allow for some ventilation at bottom and top to keep a reasonable airflow. The foam won't handle wind well unless you rig a way to tie it down, and a strong wind could tear it apart. But it will keep the sun off and the heat from building up, allowing your equipment to stay cool.

How much stuff should you leave set up night after night? The camera is probably the most sensitive, and it could be costly to experiment to see how much heat, cold, or moisture your camera can handle. If the weather is extreme, remove the camera for safety. You might also want to remove the scope, although you may have to perform a new polar alignment, or at least tweak your alignment, if you do. The longer your focal length, the more likely you are to benefit from a polar alignment check if you remove the scope. If your mount allows you to resume with just a quick realignment, you can be underway very quickly. Many mounts have a park position, and you can quickly restart from the park position even if you've removed the camera and/or scope. Simply align on a star after resuming, and if your polar alignment hasn't been disturbed, you will be good to go.

Some other tips for a good-weather-only, multi-night setup:

- When in doubt, check with the camera manufacturer to get their recommendation for leaving the camera outside in the daytime or in hot conditions.
- I've always exercised caution when leaving equipment out for several days at a time. I'd rather have to spend a little time setting back up than suffer damage from bad weather or heat.

- If your camera uses water cooling, change the water often enough to prevent stuff from growing in the water. Empty the tubing that carries the water every night.
- Since you are leaving things set up, you could take your darks into twilight while you catch a nap, and then shut everything down when it get too bright or too hot to maintain cooling. Beware of light leaks if you do flats or darks in daylight, however.
- I usually unplug the main power source for my setup when I bundle everything up for daylight. This prevents any accidental electrical problems if a sudden shower or other source of water shows up when you least expect it. Animals can chew on cables and cords, so inspect them before darkness sets in to make sure you are ready to start up safely.
- If you have an ideal location in your yard, consider whether or not you should put a concrete foundation and a pier at that spot. Having a pier, even if it only has a mount on it from time to time, will greatly simplify your setup, especially polar alignment. You'll be close to alignment each time you set up, and should only need to tweak it when you reinstall the mount. This applies as well to the following section on remote control.

Remote Control

My personal setup is a combination of the good-weather setup and remote control. I leave most everything set up from day to day while the weather is good. We don't get a lot of clear nights in Seattle, but when we get them they tend to come in streaks, usually in July and August. Even in the winter, we are more likely to get 2-4 days of good weather in a row, though it may only happen one time between October and June!

You can also do remote control using an observatory, so this section also applies to anyone using or interested in an observatory.

If you are going to control your scope remotely, the basic choice is whether to put in some kind of conduits for the cables, or whether to run the cables on the ground and take them up and put them down as needed. If you do dig a trench for conduit, use conduit that is specifically approved for exterior use. Consider putting your 110-volt power cables in one conduit, and your signal lines (camera, mount, whatever) in the other. Otherwise, you risk cross talk between power and signal, which can generate noise and make a mess of your images or scope control. Use conduit that is large enough so that if you have to pull another cable through later, you'll have room to do it. Leave a sturdy string in there in case you need to pull something else through there down the road. For example, if your next camera uses Ethernet, it will be handy if you can quickly add a cable for it.

Another issue on conduit size is whether you plan to attach your cable ends before or after you run the cables. Those 25-pin connectors are very large, and if you aren't going to add them after the fact, you'll need really large conduits.

All of these considerations may convince you that you are willing to set out your cables as needed. If so, invest in high-quality cabling. Moving it around creates a lot of wear and tear. Get cable ends with strain relief, which prevents excessive friction and bending at the connectors. Buy good connectors, too; they are bound to collect moisture and dirt. Get the type of connector that doesn't require a screwdriver. Water is the enemy of all types of cables. Even if you bring them in when it rains, dew or fog will still create problems.

The following material offers some ideas to think about if you are going to operate your camera remotely.

Focusing your CCD camera remotely takes planning. The simple solution is to add a motor to your scope's focuser, but many such setups don't give you the fine control you need for faster focal ratios. RoboFocus, however, works well. If you have a good focuser, the RoboFocus adds motorization and takes nothing away from your focusing quality. Add-on focusers that provide very fine motion control are a good solution, too. The JMI NGF-S is the budget solution, and the TCF-S from Optec is the high-end solution.

If you go with a motorized focuser (and it sure beats running outside every time you need to adjust focus), you have three approaches you can use. You can simply run a long cable between the hand controller and the focuser, and adjust focus by pushing buttons on the focuser's hand controller. Or you can buy a mount that includes focuser control, and use the same software you use to control the mount to control the focuser.

TheSky can do this with LX200, Paramount, and Astro-Physics GTO mounts. The third choice is to use a computer-controlled focuser, of which the RoboFocus and Optec TCF-S are the two available as of the time or writing. All three methods work. Computer control is more accurate and convenient for most situations. Controlling the focuser through the mount can be effective, but you are totally dependent on the quality of control the mount provides. The LX 200 control of your focuser is the least accurate; control with a Paramount, AP GTO, or direct computer control of a focuser is the most accurate. CCDSoft version 5 includes automated focusing, which is very effective with computer-controlled focusers; see the chapter on focusing for some hints on selecting the right mount and focuser for optimal results.

You can cut down on the number of cables between the mount and your control center by putting a computer at the mount. If you have an observatory, you can simply leave a computer in it. If you are setting up and taking down, then a laptop next to the mount works well. Use a Cat 5 Ethernet cable to connect the outside computer to the inside computer, and control it with remote control software such a PCAnywhere, RAdmin, VNC, etc. Software Bisque makes a suite of programs (TPoint, IAServer, IAClient, TheSky, CCDSoft) that work extremely well together, with or without Internet control. Remote opration works with or without a permanent observatory. It takes time and effort to get everything set up, but the convenience is wonderful. If you have Windows 2000 Server available to install on the observatory computer, and at least 256MB of memory to run it, you can use Terminal Services for remote control. Install Terminal Services as an application server, and choose the setting that maintains maximum compatibility for legacy applications. Two other remote control programs that won't break the budget are VNC and Radmin. VNC is freeware, and can be downloaded from:

http://www.uk.research.att.com/vnc/

RAdmin is low-cost and you can download a 30-day trial version from:

http://www.famatech.com/

If you control the mount directly with cables from a remote computer, you'll need high-quality cables between computer and mount/camera/focuser. These can't just be any old kind of cable. Electrical noise can interfere in transmission if you don't get adequate cabling. Check with the manufacturer of your equipment to see if they recommend a specific cabling arrangement for remote control. SBIG's web site fully documents what you need to do to create a cable that will give you up to about 200 feet for remote operation. If you have a cable made up, make sure the folks making the cable have a copy of the manufacturer's requirements. Print out a copy of the requirements if necessary and give it to the cable-making company. Remember to get strain relief on your connectors. Talk to them about filling the voids in the connectors with epoxy to prevent wires from working loose over time, and to keep moisture out. If you do this yourself, Goop makes a good void-filler.

A goto mount is ideal for remote operation. You can run outside and point the mount if you have the patience for it, of course. A goto mount is perfect for imaging in general. A solid goto mount takes the sweat out of putting objects on the CCD chip. I highly recommend Software Bisque's TPoint to improve pointing accuracy. You won't have to spend a lot of time finding stuff; you can spend your time imaging. Goto accuracy varies with mount quality, but TPoint will improve the accuracy of almost any mount.

Remote control also opens up the possibility of using scripts to do your imaging. Scripting is a lot like programming, and it's not for everyone. But if you use scripting, you can set up an entire night's imaging run and then get some sleep. How many imagers can actually get some rest at night? Of course, if something goes wrong, you could lose a lot of imaging time, so think in terms of getting *some* sleep while your scripts are running. You might want to check your results periodically to make sure everything is working properly.

Speaking of checking your equipment, there are probably a lot of cables attached to a lot of things. You should make generous use of masking or duct tape to keep wires under control. The worst thing that can happen is a cable getting caught between two surfaces and either shearing off or jamming the works catastrophically. Cables can also snag on protrusions and cause the mount to either work too hard or get out of alignment. When making major slewing moves, it would be great to have a simple way to observe your equipment to make sure everything is OK. You might consider putting a small, cheap video camera near your scope and keeping a monitor in the house. You can use a wireless transmitter so you have one less cable to string. If you are lucky, you can see the mount through a nearby window. Or maybe you just run outside when you do a major slew. However you do it, don't forget that your mount, camera, and scope are physical things that can go bump in the night. Be especially careful on long slews and when imaging near the zenith. Whenever the camera swings close to the pier/tripod, or crosses the meridian, trouble is only inches away.

It's easy to get caught up in the remote imaging process and forget that there is a physical system outside responding to your commands. If anything seems odd, go outside immediately to investigate! Better safe than sorry is a good policy for remote operation.

If you'll be setting up in the same location regularly, consider putting in a concrete foundation and a pier. Even if you are not going to leave the mount on the pier, the savings in setup time may make it worthwhile.

An Imaging Road Show

Taking your imaging on the road is simpler than you might expect. Simple is a relative term, however. We all get comfortable setting up in the back yard. That comfort comes from the familiarity of knowing what to expect, where to set up, what the quirks are. Dark skies beckon, and it takes a few additional tricks to take your setup on the road successfully.

The road show starts with the right batteries and the appropriate connectors to hook up your equipment. Small details loom large when you are on top of a mountain, many miles away from a hardware store. Twelve-volt connections come in a variety of configurations. For each piece of equipment, make sure you examine the connectors carefully to determine how best to connect to a battery. Pay special attention to the polarity of the connector. Some have center positive, while others have center negative. Label every 12V connector with the piece of equipment it belongs to.

Take your time to make up your travel kit. Even if you are only going on the road once or twice a year, don't just assume that you can remember everything,
and never pack at the last minute. If you routinely go on the road to image, have everything ready to go in plastic bins. Clear plastic allows you to see what's in each bin, especially at night. More bins are better than fewer. Assign a function to each bin, and don't worry about empty space. Being organized is important when you are setting up out in the middle of nowhere. Put all the camera cables and power supplies in one bin. Put all of the mount goodies in another bin. If you are tired at the end of the session, having room to spare in each bin makes it possible to throw everything together quickly.

The interior space of your vehicle determines how organized you can get. If you have a compact, you may not be able to use separate bins for everything. But if you've got a big SUV handy, take advantage of that space to keep yourself well organized.

Get a marine deep cycle battery for your road show. This is a specific type of battery that is not the same as the kind of battery you have in your car. A deep cycle battery can be fully discharged, and then recharged, without harm. You can save a little money by buying a trickle charger for it, but I prefer to have a full-scale battery charger. If you forget to recharge the battery, it pays to have a rapid charger. Many chargers can be left plugged in, and will monitor the state of the battery's charge and periodically top it off.

The greater the capacity of the battery, the more it weighs. Consider how far you have to carry the battery, then decide on how big of a battery to get. Make sure you get a battery with a built-in carrying strap. You can also get a small cart to wheel your battery around. If you will be on uneven ground, get a cart with large wheels. I use a battery that weighs in at 80 pounds. That is the largest weight I can manage safely, and at 135 amp-hours it gives me enough power for two full nights. The size battery you get will depend on your lifting abilities, but try to get at least 35-40 amp hours of capacity per night you want to image. Your requirements may vary; it depends on how much stuff you have. Consider a quiet generator, especially the small Hondas, for recharging your batteries in daylight. Solar chargers are often not fast enough to get the job done, but may be a useful (and silent) supplement.

Get at least two batteries. Use one battery exclusively for the mount, and the other battery for everything else. The mount has to track accurately, and some mounts cannot handle power fluctuations. Devices draw varying amounts of power, and can mess up a mount more easily than you would expect. The Astro-Physics mounts are among the least tolerant. If you decide to go with three batteries, then use one for the mount, one for the camera, and one for everything else. Some mounts *require* separate power from the camera, such as the Losmandy mounts. Failure to use separate power supplies could actually damage your equipment in such cases, so get at least two batteries.

The kind of tripod or pier you use will greatly affect your setup time and effort. My favorite pier for portable use has been the Astro-Physics piers, which can be set up in a few minutes, are very stable, and can handle a large or small mount as needed. You will probably have to contact a machine shop to make an adapter plate to allow you to put a non-AP mount on these piers, but they work so well it will be worth your while if you choose this route. If you use a tripod that has sharp feet that dig down into the ground, avoid concrete and very hard-packed earth for your setups. The legs will tend to skid on such surfaces, and that will ruin your polar alignment. Even if you do stick them down into soft earth, make sure you stick them in really, really well. Otherwise, you might lean on the mount at some point and push a leg down a bit further, which will also ruin your polar alignment. This is why I like the AP piers so much. They stay where you put them, unless you set up in a swamp!

Lay out your adapters and cables before your first trip, and then put everything together in your yard or garage to make sure it all works as you expect. Try everything out. Slew the mount, take some pictures, hook up any accessories, etc.

TIP: If possible spend a night imaging with your road setup in your own backyard or driveway so you can look for flaws that you can remedy before you actually hit the road.

There are two basic approaches you can use for powering AC equipment: buy 12V power supplies, or use a 12V to 110V converter. For items that use a lot of power, the 12V power supplies are more likely to give you the longest battery life. 12V to 110V converters tend to waste some of your power in the conversion process. Many devices have 12V power supplies available, or work off of wall warts (110V to 12V converters). In many cases, you can replace a wall wart with a direct connection to your battery. SBIG makes a 12V power supply for many of their cameras, and I have used it successfully in the field and in my backyard.

TIP: Make sure you pay attention to the polarity of the connectors! Using a connector of the wrong polarity can blow fuses or damage equipment.

I'm going to repeat this one more time: if you build or buy cables for your 12V equipment, make sure you are 100% clear about the polarity required by each and every device you use!!! Most devices will do no worse than burn out a fuse if you reverse polarity, but don't count on that. The little round power connectors used on most 12V devices have a center pin and an outer sleeve. One must be positive, and the other must be negative. Know which is which for every device you own, and write it down! Make labels for the connectors, too. There is nothing more frustrating than blowing a fuse (or the device itself) due to getting the polarity wrong. It's a good idea to attach a label to every 12V cable to make sure that you don't make mistakes. There is no standard about which should be positive and which should be negative (pin and sleeve). In fact, it is highly likely that you own equipment that goes each way. Label those cables!

You might assume that you would never make the mistake of switching the red and black connectors to your batteries. But rest assured that, unless you take precautions, some night you will be tired enough to make that mistake. Perhaps it will be on that third night of staying up until 5am to image; perhaps it will be when you are in a rush to get set up after dark. Sooner or later, the sparks will fly.

Make sure you have replacement fuses for every device or cable that has a fuse so that you can recover from this trauma. But also take some steps to make it harder to make this mistake. The first line of defense is procedure. Make it a habit to double-check the color of the connector in your hand, and match it to the color or sign of the terminal on the battery you are about to connect it to. Even better, set yourself up so that only one set of cables connects to the batteries; modify all of your other cables so that they plug into cigarette lighter outlets. You can buy adapters that have alligator clips on one end, and a three-way cigarette-lighter connector box on the other end. You can use additional splitters if necessary. The fewer alligator-style connectors you have to use, the better.

For the absolute best in polarity protection, buy or make a set of cables that attaches to the battery using screw-down clamps, and that has a bank of cigarette-lighter connectors on the other end. Leave this attached to the battery at all times. If you have devices that use cables that have alligator clamps, replace them with male cigarette-lighter style connectors. In addition to being a polarity risk, alligator clips have a tendency to slip off at very inopportune times.

Make sure to note the polarity you will get at the pin/sleeve end of your cables when you put the male cigarette-lighter connector on. Polarity is something you should only have to deal with one time, and it is well worth spending an afternoon standardizing your cables. If you don't believe me now, you'll understand some dark night when you are connecting cables in the dark.

Speaking of setting up in the dark, sooner or later you are going to have to do that. You won't always make it to your destination on time. Get yourself one of those lights that you can attach to your head with a headband, so that it lights up whatever you are looking at. If no one else is around, a white light is OK for setup, but you may want to put a red screen over it for times when others are around. Rubylith can be bought at many graphics supply houses. Consider putting reflective tape on the legs of your tripod, and on anything that is likely to be lying around waiting for you to trip on it (power supplies, buckets for water cooling, etc.). Even the little red flashlight you carry around will throw enough light to show you what to avoid.

Observatory Imaging

If you are in a position to create a more or less permanent setup, you have a lot of options -- too many to cover authoritatively here. The first step is to decide what kind of observatory you want: roll-off roof; clamshell dome; rotating dome, etc. The list of choices from that point goes on for about a year and a half; then you can start construction. Some things to think about as you ponder your future observatory:

- Location is everything. If you will be building an observatory on the same property as your home, consider what might happen to light sources in the future. Are there areas where new developments will or could occur? Make sure you have trees or something else high enough to block the worst of the skyglow from those directions. A rotating dome probably does this best, but you need to balance this value against the cost of a rotating dome. Also consider where your neighbors might be putting some lights, and stay away from those locations if possible, or block the light source (and potential sources) as well as you can manage. In a pinch, the cheapest light block is a heavy tarp laid over a stepladder. This is useful for one-time annoyances. Vegetative screening is also effective, but you have to plan ahead as it will take a few years to fill in effectively.

- Do you plan to simply store your scope in a dome and control it from elsewhere, or do you want to have enough room inside the dome to sit and work at a computer? This will make a huge difference in the size of the dome you will need. If budget is your top concern, build the smallest possible observatory, just large enough to house the scope and mount. You can open it up to image, and use a remote location to control everything. The purpose of such an observatory is simply to protect your equipment from the elements.

- You may want to avoid standing and working on a concrete floor. If that's the case, build a wooden deck around the pier. Make certain that no part of the deck touches the pier or its foundation. You can then walk around in your observatory without transmitting vibrations to the mount.

- Internet control of mounts and cameras is now coming into vogue. You may want to consider it for controlling your observatory even if the observatory is near your home. But it also opens up the possibility of having an observatory located away from your home, and controlling it from the comfort of your den or family room.

5   *Taking Guided Exposures*

Guiding is the key to taking long exposures with your CCD camera. It is the process of making corrections to keep your mount tracking accurately.

Historically, guiding was done by looking through an eyepiece, and using a hand controller to make small adjustments to the mount. Self-guiding cameras and auto-guiders make guiding easier, but it's not always simple.

Section 1: What Does Guiding Do?

FIGURE 5.1.1. AN IMAGE WITH GOOD GUIDING HAS ROUND, VERY COMPACT STARS.

CCD cameras enable you to automate the guiding/tracking process. Autoguiding has revolutionized imaging by allowing both film and CCD imagers to take images with long exposure lengths conveniently. You no longer have to stand over an eyepiece straining your back in order to take long exposures. An autoguider will get the job done with a minimum of fuss.

How Guiding Works

Autoguiding is a feedback loop that monitors and adjusts the pointing of your mount. Whatever the actual physical type of autoguider, the process is the same:

- Take a picture of a star every few seconds.
- If the star changes position, adjust the mount to re-center the star.

If the interval between images of the star is short enough, even an inaccurate mount can be forced to track accurately. This won't cure every bad mount, not by any means. But some otherwise troublesome mounts can be tamed with guiding.

The actual guiding process is a bit more complicated, of course. Your camera control software measures the position of a star image at intervals ranging from a small fraction of a second to 30 seconds or more. The interval is based on the brightness of the star you are using for guiding, the sensitivity of the CCD chip involved, the seeing conditions, and the ability of your mount (or an adaptive optics device) to respond in a timely and accurate fashion. This is called the guide interval or guide exposure. If the image of the star moves from one exposure to the next, the camera control software sends signals to the mount or adaptive optics unit to move east/west or north/south (or some combination of the two) to correct the pointing error. These signals are called guide corrections, and typically require a cable between your camera and the mount. Some guiders use a cable between your computer and the mount, but this is less convenient when you want to control the camera remotely since it requires an additional cable.

The key objective of guiding is to make guide corrections before the tracking errors become too large. "Too large" in this context means approximately the size of a pixel on your image CCD. Most camera control programs report errors in fractions of a pixel, so it's easy to monitor the tracking accuracy during a guiding session. Most programs also allow you to write a record of the guide corrections to disk so you can examine them afterwards.

> **TIP:** Some camera control software requires you to have the camera square (orthogonal) to the mount for best guiding results. Other software can guide even if the camera is far from orthogonal. Check your software documentation to see if it requires orthogonality.

When starting out, the SBIG cameras which contain both imaging and guiding CCD chips are the easiest way to get acquainted with guiding. The Starlight XPress cameras, which use a portion of the imaging chip for guiding, are also reasonably easy to work with. Both approaches work well with scopes that have moving primary mirrors (e.g., SCTs). If the mirror moves during an exposure, the guiding will detect this and adjust. If the mirror movement is small or slow, the image will be OK. If the movement is fast or large, the image will be lost anyway. The main disadvantage of the Starlight XPress approach is that you must double your exposure time to allow for the guiding.

The alternative to a camera with its own guiding capability is a separate camera that does the guiding. There are several dedicated guiding cameras, such as the ST-4, STV, and LISAA Guider. You can either divert the incoming light to both the guider and the main camera, or attach a second telescope (usually a small inexpensive one) to mount the guider. The light-splitting approach is called off-axis guiding. The separate scope is called a guide scope. Off-axis and guide-scope guiding tends to be more complex than integrated guiding, but it's also more flexible. The STV from SBIG comes with an optional eFinder that doubles as a guidescope for telescopes up to about 2000mm focal length, and is easier to install and use than a conventional guide scope.

Most of the current crop of camera control programs provide simultaneous control over both imaging and guiding CCD detectors in the SBIG cameras. With some packages, such as CCDSoft version 5 and Maxim DL, you can use either detector to image or to autoguide. From a hardware standpoint, both detectors are capable of imaging and autoguiding; it's up to you to decide which arrangement best suits your needs. Most of the time, of course, the larger chip is used to image, and the smaller chip is used to guide.

The basic steps involved in autoguiding are:

1. Calibrate your mount. The camera control software measures how fast the mount moves, and in which directions, when guide corrections are applied.
2. Select a suitable guide star.
3. Start guiding. The software will issue guide corrections when the guide star shifts position.

What makes the star move, you ask? Movement of the guide star typically results from sources:

- Inexact polar alignment, which causes a slow drift of the field of view as well as a slow rotation of the field. If not corrected, stars form lines whose length is determined by how far out of alignment you are.
- Periodic error in the mount's tracking rate. If not corrected, causes star images to elongate a bit as the mount speeds up and slows down by very small amounts during each turn of the worm gear.
- Random errors due to variations in gears, dirt in the gears, dirt or dents in an equatorial platform's driven surfaces, etc. If not corrected, results in non-round stars.
- Improperly set parameters (e.g., aggressiveness).

Movement can also occur from atmospheric turbulence, but this is not an error and you should adjust your guide interval so that it is long enough to eliminate the effects of turbulence. Severe turbulence could be a problem, but that is more likely to discourage imaging entirely because it makes focusing difficult.

Various complexities creep into the seemingly simple autoguiding operation:

- The precision required for successful tracking is measured in arcseconds. An arcsecond is small, only 1/1,296,000th of a circle. This is an exceptional level of accuracy. Many mounts cannot deliver the required accuracy, even *with* a guider.
- The longer the focal length of your telescope, the greater the accuracy required for accurate, successful guiding. I often recommend scopes with short focal lengths for starting out (400-750mm).
- Mounts have some degree of repeating error, called periodic error (PE). This error results from gears that are ever so slightly out of round. There will always be some PE, but a better mount will exhibit smoother PE, and less extreme movement. Many mounts have built-in periodic error correction (PEC), but if you are using an autoguider you do not *need* to use PEC, since the autoguider will detect PE and correct for it very precisely in real time. A guider will simultaneously correct for any and all tracking errors if they are small enough and/or slow enough.
- Mounts also have greater and lesser degrees of random error. This is true even of some fairly expensive mounts. Such mounts are not useful for unguided exposures, but may work well enough with guiding. Short guide intervals may provide better results with these mounts. This only works if you can find a guide star bright enough for the short intervals required. Any mount that needs guide intervals of 3 seconds or less is one that I would call problematic. It's very convenient to be able to use guide intervals of 7-10 seconds because it allows you to use dimmer guide stars. A mount that consistently delivers these longer guide intervals without causing guiding errors is an excellent mount, especially if it does so at long focal lengths (2000mm and longer).
- If the mount is even slightly misaligned from the celestial pole, this will introduce drift over time and that will cause pointing errors. As long as you are within 15 arcminutes, your mount should be able to guide correctly, but you will get much smoother results (and dramatically less field rotation on longer exposures) if you can get within a few arcminutes of the celestial pole. A marginal mount may not guide adequately if not well aligned. Some mounts (Takahashi NJP, EM series; Vixen mounts) have superb polar scopes that can get you very close. Other mounts will require drift aligning, or use TPoint for accurate alignment.
- Any mechanical looseness or flaws in the mount can contribute to pointing error. There are many potential sources of error, including backlash (slop) in the gears, end play of a worm gear, looseness in any gears inside the motor housing; looseness in gears between the motor and worm gear; bad bearings anywhere, etc.
- The tripod, mount, and telescope assembly are subject to different amounts of flexure as the weight of the telescope shifts. This can also contribute to pointing error. Light mounts and heavy telescopes are especially prone to this problem. One of the benefits of more expensive mounts is reduction or elimination of flexure.
- Telescopes with moving mirrors (such as the ubiquitous SCTs, which move the primary mirror to focus) can introduce large-scale movements when the mirror shifts over time. This can happen just from the mirror "settling in" after focusing, or from shifting that results from changing the pointing angle of the telescope during a long exposure, or when crossing the meridian. If you own one of these, the moving primary will wind up ruining a certain percentage of your images, or limiting the maximum length of your exposures. You may need to adjust your technique to accommodate this limitation, or you can lock down the mirror by various methods and attach an external focuser.
- If all of the parts of the mount aren't exactly aligned with each other, pointing error is the result. Contact the manufacturer to learn what you can do to properly adjust your mount for best accuracy.
- Turbulence in the atmosphere can cause small changes in the apparent position of the star used for reference. This makes it look as if a guiding correction is needed even when it is not. This could also mask the need for a guiding correction. The usual solution is to use longer guide intervals. This will average out the effects of turbulence.

SECTION 1: WHAT DOES GUIDING DO?

It is not a trivial task to keep a telescope pointing at the same exact object during a CCD exposure. Any given mount will have a certain level of accuracy that it can reach. Pushing a mount past this point will result in guiding and tracking errors that will at the very least reduce the quality and sharpness of your images, and at worst turn stars from round circles into lines. Later sections in this chapter explain how to assess the pointing accuracy of your mount, and how to tune your mount to optimize accuracy.

How Mounts Move

An equatorial mount has two axes, Right Ascension (RA) and Declination (Dec). The RA axis is parallel to the earth's axis of rotation, and allows the mount to track the stars. The Dec axis adds the ability to point anywhere in the sky (see figure 5.1.1). Most mounts have motorized adjustments to both axes. Some older mounts, and some lighter mounts, can only make guide corrections to the RA axis. Such mounts require a really good polar alignment. The better your polar alignment, the longer it will take for drift to occur.

The right ascension and declination axes behave in fundamentally different ways. The RA axis is always moving to keep up with the apparent motion of the stars. This motion is called sidereal, and the rate is very nearly one revolution per day. You will often see reference to some fraction of the sidereal rate, such as 1x (exact sidereal rate), .5x (one half sidereal rate), etc.

Because the RA axis is always moving, you have an opportunity to eliminate many (but not all) sources of error by simply making guiding corrections at speeds slower than the sidereal rate. For example, it is common to make RA adjustments using either .5x sidereal (slow the mount down to move it eastward relative to the stars), or 1.5x sidereal rate (speed the mount up to move it westward relative to the stars). Depending on the focal length of your telescopes, you may get better results at .75x/1.25x (long focal lengths), or at .25x/1.75x (shorter focal lengths). Some mounts offer limited choices of guiding speeds, while others offer many

choices. Check your documentation to see what your options are. For example, Astro-Physics mounts can give you .25X and .5X sidereal. Takahashi mounts can give you guide rates from 0.1X to 0.9X. Meade's LX200 mounts, unfortunately, use a 2X rate for guiding which is so fast that it makes these mounts harder to guide than necessary.

The declination axis, on the other hand, is stationary until a correction needs to be made. If there is substantial looseness in the declination axis at any point (motor bearings, reduction gears, worm end play, or worm mesh with the worm gear), the mount may literally be unable to guide adequately in declination. This results most often from backlash. If it takes too long to make up the backlash, then the guide corrections won't get made until it is too late. You can minimize declination axis problems by taking the time to do a really good polar alignment, and to minimize and properly compensate for backlash.

The result of the differences between the RA and Dec axes is that special attention must be paid to the declination axis to obtain high-quality guiding results. See the section "Assessing Autoguiding Possibilities" later in this chapter for detailed information about getting the best autoguiding results with your mount.

FIGURE 5.1.2. THE RA AXIS POINTS TO THE CELESTIAL POLE, AND THE DEC AXIS ROTATES AROUND THE RA AXIS.

Section 2: Autoguiding Hardware

It is often highly desirable to take long exposures. A longer exposure, as described numerous times and in numerous contexts throughout the book, almost always provides better images. There are factors that can limit the length of your exposure, but one critical limiting factor is the tracking ability of your mount. Only the very, very best mounts, such as the Paramount 1100 ME from Software Bisque, are capable of taking long unguided exposures with long focal length telescopes. Most mounts are limited to a minute or two unguided, at focal lengths under 750mm or so. If you want to image at longer focal lengths, or take longer exposures, the easiest solution is to use a guiding CCD camera. The guider measures movement of a guide star, and moves the mount to keep the star centered. This allows you to take exposures that are as long as you require them to be.

Figure 5.2.1 shows two images in the vicinity of Stephen's Quintet. The left image shows guiding errors. The stars are slightly elongated from lower left to upper right, and the brighter stars are doubled vertically. This demonstrates that even with guiding, all of your problems don't just disappear. The quality of your mount and the appropriateness of your setup determine how good your guiding will be. The right image shows greatly reduced guiding error, accomplished by using a more stable mount. There is still a small residual error, but the stars are rounder and less elongated.

How much better your images will be with a guider depends on the quality and state of adjustment of your mount. A mount that tracks very poorly may also make poor adjustments in response to a guider. On the other hand, short enough guiding corrections could allow you to get better results with an otherwise problematic mount. The only way to know for sure is to try. If you do elect to go with a marginal mount, expect to spend a lot of time analyzing its behavior during guiding, and experimenting with various software settings.

Guiding reduces the effects of tracking error. The physical appearance of tracking errors varies. Tracking errors usually drag star image out into a line. The line can be straight or curved, depending on the amount of periodic error versus random errors, not to mention the occasional cat rubbing up against a tripod leg. There may or may not be variations in brightness along the line, depending on how major the periodic error (which causes changes in tracking speed) is.

Guiding and tracking errors sometimes interact to cause oscillations, in which the mount bounces between two or sometimes three positions instead of settling down to make very small corrections. Figure 5.2.2 shows one of the more interesting guiding errors I have seen. The guider must have been twitching pretty badly that night. All of the bright stars have little mouse ears. The mount was oscillating between three

FIGURE 5.2.1. TRACKING ERRORS (LEFT) CAN BE REDUCED OR ELIMINATED BY USING A GUIDER (RIGHT).

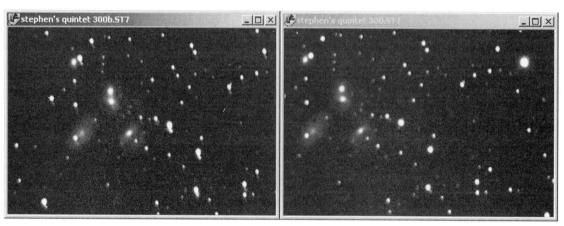

FIGURE 5.2.2. GUIDING ERRORS CAN BE DOWNRIGHT STRANGE AT TIMES.

can't take longer exposures due to the limitations of your mount at longer focal lengths, you can try taking more individual exposures, or apply a smoothing filter to the noisy background. Lack of dark frames can also cause streaking. The SBIG web site contains instructions for how to apply dark frames to Track and Accumulate exposures if you elect to use this technique.

Some CCD cameras include the ability (or allow you to add the ability) to do guided exposures using the camera itself. There are several different techniques from different manufacturers, some of which are pro-

different positions, though it spent most of the time on one of the three.

One way to take deeper exposures without guiding is to use SBIG's Track and Accumulate. This automatically combines multiple exposures. They are aligned using a star in the image. Combining images is a great way to improve signal and reduce noise, but the individual exposures must be long enough to avoid streaking caused by noise (see figure 5.2.3). The streaking comes from drift from one exposure to the next. The same problem can occur when combining (stacking) individual images. Longer individual exposures solve the problem because there is more signal, less noise. But if you

FIGURE 5.2.3. IF THE INDIVIDUAL EXPOSURES IN A TRACK & ACCUMULATE SESSION ARE TOO SHORT FOR THE SKY BRIGHTNESS, SKY GLOW WILL RESULT IN STREAKS IN THE BACKGROUND.

tected by patents and thus found only on cameras from a single manufacturer. Self-guiding cameras from SBIG and Starlight XPress are covered in this section.

This section also describes some CCD "cameras" that are not primarily intended for imaging use -- guiders. These often include a smaller CCD chip, and sometimes they have special physical or software features that enhance their ability to guide exposures.

Self-Guiding

A self-guiding camera uses the incoming light to simultaneously image and guide. There are times when self-guiding can be a major asset. When the guiding function is incorporated right into the camera, this means that the optical system you are using for imaging is also the optical system you are using for guiding. If there is any kind of change to the optical system, the guider will deal with it. Typical sources of error include periodic error in tracking, random tracking error, polar misalignment, a movement of the primary mirror, etc. As long as the change is slow and small, the guider will usually be able to deal with it.

Even a self-guider, however, will have trouble dealing with sudden changes, large changes, or certain types of changes during long exposures. A change that affects polar alignment -- poor alignment, or a change in alignment during an exposure -- affects field rotation, and a guider does not compensate for that.

The job of a guider is to recognize when a shift has occurred, and to direct the mount to adjust for that shift. For example, consider the case where you have a small polar misalignment. During a five-minute exposure, the stars will drift a small amount. The amount varies with the amount of misalignment, and the focal length of your imaging system. The larger the misalignment, and the longer your focal length, the faster the drift will occur.

The guider is constantly taking exposures of a small section of the sky. You choose a star to use as a guide star. After each exposure, the guiding software measures how far the star has drifted, and then sends signals to the mount to make it move so as to put the star back where it started.

The guiding software is usually sophisticated enough to recognize shifts that are smaller than a pixel. A star's light is spread across several pixels by air turbulence, and the software uses fancy math to figure out where the center of that spread is. This center is referred to as a centroid. These measurements can be as accurate as 1/30th of a pixel with some guiders, such as the STV.

The mount can only move in RA and Dec, so the guider will use a combination of these two movements to keep the mount pointed at the guide star. Guided exposures can go for an hour or more if you have superb polar alignment so as to avoid field rotation.

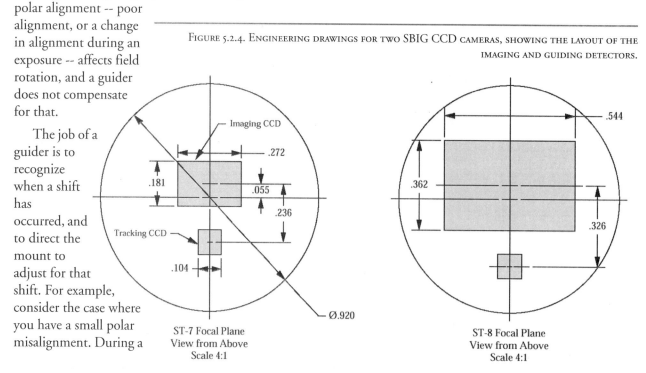

FIGURE 5.2.4. ENGINEERING DRAWINGS FOR TWO SBIG CCD CAMERAS, SHOWING THE LAYOUT OF THE IMAGING AND GUIDING DETECTORS.

The better your polar alignment, the longer the exposure you can take. With a reasonably accurate polar alignment, you can take 10-minute exposures with minimal effects from field rotation. If you require longer exposures, use drift alignment to refine your polar alignment.

If your polar alignment isn't adequate for the length of your exposures, you will see evidence of field rotation. The guide star will be at the center of rotation because the guide software is constantly centering the guide star. The portions of your image furthest from the guide star will have the largest movement.

Rotation occurs because your mount's and the earth's rotation are centered on two slightly different positions in the sky. The difference between those two positions causes drift.

Self-guiding works most efficiently with a separate guide chip, and this is the method employed in many cameras from SBIG. The efficiency comes from the fact that the guide chip is read during the exposure; no part of the signal from the imaging chip is used for guiding. Thus, using the guide chip has no adverse impact on the length of the exposure. Cameras from Starlight XPress that can self-guide do so by using the imaging chip itself. This requires doubling your exposure time to allow all areas on the chip to receive a signal.

Figure 5.2.4 shows the layout of the imaging chip and guiding chip in two SBIG cameras, the ST-7E (left) and the ST-8E (right). The larger chip is the imaging chip; the smaller chip is the guiding chip. The layout shown here is a virtual one; the guiding chip isn't physically located in the position shown. A pick-off mirror reflects light sideways to the actual position of the guide chip (see figure 5.2.5). This ensures that the light falling on the guider is as close as possible to the light falling on the imaging chip. The virtual locations of the chips are close enough that the possibility of field rotation is minimized.

The Starlight XPress cameras can use the imaging chip for guiding because of the type of CCD detectors used. The guiding software reads part of the data from the imaging chip in a sophisticated way (explained in the Starlight XPress STAR2000 section below) to perform this magic. The downside of this method is that you must take an exposure that is twice as long as an unguided exposure. The good news, of course, is that you can take much longer exposures using guiding.

SBIG Self-Guiding Cameras

Figure 5.2.5 is a photograph of the business portion of an SBIG camera (an ST-9E). This view shows the relationship between the guide chip and the imaging chip. The ST-9E chip is similar in physical size to the chip in

FIGURE 5.2.5. MOST SBIG CAMERAS INCLUDE BOTH AN IMAGING CHIP AND AN AUTOGUIDING CHIP.

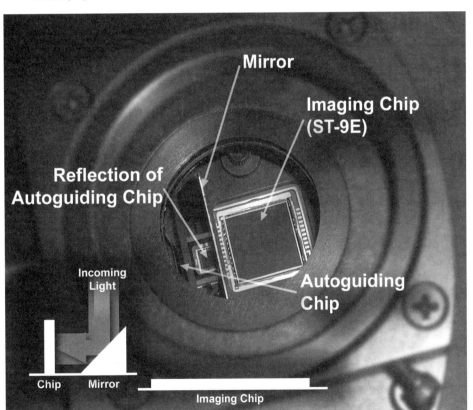

the ST-8E, but it is square instead of rectangular. The figure has a diagram at bottom that shows a side view of the light entering the camera. It reflects off of the pick-off mirror and hits the guide chip. The guide chip is standing on edge to receive the reflected light. The length of the light path to the guide chip is identical to that of the main chip, so that both chips come to a common focus.

Note that, thanks to the mirror the guide chip appears to be very close to the imaging chip. This is important for several reasons:

- If there is any vignetting (darkening away from the center) in the optical system, the guide chip will still get a good amount of illumination.
- If there is any coma or other optical aberrations in the optical system (which tends to get worse away from the center), the guide chip is close enough to minimize this problem.
- The guide chip, being close to the imaging chip, minimizes the previously discussed field rotation.

I have found the integrated guide chip in the SBIG cameras to be an extremely effective method of guiding, and I recommend it highly for starting out.

The one time where an integrated guider can be problematic is when you can't find a bright enough star to guide on. The need to keep the subject of your image on the main chip limits the amount you can move the camera around to find a good guide star. "Good" means bright enough to allow you to use a reasonable guide exposure. The higher the quality of your mount, the longer the guide exposure can be. The less stable your mount, the shorter your guide exposures must be. I have seen mounts that need guide exposures as short as half a second, and I have seen mounts that can track perfectly unguided for 5 minutes. Knowing the limit of your mount is important, but the only way to know is by actually trying various guide exposures.

Another situation where an integrated guider can be a problem: when using a filter that greatly reduces the incoming light. A hydrogen-alpha filter is a good example of this. For this situation, or when a suitable guide star simply isn't available, an external guider in a guide scope is your best option. Filters that cut the light like this usually require very long exposures (at least 10 minutes). Unguided exposures are not likely to be successful with these filters.

For all other applications, the integrated guider is extremely convenient to use. However, the longer your focal length, the greater the likelihood that, when imaging outside the general area of the Milky Way, you have some trouble putting a guide star on the guide chip. There are three things you can do to reduce the likelihood of failing to find a useful guide star:

- Use guiding software that allows you to rotate the camera and guide at any angle of rotation. This increase the area around the object where you can locate a guide star. CCDSoft v5 and MaxIm DL 3 allow you to do this.
- Use a mount that tracks well. This allows you to use dim guide stars and long guide exposures.
- Use a camera with a large chip, such as the ST-8E, so that you can shift the camera laterally to locate suitable guide stars, and still have the object of interest on the imaging chip.

Starlight XPress STAR2000

STAR2000 is Starlight XPress's autoguiding technology. Unlike dual-CCD self-guiding cameras like those from SBIG STAR2000 guides using a portion of the imaging chip -- there is no separate guiding chip. The plus is that you can guide even on moving objects, such as comets and asteroids. The minus is that your exposures must be twice as long as they would be without STAR2000, for reasons explained below.

STAR2000 relies on the structure of the CCD chips in the MX5 and MX9 families of cameras. It is not available in the HX516 camera. The MX5 and MX9 cameras use Interline Interlaced CCD chips, which are constructed with each pixel split into two vertically stacked halves. Each half can be read independently.

During the first half of an exposure, one half of each pixel continues to gather light for the exposure. The other half is read out periodically (the guide exposure time) and is used for guiding purposes. Halfway through the exposure, the situation is reversed. The other half of the pixel now gathers light for the exposure, and the first half switch to guiding duty. So half of the pixels are read at the halfway point in the exposure, and the other half of the pixels are read at the end

FIGURE 5.2.6. LONG EXPOSURES REQUIRE GUIDING.

of the exposure. This is why it takes twice as long to expose using STAR2000 -- for each pixel half, only 50% of the exposure time is actively spent gathering photons for the image. The other 50% is used for guiding purposes. Only half of the pixel area is integrating an image at any one time.

The STAR2000 can read a small area of the chip near the object you are guiding on, so it is able to do reasonably fast guide intervals. This would be about one second on a star of magnitude 11, though it will vary with the focal ratio of your telescope. One advantage of dual-chip and dedicated guiders is that they can guide at much shorter intervals when a suitably bright guide star is available.

Guiding accuracy depends mostly on the software, as with other guiders, so the accuracy is very good.

One other minus with the STAR2000 is that the CCD amplifier is turned on briefly to read the guide star image. This results in some extra light hitting the upper left corner of the image from the amplifier transistors. For optimal results, you would take a dark frame using the software that comes with STAR2000, which will do the same 50-50 exposure so that the dark frame properly matches the light frame. The glow is faint enough that on some images you can remove it using the Gradient filter (Filters menu).

Figure 5.2.6 shows two comparison photos taken from the Starlight XPress web site. The image on the left shows what happens when you take a long exposure without guiding. The star images are drawn out into lines because of a small error in polar alignment. The wiggle comes from the periodic error of the mount used to take the image. If the mount were perfectly aligned, the periodic error would still show up, but in a more compact fashion. The image on the right shows what guiding does for you: it moves the telescope in very small increments to keep up with both periodic error and drift, leaving you with a nice, sharp image (limited only by the seeing conditions and optical quality of your system).

Figure 5.2.7 shows another feature of STAR2000: the ability to guide on a moving object that is in the field of view. In this example, STAR2000 guided on a comet, so the stars form trails.

If you do not have STAR2000, you would need a guider that could be aimed independently of the imaging chip, such as a dedicated guider in a separate guidescope to guide on a moving object. See the section below on dedicated guiders for more information.

FIGURE 5.2.7. GUIDING ON A MOVING OBJECT.

STAR2000 consists of cables and an external box that you connect to your computer's serial port. During imaging, the Starlight XPress software sends guide corrections through the external box to the mount.

Dedicated Guiding Cameras

While integrated guiding is definitely convenient, there are times and situations where having a dedicated guider can be very nice. And if you want to do film imaging, most CCD cameras also work just fine as guiders. The one situation where you really want to have the guide and imaging chips using the same light path is with a Schmidt-Cassegrain or other telescope that uses a moving primary mirror for focusing. Since the moving primary can (and often does) move during imaging, sharing the incoming light allows the guider to detect this and attempt to correct for it.

If you want to use an external guiders with a telescope that focuses with a moving primary, you need to find a way to lock down the primary mirror and use an alternate focuser, such as the NGF-S from JMI or the TCF from Optec. See chapter 2 for more details on these focusers. Without locking down the primary, there is no way to guarantee that the two scopes will stay aligned.

The methods of attaching a separate guider include:

- Using a separate small telescope attached to your main telescope, and aimed at approximately the same location in the sky. This is called a guidescope. There isn't a specialized type of telescope called a guidescope; any additional telescope you use for guiding is a guidescope. If the guidescope isn't aimed at the same area of sky, you could get field rotation in your image. How close? It depends on the quality of your polar alignment. The better your polar alignment, and the shorter your focal length, the more likely you will have successful guiding on long exposures. If you do elect to use a guidescope, make sure that every aspect of it is as solid as can be. It should be very firmly attached to the main scope, and the focuser of the guidescope should have zero play in it. The idea is to everything you can to prevent that guidescope from flexing or moving with respect to the main scope. If the guidescope moves, it will ruin your exposures.

The greatest challenge to using a guidescope is getting it to sit perfectly still. Depending on the guiding ability of your guider, the focal length of the guidescope should be long enough to provide sufficient accuracy during guiding. A high-end guider like an STV, or a small-pixel camera like an ST-237, works well with a guidescope that has a fast focal ratio (f/6 or faster with the STV is best). Guiders and cameras with larger pixels will require longer focal length guide scopes. To be on the safe side, aim for a guide scope that is at least half the focal length of your imaging scope. See the section below on the STV, however, because it has unique requirements due to its small pixels and very high precision.

- Using an off-axis guide setup, with a pick-off mirror to send part of the incoming light to a guider, while the rest of the light goes to your imaging chip. These are not as simple and easy as a guidescope in use, and can be awkward at times, but they do put guider and imager in the same optical path, and can be a functional solution for a moving primary mirror. Not all telescopes have enough back focus to use an off-axis guide setup, however.

Note that you cannot use a flip mirror with a guider. The flip mirror only directs light to one device at a time, either an eyepiece (for focusing and framing) or an imager.

My own preference is to use a guidescope if I am using a separate, dedicated guider. Many imagers use simple, inexpensive 60-90mm refractors with great results. However, there's nothing to stop you from using, say, a 7" Astro-Physics refractor to guide images being taken with a 4" refractor. There are no rules other than the obvious one: the two scopes must stay rigidly aligned throughout the exposure, and must secure the camera and guider rigidly.

With cheap refractors used as guide scopes, you may need to modify the focuser, as this is often the weakest link in the chain. Many cheap refractors have good enough optics to put a clean guide star on the guider's chip, but the focusers are somewhat loose. You can solve this problem by putting some tension on the focuser to prevent it from moving. Movement can occur either as a focus change, or as an angle change if the focuser has room to flop from side to side. Large rubber bands can put tension on the focuser, but adding additional set screws to really lock it down (three equally-spaced set screws will work nicely) is one of the most effective techniques. Drill and tap through the outer portion of the focuser, and then insert three Allen-head set screws which you can tighten down once you achieve focus on the guide scope. Ideally, this will also allow you to make sure that the guide camera is perpendicular to the optical axis, and should help you prevent coma or other optical problems caused by misalignment.

SBIG STV

Once upon a time, SBIG's main guider was the ST-4. This was a stand-alone guider, able to work without being attached to a computer. The camera came with a control box (see figure 5.2.8) that provided an arcane but functional interface that told you how well the guiding was going.

The ST-4 has been discontinued, and replaced with the STV (see figure 5.2.9). The STV is more than twice as expensive, but in many ways the STV is more than twice as functional as the ST-4 was. Not everyone wants all that functionality; adding features also increases complexity. But the fact that the STV has video output, and allows you to view the status of guiding in real time using excellent visual feedback, is a big improvement for most users.

If you want the ultimate in simplicity (and can handle the arcane interface), the ST-4 comes up for sale used periodically at about a third the cost of a deluxe

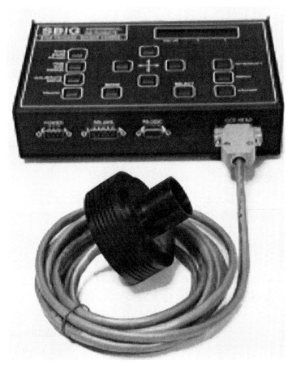

FIGURE 5.2.8. THE ST-4 WAS A VERY POPULAR DEDICATED GUIDER, OFTEN USED BY FILM IMAGERS.

STV. However, if you want the utmost in guide accuracy and functionality, the STV is my own personal choice by a wide margin. Still, if budget is the primary concern, then a used ST-4 and some time spent learning how to use the interface will deliver excellent guiding. After all, the ST-4 has been in use for years guiding for some very fussy film photographers.

The main reasons to buy an STV are:

- You need or like the video output. Video allows you to share your viewing with an audience. Planetary observing with video is extremely convenient. The ability of the STV to take a long exposure and then display it via video is a huge plus.

- You need the best guider available. The STV delivers guiding accurate to 1/30th of a pixel, which allows you to use a guide scopes with a short focal length. The optinoal eFinder works well if you are imaging at a focal length of 2000mm or less. The eFinder is an f/4 1" guidescope. That's a 100mm focal length! In fact, the STV works best with fast focal ratio guide scopes, not long focal lengths. Even a fast achromat will work well with the STV.

FIGURE 5.2.9. THE STV HAS MANY USEFUL FEATURES.

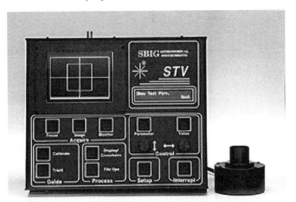

The STV is so sensitive that a guidescope with a focal length of about 500mm is enough to guide even very long focal lengths. A longer focal length just winds up guiding on the turbulence. A 500mm f/5 or f/6 refractor is ideal, and even a simple achromat will perform extremely well when imaging at 3500mm or longer.

- You need or like the analytical tools built into the STV. The STV can measure your seeing, measure the tracking performance of a mount and graph the results (see figure 5.2.10), and perform several other analytical tasks. If you have more than one mount, or even if you just love to analyze things, the STV is a great tool to have at your disposal when the time comes to determine what's wrong in tracking or guiding. You can also use the STV's graphing capabilities to help you drift align your mount.

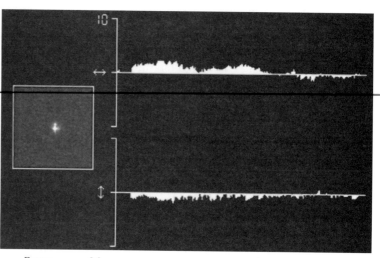

FIGURE 5.2.10. MEASURING AND GRAPHING A MOUNT'S TRACKING ERROR USING THE STV. THE TOP GRAPH SHOWS PERIODIC AND RANDOM ERROR IN RA; THE BOTTOM GRAPH SHOWS DRIFT AND RANDOM FLUCTUATIONS IN DEC.

The least important reason to buy the STV is to use it as a camera. Part of the price you pay for all of those cool abilities, especially video output, is more noise overall compared to other cameras. Still, the STV can do some nice images. Figure 5.2.11 shows planetary images taken with an STV and a 12.5" StarMaster EL.

Generally speaking, images of brighter objects have better signal to noise ratios and will be more successful. The main draw of the STV isn't as a camera. The images it takes are acceptable, but the way it can display them (video) is the real attraction.

Apogee LISAA Guider

This looks like an interesting unit (see figure 5.2.12), but until I can get my hands on one, I can't see how useful it actually is. For more information about LISAA Guider and the entire LISAA line of CCD cameras, please go to the Apogee web site:

> http://www.lisaa-ccd.com/

Look for reviews of this product on the book web site.

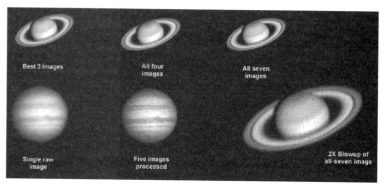

FIGURE 5.2.11. IMAGES OF JUPITER AND SATURN TAKEN WITH AN STV.

FIGURE 5.2.12. THE LISAA GUIDER

Section 3: Mount Calibration

To get the best possible results for both guided and unguided exposures, your mount should be tuned to minimize backlash and/or correct any other problems. This is critical for autoguiding. A mount that isn't tuned may under- or over-react to guide corrections. See the section "Adjusting and Tuning Your Mount" near the end of this chapter for details on getting the most out of your mount.

As described earlier in this chapter, a guider makes small corrections in your mount's position to maintain accurate tracking. Guiding will cover for small errors in polar alignment; for periodic error that results from slight eccentricity in the worm or worm gear; for non-periodic error that results from minor variations in the gears, and other sources of tracking error. The guider doesn't care what causes the error. If the guide star moves, the guider software will attempt to move it back to the center of the guide window.

If the errors causes the mount to move too far or too fast, the guider can't react in time. To deal with this, shorten your guide exposure time (limiting the number of useful guide stars), tune your mount so it tracks more accurately, or upgrade your mount.

The pixel size of your CCD detector, the focal length of your scope, and the speed at which you guide all affect the rate of movement during a correction. In addition, the orthogonality of the camera with respect to the mount's axes also plays a role with some camera control software. If the CCD pixels are not square to the mount, movement in one axis will cause a change in position in the other axis, and the rate of movement in the intended axis is smaller as well. If you want the flexibility to rotate your camera to any angle for composing your images, then go with a software package that supports this, such as CCDSoft v5. The STV and MaxIm DL 3 also support any camera angle.

Before you can autoguide, you need to perform a calibration. This calibration allows the camera control software to model your system's behavior during corrections. The software uses the calibration data to determine how long to move the mount (at whatever guide speed you are using) to make an appropriate correction. The actual rate of movement varies with your declination. The software has to move the mount more to make corrections when you are near the celestial equator, and less nearer to the pole.

If you perform a calibration at a declination near the equator, most software will scale the data for other declinations. You simply tell the software what the declination is when you calibrate, and then enter the current declination when imaging at some other part of the sky. If you choose this method, you perform the initial calibration near the celestial equator so that the scaling will be most accurate. If you calibrate too close to the pole, you lose precision of calibration and guiding closer to the equator will often be inaccurate.

You can also perform a new calibration whenever you move to a new declination. Some mounts need a recalibration when you change declination by 10 or more degrees. If you aren't getting good guiding when you move to a new position, perform a new calibration in the new area of the sky. Even very good mounts can benefit from recalibrating when you move to a new area. Some imagers recalibrate for every image at a new location as a matter of course. This guarantees the smoothest possible guiding results. The longer your focal length, the greater the likelihood of benefiting from a recalibration at each new declination.

How often should you recalibrate? It depends on your imaging session. If you move the mount across the meridian, the direction of movement in declination changes direction. For example, if you are guiding while pointing to the east, an upward correction is a move toward the west horizon. If you are guiding while pointing to the west, an upward correction is a move toward the east horizon. There is a checkbox that directs most software to flip declination corrections, and you can check this box when the mount crosses the meridian. However, for optimal results (especially with longer focal lengths and accurate mounts), a recalibration is strongly recommended. The longer your focal length, the greater the sensitivity to changes, and the more likely you are to need to recalibrate whenever you point to a new location.

I am including mount calibration examples using both MaxIm DL and CCDSoft. Both have some interesting features that make for accurate guiding, such as the ability to guide effectively even if the camera is not orthogonal to the axes of the mount. CCDSoft works cooperatively with TheSky to automatically track the current declination, so scaling of corrections is automatic. CCDSoft doesn't support all cameras, while MaxIm DL supports a large number of cameras.

The basic steps in calibration for autoguiding are similar for most camera control software packages:

1. Connect the guiding cable to the mount.
2. Move the telescope so that a suitable guide star falls near the middle of the guide chip. (More on "suitable" below.) Take images to verify the location of the guide star, and adjust position as necessary.
3. Select the star to use for calibration.
4. Adjust calibration settings, such as the exposure duration and movement time, if necessary.
5. Perform the calibration.

The first few steps are common to almost all camera control programs, and are described in a generic manner below. The software-specific steps for CCDSoft and MaxIm DL parallel these common steps. However, I have covered quite a few basic principles in the CCDSoft descriptions that also apply to MaxIm DL, such as calculating calibration times, so I suggest you read both sections even if you are only using one or the other.

Connect the Hardware

To autoguide, you need either a camera that will be dedicated to autoguiding, such as an ST-237, or a camera that has its own built-in autoguiding chip, such as an ST-7/8/9/10 or a Starlight XPress camera with STAR 2000. There is typically a cable between the camera and the mount (e.g., SBIG cameras), but mount control can also occur by a cable from the serial port of your computer to the mount (e.g., Starlight XPress cameras). Consult the camera documentation for instructions on how to connect the autoguider cable to the camera or computer. The other end of the autoguide cable attaches to your mount; consult your mount documentation for the location of the autoguider connection.

Not all mounts support an autoguider connection. Those that do sometimes require a specialized cable. Contact the mount manufacturer for information on where to buy or how to build an interface box or cable to fit such a mount. SBIG cameras come with a standard cable that fits many but not all mounts. If the standard cable doesn't fit (such as for a Takahashi mount, or a Vixen Sky Sensor-equipped mount, and others), consult the mount documentation for information about building or buying a suitable cable.

Most cameras have two additional cables: a power cord, and a connection to your computer. These vary widely, so check your camera documentation for details. Pay particular attention to limits on the length of the data cable between camera and computer. You may be able to get more distance by building a custom cable with high-quality shielded components.

Choose a Good Guide Star

Whether you are calibrating or autoguiding, and whatever software you use, selecting the right guide star is critical. The brightness and location of the star used for calibration play a major role in the success or failure of the operation. Here are some tips for choosing a good guide star. Most apply whether you are autoguiding or calibrating; a few relate specifically to calibrating.

The guide star should be bright enough to show clearly against the background. The longer your guide exposure, the brighter the stars will be. If you can't find a suitable guide star for a given exposure, increase the exposure time to see if any good candidates pop out of the background. The length of your exposure may be limited by the tracking ability of your mount, however. For calibrating, the guide star should be at least 1,000 units brighter than the background. For autoguiding, a guide star even a hundred counts brighter may work.

The guide star should not be close to saturation (4,000 units for ST-237; 40,000 units for the ST-7/8E, etc.). Random variations in brightness can easily take a bright guide star into saturation, creating false readings and invalid corrections. The guide star should be no more than 50% of the saturation level. If you don't know the saturation level for your camera, you can get it from the manufacturer, your documentation, or you can calculate it for yourself from the camera's specifications (full-well capacity in electrons divided by the gain).

Start with the guide star near the middle of the tracking CCD. This is important when you are calibrating because the software moves the mount during calibration. If the guide star moves too close to an edge, or outside the chip's field of view, calibration will fail. Most software provides a method for moving the mount to center the guide star. If you cannot put the guide star near the center for autoguiding, you can still get good results as long as the star is away from an edge during the entire calibration procedure.

The guide star must be the brightest star in the entire image when calibrating. This allows the software to find the guide star unambiguously on each exposure of the guide CCD.

No nearby stars should be close to the brightness level of the guide star. Stars that are close in brightness will cause the software to oscillate between the two stars, leading to extremely inaccurate guiding.

Make sure that another bright star doesn't move onto the CCD detector during calibration. Use TheSky or other software to visually inspect the area around your selected guide star. Check for a distance equal to 1.5 times the width of the area covered by the tracking CCD detector. See the section "Select a Guide Star" later in this chapter, which explains how to use TheSky to help you select a guide star.

Good focus is critical to accurate guiding. A small variation in focus softens the guide star enough to make its exact position less certain. Collimation is also very important for scopes that require it. Poor collimation causes off-axis aberrations. Since the guide chip is well off-axis, this results in poor guiding.

Calibration for Autoguiding with CCDSoft

To calibrate CCDSoft for your mount, telescope, and camera, open the Camera Control Panel. If it is not already open, use the Camera | Autoguide menu item to open it. This displays the Camera Control Panel with the Autoguide tab active (see figure 5.3.1).

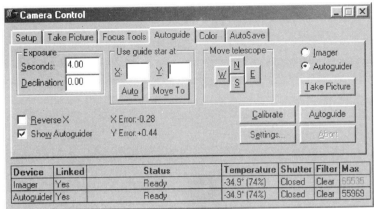

FIGURE 5.3.1. THE AUTOGUIDE TAB

Take a Reference Image

Typical autoguide exposures range from 1 to 10 seconds. Shorter and longer exposures can be used, but are rarely necessary. Guide exposures shorter than 1-2 seconds can be affected by atmospheric turbulence. Guide exposures longer than 5-10 seconds could be affected by periodic and random errors of your mount. The better your mount's tracking, the longer you can go with your guide exposures.

When the guide exposure is very short, you are measuring fluctuations in the atmosphere rather than movement that needs to be corrected. If the guide interval is too long, the guide star could move so much that corrections will be too large, and star images will become elongated in the direction of the correction.

CCDSoft can use the calibration results to scale for different declination values when autoguiding at different places in the sky. To use this feature, you must enter the current declination that the telescope is pointing to in the box labeled "Declination." An accuracy of one-tenth degree is sufficient to use this feature.

TIP: If you have TheSky installed, and are using it to point your telescope, the declination will be set automatically; you do not need to enter it manually.

The image you take as a reference image should be long enough to clearly image a star suitable for calibration. If you are using TheSky and CCDSoft together, you can determine the proper exposure based on the magnitude of the stars in the field of view. If you are

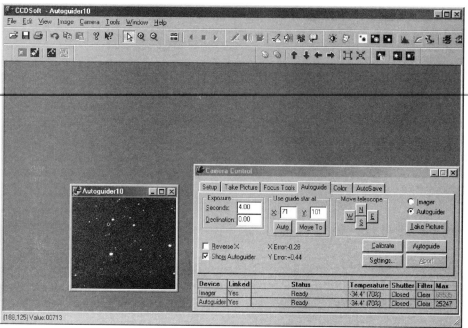

FIGURE 5.3.2. AN AUTOGUIDE IMAGE TAKEN FOR REFERENCE. NOTE THAT THERE IS ONE STAR SIGNIFICANTLY BRIGHTER THAN THE OTHERS; THIS IS A GOOD CHOICE FOR GUIDE STAR.

unsure, start with an exposure of 5 to 7 seconds. You can always adjust it upward or downward as needed.

Click the Take Picture button to expose an image. Figure 5.3.2 shows the result for the guide chip on an ST-8E camera. The guide chip is smaller than the imaging chip, so download times are short. By default, a dark frame is automatically taken and subtracted from the light frame. A dark frame is essential for calibration and for guiding. If a dark frame is not being taken automatically, go to the Take Image tab, click the Autoguide button, and make sure that the Frame type is Light, and that the Reduction type is AutoDark.

Evaluate the image to determine if the exposure time provides a suitable brightness level for the guide star. Use the criteria listed earlier in the section "Choosing a Good Guide Star."

Techniques for Locating Guide Stars

Click the Auto button on the Autoguide tab to have CCDSoft pick the brightest star to use as a guide star, or click in the image to select a guide star. A white box flashes briefly around the star you select. The coordinates of the star appear in the X and Y boxes of the Autoguide tab. The guide star must be the brightest star inside that flashing box. CCDSoft will always guide on the brightest star.

If you are using CCDSoft with TheSky, you can easily check for the presence of good guide stars in the vicinity of your current location. TheSky can project the coverage area of the CCD detector (ST-237 and similar cameras) or both the imaging and tracking CCD detectors (ST-7/8/9/10) on the map of the sky. Figure 5.3.3 shows the projection for an ST-8E camera. The central rectangle is the imaging detector; the small rectangle above it is the tracking detector. The circles show the position of the tracking detector if you rotate the camera. These circles can help you find an alternate guide star by rotating the camera, instead of moving the mount.

To set up your camera's field of view indictor in TheSky, follow these steps:

1. Select the View | Field of View Indicators menu item.
2. Click the Add button
3. Type in a description, such as the name of your camera and the telescope.
4. Select the type of indicator to use (rectangular, ST/7, or ST/8)
5. Click the Compute button
6. Click the CCD tab
7. Select the appropriate camera/CCD detector from the drop down list
8. Enter the focal length of your telescope (inches or millimeters)
9. Click the Compute button.
10. Click OK three times in the three nested dialog boxes to finish and exit.

The field of view indicator will only appear at magnifications large enough to see it. Zoom in if necessary to see the field of view indicator.

In figure 5.3.3, the pair of interacting galaxies known as Messier 51 are centered on the imaging detector, and celestial north is straight up. This assumes that your camera is oriented with north up as well. The box at the top shows that there are a few dim stars on the tracking detector, but none are very bright, and none are near the center. You could try to guide using the star at the eight o'clock position on the tracking detector, or you could move the mount to bring a brighter star onto the tracking detector. However, no star on the guider chip is bright enough to use for calibration.

A brighter star allows you to use a shorter guide exposure time, and is required for cali-

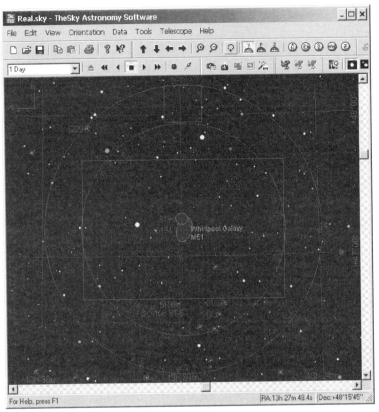

FIGURE 5.3.3. THE IMAGING AND GUIDE CHIPS OF AN ST-8E CAMERA SHOWN IN THESKY.

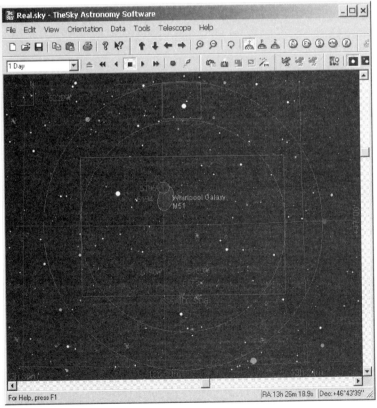

FIGURE 5.3.4. PLACING A GUIDE STAR ON THE TRACKING DETECTOR.

bration. If your mount is very stable and accurate, you can use longer guide exposure times and therefore dimmer guide stars are suitable. If your mount has poor polar alignment, larger periodic error, or suffers from random movements during tracking, brighter guide stars are more important. They allow you to use short guide exposures. This causes more frequent guide corrections and keeps your mount under better control.

Figure 5.3.4 shows the result of moving the mount to put a brighter star on the guiding CCD. M51 moves in the field of view as well since it is off-center, but it's still well within the boundaries of the imaging chip. However, M51 has been moved away from the center of the imaging detector.

This arrangement is fine if you need the brightest possible guide star for calibration or for short guide exposures. It also works if you

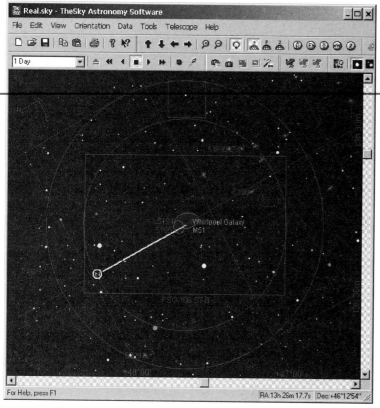

FIGURE 5.3.5. ROTATING THE VIEW TO LOOK FOR A SUITABLE GUIDE STAR.

north at about eight o'clock puts a reasonably bright star on the tracking CCD detector. You can experiment with TheSky to find the right rotation angle, and then physically adjust the camera to the new orientation. Because CCDSoft can guide accurately at any angle, even 45 degrees off of orthogonal, you have complete flexibility in framing your images or positioning a guide star.

If your mount has goto features, you can connect it to TheSky and control the mount from the computer. This allows you to move the mount to put guide stars right where you want them. Camera rotation still requires a trip out to the telescope, however.

Adjusting Calibration Settings

During calibration, CCDSoft will move the mount, and then take another picture to measure the amount of movement. The exact sequence of moves is:

- Move in a positive X direction
- Move in a negative X direction
- Move in a positive Y direction
- Move in a negative Y direction

The actual directions represented by X and Y will depend on how your camera is oriented. If the camera is oriented with north up, then +X is east and -X is west; +Y is north, and -Y is south.

An ideal mount would conclude this sequence with the guide star in the exact same position as at the start of the calibration procedure. In most cases, there will be at least a small difference in starting and ending position due to whatever backlash remains uncompensated for in the mount. In addition, unless the camera is perfectly square to the mount's RA and Declination axes, a movement in one axis will involve at least a small movement in the opposite axis. CCDSoft not only tolerates minor non-square camera placement; it can cope with a completely non-square camera position. This gives you the ability to rotate the camera to any angle, giving you much more flexibility in selecting a guide star.

intend to crop the image so that only the area immediately around M51 will be in the finished image. More importantly, if you were about to do a calibration, the bright star in figure 5.3.4 is clearly much brighter than anything else around it for a large distance. It's an ideal star for calibration. Use a very short exposure time for such a bright star, which will also speed up the calibration. In general, you should expect to move your mount to do a calibration, because putting a very good guide star on the center of the chip is the key to successful calibration. With TheSky available to position the mount, you can move, calibrate, and return to your object very quickly. Many other mapping programs can do the same job, but in different ways. Check your documentation to find out how to project your camera's chip onto the map of the sky.

If you want M51 at or near the center of your image, however, a better approach would be to rotate the camera to bring a guide star onto the tracking detector. Figure 5.3.5 shows that a rotation that puts

SECTION 3: MOUNT CALIBRATION

Table 5.1: Suggested Calibration Times

Focal length	Suggested calibration time
100-300mm (typically a camera lens)	30 seconds
300-500mm	20 seconds
500-1000mm	10 seconds
1000-2000mm	7 seconds
2000-3000mm	4 seconds

To set the time interval for calibration, as well as other related settings, click the Settings button on the Autoguide tab. Click the Advanced tab to show all available options (see figure 5.3.6). The time is measured in seconds.

In most situations, you can simply enter a time interval for each axis, and click OK. The time you enter depends on the focal length of your telescope and the pixel size of your camera. If the camera is reasonably well matched to the telescope (a camera with an image scale between 1.5 to 3.5 arc-seconds per pixel), the values shown in Table 5.1 are useful starting values for calibration time. The table assumes a 15-micron pixel size of for the guide chip. Adjust as needed for your focal length or guide chip pixel size. Larger pixel sizes require longer calibration times. When in doubt, user a longer rather than a shorter calibration move time.

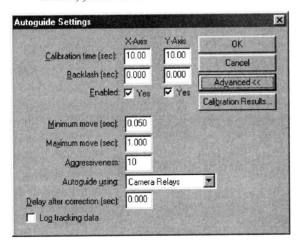

FIGURE 5.3.6. ADVANCED AUTOGUIDE SETTINGS DIALOG.

Calculating Calibration Time

To calculate a good calibration time, you need to know the image scale (arcseconds per pixel) of your guide chip. Multiply this 10 to get the number of arcseconds that correspond to 10 pixels on the guide chip.

For example consider a 10" LX200 using an f/3.3 focal reducer and an ST-7E camera. The ST-7E guide chip has approximately 15 micron pixels, and the focal length is 2500 * 0.33 = 825mm. This yields 3.75 arcseconds per pixel (see formula below). For 10 pixels this would be 10 * 3.75, or 38 arcseconds. This is the minimum angular distance you want the mount to move to get a decent calibration. The guide chip on the ST-7E is large enough that even if you move double or quadruple this distance, you are OK (in fact you'll get even better precision). I suggest that you round upward to the next arcminute as your target move.

The next step is to determine how long it would take, at your guiding rate, to move this distance.

At sidereal rate (1X or tracking speed, or approximately 15 degrees per hour), a scope moves 15 arcminutes in a minute, which is 15 arcseconds per second. The LX 200 scopes use a guide speed of 2x sidereal, so the mount moves at a rate of 30 arc seconds per second. It would thus take just two seconds to move through one arcminute and get an adequate calibration.

The formula for minimum calibration time:

$$\frac{(number\_of\_pixels\_to\_move * image\_scale)}{(15 * guiding\_rate)}$$

The formula for image scale is:

$$\left(\frac{pixel\_size}{focal\_length}\right) * 206$$

where pixel size is in microns and focal length is in mm. For example, say you want to calibrate using a generous movement of 50 pixels:

$$\frac{(50*3.75)}{(15*2)} = 6.25 \text{ seconds}$$

For a Takahashi FSQ-106, assuming 20 pixels and .5X guide rate, the calculation would be:

$$\frac{(20*5.83)}{(15*0.5)} = 15.5 \text{ seconds}$$

If you don't have your image scale handy, the following formula uses the pixel size and focal length of your setup instead of the image scale:

$$\frac{(num\_pixels * pixel\_size * 206)}{(15 * guide\_rate * focal\_length)}$$

For the FSQ-106 example, this would be:

$$\frac{(20*15*206)}{(15*0.5*530)} = 15.5 \text{ seconds}$$

Setting Advanced Parameters

The advanced tab allows you to set the following additional parameters for calibration:

Backlash compensation - Backlash exists when the motors on the mount must turn for some period of time before the gears fully engage and actually move the mount. Backlash only comes into play when you reverse direction. Backlash compensation specifies the number of seconds that CCDSoft should turn the motors at high speed. This will take up backlash when the direction of correction reverses. When guide speed in RA is less than sidereal rate (0.5x is commonly used), no reversal occurs. You should not set a time for RA backlash compensation unless you are using a guide speed of 1x sidereal or faster.

Enabled - Determines whether corrections are applied to the indicated axis. For example, if you want to avoid declination backlash entirely, and are really well aligned to the pole, you can turn off corrections to the Y axis.

Minimum move - This specifies the minimum move that CCDSoft will make. If a smaller correction is indicated, no correction occurs. When a correction of this length or greater is indicated, the correction will be made. Increase minimum move when you need to filter out small movements of the guide star.

Maximum move - The longest time to move the mount during a correction. If the mount is too aggressive no matter how much you change other settings, this will force CCDSoft to use shorter corrections.

Aggressiveness - A number from one to ten, indicating the relative aggressiveness of the corrections. With a setting of 10, CCDSoft will make the full, indicated correction. With a setting of 5, CCDSoft will make 50% of the indicated correction. With a setting of 1, only 10% of the indicated correction occurs. If the mount is over-correcting, you can reduce the amount of the correction with this setting and smooth things out. For example, when pointed near the zenith, some mounts are more responsive to corrections than when pointed at a lower elevation. You can use a lower aggressiveness setting near the zenith to compensate if your mount behaves in this manner.

Autoguide using - Determines how CCDSoft should communicate with your mount. If the camera is connected directly to the mount (or connected through the SBIG relay box), then use the "Camera Relays" setting. If you are using a special arrangement via the serial port of the computer running CCDSoft, then choose RelayAPI (serial).

Delay after correction - If your mount tends to vibrate or bounce slightly after a correction is applied, a delay can help damp this oscillation. This is usually true only of lighter mounts, but it can also occur with larger mounts that are balanced too equally on the east and west sides. A slight imbalance to the east is not only good for keeping the RA gears meshed; it also helps reduce unwanted oscillations during guiding.

Log tracking data - When checked, the log data is written to a file. See the section "Examining data from an autoguiding run in a spreadsheet" later in this chapter for information about the log file.

Table 5.1: Calibration Errors and Solutions

Calibration Problem	Solution
Error message: Star too dim. Lost during +X, -X, +Y, or -Y	The star you chose for calibration was too dim, and could not be found on the last image taken. Use a longer exposure, or choose a brighter star. A star is too dim when the brightest star in a calibration image is less than 25% as bright as in the preceding image.
Error message: Motion too small during +X, -X, +Y, or -Y move. Increase calibration time.	The calibration time was too short to move the star enough to get a valid result. Use a longer exposure time. For a valid calibration, the star must move at least 5 pixels. I recommend at least 10 pixels, however. Excessive backlash, or forgetting to turn on backlash compensation, can also cause this error.
Error message: Star too close to edge after +X, -X, +Y, or -Y move.	The guide star is too close to the edge of the image. You can reduce the calibration time to cause less movement, or start with the guide star closer to the center of the frame. This error will occur when the guide star moves to within nine pixels of the edge.
Error message: Invalid motion in X or Y axis.	This error typically occurs when: • The brightest star in a calibration image is a star other than the original guide star. The guide star should be significantly brighter than any other star in the image. • A brighter star moves into the guide frame during calibration. The guide star should be isolated from similarly bright stars by 1.5 times the width/height of the guide frame in arcseconds. • The mount is very inaccurate, with motion in X resulting in motion in Y, or vice versa. Check the Calibration Results graph to see if it provides any clues about what kind of invalid motion is contributed by the mount.
Error message: Unable to calibrate, both axis are disabled. At least one axis must be enabled to calibrate.	With both axes turned off, no calibration is possible. To enable one or both axes, go to the Autoguide tab on the Camera Control Panel and click on Settings. Click in the checkbox next to "Enabled" for one or both axes.
Problem: Guide star moves out of frame	This can cause several different error messages, depending on what happens when the guide star moves out of the image. You might see "Star too dim" if another star is interpreted as the guide star, or you might see "Star too close to edge." CCDSoft displays the images taken at each step in the calibration process; watch these images to see if the guide star is moving out of the frame, or to diagnose any other problem that doesn't have an obvious solution. Use a shorter calibration time to keep the guide star in the frame, and start with the guide star as close as possible to the center of the frame.

Calibration Results - This button shows you the numeric results of the calibration, as well as a graph of the relative movement and directions of movement. See the following section for detailed information.

Of the advanced settings, Aggressiveness is the one you are most likely to use. For your first autoguiding efforts, leave the aggressiveness set to 10. If the corrections are too aggressive, try a lower number. It is better to have too low a number than one that is too high, since the software will eventually catch up and make the necessary corrections if you are reasonably close with your polar alignment. If the aggressiveness is set too low, however, the corrections will not be able to keep up with changes and elongated stars will result.

Calibrating

Click on the Calibrate button to start the calibration process. You will see information reporting the progress of the calibration process in the Camera Control Panel.

FIGURE 5.3.7. BEGINNING A GUIDING SESSION WITH MAXIM DL.

CCDSoft will take a new reference image and dark frame, and then move the telescope four times. At the end of each move, another image is taken, and CCD-Soft compares the images to see how far the guide star has moved. If there is a problem, CCDSoft will halt the calibration and report the error. Table 5.2 lists common problems and explains how to deal with them.

Once you have calibrated the mount, you are ready to begin autoguiding. If you moved the mount to find a good calibration star, move it back to the object you want to image, and locate a suitable guide star.

Calibration using MaxIm DL

Guiding and calibration for guiding are handled on the Guide tab of the CCD control panel in MaxIm DL. Figure 5.3.7 shows the typical starting point for a guiding session. The CCD control panel is on the right, with the Expose radio button active. The window on the left is an exposure with the guide chip of an ST-8E, taken by clicking the Start button while the Expose radio button is active. The exposure confirms that the guide chip is looking at the part of the sky I expect it to.

FIGURE 5.3.8. CHANGING AUTOGUIDER SETTINGS.

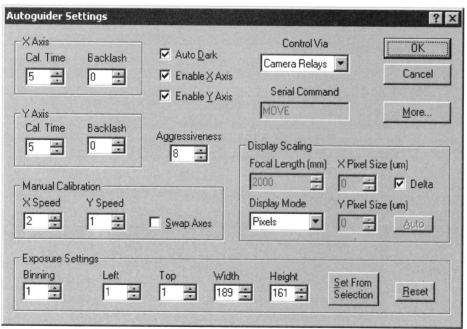

The steps in guiding with MaxIm DL are neatly organized by the three radio buttons at the left of the camera control:

Expose - Does nothing more than take an exposure with the guiding chip, for the time indicated in the Exposure box.

Calibrate - Performs a calibration. The basic steps are the same as with CCDSoft: move the scope in four directions, and measure how much it moves in each direction. The rate of speed is used to determine the duration of guide corrections when guiding.

Track - Start guiding.

As with CCDSoft, you need to set a few parameters before you perform your first calibration. Click on the Settings button in the control panel to display the Autoguider Settings dialog (figure 5.3.8). Don't be intimidated by this dialog; it has settings that you may never use no matter how many times you image.

The features of this dialog that you are most likely to find useful include:

Cal. Time (X and Y axes) - This is the same as the calibration time described in the previous section. Refer to that section for a formula for calculating the minimum calibration time for your particular telescope/guider combination. Remember to use the guider pixel size for the calculations, not the imager pixel size.

Aggressiveness - Lower numbers reduce the duration of guide corrections, which can help tame guide corrections that are too strong. Higher numbers increase the amount of correction that can occur on any one correction.

Auto Dark - When checked, automatically takes a dark frame for your guiding images. Unless you have a spe-

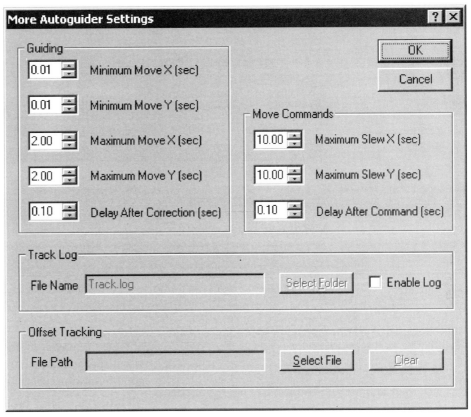

FIGURE 5.3.9. SETTING UP A TRACKING LOG FILE.

cific reason for turning this off, it should always be on. Otherwise, hot pixels in the guide images may cause the tracking software to get confused about where the guider star is. This is especially important when using a dim guide star.

Enable X and Y axes - You can turn guide corrections on or off for either axis.

The More button at upper right provides access to an even deeper level of parameters. This includes the maximum and minimum move times, which you will rarely use (see figure 5.3.9). Consider the max/min settings as a last resort to try to tame an unruly mount. The Track Log section is the most likely to be useful. You can write the guide corrections out to disk for later review. Check the "Enable Log" box to turn it on, and then specify a filename and a folder for the log file.

To get a good calibration, you should have one star in the tracking CCD image that is significantly brighter than any nearby stars. This insures that the software

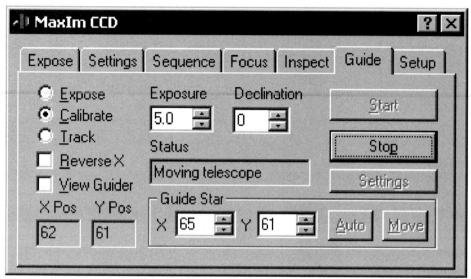

FIGURE 5.3.10. MOUNT CALIBRATION IN MAXIM DL.

This process repeats for each of the four directions. MaxIm DL stores the results internally. If you enter the declination of subsequent objects when tracking/guiding, MaxIm DL will scale the results of the calibration for other declinations.

Figure 5.3.11 shows guiding in progress. The guide star shows up in a small window if you check "View Guider," which I recommend. You can monitor the appearance of the guide star visually. This is a good way to monitor the quality of focus. Temperature shifts can affect focus, and the visual appearance and brightness of the guide star are good indications of focus quality.

will follow the star properly as it moves during the calibration process. To calibrate:

1. Put the star near the center of the tracking CCD.
2. Click the Calibrate radio button (see figure 5.3.10).
3. Click on the guide star. Verify that the star's coordinates show up in the Guide Star section (see figure 5.3.10).
4. Set an exposure time. Make sure that the guide star does not saturate. A saturated star will lead to errors.
5. Enter the current declination. You may leave Dec set to zero if you intend to re-calibrate every time you move the scope by 10 degrees or more in Dec.
6. Click the Start button.

During the calibration process, MaxIm DL performs the following operations:

1. Take an image with the guide CCD detector.
2. Move the telescope for the calibration time you set on the Autoguider Settings dialog (figure 5.3.8).
3. Take a new image to see how far the guide star has moved.

If you experience problems with calibration or guiding, the issues are the same as those listed under CCD-Soft above. Look there for suggestions that could help you resolve your problems. The most important thing is to have backlash physically adjusted and/or compensated for before you attempt to calibrate. The following sections will also give you some ideas on how to improve the quality and accuracy of your guiding.

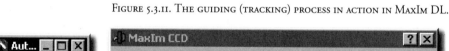

FIGURE 5.3.11. THE GUIDING (TRACKING) PROCESS IN ACTION IN MAXIM DL.

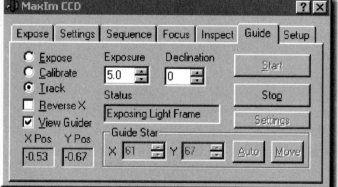

Calibration Tips

Guiding is still more art than science, and getting a few dozen hours under your belt will help you finesse some of the challenging situations you are bound to run into. Every mount responds a little differently, and time spent at your mount will teach you what is unique about your particular setup. Patience is definitely a virtue when it comes to autoguiding.

Calibrate at a faster speed

If your mount guides reasonably well, but you see some variations as you point to different areas of the sky, or change the balance of your scope, you can soften the effect of the corrections in several ways. You can reduce the aggressiveness setting, so that the camera control software only makes a portion of the correction. Or you can alter the calibration numbers manually to see if that will help. You can tell the software that your mount actually moves faster than the calibration indicated, for example, to reduce over-correction during guiding. Likewise, if your mount seems sluggish to respond, you can tell the software that your mount moves more slowly than indicated by the calibration results. How you shift the numbers will depend on the software you are using for guiding.

Another way to deal with variations in guide results is to calibrate at a higher speed than you intend to guide at. As long as your mount doesn't make large, sudden excursions, and you have a very good polar alignment, it is usually safe to make smaller, slower corrections than the situation seems to demand. If backlash is not an issue in RA, then you can calibrate at 1.0X and guide an 0.5X. If backlash is an issue, then calibrate at some fraction of sidereal, and guide at an even slower rate. Not all mounts will allow you to do this, but it is an option for some.

In general, the fewer corrections that are needed, the better. Every guide correction is a movement, and every movement carries a bit of risk with it. The better your alignment and the more accurate your mount, the better your results will be. The positive and negative aspects of your setup are cumulative. Take the time to get as many things set up correctly as possible.

Calibrate early and often

When in doubt, calibrate. If you move to a new area of the sky, and experience guiding problems, take a few minutes to calibrate. The results may vary enough to require you to do this frequently. Recalibration is the best way to make sure that you have the optimum parameters for guiding for any give part of the sky. If you are just out having a little fun imaging, then you can be more relaxed about calibration. But if you are taking 20, 30, or 60 minutes of exposures, getting the guiding exactly right is going to pay significant dividends. A guide error in a 30-minute exposure is more painful than a similar error in a two-minute exposure.

Variations on a guiding theme

Many mounts show different guiding characteristics when pointing at different areas of the sky. This goes beyond differences in calibration. For example, many German equatorials I have used tend to over-react to guide corrections when pointing near the meridian. I have found that increasing the imbalance to the east (making the mount slightly heavier to the east) helps when pointing near the meridian. This shouldn't be a large change; just increase the weight on the east side a bit when you are near the meridian. You can often improve guiding by reducing aggressiveness when pointing near the meridian. You may need to recalibrate after making these changes, however.

If you have off-center stuff on your mount, balance can change dynamically as you point to different areas of the sky. As you move the telescope the balance shifts, sometimes more than you would expect. For example, many finders are located off of the centerline of the scope. Many cameras are not evenly balanced, such as the ST-7/8/9/10 series from SBIG. Cables and water tubes can also change balance as the telescope moves. Try routing your cables first to the center of the mount, and then to their destination (camera, focuser, etc.).

Depending on your exact setup, and the capabilities of your mount, you may need to rebalance your scope as you move to different areas of the sky. Take the time to get familiar with how balance changes with your setup when pointing to different parts of the sky. Experiment with different arrangements to see how you can best balance your equipment. Make some marks (with tape or magic marker) on the mount and equipment to indicate where the correct balance points are.

Mirror flop

Scopes with moving primary mirrors (most SCTs; smaller Takahashi Mewlons, etc.) sometimes find ways to move the mirror when you are taking an exposure. And when you move across the meridian, you can get fairly serious mirror flop. The mirror never seems to move all at once, moving a little here and a little there. If you are using a guidescope, this movement can ruin your exposures because the guidescope cannot sense the change or respond to it.

The most reliable method of eliminating mirror flop is to fix the mirror in position using locking screws. This isn't for everyone; it means you have to drill some holes in the back of your scope and tap them for the appropriate screws. Don't forget a rubber tip or some epoxy or plastic on the end of the screw so it won't scratch your mirror, and don't over-tighten!!! Over-tightening could break the mirror, or flex it so that it changes its curve. If distortion occurs, it could throw off the quality of your images by quite a bit, and leave you wondering what's going on.

If you do lock down your mirror, you'll need an alternate focuser like the ones described near the end of chapter 2 (JMI NGF-S, Optec TCF-S, RoboFocus).

If you don't lock down the mirror, try guiding for a few minutes before you start to image to allow mirror shift and backlash and everything else to settle down. You can watch the numeric feedback in your camera control software to judge when the guiding has become smooth and regular. Guiding for a few seconds or minutes (whatever it takes) can even be good for setups that don't have mirror flop, as it will take up some slack or weirdness in the mount and allow you to get more consistent guiding. Some mounts may even take up to 5 minutes to smooth out, settle down, and guide reliably. I've seen mounts that required a full cycle of the worm to settle down.

Focus Drift (thermal and otherwise)

Most telescopes change focus at least a little bit if the temperature changes. Since temperature is usually dropping at a reasonable rate for about half of the night (or until the dew point is reached), it would be neat to have a handy way to assess the amount of focus change. The guide star is just such a feedback mechanism. At the start of your guiding session, when you know focus is good, observe the brightness and visual characteristics of your selected guide star. If the brightness drops in a slow, consistent manner, your focus is shifting. SCTs and other folded designs often show a greater tendency to change focus due to temperature, but almost any type of telescope will be affected.

You can spend some time experimenting to determine the amount of focus shift for your setup, and note whether focus needs to move in or out to correct for the shift when temperatures are falling. You will find that you need to make very small adjustments at a more or less steady rate. With my 4" FSQ-106 refractor, I needed to make the smallest possible adjustment approximately every 20-40 minutes while temperatures were falling.

Observe the brightness and visual appearance of your guide star to tell when you need a focus correction. If the FWHM or brightest pixel value changes, monitor the guide star to get a feel for whether or not focus has changed. If you have a motorized focuser (or, even better, a motorized focuser with digital readout), you can manually adjust focus position even during an exposure to compensate for temperature-induced shift. Such temperature-related focus adjustments are the best reason I've found to spend the extra money for a digital readout on the hand controller for a motorized focuser.

The Optec TCF-S focuser includes a temperature sensor, and once it is calibrated, you can rely on the TCF-S to change focus in response to temperature changes. If the temperature change is large, the TCF-S may require recalibration. RoboFocus also has the ability to sense and respond to temperature changes.

Section 4: Autoguiding in Action

Once your system is calibrated, you can start autoguiding. Some camera control software supports autoguiding independent of imaging (CCDSoft and MaxIm DL). Some camera control software requires you to initiate a new guiding session for each image you take (CCDOPS).

It's nice to have camera control software that will resume guiding after each image downloads. It's convenient not to have to run through the setup for the guiding session every time. This is especially true for long sequences of images. You can set up the sequence, and get some rest while the sequence executes. Guiding will resume after each image download completes.

TIP: Since some cameras have really long download times, the guide star may drift a long way while the download is in progress. To allow time for any corrections to occur when guiding resumes after the download, look for a place to enter a per-exposure delay. In CCDSoft, this is located on the Take Picture tab of the Camera Control Panel. In Maxim DL, it is located on the Expose tab.

Resuming Guiding after Downloading

Software that automatically resumes guiding after a download allows you to take a succession of images without worrying about stopping and starting autoguiding. Keep in mind that autoguiding cannot occur during image downloads if you are using a camera with an integrated guider (e.g., ST-7E). A separate guider, such as an STV or a second CCD camera in a guidescope or an off-axis guider, guides continuously without interruption during downloads.

When the shutter is closed, it closes on both the image chip and the integrated guider chip. Autoguiding is suspended until the shutter opens again. A separate camera or guider has its own shutter and can cheerfully guide for hours if necessary.

The goal when using an integrated guider is to keep that guide star within the small guide window for the length of time it takes to download. If your mount is prone to sudden movements, or if your polar alignment is poor, autoguiding will not resume, or might resume with a different star, if the guide star moves outside of the guide window.

If you are well-aligned to the pole, the guide star will normally still be visible in the little guide window after a download. If you find that the star moves more than one pixel during downloads, add a delay between images. This will allow the autoguider to make corrections before the next exposure begins. As a starting point, use a delay that is about 5-6 seconds longer than four times the autoguide exposure. For example, if your guide interval (exposure) is 4 seconds, then use a delay of about 20 seconds. The extra time allows for downloading the autoguider images and processing time. The lower your aggressiveness setting, the longer this delay must be. If you find that your particular mount requires shorter or longer delays, adjust accordingly.

A Typical Autoguiding Session

The following example uses CCDSoft version 5 to show what happens during a typical guiding session. The details will vary a bit with other software packages, but the principles are the same. A typical autoguiding session includes the following steps:

1. Adjust the position of the telescope if necessary to put a star on the autoguider detector (see the previous section in this chapter for information about suitable guide stars).

2. Take a picture with the guide chip to verify the location of the guide star, using an appropriate exposure length. If you aren't sure, try an exposure in the range of 5-10 seconds. You can increase and decrease your guide exposure based on seeing conditions and the brightness of the guide star.

3. If the brightness of the guide star is too high or too low, adjust the exposure time. You are looking for a count of at least 500-1000 above background levels, but no more than about 75% of the guide chip's saturation level (50% if you have a non-antiblooming camera). Remember that saturation level is full well capacity divided by the gain.

4. If the guide star is close to an edge, move the mount to put the star closer to the center. In CCDSoft, click on the guide star, and then click on the Move To button. A dialog opens; click on the new location, then click OK. CCDSoft will move the mount to put the guide star at the new location. You can also move the telescope manually, or using the software's N/S/E/W buttons.

5. Enter the current declination the telescope is pointing to. If you are using CCDSoft and TheSky to point your telescope, this is filled in automatically.

6. Begin guiding by clicking the Autoguide button.

When you move a large distance across the sky, or if you cross the meridian, do a new mount calibration. If you are imaging at longer focal lengths (2500mm and larger), you can calibrate before every image if necessary to get the best possible guiding. The longer the focal length, the more critical guiding becomes.

During autoguiding, CCDSoft updates a small window (see figure 5.4.1) with the latest image of the guide star. (This window is updated by default. If you do not want the update, uncheck "Show Autoguider" in the Autoguide tab.) The current position error appears near the center of the Autoguide tab, showing the number of pixels that the guide star has moved from the original position (X Error and Y Error). These numbers may be positive or negative, and are reported for both X and Y directions to the nearest hundredth of a pixel.

Note: Most programs report this error. Some, like Maxim DL, also allow you to show the error in arcseconds by entering information about your telescope.

Autoguiding will continue until you stop it by clicking the Abort button. If the star is lost (such as when the telescope is bumped), abort autoguiding and start

over. Guide exposures involve a dark frame. If you resume guiding while the imaging chip is still taking an image, you cannot change the guide exposure duration. A new exposure duration would require a new dark frame, and that can't happen since the shutter is open.

> **TIP:** If the star is lost due to excessive movement of the mount, a new star could wander into the autoguiding window. If this happens, guiding will resume but with the wrong star! You will wind up with a double image. If you see two of everything, a lost guide star replaced by a new star is the likely cause of the problem. However, double images can also occur when the aggressiveness setting is too high. With corrections that are too aggressive, the mount overshoots the target and moves between two extreme positions; the guide star never returns to center. This is called oscillation. Observe the behavior of the guide star during a guide session to determine the cause of double images.

If the CCD camera has a shutter, it will be closed when an image is being downloaded from the camera. Autoguiding is suspended while the shutter is closed, and will resume after the download is complete. If your

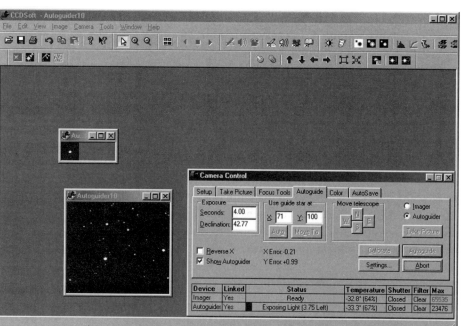

FIGURE 5.4.1. THE APPEARANCE OF CCDSOFT WHEN AUTOGUIDING IS ACTIVE.

polar alignment is good, the guide star will still be visible after image download, and autoguiding will resume. You can, as described earlier, enter a delay value in your exposures to allow for any corrections that happen right after the guide star is reacquired.

Recalibration and Guide Errors

When should you recalibrate? The simplest way to tell is to start an autoguiding session and observe what happens. Since you can autoguide separately from imaging, you can simply start an autoguiding session whenever you move the telescope to a new position. Observe what happens to the X error and Y error numbers. If the both stay within +/- 1.0 pixel, you will likely have decent results. If both stay within +/- 0.5 pixel, then the seeing is very good, and you should get excellent results. If one or both error values show a value greater than +/- 1.0, or won't settle down to the values you've been getting, either the seeing is poor, or you need to recalibrate for the new position. The need to recalibrate varies greatly from mount to mount, and only experience will show you what works best for your particular mount. The same goes for how long to observe guiding. If you mount has large, frequent periodic or non-periodic errors, you might need to observe a non-imaging guiding session for several minutes to evaluate how well it is going to work for you.

Sometimes, autoguiding will not work as you expect it to. Here are some common problems and their solutions.

Problem: No corrections occur, and the X and/or Y Error numbers continue to increase

Solution: This indicates that guide corrections are not getting to the mount. Check the Advanced Settings dialog, and make sure that corrections are enabled. Verify that the autoguide cable is connected at both ends. If you have a second cable, try it; the cable could be bad. Verify that the guide speed is set to an appropriate value. You can test connections by setting a slewing speed instead of a guide speed, such as 12X or 64x or whatever is available) and observe if the mount moves when you press the N/S/E/W buttons in the Autoguide tab. Make sure aggressiveness isn't too low.

Problem: Corrections tend to oscillate between positive and negative values (e.g., +1.3 and -0.9).

Solution: If the load on your mount is too carefully balanced, the mount will be unusually sensitive to guide corrections, and may overshoot. You want a slight but definite excess of weight on the east side. This keeps the gears meshed, and reduces the tendency of the mount to over-react to corrections. Oscillations can also occur when working near the zenith. You can reduce the aggressiveness setting to see if that will eliminate the oscillation. Sometimes the mount will simply have a resonant period similar to the exposure duration and certain guide speeds; try a shorter or longer guide exposure length or change the guide speed.

Problem: The mount suddenly makes large movements, and then returns to center as corrections occur.

Solution: A "large movement" is any sudden movement of one or more pixels. This usually results from irregularities in the mount's bearings or gears, and may limit the success you will be able to achieve with autoguiding. Shorter exposures can help by allowing you to complete a guide interval during a large movement. Contact your mount manufacturer to see if you might need to replace a worn or damaged part. Dirt or grit in the gears can also cause this problem.

If it's windy, you may not be able to guide or track effectively because the mount will bounce around. The larger and/or longer your telescope, the more likely it is to be pushed around by the wind. These deflections will show up as guiding errors. In some cases, if the movements are not extreme, you may be able to guide successfully by shortening the guide exposure so that the camera control software can detect these movements sooner and correct for them more often.

Problem: The guide star suddenly disappears.

Solution: The most common cause for this is clouds, trees, or other obstacles. Guide exposures are short enough that many stars will simply disappear if even a thin cloud moves through. If your mount is well-aligned to the pole, and tracks accurately, you can sometimes simply wait out the cloud and guiding will resume when it passes. Any exposure you were taking during the time the cloud was in the way may or may not be salvageable due to a variety of issues (bright background; tracking errors; etc.). Other obstacles are more often fatal, such as tree and buildings.

The guide star can also disappear if the mount is disturbed, in which case you may have lost your polar alignment. You might have touched your mount accidentally, or a cable may have caught on something and pinned the mount so it cannot move. Even a stray cat rubbing against a tripod leg has been known to cause trouble!

Problem: Corrections occur, but the X and Y errors are large and random

Solution: This could be a sign of over correction. The first thing to try is a lower aggressiveness setting. Try lowering the aggressiveness one unit at a time and see if you get any improvement. Stop lowering the aggressiveness when you get into the range of +/- 1.0 pixels (average seeing) to +/- 0.5 pixel (good seeing). You may also simply need to recalibrate the autoguider. Problems with your mount's bearings or gears are another potential source of trouble; consider tuning your mount if you get wild changes in the X and Y errors and simply cannot make them go away. For any given mount, there is a limit to how accurately it will track and autoguide. If tuning and/or service won't improve your results, you may need to upgrade your mount to work at your chosen focal length.

Problem: The guide star keeps moving in a specific direction.

Solution: If the move is always in declination, this is most likely the result of a polar misalignment that is fairly large (more than .5 degree). If the autoguider is able to handle the movement in declination and keep it within a trouble-free range, the only concern is field rotation for long images (varies with how much misalignment you have) or sequences of images. If the autoguider fails to make the corrections and keeps the error in a reasonable range, you can usually solve this problem by refining your polar alignment. See the section on drift alignment later in this chapter.

Problem: The guide star wanders slowly back and forth.

Solution: This happens in Right Ascension or Declination. In RA, you are usually seeing the periodic error of your mount. Increase aggressiveness until the problem is solved, or use PEC (see below). In Dec, this is usually an oscillation (see suggestions above). If your camera is oriented with North up, RA will be a side-to-side motion in the guide window. If the camera is turned 90 degrees (east or west up), then the RA motion will be up and down. Periodic error usually results in slow, steady changes that go back and forth, so it is often relatively easy for the autoguider to compensate. Some mounts have more extreme error, and you may need to use shorter guide intervals to make adequate corrections. If your mount has the ability to record periodic error and then compensate for it (usually called PEC, for Periodic Error Correction), you can use it or not, as you wish. It will not harm the performance of the autoguider if PEC and the autoguider are operating at the same time. If the autoguider is unable to compensate for periodic error on its own, use PEC to get more accurate guiding. You can use the autoguider to feed corrections to your mount when recording periodic error.

Auoguiding Possibilities

Guiding allows you to lengthen your exposures so you can take deeper, more detailed images of celestial objects. The accuracy of your mount determines how long, and therefore how deep, you can go. The more accurate your mount, the easier it will be to take long exposures.

Some mount manufacturers specify the accuracy of their mounts. This accuracy is expressed in arcseconds of periodic error. An exceptional mount can track with an accuracy of 2-3 arcseconds. A very good mount can track to about 5-6 arcseconds. A good mount will come in around 8-12 arcseconds. The average low-cost and barely photo-capable mount tracks to within 15-20 arcseconds. Mounts with even less accuracy are usually unsuitable for CCD imaging unless the focal length of the telescope or lens is exceptionally short. Guiding can improve on these performance levels.

Whatever the accuracy of your mount, the key to success is matching the focal length of your telescope to the abilities of the mount. For example, consider a CCD detector with pixels that are 9 microns square, such as the ST-7E. The angle on the sky covered by each pixel varies with the focal length of your telescope. Table 5.3 shows the relationship between focal length and image scale for an SBIG ST-7E CCD camera.

Table 5.3: ST-9E Image Scale

Focal Length of Telescope	Angle of Sky, in arcseconds
300	6.2
500	3.7
700	2.6
900	2.1
1100	1.7

If you were imaging with a 300mm camera lens, for example, you could easily tolerate 6 arcseconds of error because a single pixel covers just over 6 arcseconds. If you were imaging with a scope that has an 1100mm focal length, a single pixel would cover only 1.7 arcseconds, and the need for pointing accuracy is much higher -- almost four times higher. Guiding may give you the additional accuracy provided your mount's errors aren't too sudden, or too extreme.

In actual practice, star images cover more than one pixel due to atmospheric scattering, the same process that causes stars to twinkle, so you would have more leeway. The important concept is that the shorter your focal length, the less the need for pointing accuracy in your mount.

A short focal length alone won't solve all problems, however. Smaller or larger pixels on your CCD detector alter the numbers. As shown in Table 5.4, large pixels cover more sky for a given focal length, and smaller pixels cover less sky.

But image scale (arcseconds per pixel of sky coverage) isn't the only issue. With today's small-pixel guiders, such as the STV, focal ratio also plays an important role. A fast focal ratio delivers brighter images to the guider. This provides more potential guide stars, and allows you to use a shorter focal length. The shorter focal length covers a larger portion of the sky, also increasing the number of available guide stars.

Experience shows that, to get excellent results, your mount error when PEC is running should be no more than +/- 2 pixels. Larger errors can often be corrected, but the risk of trouble increases with the error size. For high-quality images, one and a half times the arcseconds per pixel value is a safer limit. For very long focal lengths (2500mm and up), local seeing and other conditions become more important and the calculations are not so straightforward; experience of local conditions is needed to assess the requirements for guiding.

The smoothness of the periodic error (PE) of your mount has an affect on these calculations. If the PE is very smooth, correction is more effective and larger periodic error can be handled. If the PE is rough or sudden, corrections are less effective and less periodic error can be tolerated.

Table 5.4: Focal Length, Pixel Size, and Image Scale Comparison

ST-9E (20-micron pixels)		ST-237 (6.8-micron pixels)	
Focal Length of Telescope	Angle of Sky, in arcseconds	Focal Length of Telescope	Angle of Sky, in arcseconds
800	5.1	300	4.7
1000	4.1	500	2.8
1200	3.4	700	2.0
1500	2.7	900	1.5
2000	2.0	1100	1.3

TIP: If your mount isn't providing the pointing accuracy needed by your telescope, you can either use a focal reducer to shorten the focal length of your scope, or take steps to increase the pointing accuracy of your mount, or both.

You can assess the effect of periodic error for a given mount, camera, and telescope combination quite easily. Simply misalign the mount by a significant amount (around 5 degrees), and take an image that is approximately the length of time of a single revolution of the worm gear. You can get the worm rotation time from the mount documentation or the manufacturer.

Figure 5.4.2 shows an example of a mount that guides very accurately, a Wm. Optics GT-ONE mount. The camera was an ST-8E camera on a Taka-

FIGURE 5.4.2. AN EXAMPLE OF A MOUNT/SCOPE/CAMERA COMBINATION THAT PROVIDES MINIMAL PERIODIC ERROR.

hashi FSQ-106 telescope. The mount was seriously misaligned from the pole, and a six-minute exposure taken. Note that there is very little variation from a straight line in the star trails, and the variations that do exist are smooth and slow.

Figure 5.4.3 shows a different situation. This mount, a Losmandy G-11, has frequent variations (non-periodic error) as well as a much larger periodic error. The mount from the first example could be used effectively for guided or unguided exposures. The mount from this second example would not be suitable for unguided exposures because the movements of the mount are too large and too frequent. However, this mount works reasonably well for imaging when a guider is used because the errors, while significant, are small enough and slow enough to be corrected. The random errors are fast enough to require short guide exposures, probably under 5 seconds.

FIGURE 5.4.3. A MOUNT/SCOPE/CAMERA COMBINATION WITH LARGER ERROR.

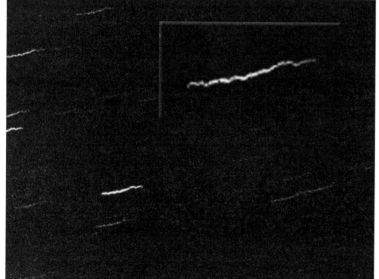

Adjusting and Tuning Your Mount

Mounts come in a variety of designs. Most mounts intended for use with film and CCD cameras involve a worm and worm gear, as shown in figure 5.4.4. There is normally one worm for Right Ascension, and another for Declination, but some excellent mounts are only driven in RA, such as the Takahashi P2Z.

A motor drives the worm gear, and there will usually be gears or belts between the motor and the worm to reduce the speed of rotation at the worm. The worm in turn drives a gear which is attached to a shaft. The RA shaft turns at sidereal rate, which is the rate at which the stars appear to rotate (approximately one revolution per day). The motor speed and gear ratios are set to track accurately at the sidereal rate. Some mounts provide a way to adjust the tracking rate, but many do not.

To track and guide accurately, a worm-driven mount must be adjusted or tuned. Before using a mount for guiding, I suggest the following procedures. Please consult your mount's documentation or the manufacturer for specific procedures on making these adjustments to your mount. Not all procedures will apply to all mounts.

Some manufacturer warranties may become void if you attempt these adjustments yourself. Check your documentation or check with the manufacturer if you have any doubts.

Things you can do to tune up your mount:

- Verify that each worm is properly seated. One end of the worm shaft often sits up against a non-adjustable stop; the other end is secured with some kind of lock nut. If the worm is held too tightly, it will bind and could be damaged over time. If the worm is too loose, the worm shaft will have end play. This results in movement that will cause guide corrections to be incorrect. Even a little end play can result in nasty problems when reversing direction. If you observe that the mount initially and briefly moves in the wrong direction when you are changing direction, end play in the worm is the usual cause. Please use caution in tightening the worm; you don't want to do any damage by over-tightening. The manufacturer of the mount may have specs on how tight it should be.

- Verify that the mesh between worm and gear is appropriate. Some mounts are made to closer tolerances, and such mounts will typically benefit from a fairly tight mesh between the worm and gear. Other mounts are made of softer metals, or to looser tolerances, and will require a looser mesh. If the mesh is too tight, the gears will bind and could overheat or become damaged. If the mesh is too loose, there will be excessive backlash. You always need at least a small amount of backlash; the amount varies from mount to mount. Some mounts have built-in backlash compensation. Many camera control programs provide software backlash compensation that serves the same purpose. During backlash compensation, the motor is run at a higher speed to take up slack in the gears. Too much compensation will result in a jerky movement; too little compensation will result in a delay before a guiding correction is effective when reversing directions. It is always better to have too little rather than too much backlash compensation. If in doubt about how tight to mesh the gears, err on the side of being too loose to avoid damage.

FIGURE 5.4.4. AN EXAMPLE OF A WORM AND WORM GEAR.

- A slow-turning motor usually has less torque (that is, less force available to move the mount). This means that a slowly turning motor is more likely to stall than one that is turning quickly. If the worm and worm gear are too tightly meshed, the slow-down to .5X sidereal that occurs during autoguiding could cause the mount to stall, or move at the wrong speed, or move jerkily as it tries to make up for missed steps.

- Very cold conditions can cause some lubricants to become stiff, and this can alter the behavior of your mount. Cold weather can also cause shrinkage of critical parts, such as motor bearing points, creating a loose condition that can causes sudden excessive sloppiness in tracking and guiding. If you need to operate your mount in very cold conditions, or if you discover increased stiffness or odd behavior in cold weather, check with the manufacturer to find out whether you will need to make modifications for extremely cold conditions.

- Some mounts will still have a large residual error even after tuning, adjustment, and PEC. A large error limits the focal length scope you can use on the mount. The larger the error (from periodic and non-periodic sources), the shorter the focal length of the scope you can use for imaging. The actual limit depends on the weight of your scope and its overall physical length as well as its focal length. Long, heavy scopes require more careful balancing and tuning for successful guiding. Focal reducers are available to help with this situation. You can also piggy-back a CCD camera with a camera lens to achieve ultra-short focal lengths.

- Test the balance of your mount in various positions. Off-center equipment, such as a finder offset from the centerline of your scope, or a camera attached piggyback fashion beside your scope, can dynamically alter the balance of your mount as you point at different areas of the sky. A mount should always be loaded so that there is slightly more weight on the east side of the mount. This keeps the Right Ascension gear train loaded at all times, eliminating backlash as a factor in RA during autoguiding (as long as the correction speed is less than the sidereal rate). If off-center equipment does cause the balance to change at different pointing angles, you can either rearrange equipment and/or add counterweights after moving the scope a significant angular distance, or you can incorporate sliding weights or some other weight-shifting system into the design of your setup. An out of balance mount can oscillate or wander unpredictably, or require a new calibration when pointing to different parts of the sky even if the declination remains the same.

- You can calibrate your setup at a higher speed than you actually use for autoguiding. This tends to mask some errors, but it isn't a cure-all. It slows down the correction process. If your mount tends to be over-responsive, it can eliminate unnecessary corrections and provide smoother autoguiding. This technique is most often useful with scopes that have focal lengths over 1500mm.

More than any other single factor, the ability of your mount to track accurately and to respond quickly and smoothly to guiding corrections determines the success of your imaging. It is well worth measuring the accuracy of your mount, and tuning and adjusting it to improve that accuracy to the greatest degree possible.

Polar Alignment

In order to track accurately, your mount needs to be aligned close to the celestial pole. Many mounts include a polar scope (also called a bore scope or bore sight) that helps you align the mount accurately. There are many different kinds of polar scopes, but most contain some kind of reticle that allow you to point the mount at specific stars to achieve alignment. Most mounts also contain provisions for aligning in either the northern or southern hemispheres.

Although you do not need perfect polar alignment if you are using autoguiding, there are some issues to be aware of with respect to polar alignment. If you are taking single images of less than five minutes, even a rough polar alignment may work if you are guiding. The lack of a perfect alignment will cause the field of view to drift and rotate slowly during the exposure. The autoguider will detect the drift portion and correct for it. The field rotation is small when reasonably well aligned to the pole, but if the exposure is long enough, field rotation will affect your images.

If the misalignment is large, the autoguider will make large, frequent corrections. The likelihood of trouble increases when this happens because the amount of correction is larger for each correction. A big change could show up in your image. The ideal correction is always no correction. The larger the misalignment, the shorter your guide exposures must be to compensate. A long focal length increases the need for precise polar alignment.

If your exposures are longer than five minutes, or if you are combining images for better signal to noise ratio or color, field rotation becomes a more important issue. You can minimize field rotation by taking the time to polar align your mount very carefully. Use the mount's polar scope or goto features for rough alignment, and then use drift alignment or a product like TPoint to get as close as you can to the celestial pole. You can do drift alignment manually, or you can use the CCD camera to assist. T-Point, a Software Bisque software tool that increases telescope pointing accuracy, also assists in polar alignment with many goto mounts. T-Point is most often used for fixed rather than portable installations, but it works for both. I use TPoint to get extremely accurat polar alignments, within an arcminute of the exact celestial pole.

Manual Drift Alignment

1. Level the base of the mount if possible. A level mount is not required, but it makes adjustment easier by eliminating interplay between altitude/azimuth adjustments and RA Dec adjustments.

2. Locate a bright star near the celestial equator and near the meridian (the midpoint between east and west). Center it in the crosshairs of an illuminated reticle eyepiece.

3. Observe the star until it drifts north or south; ignore drift to the east or west. If the star drifts north, adjust the polar axis azimuth so it points more to the east. If the star drifts south, adjust the polar axis azimuth so it points more to the west. Repeat until north/south drift becomes negligible over a 5 minute observing period (or longer if you require it).

4. Locate a bright star near the celestial equator and near either the east or west horizon. Center it in the crosshairs of an illuminated reticle eyepiece.

5. Observe the star until it drifts north or south; ignore drift to the east or west.

   If you are looking east:
   - If and the star drifts north, adjust the polar axis altitude downward.
   - If the star drifts south, adjust the polar axis altitude upward.

   If you are looking west:
   - If the star drifts north, adjust the polar axis altitude upward.
   - If the star drifts south, adjust the polar axis altitude downward.

6. Repeat until north/south drift becomes negligible over a 5 minute observing period.

Note: To determine which way is north in the eyepiece, move the telescope toward the south and note the direction the stars move; they are moving toward the north.

FIGURE 5.4.5. CALIBRATION RESULTS THAT SHOW THE CAMERA IS READY TO ASSIST WITH DRIFT ALIGNMENT.

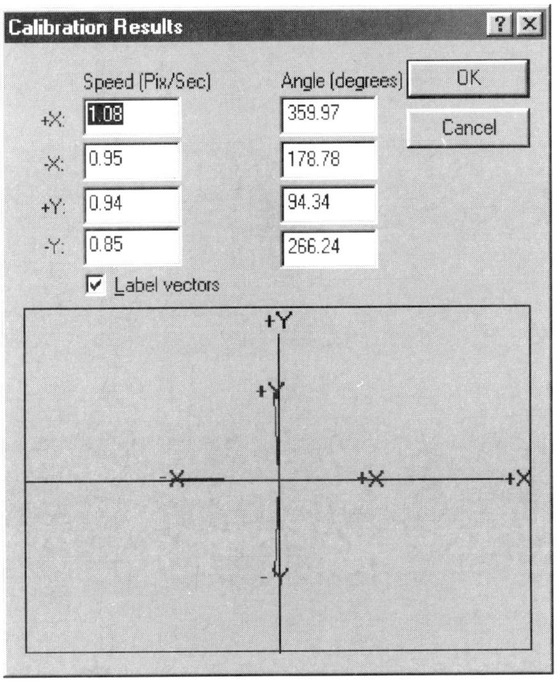

Camera-Assisted Drift Alignment (CCDSoft)

1. Calibrate your autoguider as described earlier in this chapter.
2. Verify that the camera is square (orthogonal) to the mount. On the Autoguide tab of the Camera Control Panel, click on the Settings button. This opens the Autoguide Settings dialog. Click the Calibration Results button. This displays a graph showing how the camera is oriented with respect to the RA and declination axes of the mount. Ideally, the X and Y axes of the camera will line up closely with the X and Y axes of the mount (see figure 5.4.5). The +X angle should be close to zero, and the +Y axis should be close to 90. If your camera is rotated with respect to the mount axes, it will be more difficult to get an accurate reading (see figure 5.4.6).
3. If the camera isn't square with the mount, adjust the camera and repeat calibration, then check again. Repeat until you've got everything lined up to your satisfaction.
4. Point the telescope at a star, chosen using the same criteria as for manual drift alignment.
5. Using the Autoguide tab, set an appropriate exposure time (e.g., 5-10 seconds), and click on the star.
6. Click the Settings button. If the advanced settings are not visible in the Settings dialog that pops up, click the "Advanced >>" button. Click the checkbox "Log Tracking Data." This allows you to examine your data later if you have any questions about the progress you are making toward polar alignment.
7. To measure drift, turn off corrections. Still in the Settings dialog, uncheck the "Yes" checkboxes for enabling the X and Y axes. Remember to re-enable corrections when you are done! Click OK to close the Settings dialog.
8. Click the Autoguide button on the Autoguide tab. You can now observe the Y (declination) error to measure drift accurately and quickly. You can examine the log file at any time to determine the total amount of drift versus elapsed time.
9. Once you have a measurement of the current amount of drift north or south, adjust the mount as for manual drift alignment and repeat the autoguiding process for the other axis.

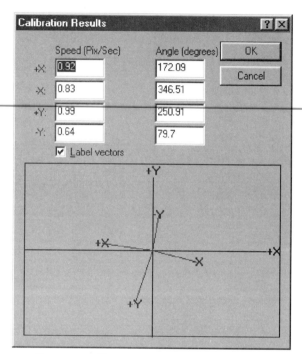

FIGURE 5.4.6. THE CAMERA AXES ARE NOT SQUARE TO THE MOUNT AXES, AND THE X AND Y VALUES OF THE MOUNT AND CAMERA ARE OPPOSITE. THE CAMERA IS NOT WELL SITUATED FOR ASSISTING WITH DRIFT ALIGNMENT.

10. For optimal results, continue drift aligning until the mount tracks for several minutes with a Y error no larger than +/- 1.0. For critical applications, continue until the mount tracks for 10-30 minutes with a minimal change in Y error.

Examining Autoguiding Data in a Spreadsheet

If your camera control software supports logging of tracking information, you can open the log file in a spreadsheet program to analyze guiding. The following example uses the log file from CCDSoft, but you can do the same thing with any log file. The columns may vary, however, so check the documentation to find out which columns are included and in what order.

Note: If you turn off corrections during autoguiding, the log file will record the periodic error of your mount. No corrections will be made, so the guide star could drift out of the field of view. Let autoguiding without corrections proceed for one full revolution of the RA worm. This is usually in the range of four to

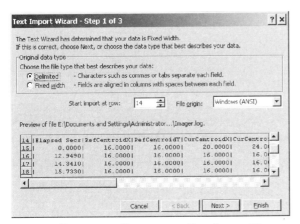

Figure 5.4.7. Step 1 of the Text Import Wizard.

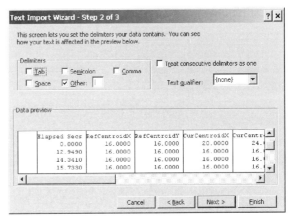

Figure 5.4.8. Step 2 of the Text Import Wizard.

seven minutes. Consult your mount documentation or contact the mount manufacturer to get this number. To turn off corrections, click the Settings button. If the advanced settings are not visible in the Settings dialog that pops up, click the Advanced >> button. Uncheck the "Yes" checkboxes to disable the X and Y axes.

To examine the data, open the log file in a spreadsheet. In Excel, use the File | Open menu item. Excel will automatically show the Text Import Wizard, a three-step process for reading the data into Excel.

1. In step 1 of the wizard (see figure 5.4.7), select Delimited as the original data type. Start import at row 14, and use Windows (ANSI) as the File origin setting. The data appears in preview form at the bottom of the import wizard. Click Next.

2. In step 2 of the wizard, choose Other as the delimiter type. Enter the vertical bar (|) as the delimiter, and set the Text qualifier as **none,** as shown in figure 5.4.8. The data columns reformat when you choose the correct delimiter character, showing solid black dividers between the columns of data instead of the vertical bar character. Click Next.

3. Step 3 of the wizard allows you to set a data format for each column. This is not necessary for the log data, so click the Finish button, and the data appears in an Excel spreadsheet.

To visualize the data, create a chart that graphs the X and Y errors. The columns you pick for graphing can vary. To visualize your mount's periodic error, for example, select the GuideErrX column (guiding error in Right Ascension if north is up). To compare the guiding errors for both X and Y directions during a typical guiding session (with corrections turned on), select both the GuideErrX and GuideErrY columns. To evaluate the actual guide error distance from the centroid of the originally selected guide star, choose the TotGuideErr column. Figure 5.4.9 shows an Excel graph of the X and Y errors with corrections turned off. This mount was troublesome, with large errors. Note the angle of the lines, which trend generally down as you move toward the right. This is a result of polar misalignment, and is not caused by mount error.

Figure 5.4.9. Graphing X and Y errors in Excel.

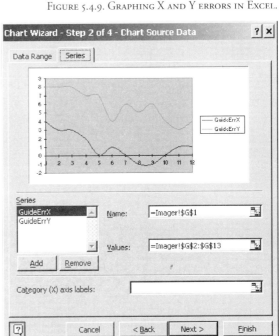

6 Increasing Image Quality

CCD cameras are sensitive devices. They respond to nearly every photon that strikes their surface.

This is very different from film, which wastes a large percentage of the photons that contact the emulsion. Some CCD chips register as many as four out of every five photons that strike them. This incredible sensitivity is part of what makes CCD cameras so valuable as research and imaging tools.

Section 1: Image Reduction

If there is a downside to CCD chips as light detectors, it is noise. The sensitivity of CCD cameras applies to unwanted photons as well as those originating deep in space. Not only are CCD cameras more sensitive to stray light, they are also very sensitive to photons outside the visual range. This includes photons radiating as heat inside the camera.

The last bit of good news is that the CCD cameras used in astrophotography use various schemes to reduce noise. There are also steps that you, as the photographer, can take to keep noise to a minimum. The techniques covered in this chapter will help you to reduce many sources of noise.

Figure 6.1.1 shows a raw and rather noisy-looking CCD image of Stephan's Quintet. This image, taken with the shutter open, is called a light frame. While figure 6.1.1 is not the most aesthetic image you'll ever see of the Quintet, it's a good example of how much noise you might find when you download an image from a CCD camera. The noise seems to overwhelm the image. If this were the only image you ever saw taken with a CCD camera, you might despair of getting a good image!

The noise can be removed from the image using a very simple technique called dark frame subtraction. Figure 6.1.2 shows the result. The heavy graininess is suddenly gone, and the dim galaxies of the Quintet show up clearly. Even faint details in the outer areas of the galaxies are visible.

Some CCD chips are noisier than others, but noise reduction is an important task in all types of CCD imaging. Fortunately, much of the noise in an image is predictable. This is called system noise, and you can remove a great deal of that kind of noise from your CCD images. True noise is random fluctuations that cannot be removed, so the focus is on removing system noise.

System noise comes from many sources, including the camera itself. Thermal noise tends to dominate, but other forms of noise occur as well, such as noise from reading the camera pixels so they can be downloaded to your computer.

Figure 6.1.1. A raw CCD image of some of the galaxies in Stephan's Quintet, showing a lot of system noise. Not all raw images are this noisy!

Most CCD cameras include some form of cooling to minimize the problems from thermal noise. At a given temperature the thermal noise is fairly predictable. To successfully remove thermal noise, accurate regulation of temperature is required. If the amount of cooling changes, you lose the ability to accurately compensate for thermal noise.

As if all of this isn't enough to worry about and deal with, the pixels on CCD chips do not respond uniformly to light. Some pixels are more receptive than others, other less receptive. Some pixels are just plain hot, and will always register as bright, even if no light is striking the chip. The behavior of each pixel is reasonably predictable. You can record pixel behavior by taking an exposure with the shutter closed. This is called a dark frame. No light is striking the CCD detector during a dark frame. Hot pixels will still be bright even without light. Thermal noise, consisting of infrared photons, will generate a certain amount of brightness in each pixel. The dark frame is a signature of the chip's behavior at a specific temperature and for a specific exposure duration.

FIGURE 6.1.2. THE APPEARANCE OF FIGURE 6.1.1 AFTER THE SYSTEM NOISE HAS BEEN REMOVED FROM THE IMAGE. EVEN A VERY NOISY CCD IMAGE CAN OFTEN BE SAVED.

For example, a given pixel might have a brightness of 500 after a 5-minute dark frame exposure at -15C. That same pixel might have a brightness of 900 after a 10-minute dark frame exposure at the same temperature. Colder temperatures and shorter exposures result in lower values for each pixel, and thus less system noise. Warmer temperatures and longer exposures result in higher values, and thus more system noise in the dark frame.

Since system noise can be almost completely subtracted from a light frame, the noise from longer exposures isn't a big deal. However, there is always some variability from one frame to the next, so cooling the camera to reduce thermal noise is always a good idea.

Dark frame subtraction is the most common and most important step you can take to improve the quality of your images. I recommend taking and using dark frames for all of your images.

System noise isn't the only noise issue you have to deal with. Your optical system may also introduce its own set of problems. Dust finds its way onto the camera's optical surfaces, as well as onto any filters you are using. This dust casts shadows on the CCD chip. Your telescope may also have some degree of vignetting. This is a darkening of the outside edges of the image compared to the middle. These kinds of problems are also predictable, and you can mathematically remove them from your images using a flat-field frame.

There is also a third type of frame, a bias frame, which you can use to get some flexibility with your dark frames. You'll learn more about bias frames after I take some time to explain dark frames. Figure 6.1.3 shows examples of all three types of reduction frames. The flat field has its characteristic brightness variations and dust shadows. The dark frame shows the effects of thermal noise, and the bias frame is a very brief version of a dark frame, used only to scale dark frames.

In this chapter, you'll learn:

- More about the common types of noise you encounter with a CCD camera.
- How to create and use dark frames.
- How to create and use flat-field frames.
- How to create and use bias frames.
- How to cut down the overall noise in your images.

CCD Chips Explained

Any discussion involving CCD imaging quickly starts to involve various technical terms, such as dark frames, flat fields, and so on. After you have been imaging for a while, these terms become second nature, but it is not at all obvious what's behind the terminology.

The need for a dark frame begins with the nature of CCD chips. CCD stands for "charge-coupled device." In technical terms, this is a semiconductor array that uses MOS (metallic-oxide semiconductor) technology, depletion storage, and data transfer by register shifting.

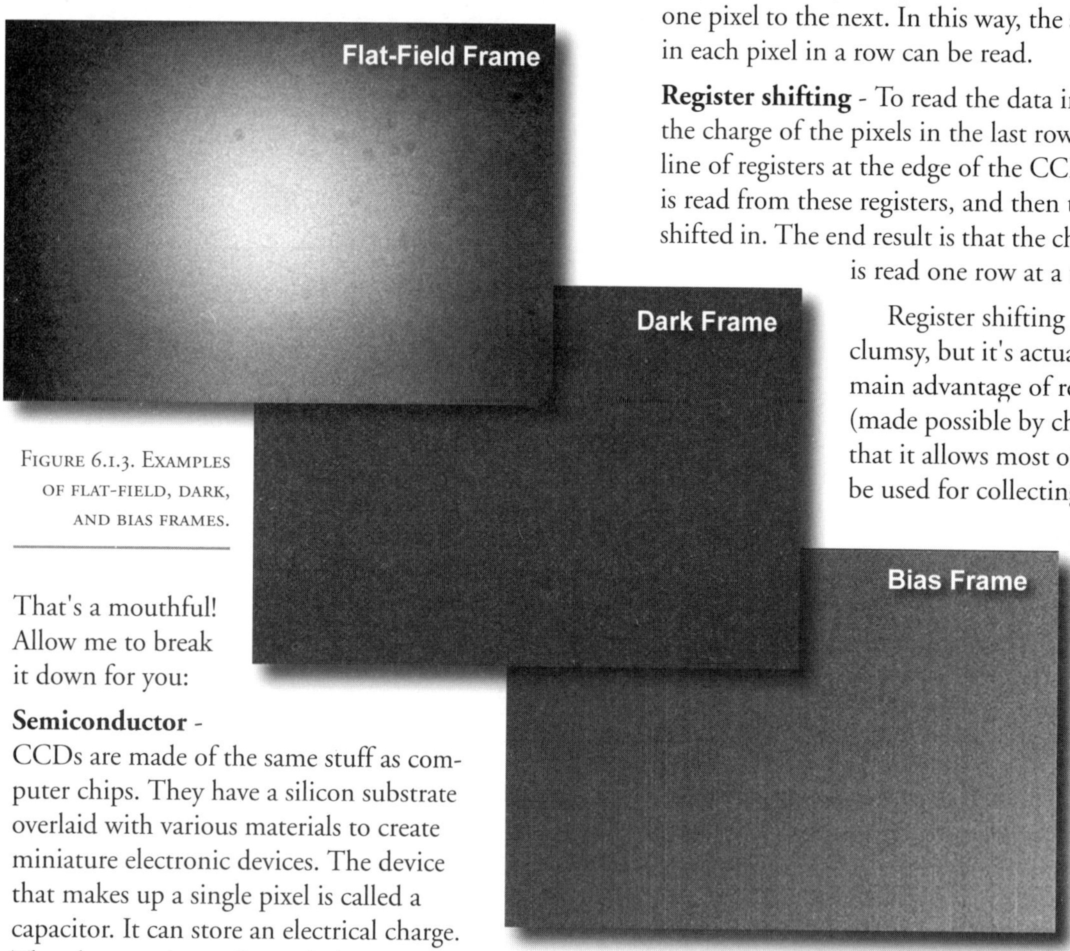

FIGURE 6.1.3. EXAMPLES OF FLAT-FIELD, DARK, AND BIAS FRAMES.

That's a mouthful! Allow me to break it down for you:

Semiconductor - CCDs are made of the same stuff as computer chips. They have a silicon substrate overlaid with various materials to create miniature electronic devices. The device that makes up a single pixel is called a capacitor. It can store an electrical charge. The photons that strike a pixel are converted into electrons, which are stored in the capacitor.

MOS technology - Metal oxides are applied to the silicon substrate to create the various electronic devices you find on a chip. This is one type of chip design; there are others (such as CMOS) that are variations on the same basic idea.

Depletion storage - As photons strike the CCD chip, most of them get converted to electrons. Electrons have a negative electrical charge, so the pixel/capacitor assumes a negative (depleted) charge proportional to the amount of light striking it. More light means more negative charge.

Data transfer - The negative charge won't do anyone any good sitting in the pixel/capacitor; we need a way to get it out of there. CCDs are referred to as "charge coupled" because the contents of one pixel can be transferred to an adjoining pixel. Adjacent pixels are coupled in a way that permits the charge to move from one pixel to the next. In this way, the amount of charge in each pixel in a row can be read.

Register shifting - To read the data in a row of pixels, the charge of the pixels in the last row is moved into a line of registers at the edge of the CCD chip. The data is read from these registers, and then the next row is shifted in. The end result is that the charge in the pixels is read one row at a time.

Register shifting might seem clumsy, but it's actually very fast. The main advantage of register shifting (made possible by charge coupling) is that it allows most of the pixel area to be used for collecting photons. A more complex method of reading pixels directly would require more circuits, thus more layers of metal oxides that would obscure some part of the pixel.

TIP: Cameras with an anti-blooming feature are less sensitive than standard CCD cameras. The circuits for anti-blooming features cover a portion of each pixel (around 30%) and block that portion of the pixel from receiving light.

Figure 6.1.4 shows the general layout of a hypothetical CCD chip with an imaging area of just 8 pixels by 8 pixels. The CCD chip has a simple image of a single star, and I have used varying brightness levels to show the amount of charge in each pixel. The brighter the pixel, the more charge it has in it. "More charge" of course means "more electrons."

There is a register row at the bottom of the CCD chip. This is where pixel information is transferred during readout, one row at a time. The lone pixel at bottom right is where the actual data is read, one register pixel at a time. The register pixels are not used for imaging, only for reading out the data in the chip.

Figure 6.1.5 shows the process of reading out the first row of pixels, which is the bottom row of the CCD chip. Starting at the top left, there are 11 steps in the process of reading out a single row of data. In step 1, the CCD chip has an image in it, and the register row and the output pixel are empty.

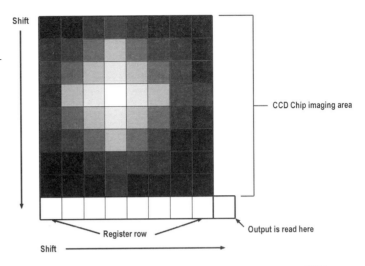

FIGURE 6.1.4. THE GENERAL ARCHITECTURE OF A TYPICAL CCD CHIP.

Moving left to right across the top of the figure, in step 2 the pixel contents have been shifted down by one row. The top row is empty, and the eight electrical charges from the bottom row of the image area are now in the register row.

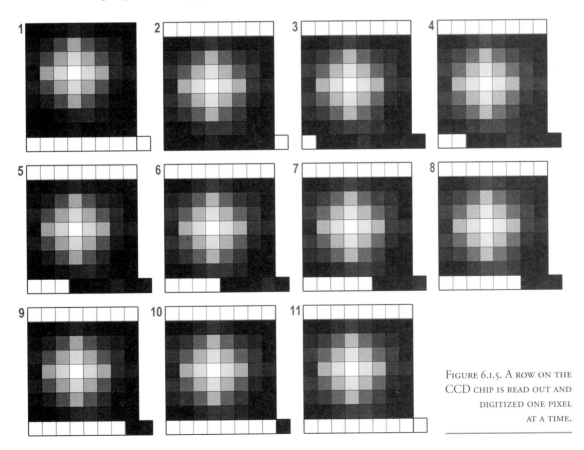

FIGURE 6.1.5. A ROW ON THE CCD CHIP IS READ OUT AND DIGITIZED ONE PIXEL AT A TIME.

TIP: All of this shifting is made possible because the pixel/capacitors are charge-coupled. This means that the charge in one pixel can be moved into its neighbor's depletion storage area. This is the key to how CCDs maintain a large image area relative to circuitry. Each pixel is an electronic connection and a miniature well that can hold a charge.

The third step shows the register row shifted one pixel to the right. The rightmost value is now in the output pixel, and can be read by the camera's electronics. Step four shows the next shift, which moves yet another different pixel's charge into the output pixel, where it can be converted into a digital value and sent on. Step five shows the third pixel being read, and so on. In step 10 the last pixel in the row is being read, and in step 11 all pixels in the row have been read.

Figure 6.1.6 shows the next steps in the progression. In step 1, the rows have been shifted downward again. Each pixel will now be read from the row, as in figure 6.1.5. Step 2 shows the third row shifted into the readout row, and the process continues until all rows have been read out.

At the end of the process, each pixel in each row of the chip has been read and converted to a digital value. The values are passed on to your computer, where they are assembled into an image.

The process of reading the charges out of the CCD chip sounds complicated, but it happens quickly and smoothly. CCD chips are also used in video cameras, where they are read out at the rate of sixty frames per second. The actual electronics in a video camera are different from an astro CCD camera, but the basic CCD chip principles are very similar. A CCD camera intended for astronomical use typically reads the values more slowly. The process of reading out the images introduces a small amount of noise into the image. Slowing down the readout process reduces the readout noise.

The science of CCD detectors runs much deeper than the simplified description above, but this should give you a basic idea of what goes on inside the camera.

Dark Frames Explained

Every CCD imager should have a bumper sticker that reads, "Noise happens." It's the reality of CCD imaging, and most of what you'll learn in this chapter will help you control the level of noise in your images. There are many sources of noise, and in this section we will concentrate on those that are inherent in the CCD chip itself. One can deal with this kind of noise by creating a dark frame, subtracting it from the actual image, and leaving behind the image data.

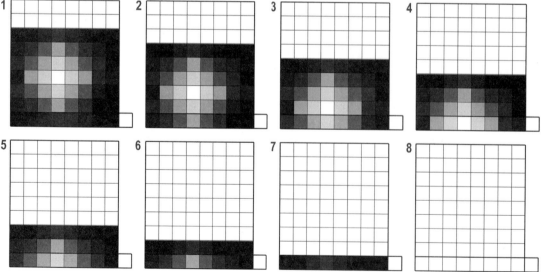

FIGURE 6.1.6. Each row of the CCD chip is read, one after the other, one pixel at a time in each row.

A dark frame happens to be exactly what its name says: an exposure of the CCD chip in darkness -- that is, with the shutter closed. If your camera does not have a shutter, you can take a dark frame by putting on the lens cap, or by holding a hat over the front of the scope. No light reaches the chip during a dark frame exposure. The only charge that accumulates during a dark frame is charge from background noise. The dark frame records it for you.

The dark frame contains nothing but system noise. This is the same noise that accumulates during a light exposure. To remove the noise from the light exposure, you subtract the dark frame from the light frame.

TIP: If you have a light leak in your camera, the dark frame won't be entirely dark. If you see a band of light on one side of the frame, you know that some light is leaking into the camera. A little black electrician's tape can work wonders for this kind of problem, or you can send the camera back to the manufacturer for a little dark adaptation. A small difference in brightness is normal for many cameras, however. If the difference in brightness does not change with the length of the exposure, then it is a normal difference that results from the time difference in reading the data from one end of the CCD chip to the other.

The dark frame is a neat solution to the noise problem, and as figures 6.1.1 and 6.1.2 show, it works. But where does the noise come from in the first place?

The pixels themselves are a source of noise. Some pixels are naturally "hot." Such pixels will read out high values, even when there is no light hitting them. These hot values are recorded in the dark frame, and when you subtract the dark frame, the hot value is removed.

For example, suppose that you are using a CCD camera that yields 16-bit values. This means that a given pixel can have a value from 0 to about 65,000. If this were an ideal world, an exposure with the shutter closed would give you a chip full of zero charge. This won't happen, however, and the dark frame will have all kinds of hot pixels on it.

FIGURE 6.1.7. AN EXAMPLE OF A PERFECTLY GOOD DARK FRAME.

> **TIP:** You can buy a camera with a better-than-average CCD chip in it. Such chips have fewer defects of all kinds. However, such premium chips demand a premium price, and you can wind up paying twice as much for a really clean Class 0 CCD chip. You can overcome all sorts of problems by careful use of dark frames, flat-field frames, and bias frames.

Figure 6.1.7 shows an example of what a dark frame looks like: evenly dark background, with a salted look throughout the frame. The white dots are hotter than average pixels. The distribution of hot and noisy pixels varies across the dark frame. Some areas of the frame have a higher density of noise; some areas have less.

The actual appearance of the dark frame will depend on your camera control software, the temperature of the CCD chip at the time you took the dark frame, the length of the dark frame exposure, and other factors. For example, figure 6.1.8 shows exactly the same dark frame as figure 6.1.7, but with different contrast settings. The data is the same; the way it is being viewed has changed. I suggest viewing your dark frames using the same contrast settings at all times. The default or auto contrast settings are an efficient way to do this. If you see a dark frame that looks alarmingly different, check the contrast settings before you panic.

Figure 6.1.9 shows an example of a dark frame that isn't thoroughly dark. There is a light gradient from top to bottom, with a soft edge toward the bottom. This appearance is characteristic of a minor light leak. If you see this effect in your dark frames, you can often solve the problem by applying electrical tape across any seams. A dark frame with a light leak is particularly nasty and useless if you are doing color imaging. You will wind up with color gradients across your images, which can totally destroy the correct color relationships in an image. The direction of the light leak can vary, as well as the extent and brightness variations caused by the leak. The light leak may only occur when the camera is oriented in a specific direction. I had a light leak problem caused by a neighbor's security light. When my telescope was pointed high to the northeast, the security camera lined up with the joint between the

FIGURE 6.1.8. THE DARK FRAME FROM FIGURE 6.1.7 WITH DIFFERENT BACK AND RANGE SETTINGS.

FIGURE 6.1.9. A 600-SECOND DARK FRAME THAT SHOWS A GRADIENT FROM TOP TO BOTTOM DUE TO A LIGHT LEAK.

cosmic ray. Cosmic ray hits can occur in any type of image: light frame, dark frame, flat-field frame, etc. You can deal with such problems by combining multiple frames, a technique covered later in this chapter.

Cosmic rays are more frequent at higher elevations. At sea level, the atmosphere acts like a cosmic ray sponge, soaking up many of them. At higher elevations, there is less air between you and the cosmos, and you are more likely to see cosmic rays.

The easiest way to show how a dark frame works is to consider what happens to a single pixel. Let's say that in a dark frame with a two-minute exposure, a particular pixel winds up with a value of 150 (out of 65,000 possible values). Then let us further assume that after a two-minute light exposure, that very same pixel winds up with a value of 350. When you subtract the dark frame from the light frame, the pixel will have the value (350 - 150), or 200. This 200 represents the true brightness of the object that you imaged on this pixel, whether it is a star or just sky background.

camera and a color filter wheel, spoiling images and dark frames.

TIP: When you are attaching a color filter wheel to the SBIG cameras (ST-7/8/9 family), you may not get a perfectly tight seal. Tape around the edge is the usual solution, although you could experiment with o-rings or gaskets between the camera and the filter wheel. The o-ring or gasket will increase the distance between the front of the filter wheel and the camera, however, which could affect your ability to come to focus.

Figure 6.1.10 shows a dark frame defect that you will encounter fairly often: a cosmic ray has struck the CCD chip toward the upper left, leaving a small arc shown magnified at right. This isn't a defect in the chip; it's just the chip responding to the energy that was contained in the

FIGURE 6.1.10. A COSMIC RAY CAN SHOW UP IN BOTH DARK AND LIGHT FRAMES.

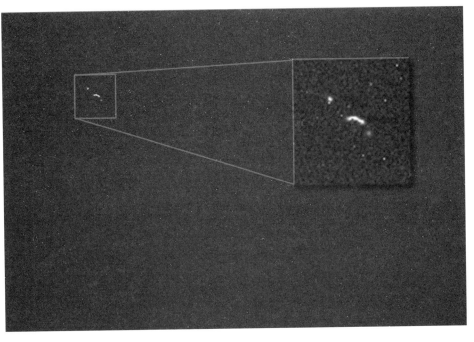

Not every dark frame looks the same as the examples above. Figure 6.1.11 shows a dark frame from a Starlight XPress MX916 camera. Note that the dark frame has a gradient, like the dark frame in figure 6.1.9 -- but in this case, the gradient is normal and not a light leak. It is a characteristic of the chip. That's the whole point behind a dark frame: it represents the reproducible noise on the chip for a specific length of time at a specific temperature. The chip in the MX916 camera isn't very noisy, even after a ten-minute exposure, but it does have some noise and images will benefit from having a dark frame.

Figure 6.1.12 shows yet another kind of gradient that is normal; it's a dark frame from a Starlight XPress MX5-C used with the Star 2000 guiding interface. During guiding, an amplifier that sits at the top left corner of the chip is turned on and off. There are two examples shown, taken with different cameras. Such variations in the appearance of dark frames are to be expected -- the appearance of a dark frame is a unique signature of each chip.

TIP: Some chip locations, and some camera designs, have minor sources of internal light that affect all frames predictably. Dark frames compensate for this problem, since the illumination is minor and repeatable. The illumination slightly reduces the dynamic range of the chip. A longer exposure will show the effect more than a short one. You should always take a

FIGURE 6.1.11. A TEN-MINUTE DARK FRAME FROM A STARLIGHT XPRESS MX916 CAMERA, BINNED 2X2.

dark frame to correct for the brightening if your camera behaves in this manner.

TIP: It's a good idea to compare your dark frames against dark frames from other users of the same camera model to make sure that what you are seeing is within the normal range. This is especially reassuring if you have a chip that tends to have more noise than other types of chips! Noise is only a problem if it can't be removed from the image, or if there is so much of it that it interferes with image quality.

FIGURE 6.1.12. TWO EXAMPLES OF DARK FRAMES FROM AN MX5-C CAMERA. NOTE THAT THERE ARE SLIGHT DIFFERENCES IN THE APPEARANCE OF EACH FRAME, BUT THE GENERAL CHARACTER OF THE FRAMES IS VERY SIMILAR, SHOWING AMPLIFIER GLOW AT TOP LEFT FROM USE OF THE STAR 2000 GUIDER.

FIGURE 6.1.13. A DARK FRAME FROM THE TC-245 CHIP IN A COOKBOOK CCD CAMERA.

Figure 6.1.13 shows a dark frame from a popular home-built camera, the Cookbook 245. This chip has noise characteristics similar to the chip in figure 6.1.8. Figure 6.1.14 shows a dark frame from a chip with larger pixels, the KAF 0261E chip in the SBIG ST-9E. This is a 10-minute dark frame.

Figure 6.1.15 shows an example of a dark frame from the Kodak KAF-1300 detector in the Finger Lakes IMG1300 camera. This is a 60-minute dark frame, yet it has relatively low noise because the camera has a very high level of cooling compared to most other cameras.

TIP: CCD chips vary radically in the amount of noise you will see in a dark frame. However, with proper cooling and dark frame subtraction, you can "clean up" the noise effectively on all of these chips and get excellent results.

How you perform the dark subtraction will vary with the camera control software you use. In some cases, such as Maxim, dark frame subtraction is combined in a single menu option with other noise reduction operations such as flat fields and bias frames. In CCDOPS, you perform dark frame subtraction separately. In Mira, you can perform dark frame subtraction on a single image, or on a set of images. CCDSoft version 5 includes tools for combining dark, flat-field, and bias frames in reduction groups that are very convenient to use. In all cases, the method behind the menus is the same: the values of the dark frame's pixels are subtracted from the values in the image or light frame. In most cases, you simply select the file that is the dark frame. The software handles the actual subtraction. This isn't always a literal subtraction. Some cameras modify the image and dark frame (e.g., by

FIGURE 6.1.14. A TEN-MINUTE DARK FRAME FROM AN SBIG ST-9E CAMERA.

FIGURE 6.1.15. A 60-MINUTE DARK FRAME FROM THE FLI IMG1300 CAMERA. NOTE THAT THIS CHIP HAS RELATIVELY LITTLE NOISE.

Here are some examples of what you will find recorded in the flat field's optical footprint:

Vignetting - This refers to having a brighter center area in the field of view, and it can range from a very slight difference to an extreme difference.

Internal reflections - These can crop up in even the finest optical systems. Sometimes they come from the optics themselves; more often, they come from various mechanical parts in the optical path. For example, a focuser with a too-shiny interior surface could add a bit of reflected light to your images. Some internal reflections are part of the characteristic footprint of a system, while others are the result of unique circumstances for a particular image. A flat field can only deal with reflections that are a fundamental, repeating part of the optical system.

Dust motes - These are lurking everywhere in the optical system -- if there is a surface, dust will eventually find its way onto that surface. The visual impact of the dust varies with its distance from the focal plane. Dust motes that are very close the CCD chip will throw small, tight, and dark shadows. Dust motes further up the optical pathway, such as on a filter a few inches in front of the CCD chip, will throw larger, more diffuse shadows.

Every optical system is unique, and the footprint changes over time. Dust motes are the most likely cause of a shifting optical footprint. Figure 6.1.16 is an example of a flat field. It shows some vignetting and quite a few dust motes, but no sign of any significant internal reflections.

adding a pedestal), and the camera control software has to take this into account.

Flat fields explained

If dark frames are the way to correct for electronic noise, then flat fields are the method for cleaning up certain kinds of optical problems.

The term "flat field" may not seem terribly descriptive. There is, however, a key concept lurking here. In essence, every optical system has a visual footprint. This footprint records the performance of the optical system down to the smallest details. A flat field is the best way to record certain aspects of that footprint. To take a flat field, you take an image of an evenly illuminated object (a white card; the sky at twilight; the inside of an observatory dome, etc.). Anything that subtracts light by casting a shadow, or adds it by a reflection, is recorded in the flat field image. The term "flat field" refers to the even, or flat, illumination of the object.

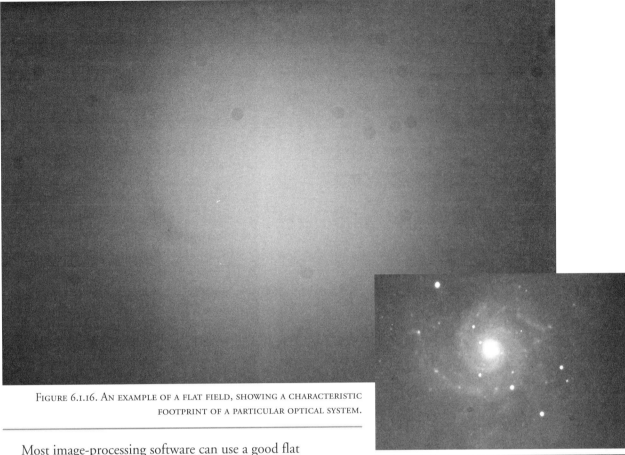

Figure 6.1.16. An example of a flat field, showing a characteristic footprint of a particular optical system.

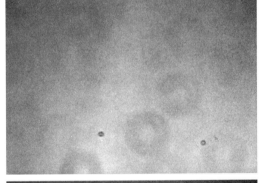

Most image-processing software can use a good flat field to clean up vignetting, reflections, and dust shadows from your images. Figure 6.1.17 shows three images. From top to bottom these are a raw image, a flat field, and the image with the flat field applied.

The raw image at the top of figure 6.1.17 has several flaws. There are two tiny dust shadows near the bottom of the image. Their small size suggests that they are sitting right on the optical window just above the CCD chip. The flat field itself shows these two shadows more clearly, as well as larger shadows from other dust further away from the CCD detector. The large shadows are probably from dust on the face of the camera, or on filters in front of the camera.

The bottom image in figure 6.1.17 shows the result of applying the flat field to the raw image. The dust shadows are gone, and the image now has a flat background that is free of any optical flaws.

Figure 6.1.17. From top: raw image, flat field, flat applied to image.

TIP: If your optical system is clean, and doesn't suffer from serious vignetting, a flat field might be optional. As you can see from figure 6.1.17, only the two really nasty dust motes sitting right on the surface of the CCD chip's optical window seriously detract from the image. It's a judgment call as to whether or not flat fields are required for a given image. The more intensely you intend to process the image, the more you will benefit from a flat field.

Obtaining a flat background is the whole idea behind the flat-field frame. The image processing software normalizes and divides the pixel values in the flat field into the image you've taken. Shadows are cancelled out; bright areas are toned down, in exact correspondence with the record of these optical flaws in the flat field.

You might be wondering just how horribly messed up the optical system can be and still deliver good images. The truth is that you could have a pretty dirty optical system, and flat fields would rescue the image fairly effectively. A clean optical system is always better because it is always better to eliminate a noise source than to try to remove it after it has gotten into an image. But it is extremely difficult and time consuming to try to maintain a perfectly clean optical system. A good flat field can handle an amazing level of dust and grime.

A gallery of really nasty flat fields follows, along with some of the images that were rescued by these flats. These images can serve as a good reference point for your own flat fields. Even if you think you're dealing with something really horrible, most of the time you can salvage the image if you handle the flat field properly. It's a good idea to avoid too-frequent cleaning of optical surfaces. Every time you clean the optics,

you risk scratching them. A flat field, properly taken, is a better alternative.

Figure 6.1.18 shows a sad case. There is an extreme hotspot at lower center. There are many dust motes on the optical window of the CCD chip. There are a number of dust motes further out, in this case on the surface of the clear infrared blocking filter sitting in the camera nosepiece. This may look like a tragically bad optical system, but I've seen much worse.

TIP: When you see a hot spot off-center as in figure 6.1.18, the CCD chip may be offset from the optical axis of the telescope. The most common cause of this is flexure in the connections between scope and camera. This can cause elongated stars or poor focus. This happens most often when you are using a 1.25" nosepiece on a heavy camera, and the obvious solution is to switch to a 2" nosepiece. On some telescopes, you will be able to purchase adapters that allow you to use all-threaded connections, and these are especially rigid. Aligning the CCD chip with the center of the optical axis is especially important with scopes that are sensitive to slight mis-collimation.

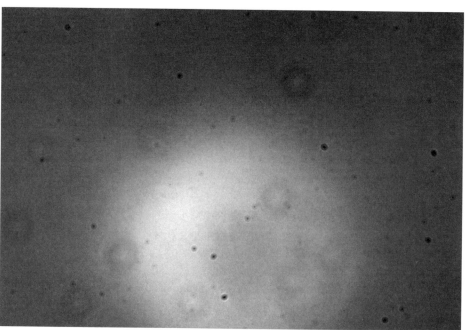

FIGURE 6.1.18. ONE EXAMPLE OF TRULY NASTY FLAT FIELD EFFECTS.

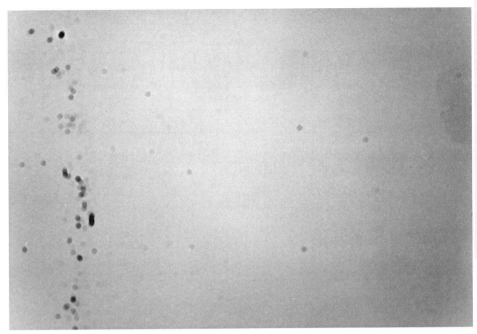

FIGURE 6.1.19. THIS FLAT FIELD SHOWS SO MANY DUST MOTES THAT EVEN A FLAT FIELD MIGHT NOT SAVE THE IMAGES TAKEN WITH THIS CAMERA. IT'S TIME FOR A GOOD CLEANING.

TIP: A 5x or 10x magnifier can be your best friend when you are cleaning the chip's optical window. Because dust motes on this window cast sharp, dark shadows, this is a key area to clean well. Hold the camera up to a bright light to see dust with the magnifier. Adjust the angle between the camera and the light source to see the dust clearly.

Figure 6.1.19 shows an optical system with a slight hot spot offset toward the top of the image. There is a line of dust mote shadows at left. They are almost certainly sitting on the chip's optical window because of their small size. There is a hint of some larger shadows cast by dust further out in the optical system. This could be a piece of lint caught inside the focuser tube, or a thread from the sleeve of your coat that got caught on a sharp edge. The diffuse nature of the largest shadows suggests that they are cast by something inside the focuser, six inches or more away from the camera. Such shadows are almost impossible to recognize in an image, but can be just seen in a flat field.

I created the line of dust during cleaning. I pushed the dust to one side instead of actually removing it from the CCD chip's optical window. When cleaning this particular location, check to make sure you've actually removed the dust, not just pushed it around. The line of dust motes in figure 6.1.19 is enough to push the limits of what can be solved with a flat field. If dust motes on the optical window are large enough, or close enough together, they can cast a shadow strong enough to interfere with the quality of your image even after a flat field is applied.

Figure 6.1.20 shows a subframe flat. Most camera control programs allow you to image with a portion of the chip instead of the full array. Take a flat to match the image subframe.

For this example, I took images of Saturn using a Barlow between the scope and the CCD camera. A Barlow tends to exaggerate dust motes, as figure 6.1.20 shows.

FIGURE 6.1.20. A SUBFRAME FLAT FIELD. THE LOCATION OF THE SUBFRAME MUST MATCH THAT OF THE IMAGE FRAME EXACTLY.

The flat field in figure 6.1.20 was taken to use with light frames of Saturn that used the same size and location of subframe. Figure 6.1.21 shows an image of Saturn. You can see that several of the dust motes are visible in the image. Since planetary images typically are heavily sharpened during processing, it is important to use a flat field to remove dust shadows. Without a flat field, the dust motes will become amplified by sharpening, as

FIGURE 6.1.23. A FLAT FIELD SHOWING A FINGERPRINT RIGHT ON THE CCD CHIP WINDOW.

shown on the left of figure 6.1.22. The image on the right of figure 6.1.22 shows the result when a flat field was applied before sharpening.

The flat field will show all kinds of dirt. Figure 6.1.23 shows a flat field from an SBIG ST-237 camera. Note that there are some distinct dark dots near the center of the image. These are the results of a fingerprint right on the chip window, resulting from an over-eager owner looking a little too deeply into his camera!

FIGURE 6.1.21. AN IMAGE OF SATURN SHOWING THE EFFECTS OF THE DUST MOTES.

FIGURE 6.1.22. THE IMAGE ON THE LEFT WAS PROCESSED WITHOUT ANY FLAT FIELD, AND IMAGE PROCESSING HAS GREATLY EMPHASIZED BOTH PLANETARY DETAIL AND DUST MOTES. THE IMAGE ON THE RIGHT HAD A FLAT FIELD APPLIED PRIOR TO IMAGE PROCESSING. PLANETARY DETAILS ARE REVEALED, BUT THE DUST IS NOT.

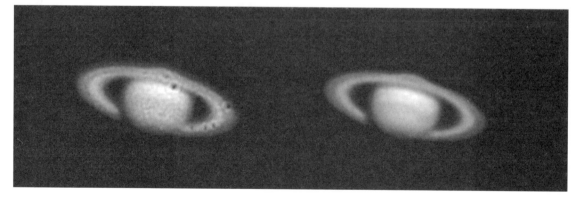

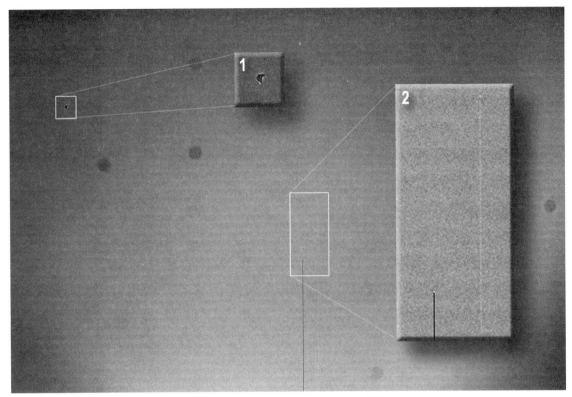

FIGURE 6.1.24. AN EXAMPLE OF FLAWS IN A CCD DETECTOR SHOWING UP IN A FLAT FIELD.

This fingerprint never got in the way of taking excellent images with the camera. The flat field took care of the problem.

Figure 6.1.24 shows a flat field taken with an engineering-grade antiblooming CCD chip. An engineering grade chip usually has serious flaws, and this one is no exception. Most manufacturers do not sell cameras with engineering-grade chips, but you can sometimes find such chips at low cost for building your own camera. The flaws are usually severe. Flaw #1 in figure 6.1.24 is the result of physical damage on the surface of the chip. Flaw #2 shows partial bad columns, one dark and one light. Data in a dead column or defect would be missing, and could not be recovered. But most software gives you the ability to repair such problems by interpolating data from nearby pixels, so even engineering grade chips are acceptable for making pretty pictures. The horizontal banding in figure 6.1.24, by the way, is not a defect but the normal appearance of a flat field for an antiblooming detector. The flat field will remove the horizontal bands from your light frames.

Even the finest telescopes will benefit from having a flat field applied to images. Figure 6.1.25 shows the flat field from an Astro-Physics Traveler and an ST-8E camera. The vignetting is the result of using an NGF-S focuser at the back of the Traveler's focuser, and it is completely handled by use of a flat field.

FIGURE 6.1.25. WIDE-FIELD SCOPES TEND TO HAVE MORE VIGNETTING THAN SCOPES WITH LONGER FOCAL RATIOS.

These flat fields reveal flaws that are a small sample of what you will see on even the most respectable optical systems. Just as you would not risk damaging the surface of a front surface mirror by cleaning it when it doesn't really require it, you can usually safely use a flat field to correct for dust, vignetting, and reflections without risking your optics or your camera. If and when the problem becomes too much for a flat field to solve, you can clean your optics.

One question I often hear is "How do I tell if I have taken a good flat field?" The value of a flat field can only be proven by applying it to your images and observing how effective it is at giving you a flat background. Many things can go wrong with a flat. The most common is a failure to get even illumination. If this happens, the flat will create a gradient in your images when you apply it. Don't confuse this gradient with gradients from light pollution; a flat can't correct light pollution gradients. To test a flat, apply it to an image taken in an area without light pollution.

Flat fields can be taken at a different temperature from your dark frames. You must, however, take separate dark frames for your flats at the same exposure time and temperature as your flats. Flats often have much shorter exposures than light frames, and often wind up being taken at the warmer temperatures of twilight.

Figure 6.1.26 shows a rather interesting flaw in a flat field. Note shadows of dust motes at left and right. A similar dust mote at top right is bright, not a shadow at all. There are two more subtle bright dust motes just to the left of center as well. These are caused by stray light entering the optical system from an oblique angle and illuminating the dust motes directly. Such a flat field will create more problems than it solves. You'll know when you have a bad flat because it will cause bright spots, dark spots, or gradients in your images instead of flattening the background.

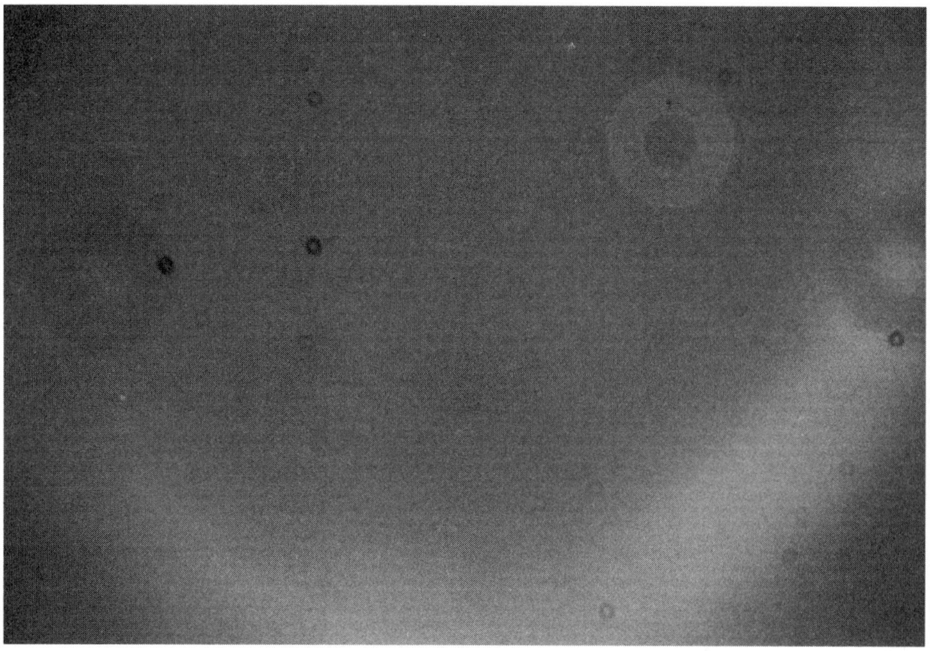

Bias frames explained

In addition to thermal noise, your CCD camera generates some noise that doesn't change with exposure time. A bias frame records this noise.

The good news about bias frames is that under many circumstances you will not need to take or apply them. The sole purpose of a bias frame is to allow you to use dark frames whose exposure time does not match the light frame exposure time. If you always take dark frames with the same exposure as your light frames, you do not need to take bias frames.

An example will make this clear. Let's look at just one pixel. If you take a zero-length exposure, you might expect that pixel to have a value of zero because there has been no time to build up a charge. That's not how it works. There is a small amount of noise that exists even if the chip hasn't had time to build up any charge from thermal noise. The bias frame records this.

For example, on an ST-8E camera, the shortest possible exposure is 0.11 seconds. For an ST-237 camera, the shortest expo-

FIGURE 6.1.26. A FLAT FIELD THAT HAS BRIGHT DUST MOTES INSTEAD OF SHADOWS.

sure is 0.01 seconds. These are not quite zero-length images, but they will do for a bias frame. Let us suppose that the bias frame records a value of 20 for one particular pixel.

Note: Many cameras add a pedestal to every image downloaded. This prevents negative values from occurring due to noise. For SBIG cameras, for example, the pedestal is 100. This means that the pixel would have an actual value of 120 when downloaded, even though it's "real" value is 20. Camera control software knows about camera pedestals, and takes them into account when performing dark subtractions, applying flat fields, etc. If you decide to manually perform these same tasks, you will need to know the value of the pedestal before you can include it in your calculations.

If you take a one-minute dark frame, that pixel might have a value of 520. Assume that the dark frame is 500 units and the bias is 20 units. If you take a two-minute dark frame, that pixel might have a value of 1020. That's 1000 units due to thermal noise, and 20 due to the bias. Doubling the exposure time doesn't exactly double the pixel value. The effect of thermal noise varies slightly from image to image, and is not necessarily linear with increasing exposure times. But for the sake of this example, we'll keep things simple.

If you were to take a one-minute dark frame and apply it to a two-minute image, the 520 units would be insufficient to remove the noise. You would still see hot pixels in your image. If you were to double the 520 units, you would wind up with 1040, which is too much. But if one first subtracts the bias (20) from the dark (520), this gives you 500 which is the actual contribution from the dark frame. If you double 500, you get 1000. Add the bias back in, and the value will be 1020, which is accurate.

CCD camera software does this for you when it scales a dark frame. The larger the difference between the dark frame and the light frame, the more important it is to have a bias frame.

TIP: The camera aboard the Hubble Space Telescope is a CCD camera, and many of the techniques we are talking about in this chapter are applied to the wide field and planetary cameras. This is done with a high degree of precision and redundancy. In the case of the Hubble bias frames, thirty-one bias frames were taken during a single functional test. In the entire suite of seven tests, a total of (7 X 31) bias images were taken, for a total of 217 bias images. Next time you feel like it's too much work to take and apply the various reduction frames, keep this in mind!

For details on how darks and bias frames are used in Hubble research, please see the following link:

http://hires.gsfc.nasa.gov/stis/postcal/quick_reports/r058/r058.html

A collection of research, reports, and white papers on reducing the Hubble's CCD data can be found at:

http://hires.gsfc.nasa.gov/stis/postcal/quick_reports/quick_reports.html

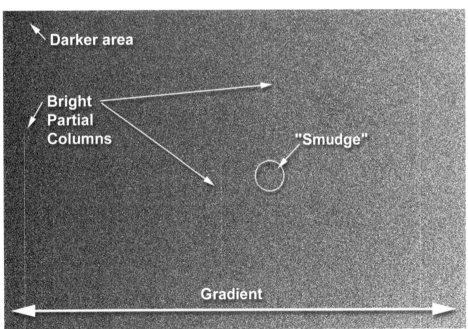

FIGURE 6.1.27. AN EXAMPLE OF A BIAS FRAME.

These links will give you a whole new perspective on your own use of dark, flat-field, and bias frames.

The bias frame contains noise from a variety of sources that are independent of exposure length. Figure 6.1.27 shows a bias frame from my own ST-8E camera, taken at a temperature of -14 degrees C.

> **TIP:** The temperature at which bias frames are taken must be the same as that used for your dark frames. If you are able to maintain a consistent temperature across multiple nights, it will assist you in re-using your reduction frames of all types in subsequent imaging sessions.

The flaws visible in figure 6.1.27 include:

- Thre is a low-level mottling effect across the entire frame. Although this mottling appears obvious in this illustration, that is a trick of processing. I have enhanced the contrast to show the variations more clearly. The brightness variations between the darker and lighter pixels are very small. The darker pixels have a value of around 93, and the lighter pixels have a value of around 113, for an average pixel brightness of 103. Most pixels are only 3.5 units above or below this average value. The very brightest pixel in the entire frame is 312 units, and the very darkest is 84 units. So although the image looks really noisy, it has very little noise in it.
- There are some slightly brighter columns. As was the case with the mottling, these columns, while definitely brighter than the average background, are not all that much brighter and thus will have little or no visible impact. The brightest column (the rightmost one) has a peak brightness value of only 122. Given that the chip can represent over 40,000 brightness values, this is a very, very tiny percentage of the entire available brightness range.
- There is a small slightly darker area at top left. This area has an average brightness value of 98, just five units below the average of the entire chip. This will be of little consequence.
- There is also a "smudge" near the center of the chip, an area that shows a slightly darker ring around a slightly brighter core. Again, the differences are small: just 1 brightness unit, on average.
- Finally, the most notable feature is a gradient from left to right. The magnitude of the variations are very small, and this gradient is typical of many cameras. It occurs because of the time delay in reading across the chip. The pixels at right have a little more time to build up thermal noise, so they are brighter. The pixels at the left edge of the frame have an average value of 100. Pixels at the far right have an average value of 107. These values are consistent with the overall average brightness value of 103, since the average of the two values is 103.5. The gradient will occur in all frames, and is automatically dealt with when you reduce the images with dark, flat-field, and bias frames.

FIGURE 6.1.28. SCLAING A DARK FRAME WITH AND WITHOUT A BIAS FRAME.

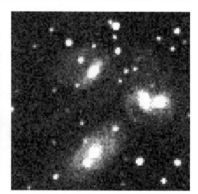

3x10   3x10 bias   1x30

Thus, despite having quite a collection of weird-looking artifacts, this bias frame makes almost no difference at all in the appearance of an image. The magnitude of the variations is very small.

Figure 6.1.28 shows what happens when you attempt to scale a dark frame without using a bias frame. The raw image in all three examples is a 30-minute exposure of Stephan's Quintet. The image at left has been reduced by scaling a 10-minute dark frame without a bias. The middle image has been reduced by scaling the same 10-minute dark frame using a bias frame. And the image at right has been reduced by using a 30-minute dark frame.

The images have all had their contrast adjusted to match each other. Note that a stream of stars resulting from the interaction of these galaxies (an arrow points to it in the middle image) is slightly less visible in the image without a bias because of the higher noise level. The image that uses a dark frame exactly matching the light frame is the best by a slight margin.

Like most electronic noise related to CCD cameras, noise in the bias frame changes with temperature: lower temperatures mean less noise. Compare figure 6.1.27, which was taken at a temperature of -14C, with figure 6.1.28, which was taken at a temperature of -32C. There is hardly a trace of noise in the colder bias frame.

So while a bias frame can help you out if you have different exposure lengths, the same is not true for variations in temperature. It's a good idea to allow your CCD camera a few extra minutes at the start of your observing session to fully stabilize in temperature. If your camera is water-cooled, the water is more effective than air in keeping temperature stabilized.

FIGURE 6.1.29. AN EXAMPLE OF A BIAS FRAME TAKEN AT -32C; COMPARE TO FIGURE 6.1.27.

Section 2: Using Dark Frames

Do you need to take a dark frame every time you image? Do you need to take dark frames right from day one? No, and no again.

The key word is *need*. A dark frame will always improve the quality of your images. If your image has horrible thermal noise, the need for a dark frame is obvious. If your image looks fairly clear, you might choose not to use a dark frame. But even if the image has very low noise to start with, the dark frame will clean up what noise there is and produce an even better image.

The longer your exposure, the more likely you are to benefit from taking a dark frame. Thermal noise builds up over time, and a longer exposure simply has more thermal noise in it.

While a dark frame will clean up thermal noise, there is also some noise added when you apply the dark frame. It will remove a huge chunk of system noise, and add in a small amount of random noise. The colder you can get your CCD detector, the less noise you will start with, and the less noise you will wind up with after subtracting the dark frame.

Maximum cooling options, if installed in your camera, will give you the most dramatic reduction in thermal noise. If you can get the temperature of the chip down to -30 Celsius or lower, you will barely be able to see the thermal noise. The new MaxCams from Finger Lakes have especially deep cooling, and I expect them to be good performers.

I make dark frames a habit. I take new dark frames almost every night I image. If I want to spend all of my time imaging, I may re-use dark frames from previous nights. There is always some level of difference from night to night, so averaging or median combining dark frames is essential.

Taking a Dark Frame

Taking a dark frame is simplicity itself: take an exposure of the same duration as your light frame, but with the camera shutter closed. If the camera does not have a shutter, take a dark frame by covering the front of the telescope during the dark frame exposure. A hat, the scope's dust cover, or even your hand on a small scope, will suffice for a cover. Most camera control programs include a simple way to take a dark frame. Figure 6.2.1 shows the MaxIm DL camera control. To take a dark frame, set the exposure time, click on the Dark radio button, and then click the Expose button. If Sequence is checked, you get multiple dark frames. I suggest taking 3-5 darks routinely, and then combining them with a median combine. This reduces noise and makes your dark frames more effective at preserving image quality.

The keys to success with dark frames are:

- Having the camera at the same temperature for the light frame and the dark frame.
- Taking the dark frame with the same exposure length as the light frame.
- Taking the dark frame near the time the light frame was taken.
- Taking multiple dark frames and combining them to reduce noise.

You can get reasonable results even if you don't follow these guidelines, but the best results will always come from matching the light and dark frames as closely as possible. The question you may have to

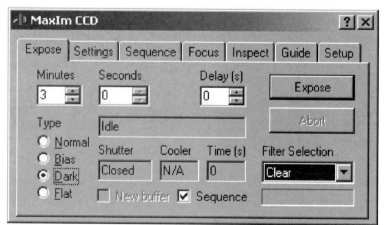

FIGURE 6.2.1. TAKING A DARK FRAME IN MaxIm DL.

FIGURE 6.2.2. AN EXAMPLE OF A DARK FRAME TAKEN AT A VERY COLD TEMPERATURE.

Figure 6.2.3 shows a dark frame taken just a few minutes earlier with the same camera. For this dark frame, the chip temperature was 0 degree C. The general appearance is much different, with a lot more noise evident. Note that the maximum pixel value is much higher, over 17,000. That is nearly one-third the maximum value. The average pixel value isn't too bad, 130.9. Although the images would have major thermal noise, you could still image at this temperature.

Although a dark frame can remove system noise, it does so at a cost. The dark frame also adds a bit of random noise. If the temperature is too high, the random noise will become noticeable. On any given night, you will get the best results with a chip temperature that is as cold as possible.

answer by actual field tests with your own camera is this: How much variation from the ideal is possible?

Let's take a moment to look at the keys to success in some detail.

Temperature and Dark Frames

The characteristics of a dark frame are highly dependent on the temperature of the CCD chip when the dark frame was taken. If the temperature is high, there will be more noise in the dark frame. If the temperature is low, there will be less noise. Figure 6.2.2 shows a dark frame taken at a very low temperature (-35C) with a 60-second exposure. There are very few bright pixels, and the overall statistics for the frame (see table 6.1) are excellent: There is a slight gradient from left to right, but this is characteristic of the camera and is not a concern. The gradient appears in both dark and light frames, so the dark frame will remove it.

The gradient results from the time delay between reading the pixels at the far left and the far right. Thermal noise continues to build up as the chip is being read, one row at a time. The longer your dark frame, the less obvious the gradient will be.

Table 6.3: -32°C Dark Frame

Average pixel value	110.7
Standard deviation	9.8
Maximum pixel value	2815
Minimum pixel value	73

Table 6.3: 0°C Dark Frame

Average pixel value	130.9
Standard deviation	55.7
Maximum pixel value	17,279
Minimum pixel value	0

You don't want to be too aggressive in setting the chip temperature. Most camera control programs will report the percentage of total cooling capacity in use. If this percentage creeps too high, the camera does not have reserve cooling capacity to deal with an above-average temperature variation. For example, on many nights, warm and cool breezes will alternate, and you want enough reserve cooling capacity to keep the chip temperature as close as possible to your selected temperature. Avoid running the chip at 90% or more of its capacity. Otherwise, a warm breeze could result in temperature variations that create noise variations, creating problems that cannot be solved with a dark frame. Keep your cooling no higher than 80-85% maximum.

TIP: A good strategy for cooling your camera is to run your cooling percentage a little on the high side at the start of your session, in the range of 90-95%. As the air cools over the next few hours, your cooling percentage will go down. This strikes a balance between the need for the coolest possible temperature, and taking only one set of dark frames during the night. If you change temperatures, you will need a new set of dark frames at the new temperature.

Exposure Length

If a dark frame has exactly the same exposure duration and temperature as a light frame, the system noise levels will be very close in both. When you subtract the dark frame from the light frame, the added random noise from the subtraction will be as small as possible. This clean subtraction of the dark frame's noise gives you a greater likelihood of a good-looking final image.

Even a very short dark frame can have unique noise components, particularly from cosmic rays striking the CCD chip. These random cosmic rays leave a bright spot or streak on the dark frame. Median combining three or more dark frames can reduce these problems. If you don't reduce them, they will leave dark spots in your final image. You can correct these by hand, but what if that dark streak is in an area of spiral arm detail? You will lose detail in your image when you apply the faulty dark frame. Median combine avoids this, but it requires multiple darks.

It is not absolutely essential to take dark frames that match the exposure times of your light images exactly. There may be times where you wind up not taking one or more dark frames. You can either use the scaling feature in your camera control program, or use a dark that is close to the same exposure length. Scaling is a better choice than a near miss, and it requires a bias frame to be accurate.

Figure 6.2.3. A dark frame with very little cooling (zero Celsius).

The noise in a dark frame (and therefore in the light frame as well) tends to increase in a predictable fashion. If a dark frame is twice as long as another dark frame, then a given hot pixel will likely be twice as bright (with proper application of a bias frame). But it won't be *exactly* twice as bright! There will always be minor variations in response for a given pixel as a result of random noise.

SECTION 2: USING DARK FRAMES

FIGURE 6.2.4. A 60-SECOND DARK FRAME EXPOSURE HAS VERY LITTLE NOISE IN IT.

Table 6.3 shows the changes that occur for a series of images with different exposure times. The images were taken with the same camera (ST-7E) and at same temperature (-30°C).

This table reveals some interesting things about the behavior of dark frames at different exposure lengths.

The average pixel value across the entire frame doesn't change much even when the exposure times vary by a factor of 10, although it does increase for very long exposures.

The standard deviation increases with exposure. This means that the variation from pixel to pixel increases with longer exposures. This is a rough measure of noise, and this means that, as you would expect, longer exposures have more thermal noise.

TIP: If you have a bright star in an image, and take your dark frame immediately after the light frame, you may see a faint after-image of the star in the dark frame. Either take the dark frame first, or take a few dummy dark frames to dissipate the excess charge from the star on subsequent dark frames. Such "ghosts" are very evident visually, so it's not hard to tell when you've got one. Retake the dark frame until the ghost goes away.

FIGURE 6.2.5. SAME CAMERA AND TEMPERATURE AS FIGURE 6.2.4, BUT A 600-SECOND EXPOSURE.

Figure 6.2.4 shows a 60-second dark taken with an SBIG ST-7E at -30°C. Figure 6.2.5 shows a dark frame with ten times the exposure length and the same temperature. The change in exposure has affected the noise level. The longer the exposure, the greater the thermal noise.

THE NEW CCD ASTRONOMY

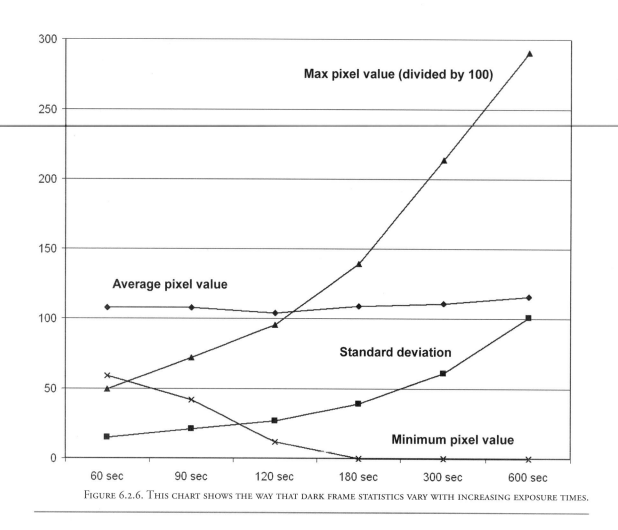

FIGURE 6.2.6. THIS CHART SHOWS THE WAY THAT DARK FRAME STATISTICS VARY WITH INCREASING EXPOSURE TIMES.

Table 6.3: Dark Frame Variations with Exposure time

Exposure (seconds)	60	90	120	180	300	600
Average pixel value	108	108	104	109	111	116
Standard deviation	15	21	27	39	61	101
Maximum pixel value	4953	7226	9541	13930	21378	29070
Minimum pixel value	59	42	12	0	0	0

The maximum pixel value increases with increasing exposure. However, the relationship is not linear. The hottest pixel in a 60-second exposure has a value of 82 for each second of exposure; the hottest pixel in a 600-second exposure has a value of 48 for each second of exposure. This means that hot pixels increase in value at about half the rate of increase in exposure time. This result will vary with different cameras.

The minimum pixel value drops with increasing exposure time. This might seem counter-intuitive, but it is true! The longer the exposure, the cooler some of the pixels will be. This contributes to the increase in standard deviation noted above. There is an increasing spread between brightest and dimmest pixels with increasing exposure times.

Figure 6.2.6 shows these conclusions graphically. The Max pixel value and standard deviation change at a rate that is close to linear, but not exactly linear. Scaling will be approximately correct, but not as good as a dark frame with the exact same temperature and exposure.

This has implications for using non-exact exposure times for your dark frames. Most image processing software will scale the pixel values in a dark frame to match the exposure times in a light frame. You must supply a bias frame to get a correct scaling (see the first section of this chapter). However, since the relationship between these values and exposure time is not linear, you won't get as good a result as you would with a dark frame having exactly the same exposure duration.

At the same time, the match may be good enough for many images. How far can you stretch exposure time variations? It really will depend on the characteristics of the chip in your CCD camera. The only way to know for sure is to experiment. However, as long as you keep the dark frame exposures longer than the light frames you scale them to, you will be close. The real risk comes when you try to scale a one-minute dark frame with a five-minute light frame.

Timing of Dark Frames

If you can take your dark frames as close as possible in time to your light frames, you are ahead in the noise-control game. Chip characteristics don't change quickly, but they do change over time. The ideal dark frame would be taken right before or after your light image. For most purposes, getting a dark frame on the same night as the light frame will be sufficient.

Combining images reduces noise, and this applies to all types of images, including dark frames. The ultimate dark frame methodology is to take one or two dark frames before your light exposure, then another one or two after, and average the dark frames. In practice, however, simply taking three or more dark frames at the start or end of the night will be more than adequate. My typical routine is to make sure that I get 3 dark frames for each exposure duration and bin mode I use during the night. For example, if I take 10-minute, 16-minute, and 30-minute images during the night, I want to get at least three dark frames at each duration, and of course at the same chip temperature, before I shut down for the night. I will typically start taking my darks as twilight makes imaging impossible, and in the above scenario that would mean about three hours spent acquiring the dark frames. By the time I start taking dark frames, I am at about 70% of cooling capacity, so I can often go for those three hours while still keeping the same temperature as in my images. Just in case, I always take the darks in mixed order: one 10-minute, one 16-minute, and then one 30-minute. Then I take at least two more sets in the same fashion. If for some reason I can't take all three sets, at least I have dark frames that are current.

Why should you take dark frames every night you image? Because dark frames vary a little from one night to the next, and increasingly more over longer periods of time. Figure 6.2.7 shows two dark frames taken with an ST-7E camera. The bottom image was taken about two months after the top image. To the eye, the two images are nearly identical.

A closer inspection reveals differences, however. The best way to compare any two images to see how they differ is to subtract one of the images from the other. In CCDSoft, use the Images | Combine | Combine Images menu item, and choose Subtract as the combination method. In MaxIm DL, use the Process | Pixel Math menu item, and set one image as image A and the other as image B, and choose Subtract as the method. In Photoshop, open both images and use the Image | Apply Image menu item, with Blending set to Subtract.

Figure 6.2.7. Comparison of two dark frames (90 seconds, -15C) taken two months apart.

Figure 6.2.8 shows the result of subtraction. Each pixel shows the level of difference between the images. If there were no differences, all of the pixels would be black. There is a subtle nearly horizontal line near the bottom that is brighter than the rest of the image, and there are plenty of individual pixel variations. The variations aren't large, however, and it would be reasonable to use a dark frame from either date. For critical use, however, the dark frame closest in time to the light frame will always be a little more accurate. And the longer your exposure, the greater the difference will be.

Table 6.4 shows the differences between the two dark frames numerically. The values for the average seem close at 109 and 107. But if you consider that the pedestal of 100 units should be subtracted to get an accurate comparison, that's a difference of 2/9, or almost 20%. The standard deviation is the same in both images. The maximum pixel value is also close, 16,229 and 16,098. The values for the difference image also indicate that the two images are fairly close: the value of the average pixel is 1.6, indicating that the pixel values in the two images are very close. The standard deviation of the difference image is 3.4, indicating that there isn't a whole lot of variation. The changes in the pixel values are all small and not very significant. Note that the variation in the maximum pixel value is only 237 units. If the temperature of the chip had been lower, these values would have been lower as well. If the exposure time had been longer, the differences would be greater.

You can perform this analysis yourself on any two dark frames. I used MaxIm DL for

Figure 6.2.8. Subtracting one image from another shows the net difference between the two images from figure 6.2.7.

Table 6.3: Long-Term Dark Frame Differences

Image	August, 1999	October, 1999	Difference image
Average pixel value	109	107	1.6
Standard deviation	49	49	3.4
Maximum pixel value	16229	16098	237
Minimum pixel value	0	0	0

Table 6.3: Short-Term Dark Frame Differences

Image	Image 1	Image 2	Image 3	Image 4
Average pixel value	208	208	209	209
Standard deviation	42	42	42	42
Maximum pixel value	7468	7452	7418	7483
Minimum pixel value	141	139	133	129

my comparison, but other software packages provide some or all of the same information.

The natural question at this point is: How much do dark frames taken on the same night vary? Table 6.5 shows the values for four dark frames, each 120 seconds long and taken at -32°C. The images are very similar.

That is why the ideal approach to dark frames is to take multiple images on the same night and median combine them.

What effect does temperature and exposure duration have on this result? Table 6.6 shows the statistics for a pair of 600-second dark frames, taken at -20°. Despite a much longer exposure, and a warmer temperature, the values are still very close because the images were taken on the same night, one right after the other.

Table 6.3: Long Exposure Comparison

Image	Image 1	Image 2
Average pixel value	301	300
Standard deviation	376	376
Maximum pixel value	41622	41636
Minimum pixel value	0	6

Generally speaking, dark frames taken closer to the time of a light frame will be closer in sytem noise levels, and will compensate for that noise more accurately. Only trial and error with your own chip will tell you how often to take a dark frame. Purists will take new dark frames every night; pragmatists might take new dark frames monthly, or even less often, until there are too many hot pixels to fix by hand.

Applying a Dark Frame

Most image-processing software makes it easy to apply a dark frame to an image. For best results, I recommend that you take more than one dark frame and median combine them before applying them to your images. The newest generation of software will do this for you, but you can do it yourself if your software doesn't include this feature. For example, in MaxIm DL, use the Process | Combine menu item in Overlay mode and with the Median radio button checked. Simply open the images, perform the combine, and save the result as your median dark frame. You need at least three images to do a legitimate median combine.

As you've already seen, even dark frames taken a few minutes apart show minor variations. By using median combine, you can get better results in the fight against noise in your images.

Figure 6.2.9 shows a raw image of the Horsehead Nebula in Orion, taken in April of 1999 in Arizona. This is a 300-second image, taken at -25C. There's plenty of noise to go around, and this image definitely needs the benefit of a dark frame. The enlarged detail at lower left shows the level of noise more clearly.

Figure 6.2.10 shows one of the two dark frames I took that night. Yes, I should have taken at least three or four but when you only get down to Arizona for one week out of the year, and you live in rain-drenched Seattle, it can be a challenge to maintain discipline.

This dark frame has a problem: a cosmic ray hit just above the center of the frame. The atmosphere absorbs cosmic rays. Since I was imaging above 5000 feet,

FIGURE 6.2.9. A RAW IMAGE OF THE HORSEHEAD NEBULA IN ORION, SHOWING NOISE THAT CAN BE ELIMINATED WITH A DARK FRAME.

FIGURE 6.2.10. A DARK FRAME FOR THE HORSEHEAD IMAGE, WITH MATCHING EXPOSURE DURATION AND CHIP TEMPERATURE.

many of my images had cosmic ray hits. Figure 6.2.11 shows what happens when you apply a dark frame containing a cosmic ray hit. The area of the hit appears darkened in the final image. The stronger the cosmic ray hit, the greater the darkening. The dark area from the cosmic ray hit is just below and slightly to the left of the star.

The best way to deal with such problems is to have those magic three or more dark frames so that you can do a median combine. I only had two, at least only two that matched my image's exposure time and temperature. I also had several 300-second dark frames taken at -28C, so I tried an experiment. I did a median combine of the four dark frames (two at -25C, two at -28C) in MaxIm DL.

To perform a manual combine, use the Process | Combine menu item. This opens the Select Images dialog box shown in figure 6.2.12. Clicked the Add All button to put all of the open images into the right-hand box, "Selected Images." To select specific images, click on them in the left frame to highlight them, and then click the double-arrow button that points toward the right (the upper one).

FIGURE 6.2.11. EFFECT OF APPLYING A DARK FRAME WITH A COSMIC RAY HIT.

TIP: Notice that I use filenames that tell me the critical information about each dark frame: time and temperature. I recommend this as a simple but effective method for maintaining some semblance of control over your dark frames. You could easily wind up with dozens and dozens of images from a single all-night session, and making sense of them the next day should be easy to do.

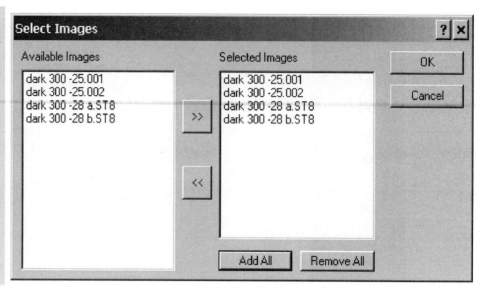

FIGURE 6.2.12. THE SELECT IMAGES DIALOG BOX.

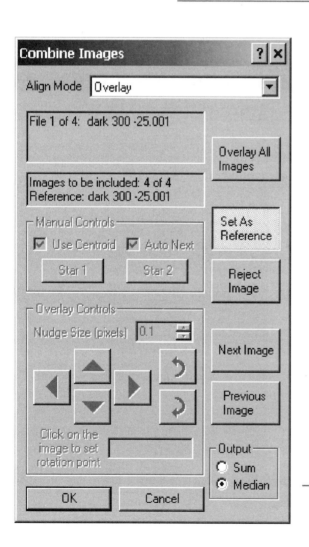

The Combine Images dialog box appears when you click OK. This dialog looks complicated. Ignore that for now. For this operation make sure that "Overlay" is selected n the Align Mode drop-down box. In the Output box, click the Median radio button. This tells MaxIm DL to overlay the images without any alignment changes, and to do a median combine on the images. Click OK. The result is a new window with the median-combined dark frame.

Depending on the size of your images and the speed of your computer, it may take a few seconds or a few minutes for MaxIm DL to work its magic. The median combine performs a mathematical median on the pixel data from the images. The median value is not the same as the average value. The median value is the value that has an equal number of values above and below it.

The main advantage of this manual method is that you can take a look at each dark frame as you open it, and confirm if it is one you want to use in the combine. Excessive cosmic ray hits, a light leak, a wrong temperature, or other flaws can be identified and the image closed before you combine. You can also have MaxIm DL do an automatic median combine.

FIGURE 6.2.13. COMBINING IMAGES IN MAXIM DL WITH A MEDIAN COMBINE.

FIGURE 6.2.14. A SMOOTH, COSMIC-RAY-FREE DARK FRAME IS THE RESULT OF A MEDIAN COMBINE.

The median calculation removes extreme values from the result. Since the cosmic ray hits are definitely extreme values, they usually disappear, leaving you with a statistically better dark frame that is cleaner than any of the single frames that were used to create it. The use of a median combine also reduces the potential problems from using dark frames at slightly different temperatures. Figure 6.2.14 shows the result of the median combine: no evidence of those nasty cosmic ray hits.

CCDOPS can also combine your dark frames, but it performs an average instead of a median combine. To average dark frames, click on the Util-

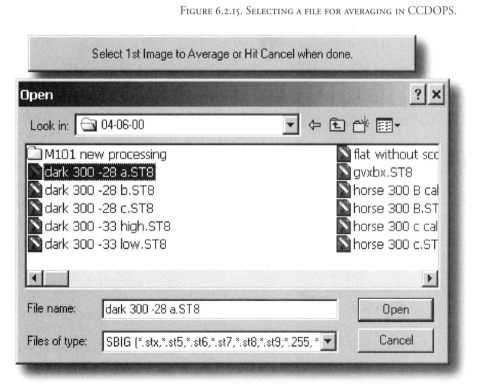

FIGURE 6.2.15. SELECTING A FILE FOR AVERAGING IN CCDOPS.

FIGURE 6.2.16. AVERAGED OR SUMMED DARK FRAMES RETAIN ALL COSMIC RAY HITS FROM INDIVIDUAL FRAMES.

ity | Average Images menu item. This opens a conventional Open dialog, as shown in figure 6.2.15. The small box above it asks you to select the first file to be averaged. Click on the filename, and then click the Open button.

This loads the first image into memory, and then the dialog appears again. Click on the next file to include in the averaging process, and click Open again. Continue in this way until you have indicated all of the files you want to average. Then click the Cancel button. **This will not cancel the operation;** it will start the averaging process. Figure 6.2.16 shows the result of averaging. Note that the cosmic rays have increased -- there are now at least 17 cosmic ray hits from all of the different images included in the average image. This is a graphic illustration of the value of a median combine for removing extremes of noise. A median combine is also very useful for light images that contain tracks from meteors or satellites.

While the median combine method of creating dark frames is preferable, the averaged dark frames will still do a credible job. The gotchas from the cosmic ray hits can be cleaned up manually in an image-editing program such as Photoshop or Paint Shop Pro.

CCDSoft version 5 takes a completely different approach. It allows you to quickly specify multiple dark frames (as well as multiple bias and flat-field frames), and it will perform averaging or median combine automatically. Figure 6.2.17 shows the Image Reduction dialog. I have created a number of reduction groups, each containing the appropriate bias, dark, and flat-field frames. There is even a separate entry for darks for the flat-field frames, which often have a different exposure and temperature from the light frame.

For each type of frame, you can set options. For example, for the dark frames, you can choose the combination method (average or median) and whether or not the darks will be automatically scaled if the exposure times do not match. The lower right portion is also very handy. It shows details about the dark frame, including the critical items exposure, temperature, and bin mode.

SECTION 2: USING DARK FRAMES

FIGURE 6.2.18. THE SET CALIBRATION DIALOG BOX

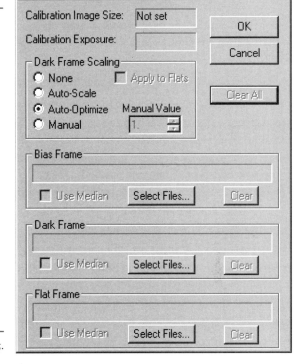

Using the buttons at top right, you can reduce open images or a folder of images. I find this to be a wonderfully efficient and clear method.

To apply a dark frame in MaxIm DL, use the Process | Set Calibration menu item. Note: The word "calibration" for this process is not ideal; the correct term is reduction. Data reduction is the process of cleaning up your data, and thus is a more appropriate term.

This opens the dialog box shown in figure 6.2.18. It has a "Select Files" button for each type of reduction frame: bias, dark, and flat field. For this example, we are only setting a dark frame, but you can include bias and flat-field frames just as easily.

It's worth a moment to talk about the radio buttons found under "Dark Frame Scaling" in this dialog. These buttons control whether and how scaling of dark

FIGURE 6.2.17. USING REDUCTION GROUPS IN CCDSOFT v5.

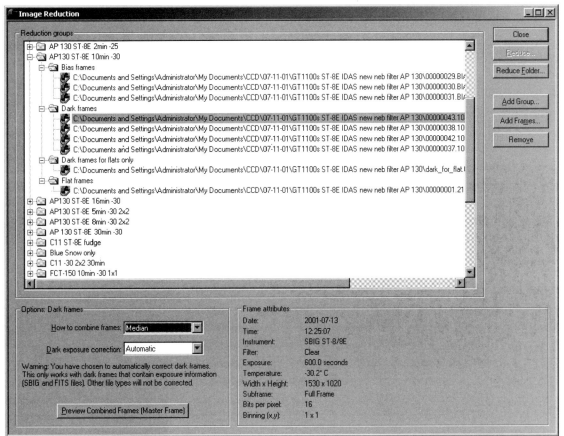

frames occurs. If you use scaling, you should include one or more bias frames with the same chip temperature in addition to the dark frame(s). Here's a breakdown of the four choices available to you:

None - Scaling is not applied to the dark frame. Use this setting when your dark frame exactly matches the light frame in temperature and exposure duration.

Auto-Scale - This adjusts the pixel values of the dark frame to compensate for differences in exposure duration. When Auto-Scale is on, you can also check the "Apply to Flats" box. This will apply the dark frame to the flat fields as well as the light image, with scaling if appropriate. If your flats have a different *temperature,* apply a dark frame to them separately.

Auto-Optimize - MaxIm DL will compensate for differences in exposure duration *and temperature.* If your exposure and temperature match, no scaling will occur.

Manual - Allows you to enter a scaling factor. If you want to double the values in the dark frame, enter a scaling factor of 2. For example, if you have a 60-second dark, and a 120-second light image, and the Auto-Scale doesn't quite give you optimal results, you could try a manual scaling factor such as 1.79 or 2.2.

To select a dark frame, click on the Select Files button in the Dark Frame section of the dialog. A special Open dialog with numerous options that you can most often ignore appears. Navigate to your dark frame(s) and click to highlight them. Hold down the control or shift keys to select multiple files. Then click on OK to load the file(s).

This returns you to the Set Calibration dialog, where the filename(s) of the dark frame(s) now appear in the Dark Frame section. If you have Bias frames and/or flat-field frames, click on the appropriate Select Files button to add them. Click OK to save.

This doesn't perform image reduction (calibration). It simply tells MaxIm DL which files to use for reduction. To reduce an open image, use the Process | Calibrate menu item. To reduce all open images, use Process | Calibrate All.

FIGURE 6.2.19. THE HORSEHEAD NEBULA IMAGE AFTER IMAGE REDUCTION (DARK FRAME ONLY).

TIP: MaxIm DL also has a Process | Calibration Wizard. This will walk you through the process of selecting the various files needed to calibrate an image. It's slower than the Set Calibration dialog, but may be easier to use until you know your way around the reduction process.

Figure 6.2.19 shows the result of applying image reduction to the Horsehead Nebula image. The enlarged detail at lower left shows the absence of the hot spots seen in figure 6.2.9. The image is still noisy, due to the short exposure, but the thermal noise has been removed.

Section 3: Using Flat-Field Frames

As described earlier in this chapter, the fundamental purpose of a flat field is to correct for certain types of noise in the optical path. In this section, you'll learn how to create flat-field frames, and to process an image using a flat-field frame. You'll see examples of how the flat field improves the appearance of, reduces the noise in, and improves the accuracy of images.

About Flat Fields

The idea behind a flat field is simple: without changing your focus position, take an image of an evenly illuminated surface so that you achieve approximately one-third to one-half the saturation level of the CCD chip. Let's break this down into plain English:

No focus position change - The flat field is a snapshot of your optical system in a particular configuration. If you change the focus position, the relative positions of the optical elements change slightly, and the shadows from any dirt or dust will change as well. Changing the focus position would make the shadows slightly larger or smaller, and the flat field would no longer match light frames. In effect, a change in focus position means that the flat field is now a snapshot of a slightly different optical system than the one you used for your image. Applying it to the image could create more problems than it solves. Small focus changes, such as those used to compensate for temerpature-induced focus shift, do not have a significant impact.

Take an image - The flat field is an image of something. That means that it must be treated like any other image. It will need a dark frame, for example. Some image processing software (e.g., Maxim DL) allows you to apply the same dark frame to both image and flat field, but I prefer to use separate dark frames for my flats. Most of the time, your flat fields will have much shorter exposures than your light frames.

Evenly illuminated surface - The flat field is often an image of a surface near the telescope. That surface needs to be evenly illuminated. Otherwise, the flat field isn't really flat. The idea with the flat field is to measure how your optical system adds and subtracts brightness from the evenly illuminated surface. When you apply the flat field to an image, the software subtracts and adds brightness to get back to even illumination. If the surface you are imaging is not evenly illuminated, then the flat field isn't flat. It's only an approximation, not a true flat field. You can get away with some unevenness, but for optimal results even illumination is essential.

Saturation level - If the flat field isn't bright enough, it won't accurately correct for uneven illumination in the optical system. If the flat field is too bright, then there is no headroom and it could over-correct. Headroom is the difference between the flat-field levels and the saturation levels. If the average value of your flat field is about one-third to one-half of the saturation values for your camera, then it's ideal.

Don't assume that the saturation value is the same as the maximum value for your camera. Those can be very different numbers. For example, a non-antiblooming chip in an SBIG ST-7E camera will hit saturation at around 40,000-45,000 ADU (analog to digital units). An antiblooming version of that camera will hit saturation around 20,000-22,000 units. For the first example, a flat field with an average value of about 15,000 to 20,000 is ideal. For the second example, a flat field around 7,000 to 10,000 units is appropriate.

Every CCD chip has a saturation value that is determined by the full charge a pixel can hold. For the ST-8E, for example, this is around 43,000 ADU. The 16-bit data stream of the ST-8E supports up to 65,000+ values, but any pixels that have a value above 43,000 ADU are in effect over-full, and have been bleeding electrons into adjacent pixels.

To calculate the saturation value for any camera, you need two numbers:

- The chip's full well capacity
- The gain in the A/D (analog to digital) converter

For example, full-well capacity in the ST-8E is 100,000 electrons. The gain is set at 2.3 electrons per ADU. A value of 1 at the computer end represents 2.3 electrons in a pixel. A value of 2 represents 4.6 electrons, and so on. To find the saturation value, divide the full well by the gain. That is 100,000 divided by

2.3, or approximately 43,500. There is some variation from camera to camera, so use a saturation level of 40,000 for ST-7E and ST-8E cameras.

For flat-field frames, aim at an average brightness that is about 35-50% of the saturation level.

> **TIP:** Always check the number of bits in the A/D conversion. The number of bits used to move the data from camera to computer determines the maximum possible value in your image. This is 2 raised to the [number of bits] power. For example, the ST-237 camera has a 12-bit A/D converter, with a maximum value of 2 to the 12th power (2^{12}), which equals 4,096. If the maximum value of the A/D converter is less than the calculated saturation value, use the maximum conversion value as your saturation value.

To calculate the saturation level for the ST-237:

- Full well: 20,000 electrons
- Gain: 4.0 electrons per ADU

To find the 100% saturation level for the ST-237, you would expect to use 20,000 divided by 4 (5,000). The 12-bit A/D converter limits you to 4,096. You would aim for a flat field with an average value of about 1300 (35%) to 2000 (50%). The newer ST-237A uses a 16-bit A/D unit, and the values are different: 20,000/0.72 = 27,000. A flat field should have an average value in the range of 10,000-14,000.

> **TIP:** If you are using SBIG's Track & Accumulate, use an average level that is much lower than the single frame saturation level. The flat field will be applied to each image that is acquired. Divide the saturation level by the number of images. For example, for an ST-7E camera and 30 images, the average value for a flat field is 40,000/30, or around 1,100 units. You can achieve this by taking a flat field with a much shorter exposure than normal. The SBIG web site has complete instructions for applying a flat field to a Track and Accumulate image here:
>
> http://www.sbig.com/pdffiles/
> flat.field.track.accumulate.pdf

Taking a Flat Field

Most camera control software includes the ability to take a flat field. A flat field is no different from any other image from the camera's point of view. It is taken with the shutter open, after all. However, your camera control software might add the word "flat" to the filename, or add a keyword to the file defining the image as a flat, so it's usually a good idea to use whatever flat-field features your software gives you.

In MaxIm DL, use the Expose tab on the camera control (see figure 6.3.1) to take a flat field. Click on the Flat radio button, and take some test exposures to determine an exposure that will give you an average level of 35-50% of saturation. Take a sequence of flats and median combine them to reduce noise.

To take a flat in CCDSoft, choose Flat as the Frame type on the Take Image tab (Camera Control panel). This disables the Reduction drop-down; you cannot AutoDark a flat field. Take a separate set of darks for your flats. There is a provision for separate darks for your flat fields in the Image Reduction dialog (see the section on Reduction Groups elsewhere in this chapter). If AutoSave is on (and you should almost always have it turned on), this will add "FLAT" to the filename, and make it easy for you to identify your flats. Regarding AutoSave, it is easier to delete bad images than to resurrect a good one that's lost!

There are several methods for making a flat field image; they range from incredibly simple to incredibly tedious. In most cases, a reasonably accurate but not

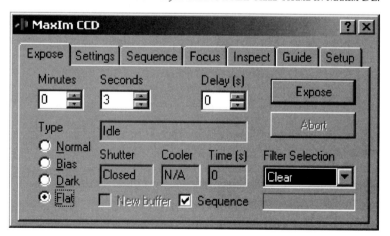

FIGURE 6.3.1. TAKING A FLAT-FIELD FRAME IN MAXIM DL.

perfect flat field will deliver excellent results. However, for high precision work such as astrometry, a carefully made, evenly illuminated flat field is essential.

This section shows you a number of different ways to make a flat field:

- The Two-Surface Diffusion Flat, which involve multiple reflections of a light source to even it out.
- The Dome Flat, which involves imaging the inside of an observatory or the wall of a nearby house, shed or other surfaces.
- The Sky Flat, which involves taking many images of the sky at twilight and averaging them.
- The True T-Shirt flat, which involves putting a T-shirt or other white cloth over the front of your scope in daylight.
- The Light Box Flat, which involves a box with a light source and a diffuser, which is placed on the end of the telescope.

The Two-Surface Diffusion Flat

The first method we will look at is a simple one: reflect light off of two surfaces to obtain even illumination. One advantage of this method is that it can be used in the field with a minimum of fuss by doing a little clever substitution for the reflecting surfaces.

Figure 6.3.2 shows the basic setup: a light source, and two reflecting surfaces. The dual reflecting surfaces act as diffusers, which is a fancy way of saying they help spread the light out evenly. Even illumination is the goal here. The first reflecting surface can even be as simple as a t-shirt being worn by one of the participants. The second reflecting surface should be parallel to the first (to help prevent any side-to-side gradient), and can be a sheet of cardboard, a wall, etc. Watch out for the light pattern on many flashlights. They often have either a central hot spot or dark spot, or some other form of uneven illumination. Uneven illumination requires more effort to smooth out the light by reflecting/diffusing it.

The more even your source of illumination, the more likely that the two reflecting surfaces will smooth it out to give you flat illumination. Your CCD camera is very sensitive, and doesn't need a lot of light to get a good exposure. You can also put a cloth over the light source to add another level of diffusion. Moving the light source during the exposure will also help to even out the illumination.

Aim for around one-third to one-half of saturation level for the average value in your flat-field exposure. A flashlight is shown here as the light source, but anything from a porch light to a string of Christmas lights bunched up can do the job. The main disadvantage to this method is that it usually takes two people to get enough distance between the two diffusing surfaces. And if your light source doesn't provide reasonably even illumination, you can still get poor results.

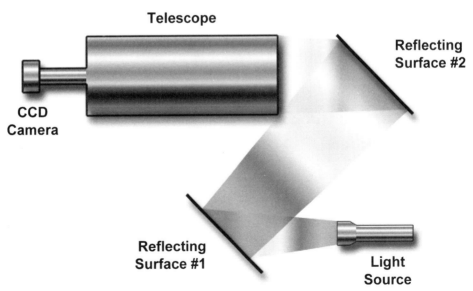

FIGURE 6.3.2. SUBSTITUTE A T-SHIRT FOR "REFLECTING SURFACE #1" AND YOU'VE GOT A SIMPLE FIELD SETUP FOR MAKING A FLAT-FIELD IMAGE.

The Dome Flat

The word "dome" doesn't just refer to the interior of an observatory; it can be any handy flat surface near your telescope that is evenly illuminated. The traditional dome flat is taken by pointing the telescope to a white painted spot on the inside of the dome, but I have used fences, house walls, the side of a shed, and so on. Any flat surface might work in a pinch. And if time is pressing or there is no other way to take a flat, even a not-terribly-evenly-lit surface may be better than not taking a flat at all. A house with clapboard siding, for example, may provide more even illumination than you expect. It will be far out of focus in the flat field, and that helps even out the brightness. Large cast shadows are more troublesome than small physical irregularities, in my experience.

If you do see irregularities in the flat field because of brightness variations in the source, try moving the scope a small amount between multiple flat-field images. Take lots of images and use median combine. The irregularities will disappear or at least be greatly reduced. If your exposures are long enough, you can also try moving the scope during the exposure. Try a slow speed rather than a fast slewing speed.

How many is "lots of images?" In order to compensate for irregular illumination or surfaces, one will need more than the minimum of three. Eight is a good low-end number, and 12 or more will often guarantee success.

The Sky Flat

The sky at twilight is a wonderful source of evenly illuminated space. There is a very, very slight gradient, but not enough to cause any trouble because you will be combining multiple images. The sky is brightest toward the sun, so aim away from the setting or rising sun when you make a sky flat.

Your CCD camera is so sensitive that, even if you can't see any stars, the CCD might. You could take your sky flats when the sky is still very bright to drown out any stars. But many CCD cameras can't take a short enough exposure to make this work. What's a CCD imager to do?

The answer is the same as my suggestion for a good dome flat: median combine. Figure 6.3.3 shows one of the sky flats from one of my own imaging sessions. It looks rather hopeless for a flat field because it is has stars all through it.

However, I took 20 sky flats as dawn approached. Before each new flat field image, I moved the scope a small amount so that the stars would be different in each image. The sky brightness is also changing fairly quickly at twilight, so you need to adjust your exposure to compensate. This is the most difficult aspect of sky flats. It takes a while to get a feel for how much to change the length of each exposure to get a similar average value for each image.

TIP: If you have a goto telescope, or if your scope is connected to planetarium software, use the software to

FIGURE 6.3.3. A SINGLE IMAGE IN A SET OF SKY FLATS. NOTE THAT MANY STARS ARE VISIBLE; THIS IS NOT GOING TO BE A PROBLEM!

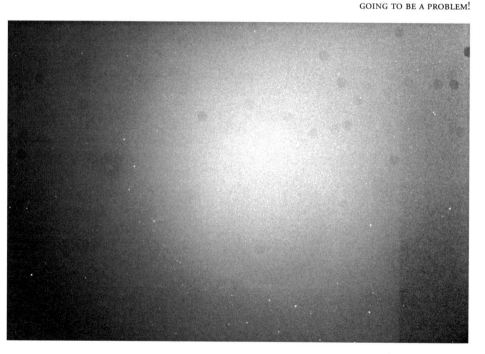

SECTION 3: USING FLAT-FIELD FRAMES

FIGURE 6.3.4. PORTIONS OF THREE DIFFERENT FLAT FIELDS IN A SEQUENCE, SHOWING HOW STARS BECOME LESS AND LESS EVIDENT AS THE SKY BRIGHTENS.

sure was shorter. The bottom third is from image #14, and the stars are barely in evidence.

Figure 6.3.5 shows the result of combining more than a dozen of the flat fields using Maxim DL's median combine. No alignment was used, just a median combine. If you look very carefully, there is just a hint of a few of the brightest stars. These can be removed using Maxim DL's Edit Pixel command on the Edit menu. Click on the Pick Up Color button and then click on a point near one of the faintly remaining stars to pick up that shade of gray. Then click and drag the color over the star until you've erased it.

Figure 6.3.6 shows the final result, with all traces of the offending stars removed.

help you point the scope at areas of the sky with dim stars. A median combine can work wonders, but give yourself a head start by using a star-poor area of the sky to make sky flats. If you have a Paramount, use the Star Search function at 8x to move the scope while you take your flats. This spreads the starlight out across the image. Any dim lines from bright stars will disappear in the median combine.

Figure 6.3.4 shows portions of three flat frames from the sequence. The top third is from the second flat I took; stars are still very evident as the sky has just begun to lighten. The middle third is from flat field #9, and the stars are less evident because the sky is brighter, and thus the expo-

FIGURE 6.3.5. THE MEDIAN-COMBINED IMAGE SHOWS ALMOST NO TRACE OF THE STARS.

FIGURE 6.3.6. NO TRACE OF THE STARS REMAINS AFTER USING THE EDIT PIXELS COMMAND.

few layers of plain white bed sheet will do the job.

The T-shirt or other fabric makes a good diffuser, although you do have to be careful about light hitting the cloth directly. I usually aim the scope downward, at a dark patch of ground, and if necessary I use a large dew shield or the shade of a nearby tree to avoid any sunlight hitting the cloth directly. If sunlight does strike the cloth, even at a very oblique angle, it can bounce around inside the scope and brighten one side or the other of your flat field, rendering it useless.

On the plus side, there are some things that you can do to get the most out of the T-shirt method. With a little practice, the following tips will allow you to get excellent flats without spending a dime:

TIP: If you wind up with too many stars in your image, then you either need to take more flat-field images before combining, or you need to wait a bit until the sky is brighter so that you get fewer stars.

The True T-Shirt Flat

I recall exactly how the inspiration struck me for T-shirt flats. I had imaged through a long winter night, and I had the best of intentions about taking sky flats just before dawn. I fell asleep and by the time I woke up the sky was too bright for flats. I was concerned that I couldn't process my images without good flats. It was too bright to take flats; it was a brilliantly sunny day.

If only I could block out most of the daylight...

A T-shirt and some rubber bands are all I needed to reduce the light level and take my flats. Figure 6.3.7 shows my "sophisticated" approach to taking daylight flats. The telescope is an Astro-Physics Traveler, a 4" APO refractor. Even the rubber bands are optional on such a small scope. For very large scopes, a

FIGURE 6.3.7. A T-SHIRT ACTS AS AN EFFECTIVE DIFFUSER.

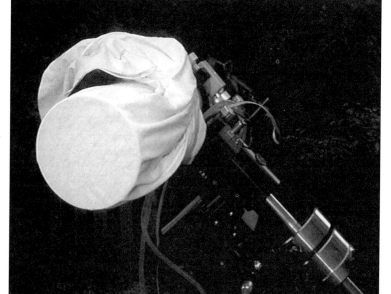

FIGURE 6.3.8. AN EXAMPLE OF A T-SHIRT FLAT.

dark on these two frames, even though they were taken with the exact same optical system. The sky flat has a brighter right-hand edge. I had been having trouble with gradients in my images after applying the sky flat. Comparing the two images shows why. The sky flat was darkening the right-hand side of the images.

I used Photoshop to subtract the T-shirt flat from the sky flat. Figure 6.3.10 shows the result of this manipulation: a gradient that is brightest at lower right. The problem was

- Use a heavy cotton cloth because you will most likely need to cut the light levels. Fold the shirt to get more layers, or use two shirts if necessary.
- Be careful to get the cloth perfectly flat across the aperture. This gives you even illumination. Use a rubber band, duct tape, or a bungee cord around the scope to keep the fabric flat.
- The surface the scope points at should have even illumination. A shady area under a tree is about right on a sunny day, and a lawn works well on a cloudy day.

Figure 6.3.8 shows an example of a T-shirt flat. There were a total of five flat-field images, and I used MaxIm DL for a median combine.

The flat in figure 6.3.8 looks pretty much like any other flat, but there's a key difference: it turns out to be a really accurate flat field. Testing shows that it models my scope's optical system very closely.

Figure 6.3.9 shows two flat field images. The top image is a sky flat, and the bottom one is a T-shirt flat, There is a clear difference in the pattern of light and

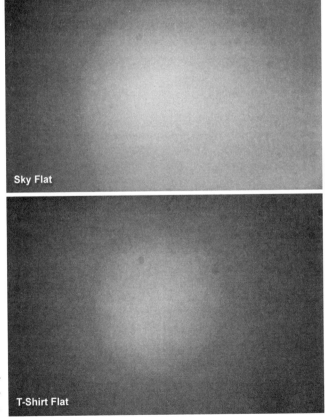

FIGURE 6.3.9. A TALE OF TWO FLAT FIELDS: ONE IS ACCURATE AND ONE IS NOT. SEE TEXT FOR EXPLANATION.

FIGURE 6.3.10. SUBTRACTING ONE FLAT FIELD FROM THE OTHER SHOWS THE DIFFERENCE BETWEEN THE TWO.

with my sky flats. My gradient problems were not from light pollution, but from a flaw in my flats.

See the Light

I have also managed to get gradients in my T-shirt flats at times, so the problem with gradients is not inherent to sky flats. With any flat, you run the risk of an oblique light source sneaking in and creating a gradient in your flats.

How you avoid the extraneous light will vary with the method you use to shoot your flats. A dew shield is one of the best tools for blocking out the extraneous light. With the T-shirt flat, you may need to make an oversized dew shield that will fit over the cloth. You can also put the cloth over your existing dew shield, but you should also have some kind of shield to block direct light from the surface of the cloth. Try pointing your telescope at different positions, and take a group of flats for each position. Compare them to each other, noting which appear to have gradients, and which do not. Apply the different flats to your images and evaluate how effective they are. When you find a position or method that gives you a clean flat, make note of what's required for future reference.

To evaluate the quality of your flat field, apply it to an image and examine the background. Be sure to image a part of the sky that doesn't have a light-pollution gradient! You should also test with an image that doesn't contain a large galaxy or nebulosity. A simple star field in the darkest area of your local sky is the best choice for a test case.

Your evaluation of the flat field will only be valid if you aren't introducing other kinds of gradients into your test image during the exposure. A really good flat field will leave you with variations of no more than 5-10 units across the test image background. In figure 6.3.10, the variation in brightness from upper left to lower right is 120 units. This doesn't

FIGURE 6.3.11. AN IMAGE OF M77 WITH ONLY DARK FRAME REDUCTION; NO FLAT FIELD APPLIED. NOTE CENTRAL HOT SPOT.

FIGURE 6.3.12. AN IMAGE OF M77 WITH THE SKY FLAT. NOTE BRIGHTER BACKGROUND AT UPPER LEFT.

I did the same for the other two images. Any differences you see in these images are solely the result of differences in the flat fields.

Figure 6.3.11 is the worst case: no flat field. A fast focal ratio (f/5) creates a hot spot in the image. Without a flat field to remove it, details on the galaxies in the image are lost in the background. You could adjust the black point to show details in M77, but then you would lose most of the dim details in the smaller galaxy closer to the center of the image.

Figure 6.3.12 shows a better but imperfect result from applying the flawed sky flat. The hot spot is gone, but, there is a new gradient. The gradient is the opposite of the gradient in the sky flat. The presence of a gradient isn't quite as problematic as the hot spot, but it interferes with optimal contrast adjustment. The processing problems are similar: optimizing the adjustments for M77 would reduce the detail visible in the smaller galaxy. One part of the background would be jet black, and the other would have a glow.

seem like much, but it's enough to create serious problems during image processing.

Let's look at a sequence of images that shows the price you pay for a missing or inaccurate flat field. Figure 6.3.11 shows an image of the area around M77. The images for this example were processed identically except for the flat field:

1. I took three images of five-minutes each with an ST-8E and a Takahashi FSQ-106.

2. I reduced the images using three different flats:
   - Dark frame only (no flat field)
   - Sky flat shown in figure 6.3.6
   - T-shirt flat shown in figure 6.3.8

3. I performed a median combine in Maxim DL for each set, leaving me with three images reduced in different ways.

4. I opened the files in Photoshop, where I adjusted levels and curves to bring out the best of the image data. Details on using these techniques are in chapter 8.

I processed each image set identically, other than the flat field differences during reduction. For example, if I adjusted the black level for one image by 20%, then

TIP: Where does the extra light come frame that causes troublesome gradients? It could come from a number of sources. The sky itself has a gradient at dawn and dusk. You might get better results with sky flats if you rotate the camera or move it to point at different areas of the sky so that any gradient will average out. Light can also reach the CCD chip via an off-axis path through the telescope. The camera itself may have a minor light leak that only shows up in the extreme brightness of daylight. Tracking down imbalances in illumination in flat fields can be challenging. But there is nothing quite as satisfying as a really accurate flat field because an even background makes image processing so much simpler.

Figure 6.3.13 shows a nearly ideal flat field result. The background is just ever so slightly brighter at lower right. The gradient is not nearly as severe as in figure 6.3.12.

The background in figure 6.3.13 is brighter than it needs to be. With no gradient to interfere, the contrast requirements are the same across the image. This makes it possible to go right to the limit on optimizing contrast. Figure 6.3.14 shows the result.

The background of figure 6.3.14 is much darker, yet no detail is taken away from the objects in the image. In fact, darkening the background enhances the faint outer detail of M77. A good flat field allows you to get the most out of your image processing.

FIGURE 6.3.13. AN IMAGE OF M77 WITH THE T-SHIRT FLAT APPLIED. NOTE THAT THE BACKGROUND IS NEARLY PERFECTLY EVEN.

Figure 6.3.15 shows four images together so you can clearly see the difference a good flat field makes.

Figure 6.3.16 shows how a good flat field can allow you to bring out faint details more effectively. The image processing for the first three images (left to right) was limited by the presence of the gradient. If I had darkened the background behind M77, I would have removed much of the detail elsewhere in the image. The rightmost example shows how much more effective processing is when you have a flat background.

Each of the images in figure 6.3.16 suffers from one limitation or another due to the uneven background.

The left image, which has no flat field applied, is

FIGURE 6.3.14. AN IMAGE OF M77 USING A T-SHIRT FLAT, FULLY OPTIMIZED. SEE TEXT FOR DETAILS.

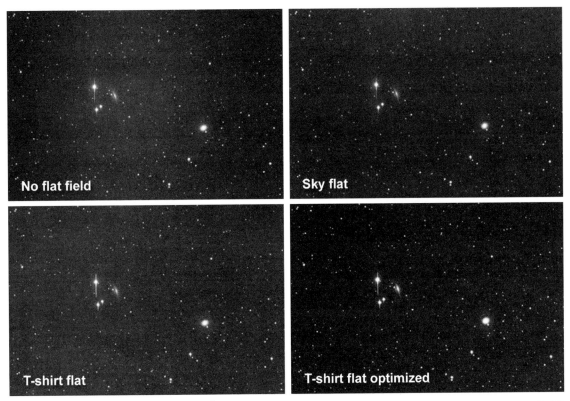

FIGURE 6.3.15. A COMPARISON OF THE FOUR DIFFERENT IMAGES FROM EARLIER IN THIS SECTION.

the worst at showing the faint outer details of M77. This is because M77 is in an area on the image where the background values are about halfway between the brightest and dimmest regions.

The next image uses the sky flat. It shows a slight improvement over the no-flat-field image, but the dim details are hard to distinguish from the background.

The third image is the T-shirt flat version before final contrast adjustments. Even without the contrast adjustments, the outer area of the galaxy shows more clearly. The flat background improves contrast between the background and all of the objects in the image.

The rightmost image shows the best dim detail of all. The contrast between M77 and the background is much better. The uniform background makes it possible to adjust image contrast without worrying about making other parts of the image too dark or too bright.

FIGURE 6.3.16. DETAIL OF M77: FROM LEFT TO RIGHT: NO FLAT, SKY FLAT, T-SHIRT FLAT, AND OPTIMIZED T-SHIRT FLAT.

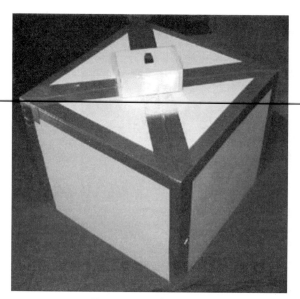

FIGURE 6.3.17. AN EXAMPLE OF A LIGHT BOX, COURTESY OF FRED HERRMANN.

Note that the final image has much better contrast following the application of the flat field. The greater the vignetting in a given image, or the greater the problem from dust shadows or other optical issues, the greater the need for a good flat field. For more information about Fred's light box, see his web page at:

http://members.home.net/mpherrmann/light-box.htm

FIGURE 6.3.18. A SAMPLE IMAGE CORRECTED WITH A FLAT FIELD MADE WITH A LIGHT BOX. COURTESY OF FRED HERRMANN.

The Light Box Flat

Rather than relying on incidental tools for taking flat fields, you can build or buy a light box to create your flats. The general idea is to provide a dim light directed toward the scope, with one or more diffusers to spread the light out and make the illumination as even as possible.

Figure 6.3.17 shows an example of a light box built for taking flat field images. The box contains a light source and a diffuser to even out the illumination. To use the box, you place it in front of your telescope, turn it on, and take a flat field image. The light box can be the simplest method of all, but there is a trick: you must get even illumination out of your diffuser. The larger your telescope aperture, the more challenging this will be. You can improve your results by using multiple light sources, and multiple layers for diffusion. One good diffusion material is translucent plastic, which when combined with four or more fairly dim light sources (such as 12-volt bulbs) can provide relatively even illumination.

Figure 6.3.18 shows an image that was corrected with a flat field created using the light box shown in figure 6.3.17 (sold by Fred Herrmann). The top image is the raw version, the middle image is the flat field, and the bottom image has had the flat field applied.

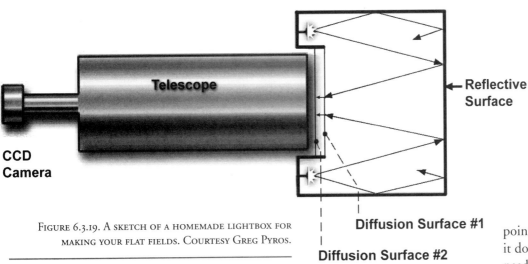

FIGURE 6.3.19. A SKETCH OF A HOMEMADE LIGHTBOX FOR MAKING YOUR FLAT FIELDS. COURTESY GREG PYROS.

You can also build a lightbox from lightweight parts. Use foam core, available at most office supply stores, and tape or a glue gun to build the box itself. Hardware stores carry a variety of translucent materials that will diffuse the light and provide more even illumination. Figure 6.3.19 shows a diagram for a lightbox designed by Greg Pyros.

The lightbox uses two diffusing layers near the telescope. A ring of dim bulbs around the outside of the box provides illumination. For field operation, use bulbs that operate on 12V. The reflective surface could be the inside of a sheet of foam core, which has a matte finish. The combination of multiple reflection paths and two diffusion layers provides even illumination that is ideal for making flat fields.

Applying a Flat Field to an Image

In this section, you will see how a flat field is applied to an image using either Maxim/DL and CCDSoft. Both products are flexible in how you can use flat fields and other types of frames for image reduction, but there are also significant differences between the two products.

Figure 6.3.20 is a raw image of the Double Cluster. The white point is set very high so it doesn't look like it needs a flat, but it does.

To get a good look at a gradient or hot spot in an image, lower the white point dramatically as shown in figure 6.3.21. This allows brightness variations in the background to take center stage. This particular image has a central hotspot; there are no visible dust motes or other optical quirks. The example shows the CCDSoft Histogram tool in action, but you can use any contrast or histogram tool in any camera control or image-editing program to do this. Be sure to use a non-destructive

FIGURE 6.3.20. A RAW IMAGE OF THE DOUBLE CLUSTER, TAKEN AT -36C THROUGH A RED FILTER.

histogram adjustment so you can put the image back the way you found it.

This technique is also good for diagnosing gradient problems. The left image in figure 6.3.22 has a central hot spot as well as a left-to-right gradient. The right image in figure 6.3.22 has pronounced hot spot and dust shadows. Dust shadows almost always require a real flat field.

The Double Cluster image does not have any serious problems, but it can be improved. The image is a 30 second exposure at -36°C. This image was taken through a Takahashi FSQ-106 refractor, a 4" f/5 instrument. The short focal length creates a very steep light cone, and typically this causes a certain amount of vignetting from center to edge. The vignetting isn't serious and it can easily be corrected with a flat field.

FIGURE 6.3.21. LOWERING THE WHITE POINT REVEALS ANY FLAWS IN THE BACKGROUND. CCDSOFT HISTOGRAM TOOL SHOWN AT LOWER LEFT.

TIP: Generally speaking, the faster the focal ratio the greater the likelihood that some degree of vignetting will occur. A fast focal ratio is most often achieved at the cost of even illumination of the field. Astrographs such as the Takahashi FSQ-106 are among the few exceptions to this. Such telescopes are specifically designed to achieve the largest possible flat, evenly illuminated field. In the mildest cases of vignetting, the center of the image will be a little hotter/brighter than the edges. In the worst cases, the edges will be substantially darker or even black, as in the right side of figure 6.3.22.

Because the image in figures 6.3.20-21 was taken at a temperature of -36C, it has very little noise from dark current. However, I did take four dark frames that I averaged and applied to the image as part of reduction. This made a small improvement to the image. There is nothing like a well-cooled CCD chip for keeping the noise down!

FIGURE 6.3.22. TWO RAW IMAGES WITH HISTOGRAMS ALTERED TO SHOW UNEVEN BACKGROUNDS.

Figure 6.3.23 shows the image open in Maxim/DL before applying a flat. The Information window is open, and I have placed the cursor roughly in the middle of the image. The average pixel value at this location is 852 ADU.

SECTION 3: USING FLAT-FIELD FRAMES

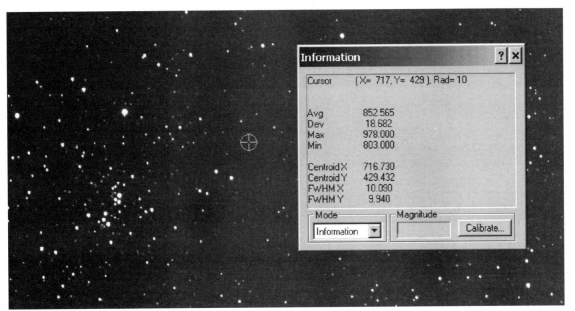

FIGURE 6.3.23. THE AVERAGE PIXEL VALUE AT THE CENTER OF THE IMAGE IS 852 ADU.

TIP: When measuring average brightness, be sure to position the cursor over an area of the sky that has no stars or other objects in it. This is especially important when working with nebulae images. Even the darkest areas of the image may contain brightness from dim areas of the nebula. Search through the image to find the darkest areas when measuring. Even if it is still a part of the nebula, at least it is the darkest part.

TIP: Variations in brightness across an image background will tend to have the most impact on color images. The brightness variations between the red, blue, and green images will generate color variations across the background when the images are combined. The even background that flat fielding creates is especially important for color imaging.

Figure 6.3.24 shows the pixel value in the top right corner of the image. It is 783 ADU, showing that the corner is darker than the center. Even though the vignetting isn't obvious in this image, the measurement confirms what we saw in figure 6.3.21: there is a slight central hot spot. The center is brighter than the edges by about 70 ADU. As small as this difference is, it will still have an impact on image processing. The image itself contains mall, subtle variations in brightness, and the variation in background brightness will hamper your ability to bring out those details. In effect, the gradient will mask the dim details in the image (in this case, that would be the dimmest stars in the clusters).

FIGURE 6.3.24. THE AVERAGE PIXEL VALUE IN THE CORNER OF THE IMAGE IS 783 UNITS, ABOUT 70 UNITS DARKER THAN THE CENTER.

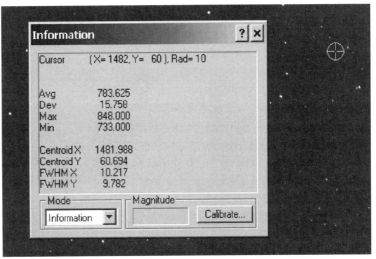

Figure 6.3.25 shows a flat field for the imaging system used to take the Double Cluster image. This is a single flat field image, and the background is somewhat noisy. There is very little dust, and what dust there is doesn't cast very strong shadows.

I took a total of four flat field images through each color filter for a total of twelve images. My intent was to do a median combine on each group of four, and to flat field the red, green, and blue images separately with matching flat fields. There was very little difference between the flat fields, however. The differences came from dust on the color filters. These filters are located relatively far from the CCD chip. The shadows they cast are weak, and do not have much impact on the image. You would need to push the image processing to an extreme degree to even see the shadow in an image. Separate flats taken through each filter are thus not always necessary. If you are after the best possible results, of course, or if you do need to push the image processing unusually far, separate flat fields will give you the best results.

FIGURE 6.3.25. A FLAT FIELD TAKEN THROUGH THE SAME TELESCOPE USED TO TAKE THE ORIGINAL IMAGE OF THE DOUBLE CLUSTER.

FIGURE 6.3.26. AN AVERAGE OF FOUR FLAT FIELD IMAGES. COMPARE TO FIGURE 6.3.25, AND NOTE THE INCREASED SMOOTHNESS. THIS INDICATES THAT THE COMBINED IMAGE HAS LESS NOISE.

Figure 6.3.26 shows the median combine of the four red flat fields. The background noise is reduced. The vignetting and dust shadows are much clearer. This shows how multiple flat fields reduce noise. The less noise there is in your flat, the less noise it will add to your image when you apply the flat. You can get

SECTION 3: USING FLAT-FIELD FRAMES

FIGURE 6.3.27. THE RESULT OF APPLYING THE FLAT FIELD.

apply it either to open images, or to an entire folder of images.

Figure 6.3.27 shows the result of applying the median combined flat fields to the Double Cluster image. Not much of a change, but you can lower the white point to verify the flat background as shown in figure 6.3.28.

Now that the image has a flat, even background, it is possible to bring out the dimmer stars without creating a "hot spot" in the center of the frame. The overall appearance is both more pleasing and more interesting. The background looks noisy when the white point is lowered too dramatically. For the final image the background level will be so low that the noise will not be visible. Figure 6.3.27 shows the contrast properly adjusted.

away with using just a single flat field, but this will leave a somewhat noisier image than when a median-combined flat is used.

To set up the flat field in MaxIm DL, use the Process | Set Calibration menu item and click on the Select Files button in the Flat Field section of the dialog. To apply the flat(s) and dark(s) to an image, open the image and use the Process | Calibrate menu item. In CCDSoft, you can right-click on an open image to apply a flat field (see example at the end of this section), or you can create a reduction group with flats, darks, etc. (see the next section) To create, manage, and use reduction groups in CCDSoft, use the Image | Reduce | Image Reduction menu item. Pick the appropriate reduction group, and

FIGURE 6.3.28. AFTER APPLYING A FLAT, THE HOT SPOT IS GONE.

THE NEW CCD ASTRONOMY

> **TIP:** When imaging nebulae and galaxies, anything that adds noise to areas of dim detail will be very troublesome. It is especially important to use median-combined flat fields with nebula and galaxy images to get the best possible signal to noise ratio in the final image. The background in the cluster image is noisy, but this is a single image. Median combining multiple images will also help to reduce the noise in the final image.

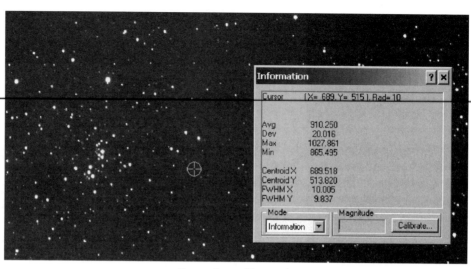

FIGURE 6.3.29. THE AVERAGE PIXEL VALUE AT CENTER IS NOW 910.

You can measure the background levels in MaxIm DL after applying the flat field. Figure 6.3.29 shows the Information dialog in Maxim/DL, with the cursor over the center of the image; the average pixel value is 910. Figure 6.3.30 shows the corner of the image; the average pixel value is 908. Mission accomplished. To measure pixel values in CCDSoft, move the cursor over a pixel and observe the brightness value in the status bar at the bottom of the program window.

You can apply a single flat-field frame to an image, or you can apply multiple flat-field frames. Both MaxIm DL v3 and CCDSoft v5 allow you to specify more than one flat-field frame, and to choose the method for combining them. MaxIm DL has a checkbox for Median combine in the Set Calibration dialog. CCDSoft offers multiple combine options when you use reduction groups.

Unlike a dark frame, which is subtracted from a light frame, the flat-field frame is applied to the light frame in a more complex manner. A dark frame represents simple offsets from the accurate value, and is therefore subtracted from the light frame. For example, if a light frame pixel has a value of 800, and that same pixel in the dark frame has a value of 500, then the corrected value of the pixel after dark frame subtraction is 300.

For a flat field, the pixel values in the light image are adjusted using a more complex formula based on the average brightness level as well as the corresponding pixel brightness in the flat-field frame. There isn't a simple relationship between the pixels in the light frame and the pixels in the flat-

FIGURE 6.3.30. THE AVERAGE PIXEL VALUE IN THE CORNER IS NOW 908.

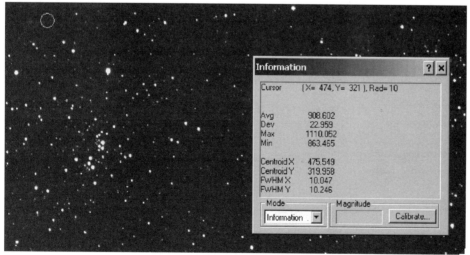

field frame; the entire image is used to effectively apply the flat field.

CCDSoft allows you to quickly apply a single flat-field frame to an image. The easiest way to do this is to right click on the image and choose Flat Field. This displays the Flat Field dialog box (see figure 6.3.31). Select the light frame in the top of the dialog, and the flat-field frame in the bottom. For both images, you can select either a currently open image, or a file. In figure 6.3.31, the light frame is a file currently open, and the flat-field frame is a file.

If you check the "Create new window" checkbox, CCDSoft will display the result of applying the flat field in a new window, leaving the original light frame untouched.

You can also quickly apply a single dark frame to an image using CCDSoft. When you right click on an image, the pop-up menu also shows an option for applying a dark frame. The correct order is to apply the dark first, and then the flat field. A right click is also the fastest way to get to the Histogram tool.

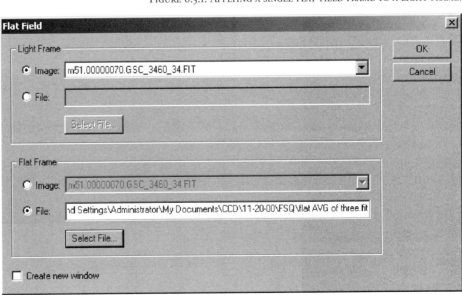

FIGURE 6.3.1. APPLYING A SINGLE FLAT-FIELD FRAME TO A LIGHT FRAME.

Section 4: Image Reduction in Action

Image reduction is the art of applying your dark frames and flat fields to your images. But before we get into the details, a short look at the lowly bias frame is in order.

A dark frame is a snapshot of your camera's instrument noise. You take one so you can subtract it from your images, giving you a cleaner image. A bias frame represents the system noise that is always there, no matter how short or long the exposure is. If you could take a zero-length exposure, there would still be some instrument noise in that exposure. The bias frame isn't exactly zero length. It's the shortest possible exposure your camera can take. This makes it a very good approximation of that residual noise that is always there.

This residual system noise, the bias, is always the same amount. If you scale your dark frames, such as halving a two-minute dark frame to reduce a one-minute exposure, you do *not* want to also scale the bias. You take an image of the bias, and you subtract it from your darks before you scale them. Actually, the software will do it for you. How you take and apply a bias is explained below.

First, I have one other quick note about image reduction. SBIG cameras include a 100-unit pedestal (that is, 100 units are added to the value of every pixel). This is not a part of the bias. This pedestal is added to all frames (bias, dark, flat-field, and light) by the camera driver supplied by SBIG. MaxIm DL, CCDSoft, and other camera control programs take the pedestal into account automatically during data reduction. If you ever attempt your own reduction, you'll need to account for the pedestal before performing any math on an image.

Taking a Bias Frame

Most camera control programs have a simple method for taking bias frames. Figure 6.4.1 shows the CCDSoft Camera Control panel. The Frame type at the center of the panel has been set to Bias. The bin mode has been set to 1x1; the temperature has been set to match the dark frames with which the bias will be used. Note that several items are automatically disabled, including exposure time and reduction. Neither of these has any meaning with respect to a bias frame.

The "Series of" setting is 3. I recommend taking at least three bias frames. These can be median combined to reduce noise. Bias frames are quick and easy, so take as many as you like. The more you take, the lower the noise will be. Combining images always reduces the noise level, and it works for bias frames just as surely as it works for every other type of frame.

Figure 6.4.2 shows a typical bias frame. It looks a lot like a dark frame, and this is to be expected. After all, a bias is simply the shortest possible dark frame.

The bias frame you see from your own camera will vary as a result of differences between cameras, camera models, manufacturers, etc. It will also vary depending on the temperature you take the bias frame at. The colder the temperature, the less noisy the bias will be. However, the temperature of the bias frame absolutely must match the temperature of the dark frame you plan to use it with. A difference in temperature will add noise that will reduce the quality of your image.

The noise in a typical bias frame is at an extremely low level. Even a nasty looking bias like figure 6.4.2 will do more good than harm!

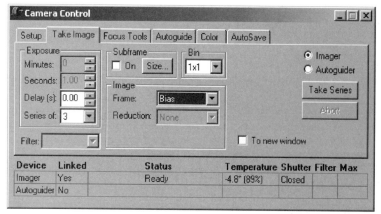

FIGURE 6.4.1. TAKING A BIAS FRAME IN CCDSOFT.

SECTION 4: IMAGE REDUCTION IN ACTION

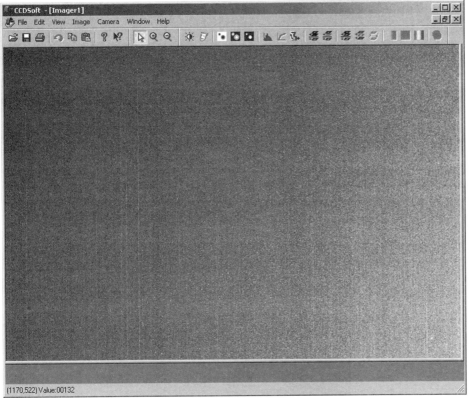

FIGURE 6.4.2. A SAMPLE BIAS FRAME.

To take a bias frame in MaxIm DL, click the Bias radio button at bottom left of the Expose tab (see figure 6.4.3). The exposure time is grayed out since a bias frame is always the shortest available exposure. Check the Sequence checkbox to take multiple bias frames. Set up the filename and numbering scheme on the Sequence tab. Filter selection is irrelevant for a bias frame.

If all of your dark frames and light frames have the same exposure duration, then you don't have to take bias frames. Bias frames are used to scale dark frames when the exposure time of light and dark frames differs.

Using Reduction Groups

Although version 3 of MaxIm DL allows you to use multiple dark, flat field, and bias frames for image reduction, CCDSoft's Reduction Groups are my favorite way of handling image reduction. The ability to create named groups means that you can keep yourself well organized. If you routinely take multiple reduction frames, the ability to organize them in groups is a great help.

The reason for taking more than one of each reduction frame is simple. Yes, a single bias, dark, or flat-field frame reduces the instrument noise in your images. But this comes at a cost. There is noise in every reduction frame. Applying the dark, flat-field, or bias frame to your images adds that noise to your image. The goal is therefore to use bias, dark, and flat-field frames that have as little noise as possible. The best way to do that is to take multiple reduction frames and median combine them.

CCDSoft's reduction groups make it really easy to combine these images in ways that reduce overall noise. You can even apply the reduction groups automatically when you are taking images with the camera. Simply select "Bias, Dark, Flat" as the Reduction type (see fig-

FIGURE 6.4.3. TAKING A BIAS FRAME IN MAXIM DL.

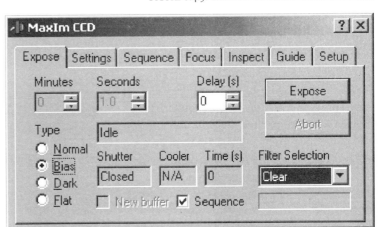

ure 6.4.4), and then choose the reduction group to use on your images.

Make sure you turn AutoSave on. CCDSoft will save two versions of each file: the raw image, and the reduced image. If an error occurs during reduction, the raw data is still saved to disk. Whenever you use the Image Reduction dialog to reduce images, the word "REDUCED" is added to the filename. If you reduce an open file, you must use the File menu to save the changes to disk.

If you have TheSky running on the same machine, the AutoSave filename will also include information supplied by TheSky about the current position of your mount. If AutoSave is turned off, only the reduced version of each image appears in a window in the CCDSoft workspace. If an error occurs during reduction, the original data is lost. My advice therefore is simple: turn AutoSave on and leave it on all the time.

In MaxIm DL, you can use the current "calibration" setting to automatically reduce each image as it is acquired. Figure 6.4.5 shows the Settings tab of the camera control. In the Auto Calibration section, select "Full Calibration." This applies the current dark, flat, and bias frames from the most recent "Set Calibration" session, and uses the various settings from that dialog.

Creating Reduction Groups

Before you can create a reduction group, first take the bias, dark, and flat-field frames you plan to include in the group. The Image Reduction dialog (Control + R) comes with two default groups, but you should ignore those and create your own groups. Figure 6.4.6 shows the default starting point. To create a new group, click the Add Group button, name the group, and you are ready to go. I suggest including the telescope name, camera name, exposure length, bin mode, and

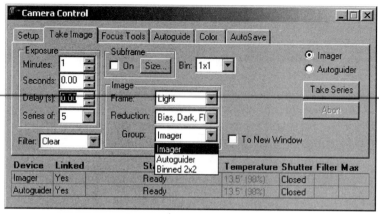

FIGURE 6.4.4. SETTING UP AUTOMATIC IMAGE REDUCTION IN CCDSOFT.

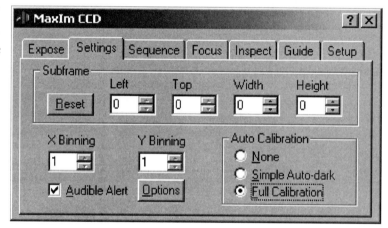

FIGURE 6.4.5. SETTING AUTOMATIC CALIBRATION IN MAXIM DL.

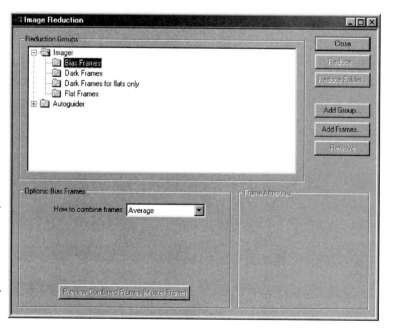

FIGURE 6.4.6. THE IMAGE REDUCTION WINDOW.

SECTION 4: IMAGE REDUCTION IN ACTION

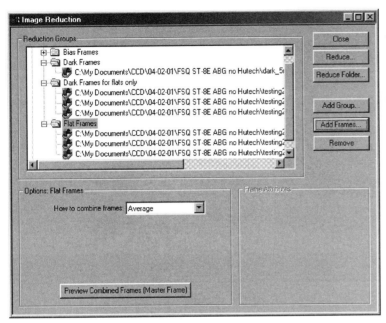

FIGURE 6.4.7. FILES HAVE BEEN ADDED TO THE SUB-GROUPS.

To add reduction frames to a group, highlight the appropriate slot (bias, dark, dark for flats, flat) and click the Add Frames button. This opens a Select Frames dialog. Navigate to the folder that contains the files you want to add. You can add more than one frame at a time; use Shift+Click and/or Control+Click to select multiple images. Click Open to add the files.

Figures 6.4.7-8 show files added to a reduction group. You can resize the Image Reduction dialog to see long filenames. Click on the plus and minus icons to view or hide the files in the slots.

Note that the bottom portion of the Image Reduction dialog changes depending on what slot is currently highlighted. In figure 6.4.7, the Flat Frames slot is highlighted. This is an option that determines how the flat frames are combined: Average or Median Combine. If you have at least 3 images, use median combine for bias, flat-field, and dark frames. It removes flaws more effectively. If you highlight an individual frame, CCDSoft will display information about the file, including temperature, bin mode, etc. (see figure 6.4.8).

temperature in the group name. This makes it easy to find a matching group when you need one. If the telescope and camera don't change, you can leave those out of the group name.

For example, typical group names would be "AP130 ST-8E -30 2x2 10min" or "C11 ST-9E -25 1x1 30min." This makes it easier to apply groups correctly. I wind up processing images long after I take them, so detailed group names make things easier.

TIP: You can use the auto save options in both CCDSoft and MaxIm DL to include this same information in your filenames as well, further simplifying file management work for all of your images.

A reduction group has slots for bias frames, dark frames, flat frames, and "Dark Frames for flats only." That last item allows you to include separate darks for your flat-field frames. Flats are often shot with different temperatures and shorter exposures than your light frames, and you can get better results by taking matching dark frames for your flats.

FIGURE 6.4.8. SETTING DARK FRAME OPTIONS.

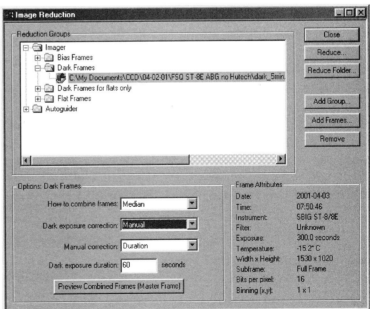

THE NEW CCD ASTRONOMY 285

CCDSoft creates master frames for the bias frames, dark frames, etc. The master frames are created automatically the first time you use a reduction group. If you add, remove, or modify files in a reduction group, the master frames are updated automatically the next time you use the group. If a file is missing, the reduction will not proceed. Remove the reference to the missing file and restart reduction.

You can create a master frame for any sub-group by clicking on the Preview Combined Frames button. A copy of the master frame appears in a window, and you can examine it to make sure it meets your requirements.

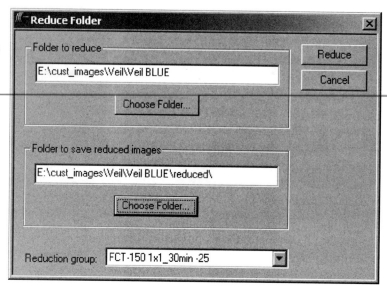

FIGURE 6.4.9. REDUCING ALL OF THE IMAGES IN A FOLDER

The Remove button acts differently depending on what you highlight in the Image Reduction window:

- If a single frame is highlighted, the frame is removed from the group. You cannot select more than one frame at a time.
- If a sub-group is highlighted, the frames in the sub-group are removed. The sub-group itself remains.
- If the group is a default group (Imager and Autoguider are the default groups), the empty group remains. If the group is one you created, the group is removed.

If you double-click on a file in a sub-group, the file opens in a new window.

TIP: To reduce a single image with a reduction group, open the image in CCDSoft and then open the Image Reduction window (Control + R). Click the Reduce button to open the Reduce Now dialog. Select the image(s) you want to reduce, choose the reduction group, and click OK.

You can also use the Image Reduction window to reduce a folder of images. This is a powerful option that allows you to get a lot of reduction done quickly. Copy the files requiring the same reduction group to a folder. Click the Reduce Folder button and set the source and destination folders (see figure 6.4.9) and choose the reduction group.

I usually arrange my files as follows:
- One folder for each night I image.
- One sub-folder for each equipment variation I use that night, such as changing cameras or scopes.
- I always use AutoSave to number files sequentially.
- Before processing images, I create folders for each object. For example, I might have an M27 folder and an NGC7331 folder, and I copy the raw files into these folders.
- For each object, I create **reduced** and **aligned** folders. They are targets for reduction and alignment operations.
- If the image involves color combination, I create subfolders for each set of component images (red, green, blue, luminance). The reduced, aligned images ultimately go into these folders.

After reduction and alignment, I am ready to sum or median combine images, color combine images, process images, etc.

Section 5: Other Tips on Cutting Down Noise

This section covers two important techniques for reducing noise in your images:

- Aligning and combining images
- Removing light-pollution gradients

Combining images improves the signal to noise ratio of your images because signal increases faster than noise. The signal increases linearly as you add exposures. The noise increases quadratically as the square root of the sum of the squares. This isn't as obtuse as it sounds! If you add five images together, the signal is fives times higher. The noise increases by a smaller amount, the square root of five (2.24). The additional exposures improve the signal-to-noise ratio by the square root of the number of exposures.

For example, assume a single image has a signal level of 10,000 and noise of 100. The signal to noise ratio is 100:1 (10,000/100). If you add five exposures with similar signal and noise levels, the signal becomes 50,000, and the noise becomes 100 * 2.24, or 224. The new signal to noise ratio is 50,000 divided by 224, or 224:1, a substantial improvement. In actual practice, the signal and noise levels of the individual images would vary somewhat, but the net result would be very similar to this example.

The most common methods for combining images are add, average, and median combine. Each has advantages and disadvantages covered in this section. Combining images can make a big difference in image quality.

Aligning Images

Before you can combine images, you need to align them. Camera control software typically provides several different types of alignment tools, each with different levels of precision and functionality:

Align centroids - Aligns several images to sub-pixel accuracy, using the centroid of a star common to all of the images. This method, although quik, does not solve problems with image rotation or scaling.

Align using multiple centroids - Uses multiple stars in each image to shift and rotate images into alignment.

Specialized alignment tools - Many software packages include advanced tools for alignment. MaxIm DL uses Auto Correlation, and CCDSoft uses object extraction.

Align groups of images - This is basically automatic alignment, where the software figures everything out for you.

Manual alignment - Has tools to manually shift and/or rotates images in whole or fractional pixels. Fractional (sub-pixel) shift and rotation are more precise.

Align Centroids (CCDSoft v5)

CCDSoft provides a simple method of aligning images using centroids. Click the Mark Centroid tool in the Astrometry toolbar (see figure 6.5.1). The cursor changes to a crosshair (see figure 6.5.2).

Click on the same star in all images to mark centroids, and press Ctrl + A to align images. The active image remains unmodified, and the other images are lined up with it. Mark Centroids does not rotate the image, it uses shift only.

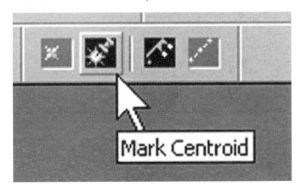

FIGURE 6.5.1. THE MARK CENTROID TOOL.

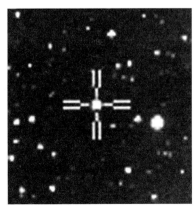

FIGURE 6.5.2. CENTROID CURSOR.

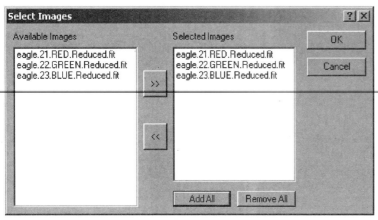

FIGURE 6.5.3. SELECTING IMAGES TO ALIGN.

Alignment (Maxim DL)

Maxim DL includes multiple image alignment options in a single tool. To align images, open all of the images and choose the Process | Align menu item. This opens the Select Images dialog, shown in figure 6.5.3. Highlight the images you want to align and click the ">>" button to put them in the Selected Images list. Click the Add All button to select all images. Click OK.

An image window with the title "Align" opens, as well as the Align Images dialog, as shown in figure 6.5.4. There are five different alignment methods available to you in Maxim DL:

Auto-correlation - Uses stars or other objects in the image to align to sub-pixel accuracy. This option provides shift only, there is no rotation. This function only works properly if the offset between images is relatively small. Occasionally it will fail utterly, in which case try the "Manual 1 star" method. Try auto-correlation for lining up planetary images.

Auto-star matching - Uses the stars in the image to align. Both shift and rotation are used as needed to align the images to sub-pixel accuracy. If you don't get a good match, try the "Manual 2 stars" method.

FIGURE 6.5.4. ALIGNING IMAGES IN MAXIM DL.

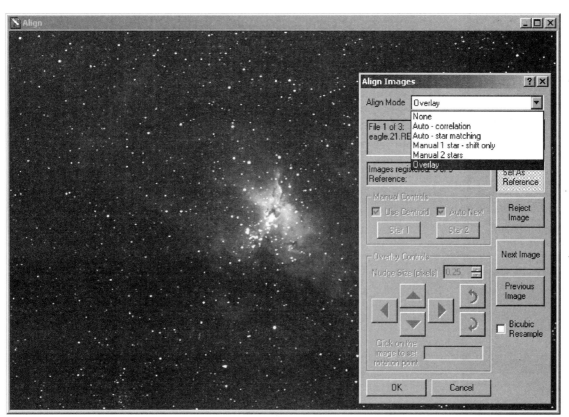

Manual 1 star-shift only - Uses one star in each image to align. You click on the same star in each image, one at a time, to set up the alignment. Uses shift only to achieve alignment. In order to speed up the alignment, turn on the Auto Next check box. Be sure to turn on the Use Centroid check box to get sub-pixel accuracy; this will cause Maxim DL to calculate the exact center of each star image for you.

Manual 2 stars - Uses two stars in each image to align. You click on the stars to tell Maxim DL which stars to use. This method performs both a shift and rotate as appropriate. Remember to turn on Auto Next.

Overlay - This is a fully manual alignment mode. You use the arrow buttons to shift and/or rotate the images while Maxim DL shows you two images overlaid, one in green and one in pink (see figure 6.5.5). You can adjust by shifting (up/down/left/right buttons), or by rotation using the curved arrow buttons. To use the rotation buttons, click on the image to set a rotation point. I recommend that you first align a star using the shift buttons, and then set that star as the center of rotation. Adjust the rotation amount so that all stars line up properly. You can use numbers as small as 0.25 for shifting, and as small as 0.01 for rotation. Overlay also works reasonably well with planetary images, but it can be difficult to find features with enough contrast to make the two-color method work for you.

I recommend that you check the Bicubic Resample box if you want the highest possible image quality. If you come across an image that would degrade the alignment or would not be useful in a combine operation, you can click the "Reject Image" button and it will not be included in the alignment. If you want to change the image used as the reference image, navigate to that image with the Next and Previous buttons,

and then click the "Use As Reference" button. If you are doing a manual alignment of any kind, you'll have to start over if you change the reference image.

If you use auto-correlation or auto-star matching alignment modes, try running the Align three times to refine the alignment as much as possible. This tip comes from various users of MaxIm DL.

The Overlay method requires a lot of manipulation on your part, but it is a good tool when you are having trouble with the automated alignment methods. Figure 6.5.5 shows how to use the Zoom mode of the Information window to magnify any alignment errors, making it much easier to achieve an accurate manual alignment. Check the corners of the image for field rotation problems; that is where they are most likely to show up. If you have pink-over-green at lower right, and green-over-pink at upper left, that is a sure sign that you need to perform a rotational adjustment.

If you find it necessary to use the Overlay method, start by aligning one of the brighter, non-blooming stars near the center of the image. Start by using larger numbers in the "Nudge Size" box. Gradually work you way down to 0.25 pixels to get optimal alignment. If you have an exceptionally sharp image, you might see

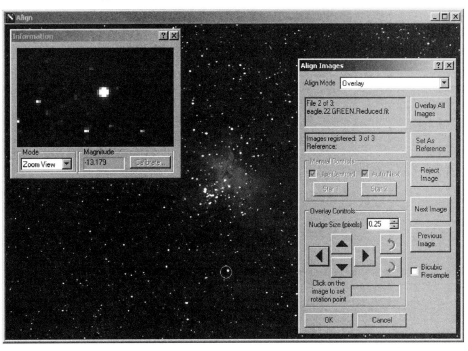

FIGURE 6.5.5. USING THE ZOOM MODE OF THE INFORMATION WINDOW DURING ALIGNMENT.

some benefit from aligning with a 0.10 nudge size, but that's too small to be useful for most images. Check the corners to see if you need to deal with field rotation. The Zoom View will often reveal very subtle field rotation, and it's worth your time to compensate for it. If you need to adjust rotation, start with very small values in the Nudge box -- something in the range of 0.05 or 0.03 should work, but you can go as small as 0.01 pixels for rotation when you have very large images such as from an SBIG ST-8E camera.

Align Folder of Images (CCDSoft V5)

CCDSoft uses a sophisticated pattern-matching algorithm, based on SExtractor (Source Extractor, software that finds the stars and galaxies in an image) to align an entire folder of images down to the sub-pixel level. It both rotates and shifts the images as necessary. It will not scale the images to align them. If your images require scaling, use Registar, covered later in this chapter. CCDSoft's alignment is more accurate than MaxIm DL's, and Registar is slightly better still.

When aligning a folder of images, only the images you want to align should be in the folder. If you put any other images in there, it will drive CCDSoft crazy trying to find a match!

The images to be aligned can have different bin modes. You can align a typical color image set, with a luminance image binned 1x1 and RGB images binned 2x2. The images will not be resized, just aligned.

If you try to align bloomed and non-bloomed images CCDSoft might not be able to align the images. It will report a "Pattern Match Error" when this occurs.

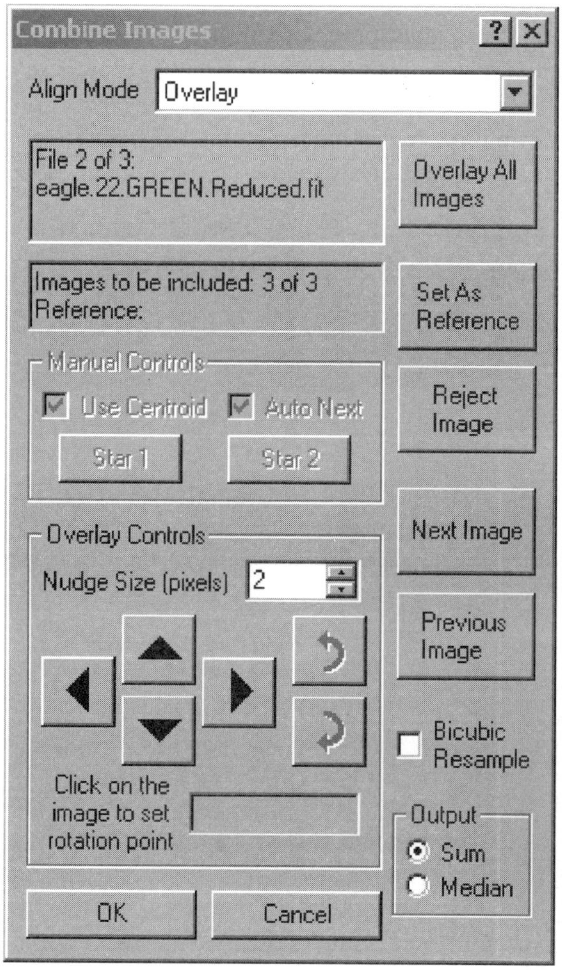

Combining Images (Maxim DL)

Combining images in Maxim DL is very similar to performing an alignment. In fact, you can do both at the same time, as the alignment controls are active when you are doing a combine. Figure 6.5.6 shows the Combine Images dialog. It adds Output choices at the very bottom right of the dialog: Version 3 is similar, but adds Average to the output options.

Other than selecting an output method, the Combine Images dialog works just like the Align Images dialog. Please see the description of the Align Images dialog earlier in this chapter for details. If you have already used Align Images before combining, you can select "None" as the Align Mode to speed up the combine process.

Combining Images (CCDSoft V5)

CCDSoft includes two different ways to combine images:

Combine Images - Combines two (and only two) images using a variety of methods. This includes Add, Subtract, Multiply, Divide, Blend, Dark Subtract, and several others.

Combine Folder of Images - Combines any number of images using averaging, median combine, or summing.

FIGURE 6.5.6. COMBINING IMAGES IN MAXIM DL

SECTION 5: OTHER TIPS ON CUTTING DOWN NOISE

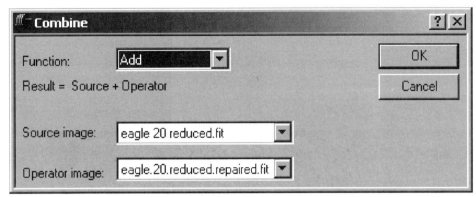

FIGURE 6.5.7. COMBINING TWO IMAGES USING THE ADD FUNCTION.

You can choose from a wide variety of combination functions when using the Combine dialog (see figure 6.5.7). The list of files for source and operator images will only include images that are the same size as the currently active image. You cannot combine images of different sizes. There is a long list of combination methods available, including add, subtract, multiple, divide, blend, logical AND, logical OR, exclusive OR, dark subtract, flat field, and add by track list.

That last option builds a flat field image according to a Track & Accumulate track list. This option creates a flat field by shifting and adding a normal flat field to itself using exactly the times and shifts from the track list. The resulting flat-field image can be applied to a Track & Accumulate image to reduce it. The flat field is not derived from the Track & Accumulate data. You take a normal flat field, and this process modifies it so it can be applied to a Track & Accumulate image. For this to work, you need a flat field whose average brightness level is equal to 50% of the camera's saturation level divided by the number of exposures in the Track & Accumulate image.

When you combine a folder of images, make sure that only the images you want to combine are in the folder. If you used Align Folder of Images to align the images previously, they are already in the output folder from that operation. The result of the combine appears in a new window.

The images to be combined must have the same size and bin mode. If necessary,

resize using the Image | Resize | Resize Folder of Images menu item.

You have three options for combining: add, average, and median combine. When the images are very clean (no satellite tracks, no serious cosmic ray hits, etc.), use the Add option to obtain the best possible signal to noise ratio. If you do have satellite tracks or other problems in your images, and you have at least three images, use a median combine. If you have large numbers of images, or a limited amount of memory, an average combine will be much faster than a median combine.

Aligning and Combining with Registar

Registar is an imaging tool that does only one job, but it does it extremely well. It aligns images to each other very accurately. This allows you to combine images effectively. Registar can also help you build mosaics; see chapter 9 for information on that subject.

Figure 6.5.8 shows a wide variety of images of the Crescent Nebula. These images were taken with different cameras and telescopes, and they mostly have different image scales and/or bin modes. I used them to

FIGURE 6.5.8. REGISTAR CAN HANDLE IMAGES OF DIFFERENT SIZES AND IMAGE SCALES AS SHOWN HERE.

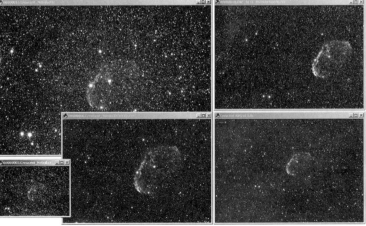

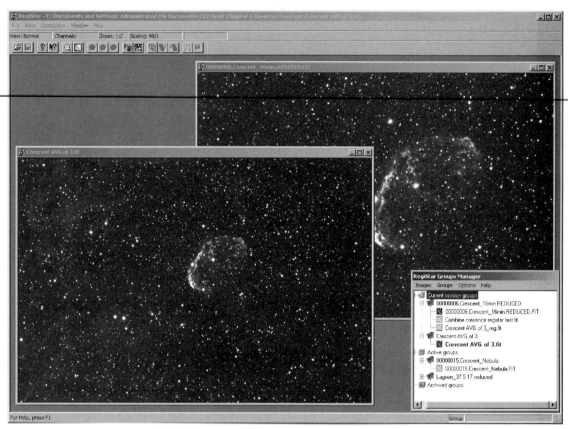

FIGURE 6.5.9. ALIGNING TWO IMAGES WITH REGISTAR.

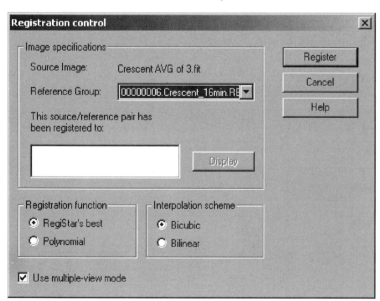

FIGURE 6.5.10. THE REGISTRATION CONTROL.

test Registar's abilities. I figured if Registar could handle such dissimilar images gracefully, it must be a pretty good alignment tool.

The short version of the results is that Registar passed with flying colors. It was able to take extremely dissimilar images and align them very accurately. The following examples show an alignment of two images. Although the images are the same size overall (they are both ST-8E images), they were taken at significantly different image scales. One image was taken with a Takahashi FCT-150 refractor at an image scale of 1.77 arcseconds per pixel. The other image was taken with a Takahashi FSQ-106 refractor at an image scale of 3.51 arcseconds per pixel. Registar can handle this difference easily.

FIGURE 6.5.11. REGISTAR SCANS IMAGES TO LOCATE STARS.

Registar takes a bit of getting used to, but it works well. The interface stumped me at first, but the product includes a detailed tutorial (you'll need it!). Start by loading the images you want to align into Registar (see figure 6.5.9). The images belong to groups. The group manager window appears at lower right in figure 6.5.9.

TIP: If you are aligning images (it doesn't matter whether you are aligning two images or 200), start by determining which image will be your reference image. As you align additional images, they will all become part of this reference image's group. This puts all of the images in one group, and when you go to combine images you will have a much easier time.

For this example, I arbitrarily chose the FCT-150 image as the reference image (at upper right in figure 6.5.9). Select one of the other images to align it to the reference image. There is only one other image here, at lower left. I clicked on that image to select it, and then clicked on the Register button in the toolbar (roughly in the middle of the toolbar). This opens the Registration Control dialog (see figure 6.5.10).

There is a drop-down list at top center that contains all current groups. At the start each image is its own group (confusing, but this is the way it works). Select the reference image from the list. Ignore the other settings in this dialog, and click the register button. Registar scans an image to identify the stars in the image. While this is happening, and if you haven't changed the default preferences, Registar will show you a progressive scan of each image. The scan reveals where Registar thinks the stars are in each image. Figure 6.5.11 shows a scan in progress. The upper portion of the image has been scanned, showing just the stars. Nebulosity disappears because Registar uses only the stars in the image for alignment.

When the scan is completed for both images, Registar calculates the alignment and displays the result in Multiple View (see figure 6.5.12). Multiple View shows both images scaled and aligned. You can use the number keys to see each image (1 and 2), or press 0 to show all images at the same time. Registar uses red and blue to show the images. Where the images overlap, content shows up as white. Where content only belongs to one image or the other, it shows up in that image's color. In figure 6.5.12, only the red of the larger image is visible. The entire small image overlaps with the larger one.

FIGURE 6.5.12. THE ALIGNED IMAGES ARE SHOWN TOGETHER.

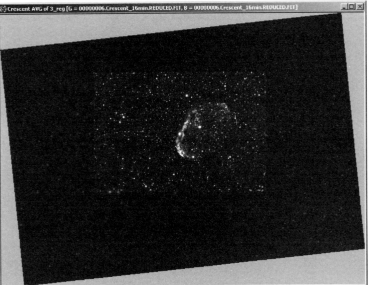

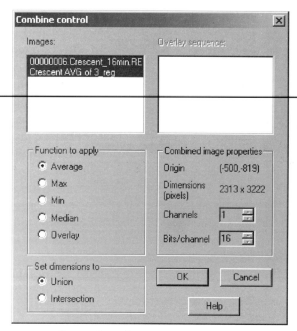

Figure 6.5.13. Choosing the images to combine.

The ability to scale images for alignment is one of the coolest Registar features. It opens up the ability to align and combine images with different image scales.

At this stage, you can either save the aligned and scaled image for use in another program, or you can perform the combination right in Registar. To com-

Figure 6.5.14. The combination doesn't have matching contrast.

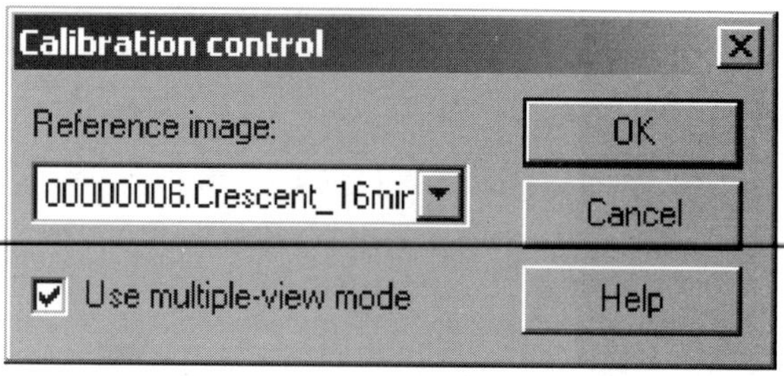

Figure 6.5.15. The Calibration control matches contrast.

bine the two images in this example, click on the Combine button to open the Combine control (see figure 6.5.13). The two main ways you can use this tool are to blend the images or to overlay the images. Select the images you want to combine from the list at upper left, set the combination method, and click OK. For this example, I chose Average and Union.

TIP: Union outputs all parts of all images, including overlapped and non-overlapped portions. Intersection outputs only the portions of the image that where all of the input images overlap.

Figure 6.5.14 shows the result of the combination. Whoops! The two images don't have similar histograms, and the combination shows this all too clearly.

If you were going to save the aligned images and combine them in another program, this wouldn't be a problem because you could adjust the histograms in the other software. But if you are going to do the combination in Registar, you need to right-click on the non-reference image and choose Calibration.

Figure 6.5.15 shows the Calibration control. Select your reference image from the drop-down list, and click OK. Registar will adjust the histograms automatically when it combines the images.

Figure 6.5.16 shows the Registar output when Calibration is included. The levels of the two images match much more closely. There is still a slight mis-match, but I find that programs like Photoshop can handle this adequately.

FIGURE 6.5.16. THE COMBINED IMAGES NOW HAVE MATCHING CONTRAST.

If you overlay with the low-resolution image on top, you'll get low resolution, as shown in the left of figure 6.5.18. If you put the high-resolution image on top, then the center of the final image will have high resolution, as shown at right in figure 6.5.18. And if you blend, such as by averaging, you wind up with an image that is in the middle of the two extremes (center image of figure 6.5.18).

Track & Accumulate

Track & Accumulate is an image acquisition method specific to SBIG cameras and software. It allows you to quickly and efficiently take multiple exposures, which are added together automatically. It is most useful with SBIG cameras that do not have a self-guiding chip, such as the ST-237.

Figure 6.5.17 shows the output from Registar adjusted for optimal display in Photoshop. Note that when Registar saved the output, it filled in the outer edges with black, not the gray you see when working inside Registar.

For the best results, use Registar to align images, and then combine them in another program that has more options for combining, such as Photoshop. Photoshop allows you to store each image in a separate layer, for example. You can apply histogram adjustments and filters to each layer as appropriate.

The method you use to combine images in Registar will affect the appearance of the final image. Figure 6.5.18 shows several different methods of overlaying the two images of the crescent. Because one of the images was scaled up quite a bit, it has a lower resolution.

In Track & Accumulate mode, all images are added together automatically. Each new image is added to the final image after it is downloaded. This means that you cannot pick and choose among your images to decide which ones will be combined.

FIGURE 6.5.17. THE FINAL IMAGE, WITH HISTOGRAM TWEAKED IN PHOTOSHOP.

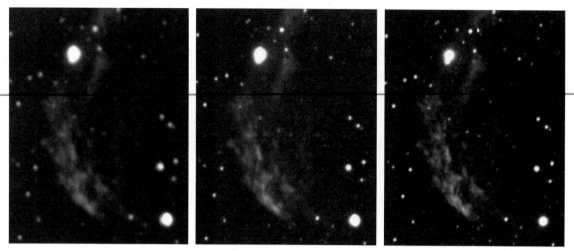

Figure 6.5.18. The combination method affects the resolution of the final image.

Although Track & Accumulate is high on convenience, it is low on flexibility. It's a great way to get started in multi-image imaging, and it can help you get more out of your short exposures. But you can get even better results by saving your multiple images to disk using your camera control program's auto save features. You can then examine the images to decide which are the best, and combine only those using the appropriate method: add, average, or median combine.

The biggest hazard with Track & Accumulate comes from short exposures. If the exposures are too short, the noise in the images lines up and creates a noisy background (see figure 6.5.19). To cure this problem, use longer exposures so that there is more signal in your images in relation to the noise.

Combining images does reduce noise, but if the noise level is very high you probably won't be able to take enough images to reduce the noise to a level where the image looks good. The brighter the sky glow, the longer your individual images must be to avoid the streaking when the images are combined.

This problem can occur outside of using Track and Accumulate, too. Any time your images have too short of an exposure, combining them will be less likely to remove enough noise to give you a good result.

Figure 6.5.19. A too-short exposure time shows excessive noise when combined.

Section 6: Dealing with Light-Pollution Gradients

FIGURE 6.6.1. THE EFFECTS OF LIGHT POLLUTION ON YOUR IMAGES CAN BE FRUSTRATING.

Even if you apply a flat-field frame to your images, you still may not get a flat background. This can happen when the flat field isn't as flat as it should be. Uneven illumination of the flat is the cause. A poor flat leaves a background gradient that prevents you from showing all of the details in the image.

A gradient often results from light pollution. Light pollution is one of the most annoying -- and common -- sources of trouble in images. Sky glow from light pollution is not evenly distributed. It occurs in what we are called light domes. These domes have a distinct gradient from base to edge. With the sensitivity and range of a CCD camera, even a very small field of view will record this gradient.

Images taken anywhere near a town or city will show a gradient in the image. The background will vary in brightness across the image in direct proportion to the level of light pollution. If you have more than one source of pollution, you can wind up with more than one gradient.

Wide-field images are the most likely to show gradient problems. They cover more sky and thus have a larger gradient from edge to edge (see figure 6.6.1). But almost any field of view will show the effect. The brighter the source of light pollution, the more pronounced the effect will be.

Figure 6.6.1 shows an example of a gradient from light pollution in my own back yard. The two most common light pollution colors are green and orange, because the lamps used for outdoor lighting commonly use elements (mercury and sodium) that emit in these colors. Such gradients can make a mess of an otherwise well-executed image. You can see in figure 6.6.1 that the Horsehead and Flame nebulae are actually clear and well defined, but they are overwhelmed by the gradient from light pollution.

FIGURE 6.6.2. THE RESULTS OF CLEANING UP THE GRADIENT FROM FIGURE 6.6.1.

Fortunately, you can work some image processing magic on gradients. The result will not be as good as an image taken from a dark sky location, but as you can see from figure 6.6.2, which is a cleaned up version of figure 6.6.1, the results can be quite good.

FIGURE 6.6.3. A GRADIENT RESULTING FROM A POOR FLAT FIELD.

There are four things you can do to reduce the effects of light pollution. Each operates at a different point in the imaging process. In increasing order of difficulty, they are:

- Use a light pollution filter to cut down the amount of pollution that reaches your camera. Filters that are good for visual use are not necessarily good for photographic purposes, however.
- Use software that detects, measures, and removes gradients.
- Create an image that has exactly the same gradient in it, by one of several available mans, and subtract it from the CCD image.
- Create an image using painting tools that has the same gradient in it. Use this image as a mask to remove the gradient. A mask is different from a selection in that you can change the mask by using painting tools.

image (the right-hand side) is black. The right edge of the galaxy (M51) sits against fairly dark sky background. The left-hand side of the galaxy, however, sits against a brighter sky. The galaxy itself looks fine, but the variation in the background brightness takes something away from the image.

If you adjust the image histogram so that the left-side background becomes black, the right side over-darkens and some of the dim portions of the galaxy are lost (see figure 6.6.5). The trail of stars flung out of the galaxy at bottom left is completely gone. The image lost detail because raising the black point high enough to hide the gradient also hides portions of the galaxy.

FIGURE 6.6.4. THERE IS A SLIGHT GRADIENT (BRIGHTER AT LEFT) IN THIS IMAGE.

Gradients Are a Problem

Even a slight gradient is a problem. Most celestial objects are faint, and a gradient makes it difficult to adjust the background to properly reveal the object details. Poor flat fields, lack of a flat field, or incorrect application of a flat field can also cause a gradient.

Light pollution gradients typically cause brightness to vary along a line. Poor flat fields usually have a radial gradient, with a hot center and dark edges. Figure 6.6.3 shows a radial gradient from a poor flat. The bottom edge of the image is also slightly brighter, so there either was also a linera gradient in the flat field, or light pollution caused a linear gradient. It's impossible to know which occurred, since the symptoms are identical.

Figure 6.6.4 shows an example of an image that has a very slight gradient from left to right. The image histogram has been adjusted so that the darkest portion of the

FIGURE 6.6.5. ABOVE: RAISING THE BLACK LEVEL ALSO REMOVES DETAIL FROM THE GALAXY.

FIGURE 6.6.6. BELOW: FIXING THE GRADIENT RETAINS MORE DETAIL IN THE DIM AREAS.

Figure 6.6.7. A light pollution suppression filter.

Figure 6.6.8. Comparing the color transmission for various types of light pollution filters.

If you fix the gradient before you finalize the histogram adjustments, you can set a lower overall black point. The image of M51 in figure 6.6.6 retains more detail. The dark, even background makes it easier for the eye to resolve details in the dim areas.

Gradients are all too common. Imaging near the zenith can help if the light domes aren't too bright. This doesn't help much in my own backyard. The lights from Seattle create a gradient that runs from my western horizon beyond the zenith. I do most of my imaging well to the east, where the Cascade mountains prevent housing developments and the light pollution that goes with them.

Using a Light Pollution Suppression Filter

The easiest way to deal with light pollution is to block it out. There are various light pollution filters available for visual observing. I haven't tested a wide variety of these filters, but there is one that I use that was designed for imaging and it works very well: the Light Pollution Suppression (LPS) filter from Hutech (see figure 6.6.7).

Lumicon, Orion, and other vendors also make light pollution filters. The Hutech filter has some features that make it well suited to imaging. The most important is that it was designed for accurate color balance.

Figure 6.6.8 shows comparison images made by Hutech comparing the color transmission characteristics of several popular light pollution reduction filters.

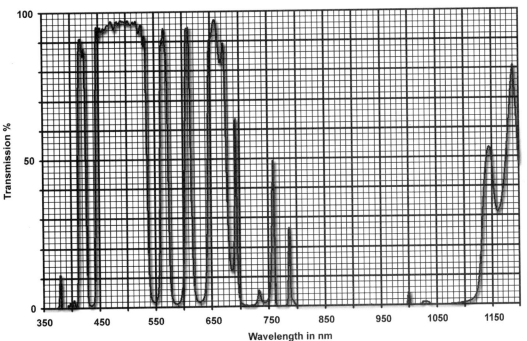

FIGURE 6.6.9. LIGHT TRANSMISSION CHARACTERISTICS OF THE LPS FILTER.

The bottom filter is the LPS. It does remove some of the light, but the overall color fidelity is good. It slightly darkens red, but blue and green are left in very good shape. It was this color fidelity that convinced me to try the LPS filter for my own personal use.

Figure 6.6.9 shows the transmission curve for the LPS filter by wavelength. Note the deep notches, where little or no light is transmitted. These correspond to various sources of light pollution. The fact that the notches are small, and the sides are steep, are what account for the good color fidelity of the LPS filter.

Figure 6.6.10 is a 10-minute exposure of an area of the sky (which just happens to contain comet McNaught-Hartley left of center) taken without the LPS filter. There is a small gradient, dimmer at upper left and brighter at lower right.

Figure 6.6.11 shows the exact same area of sky in an exposure taken just minutes later with the IDAS filter in place. This is also a 10-minute exposure. It's dimmer, because the LPS filter blocks most of the IR (infrared) light, and CCD detectors are very sensitive to IR. The good news about this is that it leads to better performance with refractors because focus is tighter without the IR. The bad news is that your exposures will be longer, about twice as long, to compensate for the missing IR energy. The best news is that the gradient from light pollution is pretty much gone. There is just the barest trace of a gradient remaining, too small to see with the eye, but measurable using software. The actual results you get will vary, depending on the amount and type of light pollution you have to deal with. The LPS filter doesn't (and can't) filter out all types of light pollution, but it does a great job with many common sources.

FIGURE 6.6.10. AN UNFILTERED IMAGE HAS A SLIGHT GRADIENT.

The downside is hardly surprising: light pollution filters require longer exposures. They filter out light, and some of the light from celestial objects is bound to get blocked at the same time.

But the upside is significant enough to make a good light pollution filter like the LPS a useful tool. My tests show that signal to noise ratios are better with the filter, and gradients are greatly reduced. I heartily recommend use of a good light suppression filter when imaging, especially from the typical suburban back yard. The penalty you pay in longer exposures (typically about 10%-30% longer if you are already using an IR blocking filter, and about double if you are not) is more than made up for with better image quality. If you are doing color imaging, you

FIGURE 6.6.11. THE FILTERED IMAGE HAS PINPOINT STARS AND NO GRADIENT.

are already using IR blocking and shouldn't require much additional exposure with the LPS. Figure 6.6.12 compares filtered and unfiltered images side by side.

If you are taking luminance images without IR blocking, the impact on exposure length will be significant. This won't hurt too much if you are using a camera with high quantum efficiency, such as one with an "E" series chip from Kodak, such as an SBIG ST-8E. On the other hand, anti-blooming chips already require longer exposures, and the lack of IR will hurt more with cameras that use ABG chips because now you have two factors requiring longer exposures. On the other hand, I have found that I can take even 30 and 60 minute exposures with an ABG camera from my back yard, and this long exposure with reduced light

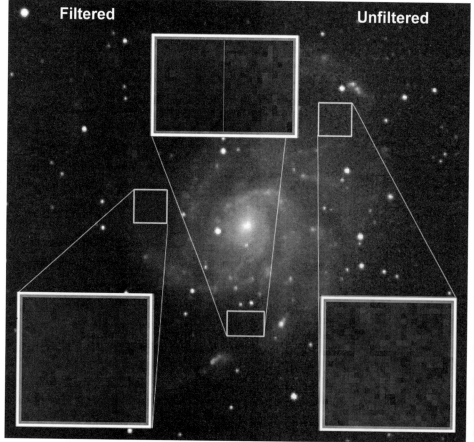

FIGURE 6.6.12. THE IDAS-FILTERED PORTION OF THE IMAGE HAS LESS NOISE IN DIM AREAS.

SECTION 6: DEALING WITH LIGHT POLLUTION GRADIENTS

FIGURE 6.6.13. AN EXAMPLE OF AN IMAGE THAT HAS HAD A GRADIENT REMOVED IN MAXIM DL. COMPARE TO FIGURE 6.6.3.

shows that while this tool is more effective, it has trouble handling a mixed radial/linear gradient such as found in the M81-82 image. I enhanced the image contrast to show that the left margin of the image has been over-brightened. The background is improved, but it's not quite even.

Mira AP offers sophisticated control over gradient removal. Mira uses mathematical modeling to clean up various kinds of gradients at various levels of complexity. You need to specify a number of factors to control the gradient detection process. To get good results, you have to understand how Mira attacks the problem, or you can use trial and error to find parameters that will remove the gradient from any given image.

Figure 6.6.15 shows the result of using Mira's "CCD Proc | Correct Background" menu item to remove the gradient from the M81/82 image. The results are very good. Mira does the best job I've seen at automatically removing background gradients, even fairly complex ones, because you can adjust settings to get the best possible result.

pollution effects is giving me some of the best quality images I've ever taken.

For more information about IDAS filters, visit the Hutech web site:

http://sciencecenter.net/hutech/tokai/lps.htm

Removing Gradients with Software

Maxim DL and Mira AP will analyze your images and remove gradients. Maxim DL implements this in the simplest possible way in version 2.x, and in a more sophisticated way in version 3. In version 2, gradient removal is poor (see figure 6.6.13) because the software expects a radial gradient (Flatten Background tool), and doesn't handle the linear gradients typical of light pollution well.

In version 3, MaxIm DL adds a new tool, Remove Gradient. Figure 6.6.14

FIGURE 6.6.14. MAXIM DL 3 DOES A BETTER JOB AT HANDLING GRADIENTS, BUT IT'S STILL NOT QUITE ABLE TO GENERATE A FLAT BACKGROUND EVERY TIME.

THE NEW CCD ASTRONOMY

The downside is that you have to adjust complex settings to get the best possible result. Figure 6.6.16 shows the dialog used to correct backgrounds.

The Column and Row terms define how complex the background gradient correction will be. The gradient from figure 6.6.3 shows a brighter area in the middle, and darker areas top/bottom, left/right. At first glance, this shows up as three zones vertically (rows) and horizontally (columns), so you might expect that Column Terms and Row Terms would both be set to 3, one for each zone. But this doesn't quite correct the problem. It's not easy to see, but there is a slight brightening along the left edge of the image. Thus, using 3 and 3 for the terms leaves you with a slight gradient from left to right (across the columns). Increasing the Column Terms to 4 tells Mira to handle an extra zone, and the result is a very smooth background, as shown in figure 6.6.15.

FIGURE 6.6.15. MIRA HAS REALLY CLEANED UP THE GRADIENT

FIGURE 6.6.16. THESE ARE THE OPTIONS FOR CORRECTING A BACKGROUND IN MIRA.

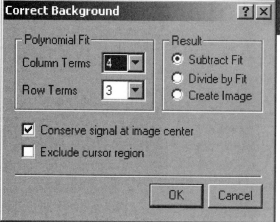

FIGURE 6.6.17. WHEN THE TERMS ARE SET TOO HIGH, CHAOS RESULTS.

If you get the terms wrong, you won't get a flat background. There will be times when you have to try different numbers to see what results you get. When you get it right, however, it really looks great. Figure 6.6.17 shows what happens when you set the terms too high -- Mira creates too many zones, and the background gets really funky. If you set the terms too low, some or all of your original gradient will remain, so I

SECTION 6: DEALING WITH LIGHT POLLUTION GRADIENTS

FIGURE 6.6.18. AN EXAMPLE OF A CREATED BACKGROUND GRADIENT IN MIRA.

recommend that you slowly increase the terms until you get a nice, even background.

You may also be wondering whether you should check a different radio button at top right of the Correct Background dialog (see figure 6.6.16). If the gradient results from lack of a flat field, use the divide option. This choice works when you did not have a flat field and you want Mira to create an artificial flat. The subtract option is effective with light pollution gradients. The rule here is a simple one. If the uneven background is a result of the telescope's optics, use Divide. If the uneven background is from light pollution or other skyglow, use Subtract.

You can also check "Create Image" and Mira will create a file that contains just the gradient. It is fun to check this option just to see what kind of gradient Mira finds. Figure 6.6.18 shows an example of a generated gradient file. This is the gra-

dient correction that Mira applied to the M81-82 image.

The ability to create a separate file with the gradient can be useful if you have a series of images without a flat field. Create an artificial flat in Mira, save it to disk, and then apply it to multiple images as a flat field.

Removing Gradients in Photoshop

The following tutorial uses files that you can download. There are two versions of the file, a small one and a large one. You can download whichever one is more appropriate to your Internet connection speed:

http://www.newastro.com/newastro/ book_new/samples/c6_veilB_big.tif

http://www.newastro.com/newastro/ book_new/samples/c6_veilB_small.tif

Figure 6.6.19 shows the image before any gradient processing. I applied linear histogram adjustments to bring out the detail in this small portion of the Veil, but I made no attempt to deal with the gradient.

FIGURE 6.6.19. AN IMAGE OF A SMALL PIECE OF THE VEIL NEBULA THAT SHOWS A GRADIENT.

This image appears to have two gradients. The arrows in figure 6.6.20 show them. As you work with the image you can temporarily adjust the histogram in Photoshop to emphasize the gradients if necessary. I do this to see what I'm dealing with in an image. You can undo the histogram changes before continuing.

The gradient from upper left is quite short, while the left-to-right gradient pretty much covers the entire image. The length of the arrows is meant to convey the approximate coverage of the gradients. As you'll see, there are several more subtle gradients in this image as well, but you can't see them easily because of the two dominant gradients. Lowering the white point further shows these subtle gradients a little more clearly (see

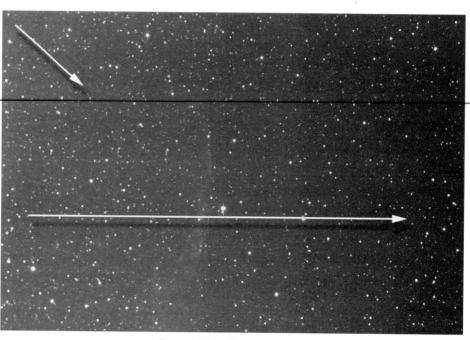

FIGURE 6.6.20. THERE ARE TWO SEPARATE GRADIENTS IN THE IMAGE.

figure 6.6.21). There is a short gradient at the top of the image, and a radial gradient near bottom center (blue in online version). The radial gradient is probably left over from a not-quite-perfect flat field.

FIGURE 6.6.21. AN EXTREME HISTOGRAM ADJUSTMENT REVEALS MORE GRADIENT TROUBLE.

Before we launch off into manual gradient fixes, it's worth a moment to examine what Maxim and Mira can do to deal with this complex of gradients. Figure 6.6.22 shows what we get from the Flatten Background menu item in Maxim DL. The gradients are reduced, but if you look carefully, the center is a little dark; the right side is the bright side of a new minor gradient, and the gradients at far left and top remain. These are minor problems, but the Veil Nebula is faint, and if you start playing with histogram settings to try to bring out details, these

FIGURE 6.6.22. FLATTEN BACKGROUND RESULTS IN MAXIM DL.

remaining problems will limit what you can do. I also tried the Remove Gradient command in version 3 of MaxIm DL, and it was better but still not completely effective with such a complex mix of gradients.

Figure 6.6.23 shows the result of applying Mira's Correct Background command. The background has minor remaining gradients. I used Column Terms set to 4, and Row Terms set to 3 to get this result. Mira overall may be a challenge to learn and use well, but its automatic gradient removal is reasonably effective.

While Mira is among the most effective gradient removal tools, you can still do better removing gradients manually. This example used Photoshop 6.0, but other image editing programs contain similar tools that allow you to do some of the same steps. If you are unable to create masks as shown in this tutorial, you may be able to create selections that perform the same function in a less convenient way.

Begin by opening one of the two veil image files. The file is saved in 16-bit TIFF format. You can experiment with histogram changes (Levels and Curves) if you want to get a feel for the image's potential, or to explore the exact nature of the gradients. When you are ready to start working on the gradients, you must convert the image to 8-bits per channel in order to work with masks. Many of Photoshop's masking and channel-based tools only work at 8-bits. Use the Image | Mode | 8-bits per Channel menu item to make this change. Some selection tools do work in 16-bit mode.

FIGURE 6.6.23. GRADIENT REMOVAL IN MIRA IS VERY EFFECTIVE.

Begin by creating a new Channel. Open the Channels palette (Window | Show Channels; see figure 6.6.24), and click on the button at top right to open the Channels menu. Choose "New Channel," and type in the name "Corner Gradient" for the new channel.

The new channel is highlighted in the channel palette, and the image appears black because there is nothing in the channel yet. Make sure that the current colors are white for the foreground and black for the background. Set the Opacity of the Gradient tool to 75%. This prevents you from painting with pure white, and allows a margin for error later when you are making corrections. By reducing opacity, you will make only 75% of the full correction when you apply this mask later on.

Using the Gradient tool with gradient set to Background to Foreground, create a short gradient at upper left. It should extend about as far as you see in figure 6.6.25. This is about where I observed the gradient ending in the original image. If it will help, you can

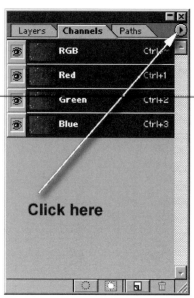

Figure 6.6.24. Creating a new Channel in Photoshop 6.0.

create guides to mark the spot where the gradient ends. Experience will help you learn the exact place to start and end the selection. To evaluate how accurate you placement is, try to correct the gradient by raising the black point using the Levels tool.

Create a new channel and name it "Horizontal Gradient." Create a horizontal gradeint, brighter on the left (see figure 6.6.26).

Load a selection using each mask in turn (Select | Load Selection menu item). Use the Image | Adjust | Levels menu item to raise the black point, which will clean up the gradient.

Figure 6.6.27 shows the Image | Adjust | Levels menu item in action. You can adjust the midpoint slider (the middle triangle below the histogram) or the black point slider (left triangle) to make the changes.

Figure 6.6.25. Above: Creating a mask using the Gradient tool.

Figure 6.6.26. Below: Creating a second mask, also with the Gradient tool.

SECTION 6: DEALING WITH LIGHT POLLUTION GRADIENTS

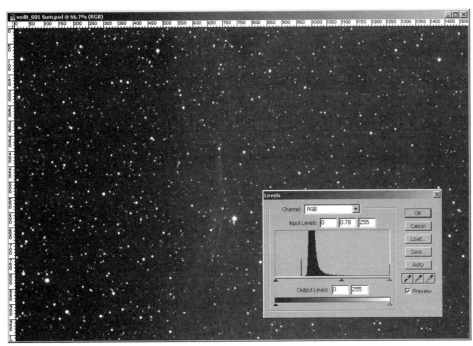

FIGURE 6.6.27. ADJUSTING LEVELS FOR THE SELECTION.

to go back and try again. If it was too large, undo and try a smaller selection.

You can correct the horizontal gradient using the same technique. Load the mask as a selection using the "Horizontal Gradient" channel. Once again, use the Image | Adjust | Levels menu item to change the midpoint so that the gradient goes away, as shown in figure 6.6.28. Or try raising the black point to see which does a better job for you. Click OK to save the change you prefer.

If you chose to use midpoint adjustments, the background remains a little bright. Clear the selection and use the Levels dialog again to raise the black point. As shown in figure 6.6.29, slide the left-most triangle

Slide the midpoint slider to the right to darken the selection until the background in the upper left corner matches the level of the background immediately around it. You can also adjust numerically; a value of 0.78 for the midpoint will be about right.

TIP: Some gradients respond better to raising the black level, some to using the midpoint slider. Try moving the leftmost triangle under the histogram to the right, and observe how effectively it removes the gradient. For any given gradient, use whichever method gives you a better result.

If your selection wasn't large enough to remove the full extent of the corner gradient, use the undo feature

FIGURE 6.6.28. REMOVING THE HORIZONTAL GRADIENT.

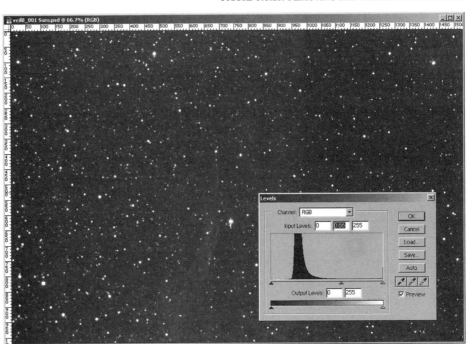

THE NEW CCD ASTRONOMY

under the histogram toward the right to raise the black point. This darkens the background. Be careful not to go past the start of the main clump of data in the histogram, or you will affect the data you want to show (the faint portion of the Veil nebula).

The next step is to use Image | Adjust | Curves to bring out more of the dim details (see figure 6.6.30). This reveals additional gradient problems. There is a slight gradient at top, and a radial gradient at bottom center. These are now the dominant problems. The Veil is extremely faint, so you need to take further steps to clean up the background if you are going to bring out as much detail as possible.

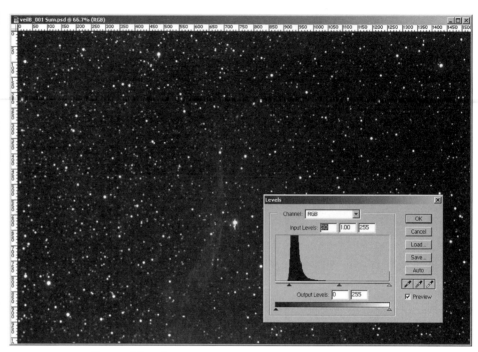

FIGURE 6.6.29. ADJUSTING THE BLACK POINT TO CORRECT THE BRIGHT BACKGROUND.

The preceeding example used masks to remove the first two gradients. You can also use selections to do the job. The primary advantage of a mask is that you can use paint tools to create one. This is more flexible than selections, which are more limited. For example, selections can only be feathered up to 250 pixels. With long gradients, this is not nearly enough feathering to remove the gradient.

FIGURE 6.6.30. BRINGING OUT MORE DETAIL USING CURVES.

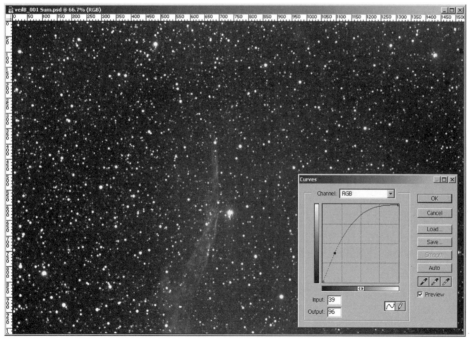

Let's attack the radial gradient just below and to the left of the image center first. Create an elliptical selection that matches the outer portion of the visible gradient. Click and drag from the center of the gradient, then hold down the Alt key so that the selection grows using the place where you clicked as the center.

SECTION 6: DEALING WITH LIGHT POLLUTION GRADIENTS

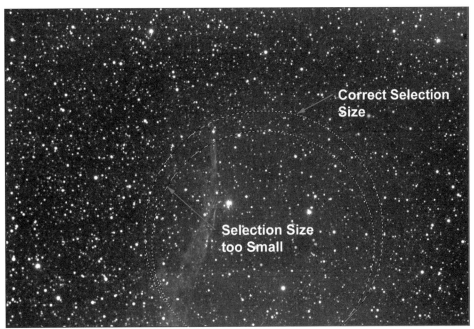

FIGURE 6.6.31. CHOOSING THE RIGHT SELECTION RADIUS TO DEAL WITH A GRADIENT.

blend between selection and background. Blending occurs on both sides of the selection boundary. The amount of feathering to use depends on the gradient. If the gradient has a very gradual rise, you should try the maximum feather of 250 pixels. If you need a larger feather, use a mask instead of a selection. If the gradient drops off very sharply, you might feather as little as 10 pixels. For any gradient, experiment until you find the sweet spot. For this example, about 80 pixels will work. To apply the feather, use the Select | Feather menu item, and enter a value of 80.

Most people have a tendency to draw the selection smaller than it should be. The natural instinct is to stop the selection at what the eye perceives as the edge of the gradient. Set the edge a little beyond this. The goal is to define the edge of what is to be dimmed, which is not the same thing as the edge of what you can see as bright. Even the very dim parts of the gradient require further dimming to match the rest of the image.

Figure 6.6.31 shows two different selections. The inner selection is where most people feel that the selection "should be." The outer selection is the one that will work.

If your selection is too small, some of the gradient will remain after the correction. If your selection is too large, you will darken some of the background

When removing a gradient with a selection, the next step is to feather the edge of the selection. This gives you a smooth

You can remove the gradient using Image | Adjust | Levels. Either slide the midpoint to the right to darken the selection, as shown in figure 6.6.32, or raise the black point. A value of 0.75 works well for the mid-

FIGURE 6.6.32. REMOVING A RADIAL GRADIENT USING THE LEVELS DIALOG.

TIP: To see a gradient clearly, use the Image | Adjust | Curves menu item. Apply a ridiculously steep curve to brighten the image. This puts the gradient into sharp relief. Click cancel when you are done.

FIGURE 6.6.33. CREATING A SELECTION WITH THE POLYGONAL LASSO TOOL.

The remaining gradient at the top of the image has a slight curve to it. Use the Polygonal Lasso tool to create a selection around the gradient. You can then feather this and use Levels to fix the gradient.

The key to success is location of the selection boundary. As always, cheat a bit to the outside. Figure 6.6.33 shows the correct position of the selection using the Polygonal Lasso.

point, but depending on the other changes you have made, a larger or smaller value may be required. Use your eye to judge when you have a smooth background.

The fall-off of this gradient is fairly steep, so a small feather is appropriate. I recommend feathering about

TIP: Observe the edge of the selection while you are making changes using the Levels dialog. If you see that you are darkening the background, the selection and/or the feather is too large. If you see a light ring outside of your selection, the selection and/or the feather is too small. You can use the Select | Modify | Enlarge/Contract menu items to change the size of your selection. Use undo and re-feather if you need to alter the feathering.

FIGURE 6.6.34. REMOVING THE LAST GRADIENT.

80 pixels. A simple way to alter the size of a selection is to Expand or Contract the selection until it's just right. Figure 6.6.34 shows the next Levels adjustment to remove this gradient from the top border of the image. You can use either the black point or the mid point to make the adjustment. Choose the one that works best.

If the image background is too bright you can make some final histogram adjustments to correct it. Figure 6.6.35 shows one way to handle the situation. There are three points in the Curves dialog. The bottom point darkens the dimmest areas in the image. This hides any remaining imperfections in the background. The next point brightens the pixel values that are just brighter than the background. This would be the portion of the Veil visible in the image, and a slight boost in brightness will make this area stand out a bit better. The third point from the bottom is an anchor, and it prevents the other two changes from impacting the brighter pixels in the image. Without this third point, the effects of the second point would extend further into the bright range of values, emphasizing stars instead of nebulosity.

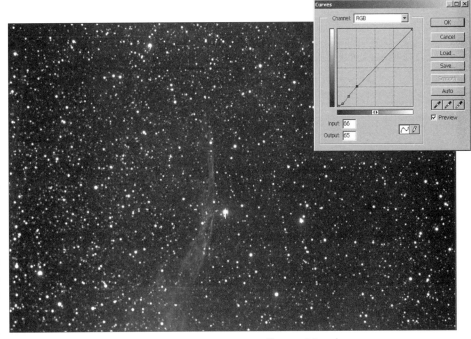

FIGURE 6.6.35. ADJUSTING THE HISTOGRAM.

I would also suggest a subtle Gaussian blur of the background. Use Select | Color Range to control how much of the background you blur. Hold down the Shift key and/or use the Fuzziness setting to control the extent of the selection. Contract the selection by one pixel to make sure not to include the Veil in the blurring. Figure 6.6.36 shows my final version of the image. Given the extensive gradients, this is a good salvage job.

FIGURE 6.6.36. THE FINAL VERSION OF THE VEIL IMAGE.

Figure 6.6.37. Blurring the image to extract the gradients.

Removing Gradients with Subtraction

If it weren't for the stars, nebulae, and galaxies in your images, you already have all the information you need to correct gradients. If there were a way to remove those stars, galaxies, etc. you could subtract the gradient from the image and get an instant correction. There are ways you can do this. One method uses cloning to cover up the bright stuff in the image. The other uses Photoshop's Dust and Scratches tool to remove most stars.

The idea is to get rid of everything except the gradients:

- Make a copy of your image
- Get rid of the stars and large objects in the copy
- Blur it like crazy
- Subtract the result from your original.

Figure 6.6.37 shows a blurred version of the original Veil image (before any corrections to the gradients). This blurred version shows the various gradients in the image, but it also shows that the bright stars and groups of stars, and even the small portion of the veil itself, are still present in the blurred image. I blurred using a Gaussian blur of radius 30, which is about as far as you ever want to blur. Further blurring reduces the contrast to a point where the gradient won't be strong enough to get the job done. Remove stars so that a Gaussian blur of no more than 15-20 will give you a smooth result.

Figure 6.6.38. Above: The clone tool was used to remove bright stars and extended objects.

Figure 6.6.39. Below: There are still enough stars to allow parts of the image to show despite the blur.

You can use the Clone tool to cover up bright objects with bits of matching background. When you blur, you don't have the problems visible in Figure 6.6.37. How heavily should you use the Clone tool? Very heavily. For best results, set the opacity of the Clone tool to 50%. This allows you to clone background stars without causing problems later on. Pick up background repeatedly and clone only very close to where you pick up, or you will alter the gradient.

Start by stamping out the brighter stars and objects. Figure 6.6.38 shows the result. Figure 6.6.39 shows that it's not quite enough even with a radius 30 blur.

A little more time with the Cloning tool gives us an image virtually devoid of even slightly bright stars, as shown in figure 6.6.40. This is what you should aim for to get optimal results.

A radius 20 Gaussian blur of figure 6.6.40 removes stars and leaves only the gradients (see figure 6.6.41). The background isn't as even as I would like it to be, mostly a result of the original image having so many stars in it. The more stars you have, the harder it is to get a smooth background.

TIP: If you have too many bright stars, too many stars, or too large of an object to get a reasonably accurate and even blur, there is an alternate solution. If the gradient problem is one that repeats as you move slightly off the target image, you can take another image, one that has more background in it, and use that to create the blurred image for subtraction. If the image was taken very far from the location of the problem image, however, it's likely that it won't have the same gradients. In that case, try the Dust and Scratches trick, later in this section.

The next step is to subtract this gradient from the original, leaving a corrected image with a flat background. Open both the original image and the blurred image in Photoshop. For convenience, I have made versions of these files available for download. There is a large pair and a small pair for faster download:

http://www.newastro.com/newastro/book_new/samples/c6_veilB_big.tif

http://www.newastro.com/newastro/book_new/samples/c6_veilB_blur_big.tif

http://www.newastro.com/newastro/book_new/samples/c6_veilB_small.tif

http://www.newastro.com/newastro/book_new/samples/c6_veilB_blur_small.tif

FIGURE 6.6.40. THE CLONE TOOL HAS BEEN USED MORE EXTENSIVELY, LEAVING NO BRIGHT OR MEDIUM-BRIGHT STARS IN THE IMAGE AT ALL.

FIGURE 6.6.41. A GAUSSIAN BLUR OF FIGURE 6.6.40.

Open the original and blurred images in Photoshop. Make the original image the active image. Use the Image | Apply Image menu item to open the Apply Image dialog (see figure 6.6.42). Select the blurred version of the image in the Source drop-down list. The Layer should be set to Background, and the Channel to Gray. Make sure that Invert is unchecked.

Choose Subtract as the Blending mode. Verify that Scale is 1, and that Offset is 0. I recommend opacity set at 90-95%. This means that slightly less than the full blurred gradient will be subtracted from the image. This is a compromise that leaves a small amount of gradient, but this helps avoid a

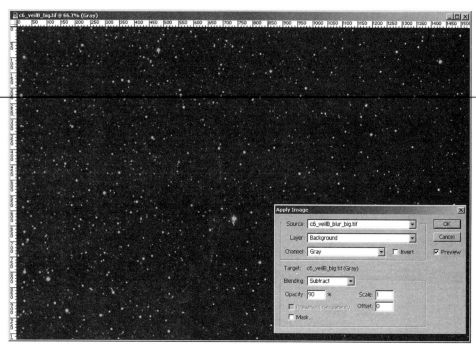

FIGURE 6.6.42. APPLYING THE BLURRED IMAGE TO THE ORIGINAL IMAGE USING SUBTRACT.

totally dark background. If you subtract too heavily, the background will be pure black and the contrast between the object and the background will often be too high. Feel free to experiment with opacity from about 80-100% to see how you like the results.

TIP: Experiment with Difference, Divide, and Screen to apply the gradient to the image. They offer alternatives that may sometimes be more pleasing to the eye.

Even with a 90% application of the background, as shown in figure 6.6.42, you will often find that your object is still too dim. This is not a crisis; as long as you haven't subtracted too much, you can use Image | Adjust | Curves to bring out

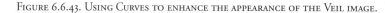

FIGURE 6.6.43. USING CURVES TO ENHANCE THE APPEARANCE OF THE VEIL IMAGE.

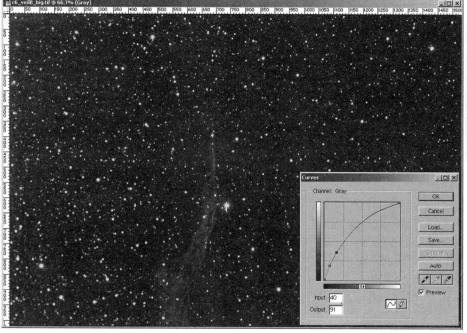

SECTION 6: DEALING WITH LIGHT POLLUTION GRADIENTS

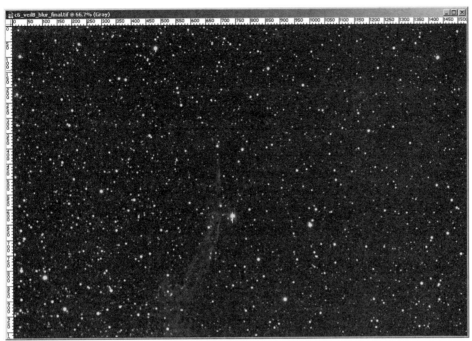

FIGURE 6.6.44. THE FINAL RESULT OF THE VEIL IMAGE.

Layers palette, and choose Duplicate Layer). All of the following operations should be done on the new layer. Make sure the new layer is above the original.

Use the Filter | Noise | Dust & Scratches menu item to open the dialog shown in figure 6.6.45. Normally, you would use fairly small numbers to remove very small defects in a scanned image. But with the right settings, you can actually remove nearly all of the stars from an image with this tool.

If you are following along with the Veil image, the settings in figure 6.6.45 are a good starting point.

dim details. Figure 6.6.43 shows a two-point curve that nicely brings out the Veil without over-brightening the stars or over-emphasizing background. The upper point slightly restrains the curve, and prevents the already bright stars from becoming too bright.

Figure 6.6.44 shows the final appearance of the Veil nebula with the gradient removed.

The biggest obstacle to use of the blurring technique for gradient removal is images that have large, extended objects in them. It can sometimes be simply impossible to blur adequately with such images. For example, an image of M33 that occupies 80% of the field of view is just too large to clone out of existence without completing messing up the gradients.

If you simply have too many stars in your image, however, there is one more technique that works very well to remove them. This technique, which uses features only available in Photoshop version 6, involves using the Photoshop Dust and Scratches tool to remove most stars, and then blurring the result. This method also introduces an alternate method for blending the image and the gradient-removal image.

Start with the original image. Copy the original image into a new layer (right click on the layer in the

FIGURE 6.6.45. SETTING DUST & SCRATCHES VALUES.

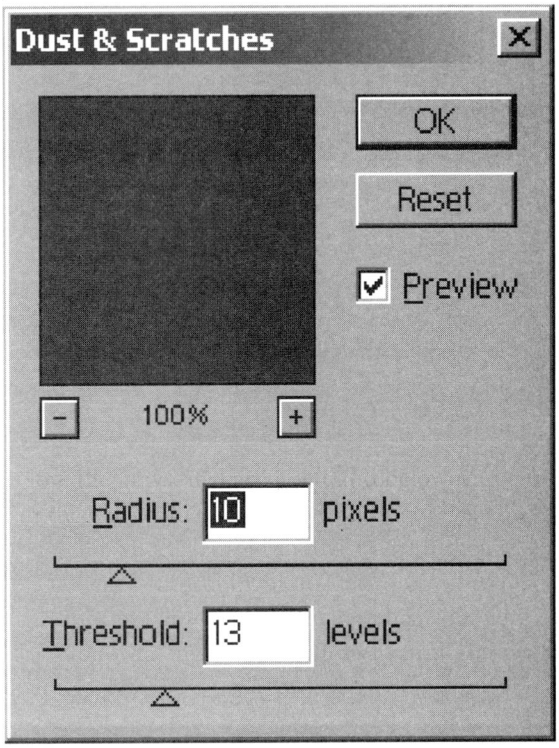

THE NEW CCD ASTRONOMY   317

Figure 6.6.46. The stars have been removed from the image using the Dust & Scratches tool.

Figure 6.6.47. The Clone Stamp tool does final cleanup.

The radius of 10 pixels will remove all but the very brightest stars, and the threshold of 13 will give you about the right level of aggressiveness in seeking out stars in the image.

The result of the Dust and Scratches tool is a strange-looking image that has only the barest hint of most of the stars (see figure 6.6.46). The Veil remains, and is actually fairly clear although the image overall has become a little blurred. That's the direction we want to go anyway.

Since there are still some slight remnants of the brightest stars, and because

SECTION 6: DEALING WITH LIGHT POLLUTION GRADIENTS

FIGURE 6.6.48. THE BLURRED IMAGE SHOWS THE GRADIENTS.

the veil remains as well, use the Clone Stamp tool to even up the background. Figure 6.6.47 shows the result. This combination of Photoshop's Dust and Scratches tool with the Clone Stamp tool is overall the most effective way to remove bright stuff from an image prior to creating a blurred gradient.

The next step is to blur what's left of the image. I used a Gaussian blur of radius 17, and the result in figure 6.6.48 is the smoothest gradient so far. This smoothness is what makes the Dust and Scratches technique such a good one.

Go to the Layers palette and click on the Opacity percentage. Use the slider to set it about 85-90% (see left image in figure 6.6.49). This is the same as the Apply Image opacity in the previous method.

Next, click the blend method list (see right image in figure 6.6.49) and choose Difference. This will perform a subtraction.

As shown in figure 6.6.50, the gradient simply disappears, leaving a nice flat background. As with other gradient removal methods, your final step is to use Image | Adjust | Curves to boost the dim details. You can use a curve similar to figure 6.6.43, but this method may allow you to be a bit more aggressive in bringing out details, so you may find that you can use a steeper curve.

Figure 6.6.51 shows the result of applying Curves. This method did the best job

FIGURE 6.6.49. USING BLENDING TO APPLY A LAYER TO THE LAYERS UNDER IT.

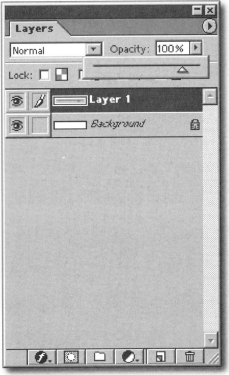
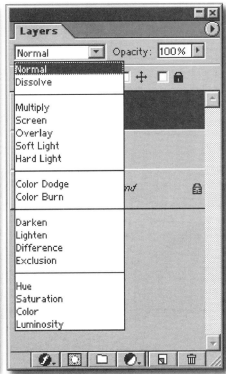

THE NEW CCD ASTRONOMY

overall in bringing out the dimmest details of the Veil, and I recommend it highly. The only time this (or any other non-manual) gradient removal won't work is if you have a very large object masking the gradients.

Light pollution can create problems with your images, usually in the form of one or more gradients. Long exposure, and/or sufficient multiple exposures, yield a good signal to noise ratio and are more likely to give you good results after removing gradients. If you see excessive noise after gradient removal, the image didn't go deep enough to successfully overcome light pollution.

FIGURE 6.6.50. THE RESULT OF BLENDING THE GRADIENT LAYER INTO THE IMAGE LAYER.

If you are faced with gradients in a color image, you can usually get the best results if you process the gradients in each channel separately. This allows you to isolate the gradients unique to each channel. If you work carefully, and get a very even background in each color layer, you will get a flat, black background in the color image. That is how I fixed the Horsehead image shown at the start of this chapter.

FIGURE 6.6.51. THE RESULT OF USING CURVES TO BRING OUT DIM DETAILS.

You can also attempt to work with all of the color layers together, using one of the techniques here. The manual method is the least likely to be successful when working with all colors at the same time. It can be very hard to judge the location of the gradients, and you often wind up making one gradient worse while you fix another one. If you have painting skills, you could try painting a gradient, but usually for a color image the Dust and Scratches method will work best.

7 Color Imaging

PART THREE: ADVANCED IMAGE PROCESSING

Color adds a dramatic dimension to images of celestial objects.

A black and white image can convey a lot of detail about an object, but color always tells you more. The blue color of hot young stars, or the blue-green emissions of excited OIII atoms, tells us something about the nature of a celestial object that a black and white image cannot do.

Section 1: Principles of Color Imaging

Note: Please see the online version for full-color images:

http://www.newastro.com/book_new/default.asp

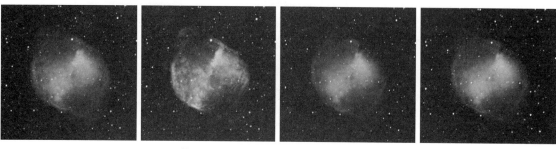

FIGURE 7.1.1. M27 LOOKS DIFFERENT WHEN IMAGED IN DIFFERENT COLORS OF LIGHT.

Figure 7.1.1 shows four images of M27. From left to right, these images are taken in full color; red light; green light, and blue light. The Dumbbell looks very different in each color of light. Different processes and structures are emitting light in each of those colors, giving us information about the nature of the nebula.

In addition, color imaging is without question a lot of fun. It's a great feeling to put together your first color image, and this satisfaction is a big part of color imaging for many CCD imagers.

However, most CCD cameras are not inherently color cameras. The CCD chip is a matrix of pixels, and each pixel is sensitive to all visible colors of light (as well as infrared and ultraviolet). There are a few one-shot color cameras, such as the LISAA from Apogee and the MX7-C from Starlight XPress, which use micro-filters on the chip itself to direct light of different colors to different pixels.

Color filters

In order to get a color image with most CCD cameras, you must place a series of color filters in front of the camera, take images through the filters, and then combine them afterwards. The filters are most commonly put into a motorized wheel or bar. This allows the camera control software to select the filter for an exposure.

The color filter wheels supplied with SBIG and most other cameras provide four filters and sometimes a few empty slots for other types of filters. The standard color filters supplied with most filter wheels are red, green, blue, and clear. For the last several years, color filters have incorporated infrared blocking (IR blocking) as well. Almost all of the color filters sold for use with CCD cameras are interference filters, which admit a very precise band of wavelengths of light. Interference filters block out unwanted wavelengths very effectively. They use very thin coatings on a glass substrate. The coatings are thinner than a single wavelength of light. As light passes through the filter, most of the light gets relfected or trapped by the coating and never makes it through. Only light of the desired wavelengths can pass. Thus the name "interference filter." The filter causes unwanted wavelengths to interfere and cancel.

The IR blocking layer is necessary for color imaging. CCD chips are very sensitive to IR. If IR were allowed to pass through, it would contaminate the color images. The clear filter in many filter sets does not have IR blocking, since infrared light can help increase the detail in your luminance images. However, if you have a refractor or any other type of telescope that has chromatic focus shift, you may find that using an IR filter for clear exposures will give you smaller, tighter star images. The SBIG clear filter does not use IR blocking. Hutech offers both IR blocking and non-IR blocking clear filters. If you use an IR blocking clear filter, you will need to take longer exposures to compensate for the blocked infrared.

SECTION 1: PRINCIPLES OF COLOR IMAGING

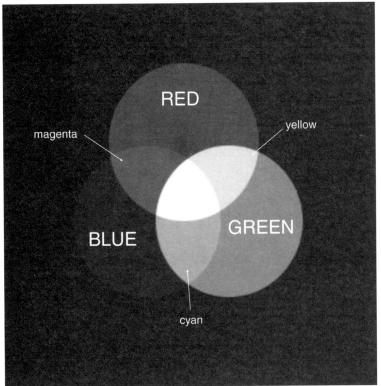

FIGURE 7.1.2. USING RED, GREEN, AND BLUE TO CREATE OTHER COLORS.

other colors in the spectrum. For example, red plus green creates yellow, and blue plus red creates magenta. Mixing various shades creates further in-between colors, such as red and a darker green to create orange. Combining all three colors in equal amounts creates white.

Figure 7.1.2 shows the three basic colors and how they combine in equal proportions to form magenta, cyan, yellow, and white. When imaging, you can use just the RGB filters, or add in a clear filter to increase image resolution (RGB and LRGB methods, both described in detail later in this chapter).

Figure 7.1.3 shows an image of the Trifid Nebula broken down into its component color images. From upper left, moving clockwise, the images shown are the red, green, blue, and RGB combined images. Each image was taken separately, and then combined to create the color image.

Most CCD detectors are not equally sensitive to all wavelengths of light, so you don't necessarily take exposures of the same duration through each filter.

If you have color filters that do not include IR blocking, you can use a single IR blocking filter on the nosepiece of your camera to achieve the same results. In that case infrared light will not pass through the clear filter, so longer clear filter exposure times will be necessary.

The most common method for taking color images is to take component images in red, green, and blue, and then combine them. Just as the eye synthesizes color images using red, green, and blue receptors, you can create color images using these same colors. The varying proportions of the three colors create all of the

FIGURE 7.1.3. THE TRIFID NEBULA'S COMPONENT COLOR IMAGES.

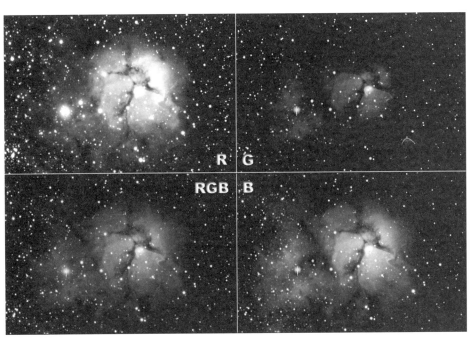

THE NEW CCD ASTRONOMY

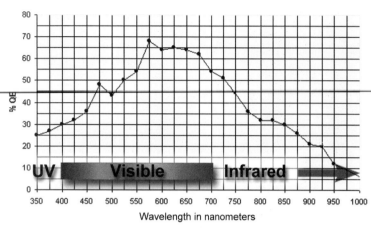

Figure 7.1.4. CCD detectors respond differently to different wavelengths of light.

Response Curves

Figure 7.1.4 shows the response curve for an ST-8E's detector, the Kodak KAF-1602E chip. The graph measures the quantum efficiency of the detector, showing how efficiently it turns photons into electrons. Different CCD detectors have different response curves. The response of the 1602E peaks in the visible range (roughly 400 to 700 nanometer wavelengths), and falls off toward the ultra-violet and infrared portions of the spectrum. Note, however, that the CCD detector is sensitive to both infrared (IR) and ultraviolet (UV) light. Figure 7.1.4 illustrates why it is important to use an IR blocking filter when imaging in color. Many CCD detectors are very sensitive to these wavelengths of light. The 1602E, for example, is 50% efficient at capturing near-infrared photons.

Because of the variation in response with wavelength (color) of light, most CCD detectors require different exposures for red, green, and/or blue. One exception to this rule is the ST-237 camera from SBIG. With the color filter wheel installed, this camera uses a 1:1:1 ratio of exposures.

You may find that you need a slightly longer blue exposure with any camera if you image away from the zenith. The atmosphere scatters blue light, and the lower the elevation you are imaging at, the longer you need to expose to get enough blue. This is called atmospheric extinction. It is most important to take it into account when imaging below about 45 degrees of elevation. Since the air is more turbulent below that elevation, you are more likely to image above that elevation than below it. The far-south summer sky, including Sagittarius, is very low for most northern observers and you are most likely to need to take blue extinction into account when imaging that area. For objects to the east and west, you can simply wait until they are high in the sky.

If you image at about 45 degrees of elevation, add 5-10% to your blue exposures. At 40 degrees, add 10% to blue. At 30 degrees, add 15-20% to blue and 5-10% to green. At 20 degrees, and add 40% to your blue exposures and 20% to your green exposure. This should keep you relatively balanced color-wise. However, at such low elevations, expect more turbulence and light pollution. You may also need to at least double all of your exposure times just to compensate for the general brightening nearer the horizon at light polluted locations.

Other factors can also affect color component exposures, including:

- Natural variations in sensitivity from one chip to the next.
- Light pollution adds light at various wavelengths. The color of your light pollution depends on the source of the pollution. Sodium vapor lights, for example, emit strongly in the green portion of the color spectrum. Mercury vapor lights emit strongly in the orange. If light pollution from such sources is a problem for you, a light pollution suppression filter can remove a significant portion, but not all, of the light pollution and restore a better color balance. I recommend the Hutech LPS filter for this task.

Exact exposure times for color images are not critical for non-metric use (that is, taking pretty pictures). Camera control programs provide tools for adjusting color balance after you take your component images. The most important thing is to get long enough component exposures for each color so that you get good signal to noise ratios in each image. See chapter 3 for details on the effects of signal to noise ratio on image quality.

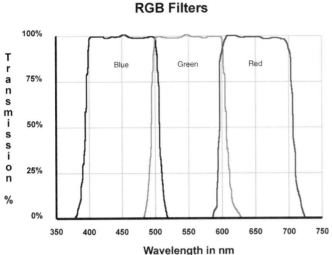

FIGURE 7.1.5. EACH FILTER TRANSMITS LIGHT OF A CERTAIN COLOR.

Graininess is the visual appearance of poor signal to noise ratio. It means that your exposure is too short. You can overcome this either by taking longer component exposures, or by combining multiple images for each color. Either technique will reduce noise effectively. See chapter 6 for details on aligning and combining multiple images.

Graininess in the dimmest portions of an image can be cleared up by smoothing techniques, covered in chapter 8. When working with color, apply image processing *after* you combine the RGB component images to create a full-color image. The exception to this is gradient removal; you'll get the best results if you correct for any gradients before you combine colors. If you do LRGB images, you will perform all histogram adjuments to the luminance layer only.

If you try to process a color image before you combine colors, you will usually disturb the color balance. If you need to use image processing prior to combining, test the results of your processing frequently to see what effect they have on color balance.

Color Filters Explained

Color filters work by passing only light of a certain color. Light is composed of various wavelengths, and each filter in an RGB (red, green, blue) filter set passes light of a very specific group of wavelengths. Figure 7.1.5 shows a hypothetical example. For each color, there is a band of wavelengths that are passed nearly 100%, and outside that band the filter passes only a very small amount of light.

For example, the blue filter passes just about 100% of the light with a wavelength between 400 and 500 nanometers. From about 360nm to 400nm, a small amount of light gets through, but all other wavelengths are blocked. The green filter passes light from 500nm to 600nm, and the red filter passes light from 600 to 700nm.

The passbands for the three filters in figure 7.1.5 are arbitrary. The actual passbands of real-world filters will vary. For example, some filter sets will have the passbands shoulder-to-shoulder, right at the 100% transmission points, as shown in figure 7.1.5. Others will pass from color to color at the 50% transmission point, or at some other point that the filter designer feels is best for the intended use of the filter. Figure 7.1.6 shows the transmission curves for the filters that SBIG uses in their cameras. The blue and green curves overlap significantly, while the red and green curves hardly overlap at all. This is by design; the filters are matched to the emission characteristics of the elements that predominate in celestial objects. These include the emissions from OIII, H-alpha, SII, and other elements and ions.

FIGURE 7.1.6. ACTUAL COLOR CURVES FOR REAL-WORLD COLOR FILTERS.

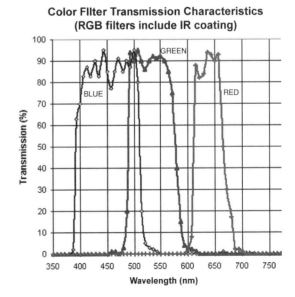

Figure 7.1.7 shows a different type of filter, a narrow band filter. This particular example is a hypothetical filter that passes only the light at a wavelength of approximately 657nm, the light of hydrogen-alpha emission. This light is emitted by many celestial objects, including many nebulae. A filter that passes only this narrow range of wavelengths will exclude other wavelengths, including most of the light from stars, and pass the light from hydrogen-alpha sources such as nebulae.

Other common narrow-band filters include SII (ionized Sulfur), Hydrogen beta, and OIII (triply ionized Oxygen). You can even use such filters to create false-color images, as shown in figure 7.1.8. Instead of using red, green, and blue filters to make the color image, Arnie used H-alpha, SII and OIII filters. The images were combined in the usual manner, with H-alpha substituted for red, OIII for green, and SII for blue.

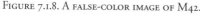

FIGURE 7.1.7. A NARROW BAND FILTER, PASSING ONLY A VERY SMALL RANGE OF WAVELENGTHS.

Luminance Layers

The red, green, and blue images are often combined with a full-spectrum image called a luminance layer. The luminance image has light from all wavelengths in it. This might include IR if a clear filter without IR blocking is used, or it might not include IR. Whether to include IR depends on how well your telescope focuses infrared, and on your intent in making the image. The IR can add detail, or it can detract from the image by washing out color, or it can enhance the image by increasing the brightness of dim areas in shorter exposures. There is no one right answer when it comes to including IR in the luminance image.

The use of a luminance layer (L) is also optional. If you do take a luminance layer, you can apply image processing to the luminance layer without affecting the color layers. You can also extract the luminance layer from a pure RGB image using programs like Photoshop.

Probably the most common use of an L layer is to allow you to use binning on the color layers. Figure 7.1.9 shows images of M27 built with various binning modes for the luminance and color layers, and combined in different ways. Clockwise from top left:

- Luminance and color layers binned 1x1 (no binning). This provides a sharp image and good color. Highly recommended, as it will provide the sharpest detail and excellent color. Requires additional imaging time relative to 2x2 binned color components.

FIGURE 7.1.8. A FALSE-COLOR IMAGE OF M42.

- Luminance 1x1 and color layers 2x2, combined in MaxIm DL. Star images are fatter, and detail is less sharp overall. Not recommended.
- No luminance layer, just RGB images unbinned (1x1). Due to excellent S/N, looks just as good as the top left image, with slightly better color saturation. Recommended, especially for bright subjects.
- Luminance 1x1 and color layers 2x2, combined in Photoshop. Star images are very similar to top left image, but detail is slightly less sharp overall. Recommended when you have limited time to capture your data, since binned color images require less time relative to 1x1.

The main advantage of binned color components is that they take less time to acquire. In the examples in figure 7.1.9, the all-1x1 image at top left consists of 10-minute luminance, and 10:10:16 minutes of RGB color images. For the image at bottom left, which combines 1x1 luminance and 2x2 color, the color images were 5:5:8 minutes. They took half as much time to acquire, yet provide a good result.

If you do take 2x2-binned color, you want to avoid simply resizing those images to match the luminance and combining. As the top right image in figure 7.1.9 shows, this can result in stars that are bloated due to the binning. To get small star sizes and sharp details when

FIGURE 7.1.9. THE SHARPNESS OF A COLOR IMAGE DEPENDS BOTH ON THE SOURCE IMAGES AND THE COMBINATION METHOD.

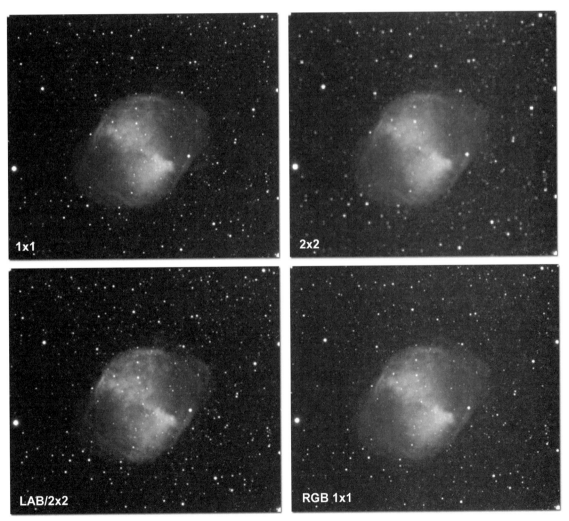

you bin your color shots, follow this procedure using Photoshop or a similar image editor:

1. Align all of the images for the LRGB layers.
2. Combine the 2x2 R, G, and B layers using whatever method or software you prefer.
3. Load the low-resolution color image into Photoshop. Resize if you haven't already done so. Verify that it is in 8-bit color mode.
4. Convert the image to LAB color model (Image | Mode | Lab Color menu item).
5. Open the Channel palette (Window | Show Channels)
6. Make the L (luminance) layer the active layer by clicking on it.
7. Load the high-resolution (1x1) luminance layer into Photoshop.
8. Make all processing adjustments to the luminance image, including gradient removal, sharpening, smoothing, histogram adjustments, etc.
9. Convert the luminance image to 8-bits per channel if necessary.
10. Make sure that only the L layer is active in the color image. Copy the luminance image to the clipboard, and paste it into the color image's L layer.

Convert the image back to RGB color model, and make any final adjustments required. You should see a color image with nearly the full resolution of the 1x1 luminance image, despite the fact that the color is at a lower resolution. This works because your eye uses luminance data to see fine detail.

CMY Filters

Another type of color imaging is called CMY. It uses cyan, magenta, and yellow filters instead of red, blue, and green. Each of the filters in a CMY filter set passes two primary colors instead of one. The cyan filter passes blue and green. The yellow filter passes green and red. The magenta filter passes blue and red.

The RGB color method is an additive method. The individual colors are added together to make the final image. The CMY color method is subtractive. The cyan image contains blue and green data. The magenta image contains blue and red, and the yellow image con-

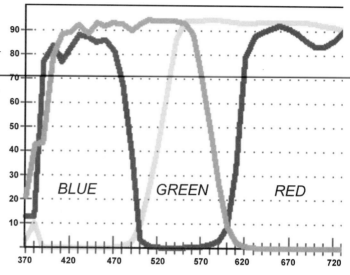

FIGURE 7.1.10. CMY FILTERS EACH PASS TWO PRIMARY COLORS.

tains green and red. Figure 7.1.10 shows response curves for CMY filters sold by Optec. Note how each color curve includes two of the primary colors.

To build a color image from CMY data, the images through each filter are subtracted from one another in specific ways to generate red, green, and blue data to create an RGB image. In theory, you could generate blue by first adding the cyan and magenta data, and then subtracting the yellow:

(blue + green) + (blue + red) - (green + red) = blue

I say "in theory" because as a practical matter, the passbands of a given filter set are not perfectly equal. To get accurate color, you would need to modify the above formula to take differences between the filters into account. For detailed information about this and many other image processing techniques, please see *The Handbook of Astronomical Image Processing* by Richard Berry and James Burnell.

CMY imaging can deliver good results, as shown by my image of M101 in figure 7.1.11. But I have found that CMY filters are harder to work with, and they do not save any time in the long run. I continue to prefer RGB filters for my color imaging. I use the Hutech LRGB filter set for refractors, which includes IR blocking on all four filters, including clear. For other telescope types that do not have chromatic focus shift,

Hutech and other manufactures provide LRGB filter sets with clear filters that do not block IR. In addition, some camera control programs do not provide support for combining CMY images.

One important point to make about CMY filters is that they can be a good choice for emission nebulae, particularly planetary nebulae. One of the prominent emission lines is OIII, which lies right on the boundary of blue and green. This means that it can literally fall between the cracks with some RGB filter sets. The cyan filter in the CMY set covers this area like a blanket, and you are assured of capturing all of the data from this emission line. If you are serious about planetary nebulae, you can use CMY filters to give yourself guaranteed coverage of this emission line.

As with RGB filters, you can also use a high-resolution luminance layer with low-resolution CMY color images to speed up your color imaging. However, for optimal results, using all high-resolution images makes for a slightly better final image.

You can do scientificly accurate measurements with color filters by using a UBVRI filter set. This is a five-filter set that contains filters in the Ultraviolet, Blue, Visible (green), Red, and Infrared bands. The bands are defined in various ways, so if you plan to contribute your data somewhere, find out which flavor of the UBFVRI filters they require you to use. These filter sets are available from Custom Scientific and Optec.

FIGURE 7.1.11. A COLOR IMAGE OF M101 TAKEN WITH CMY FILTERS.

Section 2: Using a Filter Wheel

Most CCD cameras are monochromatic -- they react to all colors of light (and even infrared and ultra-violet light) at the same time. By placing color filters in front of the CCD chip, and taking multiple exposures, it is possible to combine exposures taken through the filters to build a color image. See the first section of this chapter for details.

Although hardware exists which allows you to manually place colored filters into the light path, by far the most common technique is to use a motorized wheel or bar to make the filter changes automatic.

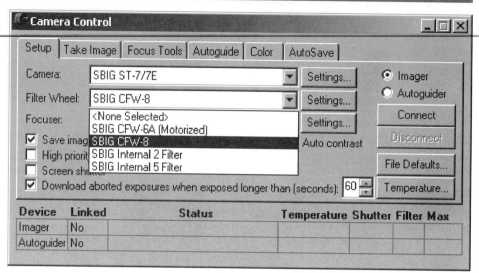

FIGURE 7.2.1. SELECTING THE TYPE OF FILTER WHEEL INSTALLED USING CCDSOFT.

In order to take color images automatically, you must have a color filter wheel attached to your camera. Both CCDSoft and MaxIm DL support a variety of filter wheels, with MaxIm DL supporting the larger number of wheels.

CCDSoft and MaxIm DL will be updated to support additional cameras and/or filter wheels as they become available, so check with the software makers to get the latest list of supported filter wheels (and bars).

Selecting a Filter Wheel

To select a color filter wheel in CCDSoft, first select the camera type using the Setup tab on the Camera Control Panel. If the control panel isn't visible, use the Camera | Setup menu item to open it.

Figure 7.2.1 shows the drop-down list of filter wheels available; the contents of this list will vary as additional filter wheels are added.

Note that there is a "Settings" button next to the list of filter wheels. After you select the type of filter wheel, you can use the Settings button to select the names of the filters installed in the filter wheel. If you are using the default filters in the default locations, as installed at the factory, you do not need to make any changes to these settings.

Figure 7.2.2 shows how you select a filter wheel in MaxIm DL. The choice is made on the Setup tab of the camera control. MaxIm supports a long list of different devices, and the author, Doug George, adds new devices as they become available. To enter the names of the filters, click on the Filter button to the right of the drop-down list.

When you have selected and configured your camera and filter wheel, connect to your camera as you usually do. In CCDSoft, click the Connect button. In MaxIm DL, click the Restart button. Your camera functions the same as it does without a filter wheel and now you can also access the color-specific features of your camera control software.

Automating Color Imaging (CCDSoft)

Once you have set up and connected to your camera and filter wheel in CCDSoft, you will normally use the Focus Tools tab and, optionally, the Autoguide tab before you are ready to take color images. If you use

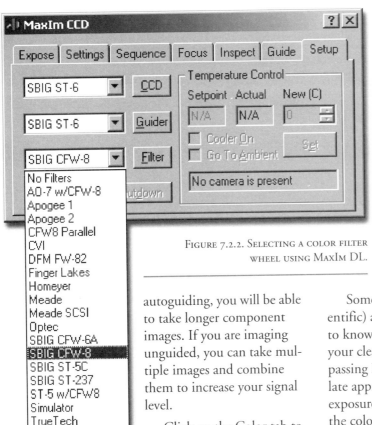

FIGURE 7.2.2. SELECTING A COLOR FILTER WHEEL USING MAXIM DL.

The Color tab shows the current settings for each filter. To change the settings for any filter, click the Edit button below the settings for that filter. Figure 7.2.4 shows the dialog that appears when you click the Edit L (luminance) button; the other three dialogs are identical in all but name.

Color filters reduce the amount of light reaching the CCD chip. They usually also incorporate an IR blocking filter. Many color filter sets include a clear filter that is parfocal with the color filters. You can use the clear filter for focusing, and for luminance images if you take LRGB image sets. See section 3 of this chapter for information about different types of color imaging.

Some clear filters pass infrared (SBIG, Custom Scientific) and some do not (some Hutech sets). You need to know whether your clear filter is passing IR to calculate approximate exposure times for the color filters. I always determine my color exposures relative to the clear exposure. If your clear filter doesn't use IR blocking, the exposure times

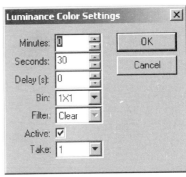

FIGURE 7.2.4. CHANGING SETTINGS FOR THE LUMINANCE IMAGE.

autoguiding, you will be able to take longer component images. If you are imaging unguided, you can take multiple images and combine them to increase your signal level.

Click on the Color tab to display the color-imaging interface shown in figure 7.2.3. This interface automates the process of taking multiple exposures using different filters. You can use up to four filters. To change the behavior for any filter, click on the Edit button for that filter.

The four columns on the color tab correspond to the four types of filters commonly used for color imaging: luminance (clear), red, green, and blue. If you have different filters installed and configured, the filter names may be different from what appears in figure 7.2.3.

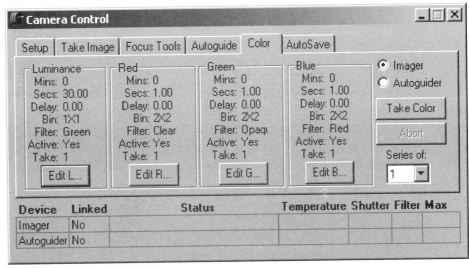

FIGURE 7.2.3. THE COLOR TAB ON THE CAMERA CONTROL PANEL.

using the same bin mode are usually longer for color than for luminance. With the SBIG RGB color filters and a clear filter that passes IR, start with a red exposure length that is approximately double the luminance exposure at the same bin mode. If the luminance exposure is binned 1x1 and the red exposure 2x2, start with the same exposure as the luminance image. If you are also using an IR blocking for the clear filter, then cut the suggested color exposure times in half. Blue and green exposures are set proportional to the red exposure using the manufacturer's guidelines.

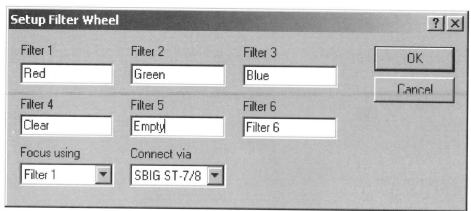

FIGURE 7.2.5. IF NECESSARY, CHANGE THE NAMES TO REFLECT THE FILTERS AND FILTER ORDER YOU HAVE INSTALLED IN YOUR FILTER WHEEL.

If time is limited, use 1x1 binning for your luminance image and 2x2 binning for your color images. For higher resolution and longer exposure times, use 1x1 binning for all images.

Note: If 1x1 binning provides fewer arc-seconds per pixel than is optimal for your equipment and condtions, substitute appropriate binning modes. For example, when using an 8" SCT at f/10 with an ST-7E, you might use 2x2 binning for luminance, and 3x3 binning for color filters. 2x2 gives you 1.85 arcseconds per pixel. 1x1 binning would give you 0.93 arcseconds per pixel, and the seeing usualy won't support that.

FIGURE 7.2.6. SETTING UP THE EXPOSE TAB FOR RED EXPOSURES.

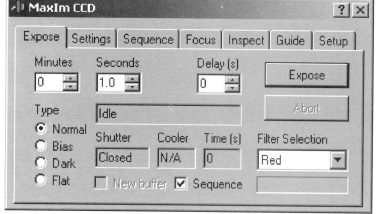

If no color filter wheel is attached, the filter dropdown will either be gray, or will only allow Clear and Opaque. Even though you won't be able to take color images without a filter wheel, you can make use of the Color tab for shooting sequences of images at different exposures times and bin modes. For example, if you are unsure of the optimal exposure that will balance blooming and depth of exposure, you can use the color tab to take several exposures of different length which you can then compare and evaluate.

When the Active checkbox is checked, the filter will be active. If unchecked, the filter position will be bypassed when the imaging sequence occurs.

The Take value determines the number of images to take with this filter. For example, you can elect to take three luminance images and one each of red, green, and blue. This constitutes a single image set. You can use the "Series of" dropdown list on the Color tab to take multiple image sets.

Automating Color Imaging (MaxIm DL)

The newest version of MaxIm DL comes with very usable color imaging features, and it supports more cameras and filters wheels as of the time of writing. You specify the model of color filter wheel you are using, tell Maxim what filters it contains, and then you are ready to image.

If you haven't previously set up the filters, click the Filter button to display the current filter names. Type in the actual names of the filters you have installed. Figure 7.2.5 shows the defaults for the CFW-8. If you have a different filter wheel installed, check the documentation for the names and order of the installed filters.

To take an automatic sequence of RGB images in version 2.11 of MaxIm DL, go to the Expose tab and check the Sequence checkbox (see figure 7.2.6). Make sure you also set the Filter Selection to Red. This takes care of the red exposure; now switch to the Sequence tab where you will set up the green and blue exposures.

TIP: It is all too easy to forget to set up the red, green, and blue exposures properly, since they exist on two different tabs. The most common mistake is to forget to change the filter to Red on the Expose tab. The next most common mistake is to forget to set the Green and Blue filters on the Sequence tab. Also be sure to check "Enable Tricolor" and also check "Different Exposures" for cameras that require a longer blue exposure.

If you are using a set of filters that requires different exposure durations for different filters, you must also check the "Diff. Exposures" checkbox (see figure 7.2.7). This enables the Minutes and Seconds controls for the Green and Blue filters. Set exposure times as recommended by the manufacturer of the filter wheel when used with your CCD camera.

TIP: Most exposure recommendations are expressed as ratios, with red arbitrarily expressed as 1. For example, the recommended exposures for the ST-7E and ST-8E cameras are 1:1:1.6. This means that the red and green filters should have the same exposure time, and the blue filter should have an exposure time that is 160% of the red exposure time. If you are using a one minute red exposure, then use a one minute green exposure and a 1:36 blue exposure (one minute and 36 seconds). I would typically round the blue exposure to 1:30, or 1:45 if imaging below 60 degrees of elevation (to compensate for blue extinction).

When you have properly set up the filters, exposure times, and number of exposures, return to the Expose tab and click the Expose button to begin the automatic color sequence.

Version 3 of Maxim uses a completely new interface for exposure sequences. The new interface allows you to set up to 16 separate exposures with different filters and exposures times for each.

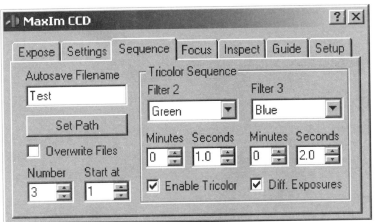

FIGURE 7.2.7. SETTING UP THE SEQUENCE TAB FOR THE GREEN AND BLUE EXPOSURES.

The Color Imaging Process

As with any CCD image, color imaging demands an accurate mount. Autoguiding is not a requirement, but it simplifies the color imaging process and makes it easier to obtain high signal to noise ratios. If you do not have an autoguider, and wish to do color images, a very good polar alignment is essential. It takes a long time to capture the necessary images, and a good polar alignment will minimize the amount of drift and field rotation. In addition, the smaller the amount of drift due to polar misalignment, the longer your exposures can be. With a poor polar alignment, your subject might even drift out of the frame before you have collected enough images to create your color image. Autoguiding follows movement accurately, although if polar alignment is poor you will need to remove field rotation from the images before you combine.

FIGURE 7.2.8. ABOVE: AN IMAGE OF M101 WITH POOR SIGNAL LEVEL.

FIGURE 7.2.9. BELOW: AN IMAGE OF M101 WITH AN EXCELLENT SIGNAL TO NOISE RATIO.

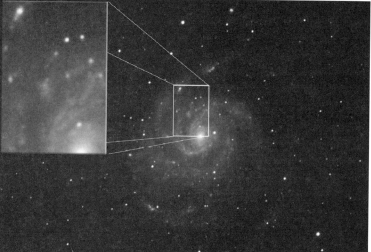

If your exposures are too short, you will see graininess in the image as shown in the image of M101 in figure 7.2.8. Grain tends to be a more serious problem in color images. When you combine such images to make a color image, you wind up with obvious dots of color instead of a smooth blend. To get good signal levels for your color images:

- Take longer exposures
- Use a faster focal ratio
- Take and combine multiple images.

All of these techniques will improve the signal level with respect to the noise, as described in chapter 3. The dimmer your subject, the longer your exposures need to be to achieve a good signal to noise ratio.

Figure 7.2.9 shows an image of M101 that has a much better signal to noise ratio. I combined multiple images to improve the signal to noise ratio. The same portion of the galaxy (rotated 90 degrees counterclockwise) has been enlarged to show that there is much less noise in this image. This image is a median combine of four exposures. Combining multiple images in each color is one of the most effective ways to get really good component images for your color images.

If your red, green, blue, and luminance component images end up with poor signal levels, the background in the final image will be excessively grainy, as shown at the top of figure 7.2.10 (M81). The bottom image (Lagoon Nebula) shows almost no grain whatsoever. This is due to excellent signal to noise in luminance and color components. The M81 image needed longer exposures in all components, while the Lagoon image has sufficient exposures in all components.

Figure 7.2.11 shows details of the two boxed areas from figure 7.2.10. In the M81 image, color graininess is pronounced. In the Lagoon image, the noise, and therefore the grain, is insignificant. The only "secret" to smooth, clean-looking color images is long total exposure time. If you don't have an ABG camera, take many multiple exposures and combine them to achieve the same result.

You can improve the appearance somewhat by blurring/smoothing the grainy areas. Use different amounts of smoothing based on the level of graininess in each component image. If you are taking LRGB images, you can take five or more luminance images to increase the signal level where it matters the most. The eye judges sharpness mostly by luminance. If some areas are clear and some are noisy, you can use image editors such as Photoshop to select only the noisy areas for smoothing. See chapter 9 for details.

SECTION 2: USING A FILTER WHEEL

FIGURE 7.2.10. THE UPPER IMAGE HAS MORE NOISE, AND THE LOWER IMAGE LESS.

FIGURE 7.2.11. THE NOISE IN THE UPPER IMAGE MAKES THE IMAGE LOOK GRAINY.

THE NEW CCD ASTRONOMY 335

Color Imaging Steps (CCDSoft)

Once you have the appropriate settings for each filter, click the "Color Series" button on the Color tab to start the color sequence. A complete color imaging session includes the follow steps:

1. Verify settings for your camera, autoguider, and color filter wheel as appropriate. (Setup tab, Camera Control panel)
2. Turn AutoSave on, and select the folder where you want the component images stored.
3. Focus in the usual manner. (Focus Tools tab)
4. Frame your subject using test exposures. Determine your luminance exposure duration. (Take Image tab)
5. If you are autoguiding, calibrate if necessary and start autoguiding. (Autoguide tab)
6. Choose your Image Reduction options. (Take Image tab)
7. Adjust the settings for each filter on the Color tab.
8. Select the number of color image sets using the "Series of" drop-down list.
9. Click the Color Series button to begin acquisition.

Color Imaging Steps (MaxIm DL)

Once you have the appropriate settings for each filter, click the "Color Series" button on the Color tab to start the color sequence. A complete color imaging session includes the follow steps:

1. Connect cables to your camera, autoguider, and color filter wheel as appropriate.
2. Make sure that the Clear filter is the filter used for focusing (Setup tab, Filters button).
3. Focus in the usual manner.
4. Frame your subject using test exposures. Determine your luminance exposure duration.
5. On the Exposure tab, make sure that the Clear filter is selected.
6. Go to the Sequence tab, and click "Set Path" to select the folder where you want the luminance images stored. Type in a filename in the "Autosave filename" box.
7. Also on the Sequence tab, make sure that "Enable Tricolor" is not checked.
8. On the Expose tab, check the Sequence checkbox. Click the Expose button to take the set number of luminance exposures.
9. When luminance exposures are completed, go to the Expose tab, and select the Red filter and set the appropriate exposure time (usually 1x or 2x the luminance exposure time).
10. Go to the Sequence tab, and click "Set Path" to select the folder where you want the component images stored. Type in a filename in the "Autosave filename" box.
11. Also on the Sequence tab, check "Enable Tricolor." If necessary, also check "Diff. Exposures."
12. Also on the Sequence tab, set the number of exposures, and the starting number for the sequence.
13. Also on the Sequence tab, set appropriate exposures, if necessary, for the green and blue exposures.
14. If you are autoguiding, calibrate if necessary and start autoguiding.
15. Click the Expose button to begin acquisition of the color images.

Color Imaging Guidelines

You may need to monitor several things while the color series is in progress:

- Is the delay you set adequate? Does it allow the guide star to get close enough to the 0,0 position before the next image starts?
- Is your polar alignment good enough to keep the guide star visible after image downloads?
- Is autoguiding working properly? Look for star trails, multiple images, or other signs of trouble on the first few images.
- If you are taking three images of each component, make sure all three are good before you move on. You need at least three to take advantage of median combine.
- Is the guide star sufficiently bright when the color filters are in the optical path? For example, a red star may not be bright enough when the blue filter is in place. If so, choose another star, or increase

exposure time for autoguiding. The same autoguiding exposure time is automatically used for all filters in the current generation of camera control software. If you require different autoguiding times, you can use the scripting capabilities built into many camera control programs to provide this functionality.

- Does the guide star become too bright with any of the filters? If you use a blue star as a guide star, for example, it may be fine with a red filter, but excessively bright (saturated) when the blue filter is used. Saturated stars don't have good centroids, and can cause poor guiding.
- Is the guide interval appropriate? Does the mount drift too far during the guide interval, leading to trailing of star images?

Figure 7.2.12 shows what can be done with good optics, color imaging, and long exposures to generate good S/N. The image of the core of M31, the prominent galaxy in Andromeda. It was built up from a luminance image plus one set of red, green, and blue images. To learn how to combine component images into a color image, see the next section of this chapter.

FIGURE 7.2.1. AN EXAMPLE OF A COLOR IMAGE BUILT UP FROM LUMINANCE, RED, GREEN, AND BLUE IMAGES.

Section 3: Color Combining

FIGURE 7.3.1. AN IMAGE OF M16 TAKEN WITH AND COMBINED IN CCDSOFT.

You can combine red, green, and blue images into RGB color images with CCDSoft, Maxim DL, and other programs. You can also create LRGB images by including a luminance image in the combination.

In this section, you'll learn how to create color images with CCDSoft and MaxIm DL. CCDSoft provides a more flexible preview, but Maxim DL includes the ability to set a common white point for the component images for overall easier color combining.

You'll need a color filter wheel attached to your camera to take the component images. If you don't have a color filter wheel, sample color images for practice are available for download from the book web site.

http://www.newastro.com/newastro/book_new/samples/c7_m16_rgb.zip

http://www.newastro.com/newastro/book_new/samples/c7_m16_lum.zip

If your component images are not already aligned, how you start depends on the software you are using. In CCDSoft, alignment is not available during a color combine. Copy the component images to a folder and align them with the Image | Align | Align Folder of Images menu item (or Ctrl + Shift + A). Maxim DL includes a button to align images in the color combination dialog, so you do not need to do alignment before combining. If you have taken multiple images in red, green, and blue, you must align and combine those first before you do the final color combine in Maxim DL or CCDSoft.

Figure 7.3.1 shows an image of M16, the Eagle Nebula. The component images were taken with CCDSoft's Camera Control panel, using the Color tab. The images were combined with the Color Combine dialog. The result is too strong in the red to be technically correct, but the deep red saturation is visually appealing. You will see a wide range of interpretations in color combining. There is no one right way to do color. You can go for technical accuracy, or you can go for drama, or any combination of the two.

There are two basic types of color combine: RGB and LRGB. An RGB combine includes three images: red, green, and blue. These images combine to form a full-color image, as explained in section 2. An LRGB combine also includes a luminance image taken either without a filter or using a clear filter. To speed up the color imaging process, you may take the RGB images binned 2x2, while the luminance is binned 1x1. Luminance is the dominant factor when the eye perceives detail. However, for maximum detail, you can take the color exposures binned 1x1. If the component images are different sizes because of different bin modes, CCDSoft will automatically resize them for you. MaxIm DL requires images at the same size and scale.

Color Combining in CCDSoft

To combine an aligned set of component images into a full-color image, open the images in CCDSoft and use the Image | Color | Combine Color menu item (or Alt + K). This opens the Color Combine dialog, shown in figure 7.3.2.

The Color Combine dialog includes the following tools and features:

Red, Green, and Blue drop-down lists - Select the component images from the list of available images. You must have images open in order to see images in the drop-down lists. You may leave any of these set to "None" if appropriate. You can also use the "wrong" color to create special effects or false-color images.

Luminance drop-down list - Select the luminance image from the list of available images. You must have images open in order to see images in the drop-down list. You may leave the luminance image set to "None" if you are creating an RGB combine using only red, green, and blue component images. You can also use color images for the luminance channel, such as using a red image for a dominatnly red emission nebula (e.g, Eagle Nebula, or a blue image for the Witch Head Nebula.

Combine - Click this button to create the final color image. The Color Combine window stays open, so you can create more than one color version.

Close - Closes the Color Combine window.

Show Preview - Opens a preview window. All changes you make to the component images affect the preview. You can resize the preview window as needed. If a component image covers the preview, rearrange the location of the component images so the preview remains visible. You can use almost any tool or command in

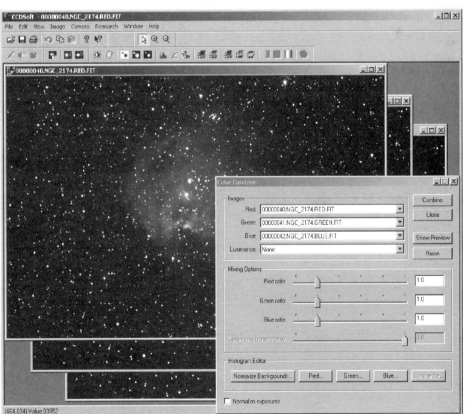

FIGURE 7.3.2. THE CCDSOFT COLOR COMBINE DIALOG IN ACTION.

CCDSoft on the component images. However, when making changes during an RGB combine (no luminance image), the more changes you make to the component images, the more difficult it will be to maintain a true color balance. You can move, scroll, zoom, and resize the Preview window as needed while you work.

Reset - Resets the controls in the Mixing Options section of the Color Combine dialog.

Red, green, and blue ratio sliders - Sets the contribution of the red, green, and blue images. By default, the ratios are set to 1.0 so that each color is equally represented. If your exposures match the recommendations of the filter-wheel manufacturer, leave these sliders at or close to the default value. For example, the SBIG CFW-8 filter wheel exposure ratio is 1:1:1.6. An example of correctly balanced exposures would be 60:60:96 seconds. If you have longer or shorter exposures, use the sliders to adjust color. This provides balanced color even when the component images are not accurately balanced.

Luminance ratio - Determines how much of the luminance exposure is included in the final LRGB image. The default value is 1.00, or 100% of the luminance. If the combined image in the preview window appears over-exposed, you can reduce this value. Alternatively, you can adjust the luminance histogram to achieve the same result.

Normalize Backgrounds - Clicking this button sets the background levels (black point) using the "Strong" option of the Histogram tool. This sets the background to black, and is usually the best option for color combining. A bright background reveals slight variations in the background color and reduces the quality of the color image. See chapter 8 for details on histogram adjustments.

Red, Green, Blue, and Luminance buttons - These buttons bring the appropriate Red, Green, Blue, or Luminance image to the front, and open the Histogram window for that image. You can then adjust the histogram of the image. Changing the black point will affect the brightness of the background. Changing the white point will change the amount of color or luminance detail that shows up in the final full-color image. Adjusting the histogram is an alternative to using the ratio sliders. Unlike the ratio sliders, adjusting the white point can hide or reveal details in the image.

Normalize exposures - Adjusts the exposures of the blue and green images so they match the red exposure time. For example, if the red exposure is 60 seconds, the green exposure is 90 seconds, and the blue exposure is 120 seconds, checking this checkbox will scale the blue and green images so they act as if they are also 60-second exposures.

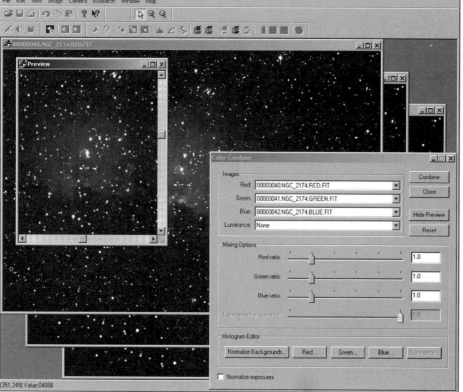

FIGURE 7.3.3. THE PREVIEW WINDOW IS OPEN.

SECTION 3: COLOR COMBINING

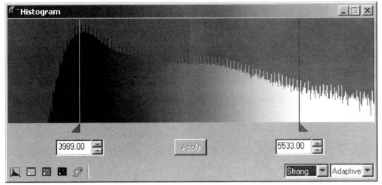

FIGURE 7.3.4. THE BLACK POINT IS SET TO THE "STRONG" VALUE.

Combining Color Images: RGB

An RGB color combine requires a reduced and aligned set of red, green, and blue images. To follow along, download and unzip this file with RGB images in it:

http://www.newastro.com/newastro/book_new/samples/c7_ngc2174.zip

Open the images in CCD-Soft v5, then open the Color Combine dialog and select the images for each color. Click the Show Preview button to monitor the results using the current settings (see figure 7.3.3). The size and location of the Preview and Color Combine windows are adjustable.

In figure 7.3.3, the Preview window doesn't show much color. With other images, it may show too much or just enough. Use the various controls in the Color Combine dialog to change the preview until you are satisfied, then click the Combine button to create the output image. You can click Combine more than once to create different versions.

The Preview window updates automatically. If the background in the Preview window isn't dark, the black points of the images are set too low. click the "Normalize Backgrounds" button.

This sets the black point close to the peak value at the left end of the histogram (see figure 7.3.4). Note that the black point (the left slider below the histogram) is close to the peak of the histogram curve at the left side.

You can set the black point yourself by selecting Strong or another value from the left of the two drop-down lists in the lower right corner of the Histogram tool. For a complete discussion of the histogram tool, black point, white point, etc. please see chapter 8.

Normalizing the backgrounds will usually bring out more color and darken the background, as shown in the Preview window in figure 7.3.5.

FIGURE 7.3.5. NORMALIZING BACKGROUNDS OFTEN IMPROVES IMAGE CONTRAST.

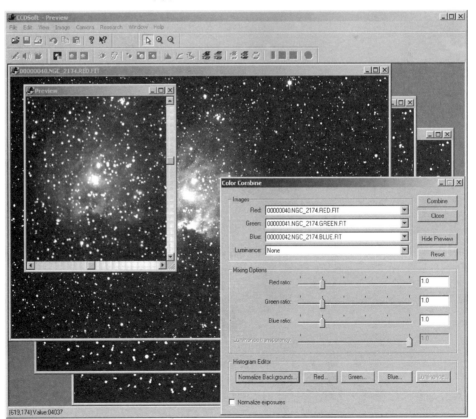

THE NEW CCD ASTRONOMY   341

If the color balance is not correct, it is usually due to incorrect exposure ratios. Use the color ratio sliders to balance the color. For example, if the image is too blue, you can slide the Blue ratio slider to the left to decrease the contribution of the blue image. Conversely, if there is too little blue in the image, you can slide the Blue ratio slider to the right to increase the contribution of the blue image.

If your component images do not have the recommended exposure ratios, you can search the web for color images of the object in your images and use them to determine the correct colors. You can then use the ratio sliders to adjust the contribution from each component image to match the desired result.

If you look closely at the central star in the Preview window in figure 7.3.5, you may notice that the star isn't well defined. If this were a monochrome image, it would be a simple matter to adjust the white point so that the bright area around the star is no longer merged with the star.

> **TIP:** When working with a color image, if you adjust the white point for a component color image, you will always alter the color balance in the combined image. Make such changes with caution!

If you move the white point higher for the red image, for example, this darkens the red image and reduces the contribution of the red component to the color image. There are two things you may need to do if you adjust the white point in a component image:

- Use the color ratio slider for the color you changed to rebalance color in the image. For example, if you move the white point for the red image higher, this will darken the red image. Move the Red ratio slider to the right to increase the contribution of the red component until color is rebalanced.

- Adjust the white point of the other two color component images to compensate. Observe the changes in the Preview window to get feedback on your changes.

FIGURE 7.3.6. ADJUSTING THE HISTOGRAM OF THE RED COMPONENT IMAGE.

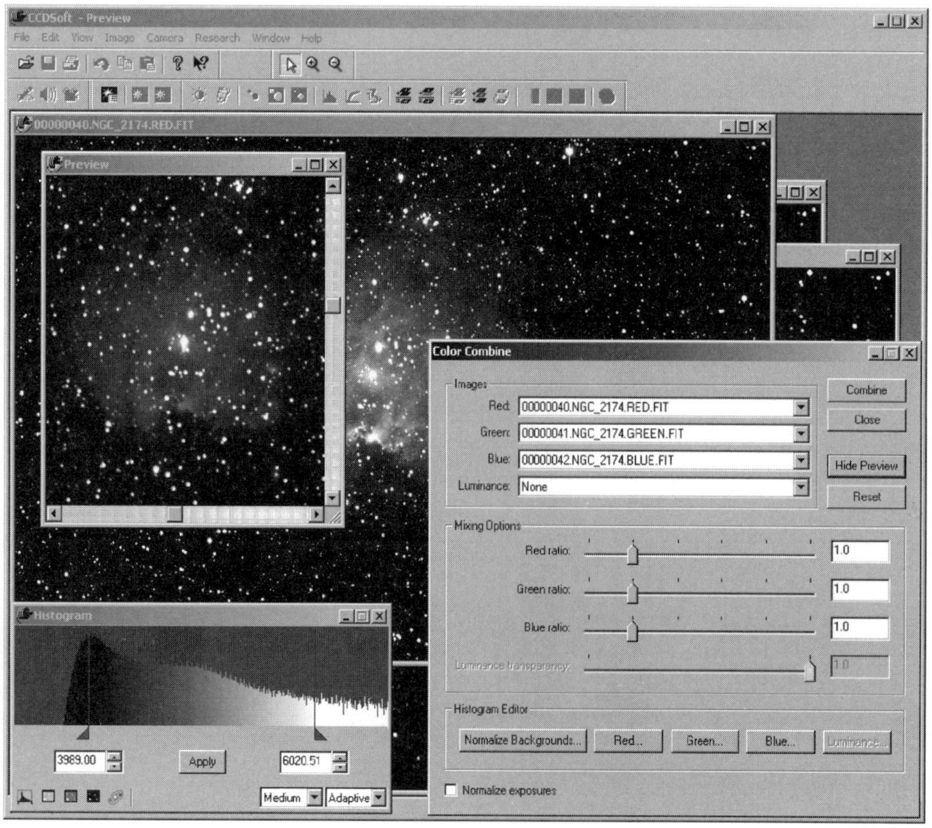

Both methods work. Which one you choose depends both on your personal preference and on the effect each has on the image. Changes to the ratio sliders change the color balance only. Changes to the white point affect the brightness and contrast of the component image. The actual results for each technique will vary from image to image, so experiment to see which method works better for a given image.

Figure 7.3.6 shows an example of the changes you can expect from adjusting a component image's white point. Clicking the Red button in the Histogram Editor section of the Color Combine dialog makes the red image active, and opens the Histogram tool.

SECTION 3: COLOR COMBINING

In figure 7.3.6, the white point for the red image has been moved to the right (higher value). This has darkened the red in the Preview window (compare to figure 7.3.5).

TIP: To darken any image, including a component color image, raise the white point. To lighten an image, lower the white point. Darkening a component image is especially useful when that color has areas that are saturated (100% of the color). In most cases, you want to avoid saturated color.

When the blue and green white points are raised to compensate for the change in the red histogram, the result is an accurate color rendition of NGC 2174, as shown in figure 7.3.7. There is some blooming on a few stars, and some red and green stars are visible at the bottom edge due to alignment differences between the images. The blooming can be cleaned up in an image editor, such as Photoshop or Picture Window Pro (see chpater 9 for details on removing blooming). The area where the images do not overlap, which has various false colors in ti, can be removed by cropping.

Combining Color Images: LRGB

An LRGB combine includes a luminance image in addition to the red, green, and blue images. It is similar to an RGB combination in CCDSoft.

Both image types yield excellent results, but there are differences. The main advantage of an RGB image is that it is composed solely of color component images. Colors will be richer. An LRGB imgae includes an unfiltered full-spectrum image, which may wash out the colors somewhat. You'll find some tips for dealing with that shortly, however.

The main advantage of an LRGB image is that you have a component image (the luminance layer) that is solely devoted to capturing fine details. Because of this, you can save some time and capture your color components using binning. This allows you to take color component in less time.

Another nifty use for the luminance layer is to use a color component, or an average of a color and a luminance component, to create a stronger impression of a specific color. For example, you could make an RRGB image of NGC2174 by using a red image in the luminance layer. Typically, you would take 1x1 red images as well as 2x2, and use the 1x1 in the luminance layer, and the 2x2 in the red layer.

FIGURE 7.3.7. THE FINAL COLOR COMBINE OF NGC 2174.

The key issues to pay attention to for an LRGB combination are:

- If the luminance component washes out colors, you can move the Luminance ratio slider to the left to reduce the contribution of the luminance image to the full-color image.
- The luminance image will not have any impact on color balance. You can balance color in the same way you would for an RBG combine.

THE NEW CCD ASTRONOMY

- Once you have good color balance, make all of your editing and histogram changes to the luminance image. This allows you to control the appearance of details in the full-color image without disturbing the color balance. This is a very important advantage of LRGB combine over RBG combine.

Figure 7.3.8 shows an example of an image where the luminance image is too bright (the white point is set too low). The image is of M16, the Eagle Nebula, but you can't tell because the core is too bright.

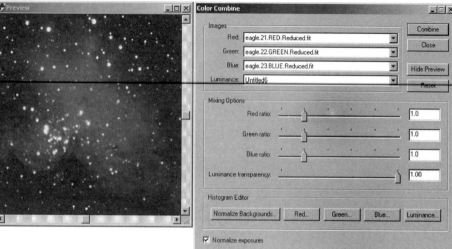

Figure 7.3.8. An LRGB combine with too much luminance contribution.

One solution to the problem is to move the Luminance transparency slider to the left. Figure 7.3.9 shows the result. The brightness of the image is reduced, and the details at the core of the nebula are clearer. You could also make this change by altering the white point using the Luminance button in the Histogram Editor section. Which method you use depends on personal preference; there is no single right way to do it.

Figure 7.3.10 shows the result of clicking the Combine button. The image has nice details in the core of M16, but the surrounding nebulosity is not visible.

You aren't limited to using the buttons and controls in the Color Combine dialog when you are combining component images to make a full-color image. You can also use any of the available CCDSoft tools to

Figure 7.3.9. Adjusting the Luminance transparency slider solves the problem.

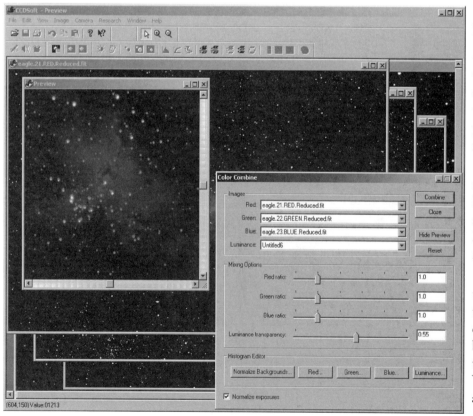

SECTION 3: COLOR COMBINING

FIGURE 7.3.10. THE RESULT OF AN LRGB COMBINE.

ferent image would require different settings. The settings also depend on the histogram of the luminance image at the time you open the Tune Brightness & Contrast tool.

Use the Preview window to see the impact of your changes. Your changes might make the component images look too dark or too bright. The primary concern in a color combine (RGB or LRGB) is to get the color balanced and the details visible. You can move, scroll, zoom, and resize the Preview window as needed while you work.

make changes to the component images. This is useful when you are doing an LRGB combine, since you can make changes to the luminance image without affecting the color balance.

Note: When doing an LRGB combine, all histogram and editing changes should normally be made to the luminance image. Changes to the red, green, and blue images would affect the color balance.

In figure 7.3.11, the overall brightness of the luminance image has been reduced by 24%. The contrast is reduced by 15%, and the Contrast pivot point has been increased by 100%. These adjustments reveal additional dim details effectively for this image. A dif-

FIGURE 7.3.11. USING THE TUNE BRIGHTNESS & CONTRAST TOOL.

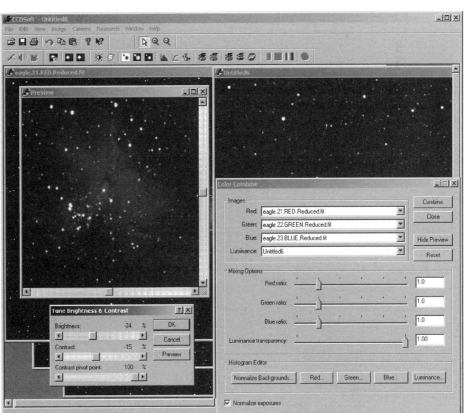

THE NEW CCD ASTRONOMY

For example, the changes to the luminance image using Tune Brightness & Contrast are shown in figure 7.3.12. Note that the luminance image is dark, too dark to be used alone. But when combined with the red, green, and blue component images, the result is just right. Figure 7.3.13 shows the results of the Tune Brightness & Contrast changes. The nebulosity is now much more extended, yet the details at the core of the nebula are still clearly visible.

There are two reasons why such a dark luminance image works properly in this example:

FIGURE 7.3.12. THE CHANGES TO THE LUMINANCE IMAGE DO NOT HINT AT THE EFFECT OF THE CHANGES ON THE FULL-COLOR IMAGE (SEE FIGURE 7.3.1).

FIGURE 7.3.13. THE FINAL RESULTS OF THE LRGB COMBINE.

- The full luminance image is so bright that it overwhelms the color components.
- The luminance image contains extensive detail of the dim nebulosity around M16

The level of detail hidden in the image is revealed when a non-linear histogram adjustment is applied to the image using a Mid-range Brighten, as shown in figure 7.3.14.

Color Combining in MaxIm DL

Combining the color images for M16 in MaxIm is more straightforward in some ways, more complex in others. MaxIm cannot combine images taken with different bin modes, but you can align, normalize, and color balance images more easily than with CCDSoft.

I recommend that you do as little processing as possible on the component images, especially the red, green, and blue components. You can easily upset the color balance by over-processing the component images. Processing a luminance image is much safer. Whenever possible save your adjustments for the final color image so you can see what impact the changes have on the finished product.

> **TIP:** The ability to process the luminance image independently of the color components is a significant advantage of LRGB imaging versus RGB imaging. However, RGB images tend to have richer color, so you should choose the color combination method based on what is more important to you: color (RGB) or details (LRGB). If you want the best of both worlds, either shoot your LRGB with the color images binned 1x1 instead of the more usual 2x2, or take longer and more RGB images to reduce noise and improve color accuracy.

FIGURE 7.3.14. BRIGHTENING THE IMAGE REVEALS THE LEVEL OF NEBULA DETAIL VISIBLE.

To begin, load the images you want to color combine (see figure 7.3.15). This example uses LRGB, so it has one luminance image, and three color images. The color images are binned 2x2, and the luminance image is binned 1x1. The advantage of this technique is time: you can use shorter color exposures if you bin 2x2. The eye is less sensitive to detail in color, and thus the luminance will be the dominant factor in determining resolution. MaxIm DL's LRGB combine creates images with lower resolution than other tools when using 2x2 binning for color. MaxIm allows the low resoltuion of the binned images to affect the luminance layer. Color combine in Photoshop and most other programs, however, maintains the full resolution of the L layer if done properly.

MaxIm DL cannot combine images shot with different bin modes, so you must first resize the red, green, and blue images to match the size of the luminance image. The L image was taken with an ST-8E camera in 1x1 bin mode, so it is 1530 pixels wide by 1020 pixels high. The 2x2 binned color images are 765 x 510 pixels.

To resize the smaller color images, highlight the red image and use the Process | Resize menu item. This displays the Resize dialog shown in figure 7.3.16. In the

New Image Size section of the dialog, set the Width to 1530, and the Height to 1020. The images were not taken with MaxIm DL, so ignore the current and new pixel dimensions -- they are calculated incorrectly because the images were not taken with MaxIm DL. Click OK to resize the image, and then resize the green and blue images.

As it happens, the background levels on the red, green, and blue images don't match up. There is a small difference between the red and green images, but the blue image was taken as the sun was starting to rise. The blue background level is high enough that it will create a color imbalance if it is not corrected.

The MaxIm DL color combination dialog has a checkbox for equalizing the color image backgrounds, but it has been my experience that you can get some-

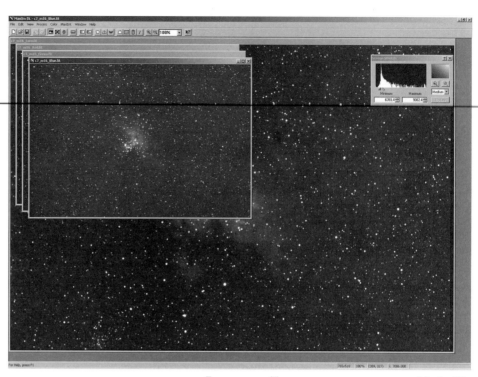

FIGURE 7.3.15. THE FOUR IMAGES LOADED INTO MAXIM DL.

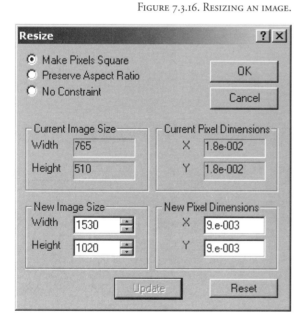

FIGURE 7.3.16. RESIZING AN IMAGE.

what better results by manually normalizing the background levels. You do not need to do the manual normalization, but it is the most accurate method.

The idea is to measure the average background level in each image, and then to subtract a value from all pixels such that the average background level drops to about 50 units. If any of the images have a gradient, you want to make sure you choose the darkest portion of the gradient for measuring the average background level. You can then treat the gradient separately, in a program such as Photoshop if necessary. See chapter 6 for information about removing gradients.

To measure the background level, open the Information window (View | Information menu item). The cursor changes to a circle with cross hairs. Pass the cursor over the darker background areas, and find the area with the lowest overall brightness. Figure 7.3.17 shows the MaxIm DL version 2.x information window and the simmest portion of the red image: the lower left corner. The average brightness in the area of the cursor is 1000.830 ADU. The blue image has the most pronounced gradient, and that gradient is darkest on the left side (with a lesser degree of darkening on the right edge as well), so the lower left corner is a good choice for all of the images. Figure 7.3.18 shows the gradient in the blue component. To check for gradients, set the black point so that the dimmest portion of the background is black, and lower the white point until the gradient shows up clearly.

Passing the cursor over this same area in each of the red, green, and blue images shows the average background levels shown in table 7.1. Depending on the exact position of the cursor, your results may vary a bit from these numbers. See section 4 of this chapter for details on manual normalization.

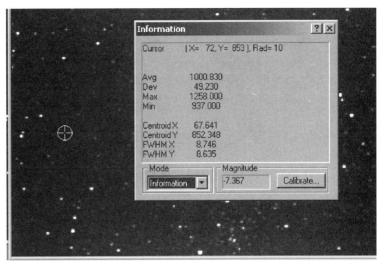

FIGURE 7.3.17. EXAMINING THE AVERAGE BACKGROUND LEVEL.

Table 7.1: Normalizing Backgrounds

Image	Average background	Amount to subtract
Red	1000	950
Green	915	865
Blue	6770	6720
Luminance	608	558

FIGURE 7.3.18. THE BLUE IMAGE HAS A GRADIENT WITH THE LEFT EDGE DIM.

To manually normalize the backgrounds, use the Process | Pixel Math menu item (see figure 7.3.19). Since all you want to do is subtract the same value from each pixel in the image to get the background levels close to each other (around 50 units, except for gradients), you must set the Operation to none and use a minus sign in the "Add Constant" box. For each of the images, use the number in the right-hand column of the table above for the subtraction. To check your results, pass the cursor over the each image while the Information window is open. The average value should be close to 50 in the area where you originally measured. If it is not 50 at other locations, that indicates that some gradient removal should take place at some

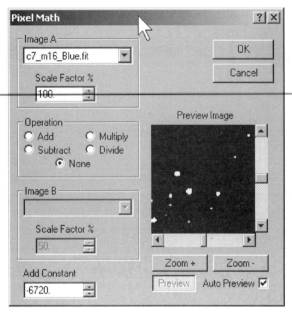

FIGURE 7.3.19. SUBTRACTING A VALUE FROM ALL PIXELS USING PIXEL MATH.

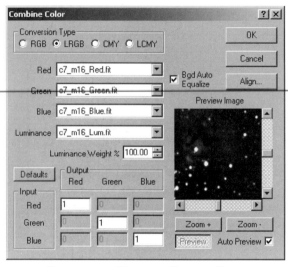

FIGURE 7.3.20. USING THE COMBINE COLOR DIALOG.

point in the overall processing. See chapter 6 for details on removing gradients.

To combine the component images into a full-color image, use the Color | Combine Color menu item to display the Combine Color dialog (see figure 7.3.20). This example show an LRGB combine. If necessary, check the LRGB radio button in the Conversion Type area to set the type of color combine. If, as is often the case, MaxIm DL doesn't guess the correct filenames, use the drop-down lists to set the files for red, green, blue, and luminance images as shown in figure 7.3.20.

Make sure that the luminance weight is set to 100% to start with, and that the Input and Output levels are all set to 1 at the bottom of the dialog. If necessary, you can change these settings to balance color, brightness, and saturation by raising/lowering these numbers.

TIP: The preview image is helpful, but it is not always 100% accurate. Don't fine tune your adjustments based only on the preview in any of the dialogs in MaxIm DL.

In this case, click OK to generate a color image. Evaluate this image, not the

FIGURE 7.3.21. THE RESULT OF THE COLOR COMBINE.

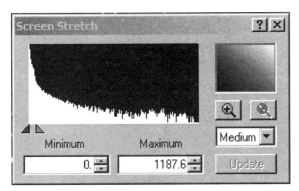

FIGURE 7.3.22. THE DEFAULT HISTOGRAM FOR THE COLOR IMAGE HAS A WHITE POINT SET TOO LOW.

Figure 7.3.21 shows the result for the color combination. The image is too saturated in red -- not at all like the preview, which is why I recommend using an image rather than the preview to make adjustments.

If this saturated look happens for you, the most frequent solution is to adjust the histogram of the image using the Screen Stretch window (at upper right in figure 7.3.21, and shown in detail in figure 7.3.22).

To reduce saturation, move the white point higher. Click on the green (rightmost) triangle and drag it to the right. For fine adjustments, use the up and down arrows in the Maximum box. Figure 7.3.23 shows the result of raising the Maximum setting to 2262. The central area is better, but not the outer area.

One solution is to use Digital Development, which will bring out dim details and sharpen the image at the same time (see figures 7.3.24 and 7.3.26). This is certainly a fast solution, but not quite ideal. You could also apply digital development to the luminance image prior to color combining, but this doesn't always (or even often) give you the simple, elegant result you are probably looking for. The dim areas are still too dim, and the color is still off due to the blue gradient.

preview, to make final decisions about the settings in the dialog box. If the image looks good, you're done. If it doesn't, you can go back and change the luminance weight and/or the input/output values, as follows:

- If the image lacks color (saturation), reduce the luminance weight.
- If one color dominates, you can reduce the contribution from that color by using a number smaller than 1 in the input/output area. Small changes in numeric values result in big changes in color. Use numbers close to 1 to start with. For example, if the image appears too blue, try using 0.95 instead of 1 for Blue.

Once you have manually normalized the backgrounds of the component images, you can still check the "Bgd Auto Equalize" checkbox to have MaxIm DL refine your work. The background auto equalization works best when the backgrounds are already close; don't rely on it to make major changes. And gradients can fool the automatic feature, so if one or more images suffer from gradients, you should generate two color images: one with, and one without auto equalization.

FIGURE 7.3.23. RAISING THE WHITE POINT REVEALS DETAILS IN THE BRIGHT PORTION OF THE NEBULA AT THE CENTER OF THE IMAGE.

MaxIm DL has two tools for fixing background problems. The Process | Flatten Background menu item works as an artificial flat field, and not so well as an anti-gradient tool. Figure 7.3.25 shows what happens when you apply the Flatten Background command to a gradient in an image of a nebula or galaxy. The Process | Remove Gradient tool is a better choice.

FIGURE 7.3.24. ABOVE: DIGITAL DEVELOPMENT DEFAULT SETTINGS.

FIGURE 7.3.25. RIGHT: DIGITAL DEVELOPMENT BRINGS OUT DIM DETAILS.

FIGURE 7.3.26. BELOW: DIGITAL DEVELOPMENT BRINGS OUT MORE DETAIL, BUT IT ALSO BRINGS OUT THE COLOR IMBALANCE DUE TO THE GRADIENT IN THE BLUE IMAGE.

Digital development works best on monochrome images, or on the luminance component of color images. Figure 7.3.27 shows the default digital development results for just the luminance image. Often, if you apply digital development to the color components, the color gets out of control, and even using the Color | Color Balance menu item can't fix it (see figure 7.3.28). For normal use, however, the Color Balance tool is marvelous.

The best solution is to take the results of the color combine (figure 7.3.23) and to do final processing in an image-editing program such as Photoshop or Picture

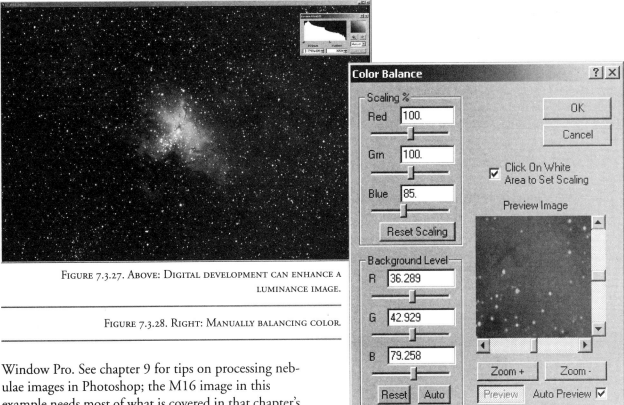

FIGURE 7.3.27. ABOVE: DIGITAL DEVELOPMENT CAN ENHANCE A LUMINANCE IMAGE.

FIGURE 7.3.28. RIGHT: MANUALLY BALANCING COLOR.

Window Pro. See chapter 9 for tips on processing nebulae images in Photoshop; the M16 image in this example needs most of what is covered in that chapter's section on Nebulae processing. In addition, the blue channel has a pretty severe gradient that is actually easy to fix; see the last section of chapter 6 for tips on removing gradients.

Figure 7.3.29 shows the final result of some Levels, Curves, and gradient fixes -- not bad for an image whose blue component was nearly ruined by the rising sun!

To perform an RGB combine in MaxIm DL, simply check the RBG radio button in the Color Combine dialog. This disables the Luminance features, and you can do your color combine using only red, green, and blue images.

FIGURE 7.3.29. THE FINAL M16 IMAGE AFTER LEVELS, CURVES AND GRADIENT FIXING IN PHOTOSHOP.

Section 4: Advanced Color Combining

Although many camera control programs include color combining tools, there are some real advantages to doing most of the steps yourself. If you have any problems in your images, you have a chance to deal with them as you step through the process.

For example, if you took your images on different nights, and wound up with a higher background level in one or more images, manually adjusting background levels will often give you a better result than the automatic tools will. Automatic tools inevitably make some assumptions about how things work, and your images won't always line up with those assumptions.

The following example shows how to do a manual combination using MaxIm DL and Photoshop. Other camera control and image editing programs will have similar tools that will also allow you to do a color combination manually. Manual combining isn't the easiest approach but it does allow you to utilize the features of the appropriate software for each step. The more tools you have, the more likely you are to find a way to optimize the process.

You can do a very good color combination using MaxIm DL and Photoshop. MaxIm DL is very effective at preparing the individual RGB images, and Photoshop is very good at merging them (and at adding a luminance image, too). Let's walk through what I consider the ideal multi-tool approach to color combining, with a heavy reliance on the tools in Photoshop.

The process starts with three image files: red, green, and blue. There are many ways to arrive at these three files. You could take three single images through red, green, and blue filters using long exposures. Or you could take ten each of red and green, and sixteen of blue, and sum each set of images. Or you might take two each of red, green, and blue, with exposure times of five, five, and eight minutes, and add each pair. If you are unable for any reason to get the proper exposure ratios for the color images, you can use Photoshop's Levels tool at the end of the

combination process to balance the color manually by altering the white point of the appropriate images.

For this example, you can download a trio of images of a portion of the California Nebula. They are binned 2x2 and taken with an ST-8E camera and a Takahashi FCT-150 refractor. Download them in a ZIP file from:

http://www.newastro.com/newastro/book_new/samples/c7_ca_neb1.zip

A luminance image at 1x1 binning is also used in the following examples. A ZIP file with the luminance image can be downloaded from:

http://www.newastro.com/newastro/book_new/samples/c7_ca_lum.zip

Normalizing with MaxIm DL

Normalization can mean many things, but when working with images it usually means doing something that gives the images a common brightness level. In this example, we want to normalize the backgrounds of the three images. Due to various factors, such as light pollution and exposure time, the backgrounds of color images may vary. Some color combining methods will do this normalization for you, but it's handy to know

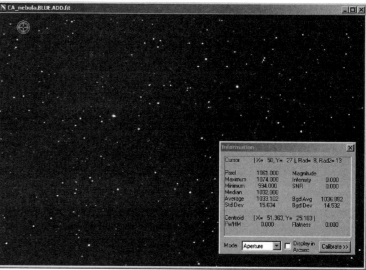

FIGURE 7.4.1. MEASURING BACKGROUND LEVELS.

how to do it yourself. There are times when you want very tight control over color combination, or when you use multiple tools and need to do this manually.

The process starts by measuring the background level of the images. MaxIm DL makes this easy, but you still have to be careful to measure the background in the right place. I prefer to measure the background at its darkest point. This might be a corner of the image that does not have nebulosity, as in this example, or simply the dark end of a gradient that you plan to fix later. The MaxIm information window (see figure 7.4.1) displays a wealth of information about an image. We are interested in the Median or Average values. In figure 7.4.1, the top left corner of the blue image turns out to be the darkest portion of the background, with an average value of 1033.

Use the same general area to measure the background level for all three images. Even a slight movement of the cursor will give you somewhat different readings, and if they do then round down. In this example, although the average value was 1033 at the point shown at upper left of figure 7.4.1, this varied from 1030 to 1035 over even a short distance. Make a note of the numbers for each image.

A good value for the darkest portion of the background is 50 ADU (analog to digital unit). To get to this value, you need to subtract 980 from every pixel in the image. In MaxIm DL, use the Process | Pixel Math menu item (see figure 7.4.2). Make sure the Scale Factor is set to 100%, and that the Operation is set to None. In the Add Constant box, enter the average value less 50 ADU as a negative number to indicate subtraction instead of addition (-980). Click OK to subtract this value from every pixel in the image.

To verify the change, pass the cursor over the same area of the image where you took your measurement. The average value should be close to 50. Repeat this process for the other two images, using the same area of each image for the measurement. Save the images. If you want to download already-normalized images, you may do so from this location:

http://www.newastro.com/newastro/book_new/samples/c7_ca_neb2.zip

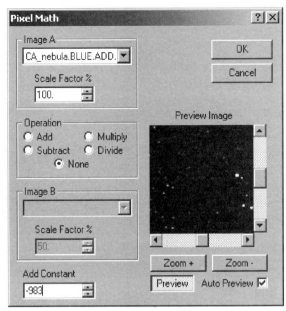

FIGURE 7.4.2. SUBTRACTING A CONSTANT FROM THE IMAGE PIXELS.

Color Combine in Photoshop

Photoshop can combine 8-bit grayscale images into an RGB image. Open all three grayscale images in Photoshop. If you don't already have Eddie Trimarchi's FITS Photoshop plug-in, it's incredibly useful and I suggest that you download it from his web site at:

http://users.fan.net.au/~eddiet/indexh.html

If you aren't using this FITS plug-in, you must save the red, green, and blue images as 16-bit TIFF files in MaxIm DL so you can open them in Photoshop.

Whether you open the images as FITS or TIFF, they are 16-bit images. Photoshop will show them as mostly or completely dark, similar to figure 7.4.3. This is normal. Photoshop is trying to show the full range of brightness values -- all 65,000 of them. Since there are only 256 brightness values available, and all 65,000 levels are being shoved into that small bucket, almost nothing in the image is visible.

Camera control programs like CCDSoft, Astroart, and MaxIm DL automatically set temporary, non-destructive black and white points when you open an image. In Photoshop, you'll need to do that manually. Black and white point changes in Photoshop are permanent. Use the Image | Adjust | Levels menu item to open the dialog shown in figure 7.4.4. Drag the rightmost slider below the histogram toward the left until you start to see the nebulosity in the red image. You are lowering the white point as you move the slider. Adjust the red image first because it has the brightest nebulosity. There is no nebulosity in the green image, and just the barest amount in the blue. Note the numeric value in the right-hand box above the histogram. It's 14 in this example, but any similar value would be fine. There is no one right setting; anything close will be acceptable.

TIP: Generally speaking, you don't want to make too much of an adjustment to the white point all at one time in Photoshop. The value of 14 for this image is about as low as I go. If you need to go further, do it in two or more steps so that you have better control over the white point.

To keep the three color images balanced, make the same adjustment to the green and blue images. Instead

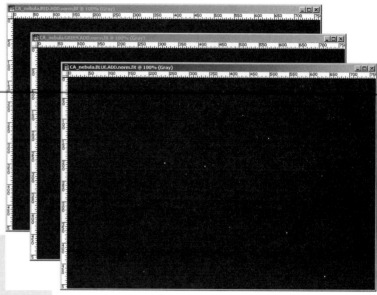

FIGURE 7.4.3. OPENING 16-BIT IMAGES IN PHOTOSHOP SHOWS LITTLE OR NO DETAIL.

of moving the rightmost slider, you can simply enter the same value in to the right-hand box above the histograms. If the color images are not already balanced, you do not need to be precise.

The white point needs to be lowered a bit more, so select the red image again and open the Levels dialog. Lower the white point to brighten the nebulosity. You don't want to try to brighten it completely; there are better tools for that. The goal is to lower the white point enough to show the middle brightness levels in the nebulosity. As you can see in figure 7.4.5, I found this to be at about a value of 180 for the second application of Levels. This is brighter than I would do if it were a monochrome image. For a monochrome image, it is safe to make most of the final brightness changes using other tools.

FIGURE 7.4.4. SETTING THE WHITE POINT FOR THE RED IMAGE.

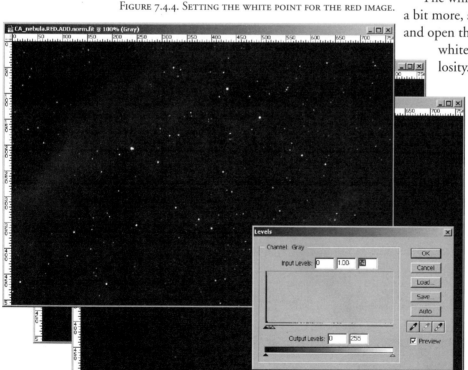

FIGURE 7.4.5. APPLYING LEVELS A SECOND TIME.

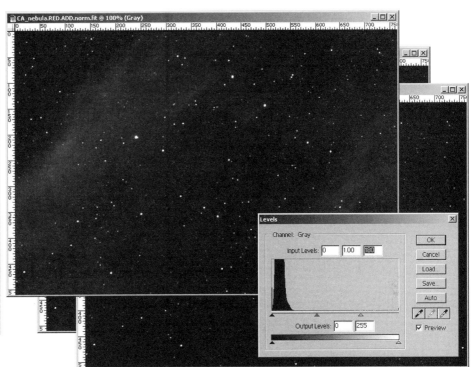

TIP: If you were processing a luminance image, you would use the Image | Adjust | Curves menu item to boost the contrast of the dim portions of the nebula. However, Curves is a non-linear change, and it is difficult to control across all color images. It is better to combine these three images into an RGB image, and then make the final brightness changes to the combined image.

For color you need to convert to 8-bit data, which has a much smaller brightness range. It makes sense to do as much brightening as possible before we convert to 8 bits. Make this exact same change to the green and blue images as well.

Before you can combine these three images into a single RGB color image, you must convert them all to the 8 bits per pixel format. Photoshop cannot combine 16-bit images into a color image. Use the menu arrow at the top right of the Channels palette to open the fly-out menu shown in figure 7.4.6. Click on the Merge Channels menu item.

This opens the Merge Channels dialog, shown in figure 7.4.7. The default mode is Multichannel. Click on the list and select RGB Color instead. The default number of channels for RGB Color is 3, and this is the value we want. Click OK to continue.

FIGURE 7.4.7. SETTING RGB COLOR MODE.

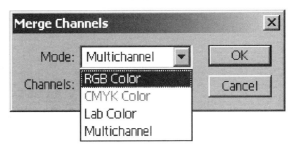

FIGURE 7.4.6. MERGING CHANNELS.

FIGURE 7.4.8. CHOOSING THE FILE FOR EACH CHANNEL.

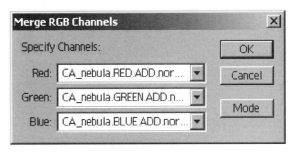

THE NEW CCD ASTRONOMY 357

The Merge RGB Channels dialog appears (see figure 7.4.8). The open images show up in the Red, Green, and Blue lists. Click on the appropriate filename for each color. Click OK when you have selected the proper file for each color. The three images disappear, and are replaced by a single color image. The color will be dim at this point, because we deliberately chose to leave some of the histogram adjustments for after the RGB combination. Figure 7.4.9 shows the result of the RGB combine.

The Channels palette shows an RGB channel, and individual channels for Red, Green, and Blue. You can make changes to all three channels by clicking on the RGB channel. To change just one channel, click on that channel to select it. For this example, we will be making changes to all channels at one time.

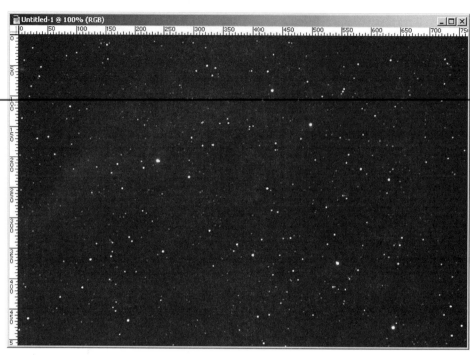

FIGURE 7.4.9. THE RESULT OF THE RGB COMBINATION.

FIGURE 7.4.10. BRIGHTENING THE IMAGE USING CURVES.

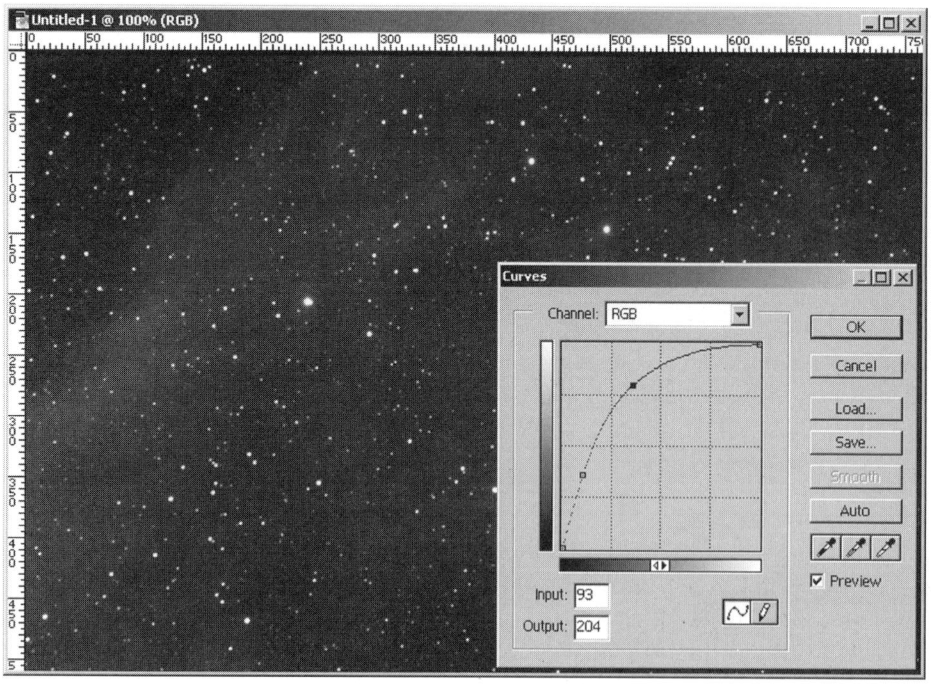

TIP: Light pollution can create a need to change a single channel. If it gives the image a green cast, raise the black point for the green channel to balance the background. This isn't necessary in this case because we normalized the backgrounds earlier.

Use the Image | Adjust | Curves menu item to open the Curves dialog. Click and drag on the diagonal line to brighten the image. Figure 7.4.10 shows how I chose to brighten the image. I added two points. The leftmost point brightens the dim areas, and the upper point prevents over-brightening of the stars and bright portions of the nebula.

I felt that the image needed even more contrast, so I applied Curves a second time using a completely different type of curve (see figure 7.4.11). The lower left portion of the curve slightly dims the background, and the upper right of the curve adds some brightness to the very brightest portions of the curve. This enhances the overall contrast of the image, and gives it a little more punch. You might prefer the image in figure 7.4.10 to figure 7.4.11; there is no one right way to process an image.

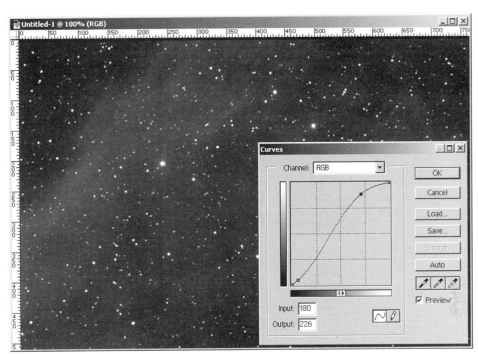

FIGURE 7.4.11. FURTHER ADJUSTMENT OF THE IMAGE.

Making an LRGB image

The color image in figure 7.4.11 is a nice image, but it was binned 2x2 so the resolution is limited. Binning color images is useful when you combine the color images with a high-resolution 1x1 (unbinned) luminance image. Figure 7.4.12 shows the high-resolution image of roughly the same portion of the California Nebula, taken three days after the color images.

The high-resolution image shows more detail and has smaller star images than the binned color image. If this isn't clear in the figures, it will be very clear when you download the files and examine them.

FIGURE 7.4.12. A HIGH-RESOLUTION IMAGE OF THE CALIFORNIA NEBULA.

The luminance image is already in 8 bits per pixel mode. If you had opened one of your own images, you would need to finish processing it and convert to 8-bit mode before making the LRGB image. The luminance image is a grayscale image, and needs to be converted to RGB color. Use the Image | Mode | RGB Color menu item to make this change.

Select the color image you just built, and resize it to match the luminance image. The luminance image is 1530x1020 pixels. Then select the entire color image, and copy it to the clipboard. Paste the color image into the luminance image.

TIP: If the color image pastes as black and white, you forgot to convert the luminance image to RGB. Undo, convert to RGB color, and perform the paste again.

The color image will be Layer 1 above the background in the Layers palette. Click on the Opacity drop-down list in the Layers palette and set the opacity of the pasted color layer to 50% (see figure 7.4.13).

This shows both layers at the same time. The images do not line up. This is to be expected when they were taken several days apart (see figure 7.4.14).

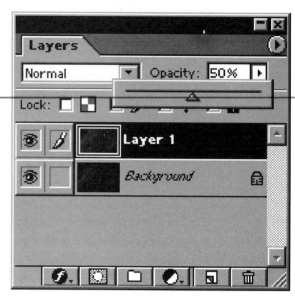

FIGURE 7.4.13. SETTING THE OPACITY OF THE COLOR LAYER.

Now that you can see the stars in both images, align the images using the Edit | Transform | Rotate menu item. However, I suggest that you first shift the color image in such a way that a star fairly close to the center of the color image lines up with the same star in the luminance image (see figure 7.4.15). This makes it easier to keep track of the rotation as it progresses. For troublesome images, use a corner star for greater precision.

Now use the Edit | Transform | Rotate menu item to begin the rotation. Move the cross-hairs that mark the center of the rotation to the aligned star. If necessary, use the Ctrl + "-" keys to reduce the image size so you can see the corners of the color image. Click near the corners and drag to perform the rotation. In Photoshop 6, you can type in numeric values interactively and observe the results for a more exact rotational alignment.

FIGURE 7.4.14. THE TWO IMAGES ARE NOT QUITE LINED UP.

SECTION 4: ADVANCED COLOR COMBINING

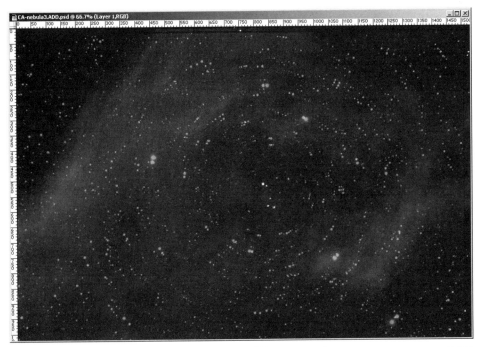

FIGURE 7.4.15. ALIGNING A STAR IN BOTH IMAGES.

some background in the image window is helpful. It gives you some room to maneuver. When you are satisfied with the alignment, double click the color image to complete the change. Set the opacity of the color image back to 100%.

Although the two images are now lined up, all you can see is the color image. The next step is a neat trick that allows you to use the color from the color image, and the high-resolution of the luminance image, to create the final image. Click on the other drop-down list on the Layers palette (refer to figure 7.4.17) and choose "Color" near the bottom of the list. This must be done while the color image is the active layer.

TIP: Photoshop has limited alignment capabilities. For a more precise alignment, you can pre-align the images in MaxIm DL, CCDSoft, Registar, Mira, or other software programs that provide such tools. These other programs will do sub-pixel alignment and can provide a more precise alignment than Photoshop alone. If you do the alignment in Photoshop, use Ctrl + "+" keys to enlarge the image so you can see the overlapping stars in detail. Use Ctrl + "-" keys to zoom back out.

Figure 7.4.16 shows the images lined up, with the stars very close to exactly overlapped in both images. This image also shows the corner handles clearly, and it illustrates why shrinking the image scale a bit to show

FIGURE 7.4.16. LINING UP THE IMAGES USING ROTATION.

THE NEW CCD ASTRONOMY   361

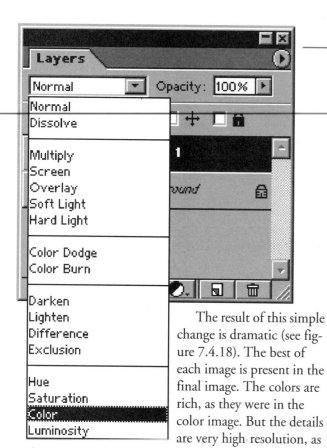

FIGURE 7.4.17. SETTING THE BLEND METHOD FOR THE COLOR LAYER.

these dim areas a bit (lower left point on the curve). I also found that I could brighten the brighter portions of the nebula still more to increase the contrast yet again (top point on curve). The final result appears in figure 7.4.20.

TIP: It is very important to note that after you combine the two images, all color changes must be made to the color layer, and all histogram adjustments must be made to the luminance layer.

Color Processing Guidelines

Many of the techniques outlined for monochrome image processing in Chapter 9 also apply to color imaging. The most important guidelines is to apply your histogram adjustments and sharpening to the luminance image or layer only. Smoothing is sometimes applied to the color layer to remove color noise, but it can also be applied usefully to the luminance layer.

If you want to live a little closer to the edge, try applying histogram adjustments to one color layer at a time. This is a very powerful way to modify the color balance of an image. It is also a way to get your color

The result of this simple change is dramatic (see figure 7.4.18). The best of each image is present in the final image. The colors are rich, as they were in the color image. But the details are very high-resolution, as they were in the luminance image. This method combines the best of both approaches, and there is very little difference between this approach (low-res color, high-res luminance) and using all high-res images for red, green, blue, and luminance. The low-res color images can be taken in half the time of the high-res 1x1 color images, saving considerable time in image acquisition at a small cost in final detail level.

I felt that the very dimmest portions of the nebula in figure 7.4.18 were too dim. I used the Curves tool (see figure 7.4.19) to boost

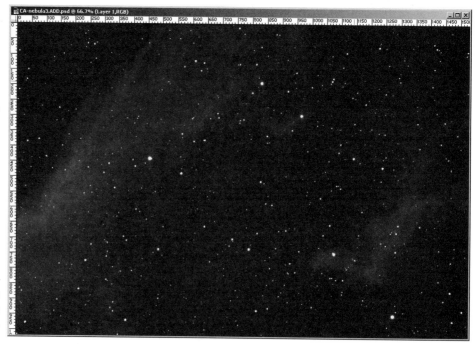

FIGURE 7.4.18. THE COMBINED IMAGE SHOWS THE BEST OF BOTH COLOR AND LUMINANCE IMAGES.

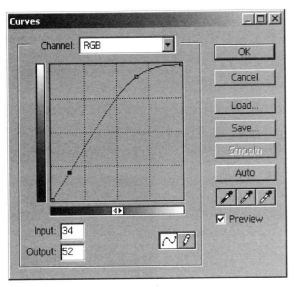

FIGURE 7.4.19. APPLYING FINAL HISTOGRAM ADJUSTMENTS.

badly out of balance. A common trick is to adjust the green layer to compensate for light pollution. Many images have too much green because of light pollution.

White point adjustments are also powerful but difficult to do . Step lightly and experiment to see what happens when you change each white point. Some products, like Maxim DL, can automatically find a common white point for you (a checkbox in the Color Balance dialog). Finding the white point manually is hard. Most of the time you are better off to simply use the LRGB method outlined earlier. With LRGB images, the key is setting a good black point in the component images. You can then control color balance after you do the original combine.

Normalize color backgrounds before you combine colors, either by carefully setting matching visual levels or by performing pixel math in programs such as Maxim DL or Astroart.

Try blurring just the color layers (in Photoshop, make the RGB layer active before you do this). This can reduce color noise (mottling), smooth out color fringes on stars, and mask slight alignment errors. Sharpening should only be applied to the luminance layer.

Avoid non-linear histogram adjustments to single layers whenever possible. Non-linear changes can have profound effects on color balance, to the point where you may not be able to correct the problems. Minor non-linear adjustments can work, but you will be safest if you make non-linear adjustments to the entire image. Use linear adjustments (Levels in Photoshop) for individual colors whenever possible.

There isn't one specific color balance that is guaranteed to be absolutely correct. Compare the M16 images in this chapter. The one done in CCDSoft is rich and red, while the one done in Maxim DL has a nice contribution from blue light in the brightest section. Both are valid interpretations; it all depends on what you are trying to show, and what pleases your eye.

FIGURE 7.4.20. THE FINAL IMAGE OF THE CALIFORNIA NEBULA.

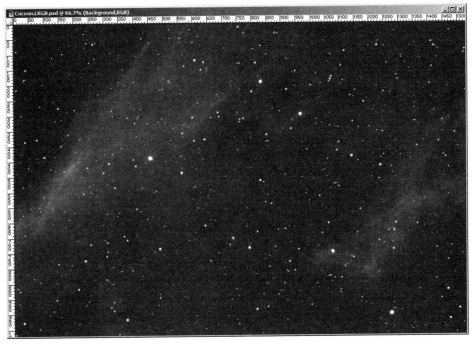

8 Image Processing Fundamentals

Image processing is what most people look forward to when they get into CCD imaging.

While it's true that some real magic occurs with image processing, there's nothing as good as starting out with a quality image. You will get the best results with image processing if you start out with an image that has a long exposure plus good focus and accuracte tracking.

Section 1: Selecting Data to Display

If a CCD image has a good signal to noise ratio, image processing will be more effective. You can't make a bad image into a good one, but you can make a good image into a great one.

The basic steps in image processing are:

- Select the portion of the data you want to display using black and white points (histogram). Exclude pixels outside the brightness range that is of interest to you. In camera control programs, this is a non-destructive change that is useful for evaluating image quality. Image editors such as Photoshop make permanent changes to the histogram.

- Stretch or otherwise modify the brightness of portions of the data to emphasize what is important to you. This is non-linear histogram stretching. It involves emphasizing certain brightness zones, and de-emphasizing others. Non-linear histogram changes permanently alter the image. See Non-linear Histogram Adjustments below for details.

- Enhance the appearance of the image with sharpening, smoothing, and other tools. Deconvolution is an exception, and must usually be done after reduction and before other image processing.

These steps are all you need to create a good-looking result. Advanced image processing techniques are most effective when used to get more out of a good image, not to rescue a poor image. Good data comes from long exposures or from combining shorter exposures. See chapters 3 and 7 for more information.

Data versus Image

Depending on the capabilities of your CCD chip and the camera hardware, the number of illumination levels in your images will vary. For a 12-bit camera, the largest possible number of levels is 4096. For a 16-bit camera, the largest possible number of levels is 65,536.

The actual number of levels may be different from these values, however, depending on the capabilities of your camera. The photons striking the CCD chip are converted into electrons. The hardware in the camera counts these electrons, and if one pixel has twice as many electrons, then it will be twice as bright in your image. Each pixel has a full-well capacity, which is the maximum number of electrons it can store. If a pixel exceeds this capacity in a non-antiblooming camera, electrons flow into adjoining pixels, resulting in blooming. If a pixel exceeds about 50% of full well in an antiblooming camera, special circuitry will bleed off electrons to prevent reaching full well. An antiblooming camera prevents blooming by draining off electrons before they can flow from one pixel to another.

Every photon doesn't get converted into an electron. CCD detectors do not operate at 100% efficiency. The most efficient CCD chips convert up to 90% of the photons striking them into electrons. CCD chips also have varying sensitivity to light of different colors. For example, the Kodak E chips, found in the SBIG ST-7E and other cameras, has more sensitivity to blue light than previous versions. Most chips have peak sensitivity at a particular color, with progressively less sensitivity as you move away from that central peak.

Let's look at an example of the brightness levels in a hypothetical CCD camera. Suppose the pixels can hold a maximum of 50,000 electrons. Assume further that

FIGURE 8.1.1. A VERY SMALL HYPOTHETICAL CCD CHIP WITH A 5X5 ARRAY OF PIXELS.

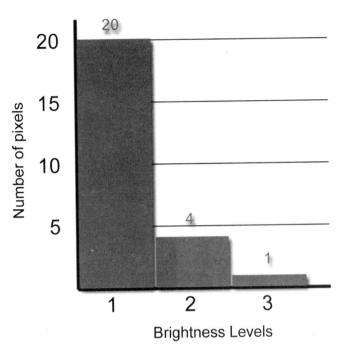

FIGURE 8.1.2. A GRAPH OF THE BRIGHTNESS LEVELS IN THE MINI-CHIP FROM FIGURE 8.1.1.

The bottom line here is that for just about any camera, the number of levels in the output is enormous, involving many more brightness levels than the human eye can distinguish. Since you can't display all of the brightness levels at one time, it is up to you to determine which levels will be included in your image.

Histogram Adjustments

A histogram is a graph of the brightness values in an image. Figure 8.1.1 shows a simple test image, five pixels wide and five pixels high. Think of it as an image of a single star taken with the world's smallest CCD camera. I've used an ultra-small image to simplify, and to simplify even further there are just three brightness levels: white, gray, and black. A quick count shows that there are 20 black pixels, four gray pixels, and one white pixel. A black pixel has a value of 1; a gray pixel has a value of 2, and a white pixel has a value of 3. In real life, of course, an image would be at least several hundred pixels on a side, and would have from 4,000 to 65,000 brightness levels.

Figure 8.1.2 shows a graph of the counts for each brightness level. The number of pixels that have a given brightens level increases in the vertical dimension, and the brightness levels start with 1 (black) on the left and goes to 3 (white) on the right.

The graph in figure 8.1.2 is called a histogram. The shape of the histogram in figure 8.1.2 is typical of most CCD images: lots of dark, and a little bit of light. If we take a typical CCD chip, and graph the brightness levels from a typical exposure, the shape of the curve will be similar. Figure 8.1.3 shows an image of the group of galaxies known as Stephan's Quintet. The image has many stars, a lot of dark-sky background, and a few small galaxies. It was taken with an ST-8E binned 2x2 on a Takahashi FSQ-106 refractor. To download:

http://www.newastro.com/newastro/booknew/samples/c8_squint.zip

Figure 8.1.4 shows a histogram of the brightness levels in the Stephan's Quintet image, using the Screen Stretch dialog of MaxIm DL. The general shape of the histogram is similar to the original example. The high values are on the left (dark pixels), curving down toward lower values as you move to the right (brighter pixels). The brightness values range from zero (far left)

you are using a camera that supports 16-bit data transfer. The maximum meaningful value is only 50,000 (the highest possible number of electrons), not 65,536 (the highest possible value in a 16-bit system). Further, the hardware in your camera may not translate one electron to one unit of brightness (called an ADU, for "analog to digital unit"). The designers of the camera work under various constraints, physical and electronic, that may require that the conversion ratio be different. For example, if the camera above has been set to deliver at a ratio of 1.5 electrons per brightness unit (1.5 is called the gain), the actual range available to you would be still smaller, around 33,000 units.

TIP: You may see individual pixel values above this number when very bright stars are in the field. These stars fill pixels to overflowing if the exposure is long enough, and this can lead to false maximum values.

You can obtain information from your camera manufacturer that will tell you the actual full-well and gain values for your camera. You need this information to create an appropriate flat field image for your camera (see chapter 6 for details).

FIGURE 8.1.3. A TYPICAL ASTROPHOTO, SHOWING THE GALAXY GROUP STEPHAN'S QUINTET.

to approximately 65,000 (far right). Only the brightness levels between the red and green triangles are actually displayed in the image, however.

There are "Minimum" and "Maximum" numbers at the bottom of the screen stretch window. They define which brightness levels are visible. The Minimum value is set to 2600. This means that all pixels with a brightness level less than 2600 will be rendered as pure black. This is called the black point, but it has different names in different software packages. The Maximum value is set to 3971; all pixels with brightness levels higher than this value will be rendered as pure white. This is called the white point. The black and white points are called the contrast settings.

The difference between these two values is 1371. This is the range of values that are rendered in the image. (In fact, CCDOPS doesn't set the white point directly. You define the white point in CCDOPS by defining the range.) Only pixels within this range of values show up as shades of gray. Depending on the number of shades of gray your monitor can show, the actual number of values will be different. The typical monitor can show 256 shades of gray, and those 256 shades will be selected so as to spread out over the total of 1,371 values. 1371 divided by 256 is 5.35, so there will be 5.35 pixel values showing up at each shade of gray.

Figure 8.1.5 shows the histogram for the same image in Photoshop 6. The same general shape is evident, but the exact shape of the curve is different. The Photoshop curve is smoother. The small spikes shown in the MaxIm DL version are not visible here. Different software packages round off the actual numbers differently. Photoshop tends to smooth out the curve, while MaxIm DL tends to emphasize every bump in the curve.

Whatever software you use to manipulate your images, it will probably have some features that allow you to view and modify the histogram of an image. The details of the histogram's appearance will vary from one software package to the next, but the important part is getting familiar with how your software allows you to modify the histogram.

FIGURE 8.1.4. A GRAPH (HISTOGRAM) OF THE BRIGHTNESS LEVELS FOR THE IMAGE IN FIGURE 8.1.3.

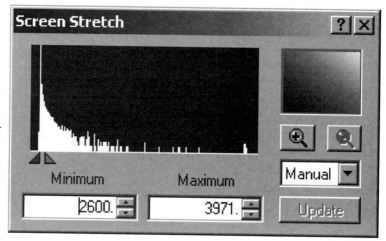

SECTION 1: SELECTING DATA TO DISPLAY

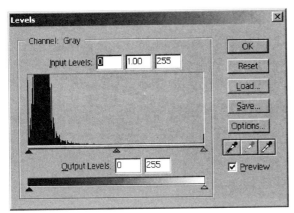

FIGURE 8.1.5. THE HISTOGRAM FOR FIGURE 3 AS DISPLAYED IN PHOTOSHOP 6.0

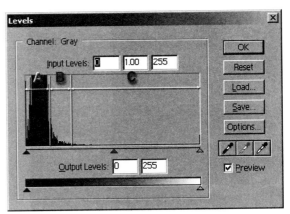

FIGURE 8.1.6. THE IMPORTANT SECTIONS OF A TYPICAL HISTOGRAM FOR AN ASTROPHOTOGRAPH.

The typical CCD image histogram has a lot of dark pixels, but they are uninteresting: they make up the sky background (area A in figure 8.1.6). There is a small shelf of dim pixels (area B) where the most interesting information is. Examples include nebulosity and dim areas in the spiral arms of galaxies. Finally, there is a long, thin area where the bright pixels in all the different stars live (area C).

Since we judge the brightness of a star more by it's size than by its actual brightness in the image, area C is also not especially interesting. What interests us most is area B. The image in figure 8.1.3 has already been adjusted to emphasize this portion of the image. That's why pixels below 2600 are safely all black, and pixels brighter than 3971 are all white. They just aren't interesting enough to be included.

Figure 8.1.7 shows the histogram of the image as it was when it was first downloaded from the camera. The general shape is similar. There are lots of dark pixels, as shown by the thin peak at left. The bright pixels are spread out over a long range of brightness values, as shown by the long thin curve to the right of the peak. There are also some interesting differences:

- There is a blank area to the left of the spike of dark values. This is a result of sky glow.
- The spike of dark values is much thinner because we are showing the full range of brightness values, not only the ones selected for display.
- The shelf, area "B," is so small that you can barely see it.

- The curve from high values to low values is very short.

In other words, it looks like a typical histogram, but it's all scrunched up toward the left, with a gap at far left that represents sky glow.

These characteristics tell us important things about the original image:

- There aren't any truly black areas in the image because of the sky glow. The darkest values in the image are not actually black.
- Of the full range of values in the image, only a very small portion of them are "interesting" to the eye. They are compressed into a very small range of brightness values.

FIGURE 8.1.7. THE UNADJUSTED HISTOGRAM FOR THE STEPHAN'S QUINTET IMAGE.

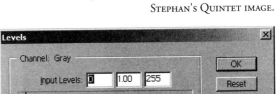

THE NEW CCD ASTRONOMY 369

Figure 8.1.8 shows the appearance of the Stephan's Quintet image if the full range of values in the image is displayed at one time. The galaxies are completely lost. They are too dim to show up when we include the full range of brightness values.

Figure 8.1.8 is typical of the data you download from a CCD camera. Most camera control programs will automatically select a range of values to display so you can see what's in your image. This automatic setting will usually be reasonably good, but it will seldom be perfect. You can make manual changes to the histogram to eliminate the data you aren't interested in, and to call attention to the data you are interested in.

You can turn off automatic histogram settings in most camera control programs. If you do your image editing with programs like Photoshop or Paint Shop Pro, you will see images like figure 8.1.8 most of the time. You won't even know what's in your image until you adjust the histogram settings.

Histogram Adjustment in MaxIm DL

MaxIm DL automatically adjusts the histogram when you open an image unless the screen Stretch window is set to manual (see figure 8.1.4). MaxIm has a drop-

FIGURE 8.1.8. THE FULL RANGE OF VALUES IN THE STEPHAN'S QUINTET IMAGE IS TOO LARGE FOR THE HUMAN EYE TO PERCEIVE ALL AT ONCE.

down list of automatic histogram adjustments. The medium setting works for most deep-sky objects.

Maxim also includes automatic histogram stretches for planetary and lunar images. Unlike deep-sky images, which concentrate the data in the dim brightness levels, planetary images often have useful data across the full range of brightness levels, with most of the interesting detail closer to the bright end (right) of the histogram.

Although MaxIm's automatic adjustment is good, you can usually tweak it to get better results. Figure 8.1.9 shows the appearance of the Stephan's Quintet image using the Medium automatic setting. The details around the galaxies can be seen somewhat clearly, but the background is too bright. The bright background reveals slight brightness irregularities in the background. The irregularities are small enough that a change in the histogram will hide them. If they were more intense, gradient removal would be required (see chapter 6).

You can change the values for maximum and minimum either by typing in new numbers, by moving the red and green triangles, or by using the up/down arrows at the right of the Minimum and Maximum boxes. If the image background is too bright, darken it by

FIGURE 8.1.9. THE AUTOMATIC MEDIUM SETTING IN MAXIM GETS CLOSE, BUT DOESN'T QUITE NAIL THE OPTIMAL SETTINGS.

FIGURE 8.1.10. A MORE BALANCED VERSION OF THE STEPHAN'S QUINTET IMAGE.

increasing the black point (increase the Minimum value). If the background is too dark, with inappropriately sharp edges on galaxies and nebulae, lighten it by decreasing the black point.

If the bright details in the image are too white, and don't show much detail, increase the number of brightness levels included in the visible image by raising the white point (increase the Maximum value). If nebula or galaxy details are too dim, try lowering the white point.

To improve the appearance of the image in figure 8.1.9, raise the Minimum value (black point) to darken the background. To get a smooth edge on the galaxies, raise the white point. This will help to hide the noise (grain) in the dimmest portions of the galaxies. Moving the white point further out (increasing the Maximum value) also helps darken the background, so the settings interact. You'll need to get a little experience adjusting them to work out which setting would be best to use on any given image. Generally speaking, set a black point that provides a not-quite-black background, and then adjust the white point so that the effects of noise (graininess) are minimized. You can go back and forth between these two adjustments, making small changes, until you find the sweet spot for both.

Sometimes it helps to think of your image as "data" instead of "image." The raw data has a huge amount of information in it. The final image is based on the choices you make about how much of that data to display. You can, for example, use a low Maximum value to show as much detail as possible in the fringes of the galaxies. This occurs at the cost of more graininess in these areas. A low black point brightens the background, which can make it easier for the eye to see small differences in brightness close to the background level. Or you could use a high Maximum value to hide the grainiest parts of the data. at the cost of hiding some of the dimmer portions. The choices are yours to make, and there are no right and wrong ways to present the data in the final image. I generally strive for balance in images that I want to look good, while a low white point can help show dim details when that is your number one concern.

Figure 8.1.10 shows contrast settings that reveal faint detail but keep the background dark. These settings bring out the faint streams of stars flung off in the galaxy collision. The sky background is now dark, and the faint details in the galaxies stand out well.

Histogram Adjustment in CCDOPS

Figure 8.1.11 shows the appearance of the Stephan's Quintet image in CCDOPS prior to any manipulation. Note that the Auto checkbox is checked, so the software has attempted to find the optimal settings for the image. As with MaxIm, the result isn't quite optimum. The background is brighter than it should be, and the dim galactic details are enhanced so much that the galactic cores are too bright.

The settings in the Contrast dialog box are different than those in Maxim. There are still two numbers, and you can change them either by typing new numbers in directly, or by using the up/down arrows. For figure 8.1.11, the values are:

Back - 2558
Range - 668

These would be equivalent to the following values in MaxIm DL (the automatic MaxIm DL values are in parentheses):

Minimum - 2558 (2253)
Maximum - 3226 (3842)

The background is a little darker than it was for the Maxim defaults. The galaxy cores are bright because the white point is set so low.

Figure 8.1.12 shows the Back and Range settings I preferred for this image: Back set to 2576, and Range set to 1674. This equates to MaxIm Max/Min settings of 2576/4250.

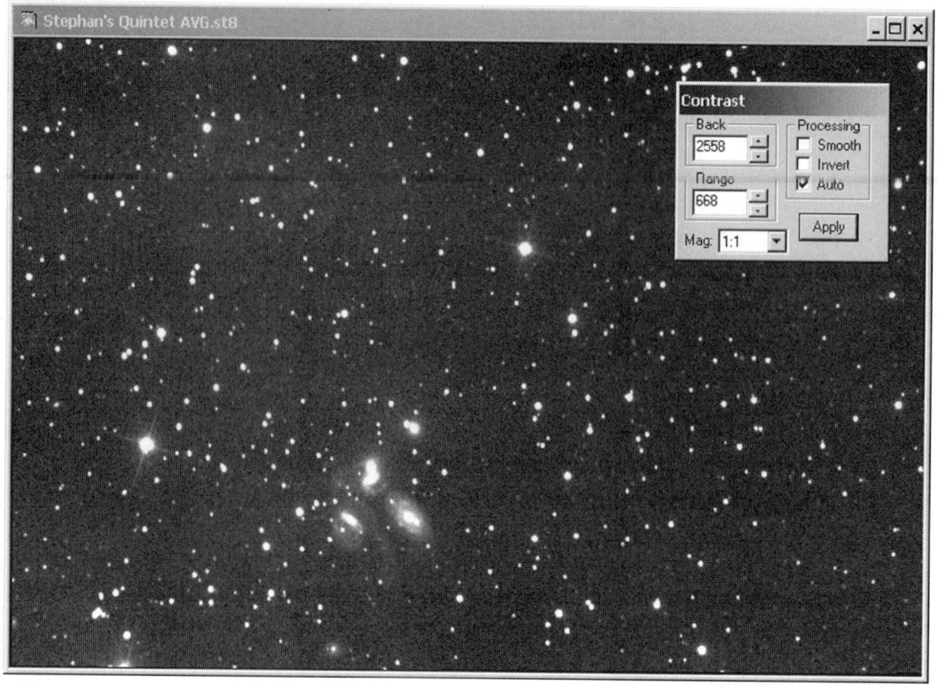

Note that these numbers are different than the numbers I came up with in MaxIm. One's preferences and sense of light and dark are subtle, changeable, and not very repeatable! There truly is no one right way to display the data in image form. The choice is yours, and as these images show, you may not always make the same choices. For many

FIGURE 8.1.11. THE AUTOMATIC LEVEL SETTING IN CCDOPS ALSO GETS CLOSE, BUT STILL DOESN'T GIVE THE MOST VISUALLY PLEASING BALANCE BETWEEN LIGHTS AND DARKS.

images, you may want to have two or more different versions of the image. Perhaps one will show a deep black background and look pretty, while another will show the noise in the background but reveal subtle details of structure in a nebula. Or in another image you might adjust the black and white points to show something besides the main subject, thereby horrendously overexposing or underexposing the main subject. When imaging M57, for example, there is a faint galaxy in the background of most images. The settings you need to show M57 clearly are usually very different from those needed to show the background galaxy (IC1296) clearly.

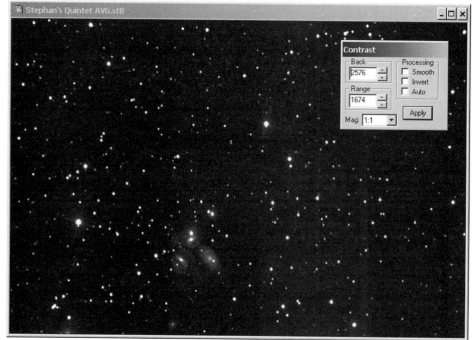

FIGURE 8.1.12. A MANUALLY ADJUSTED CONTRAST SETTING PROVIDES A MORE PLEASING RESULT.

TIP: When you need to use two different histogram settings, you can create two different versions of your image, each optimized for a particular subject. Then use Layers and Masks in Photoshop to blend them. See chapter 9 for details.

When you set the black and white points, you are choosing the portion of the data you want to display. The shorter the distance between these two points, the grainer your image will be, but more dim details will be visible. The greater the distance between these two points, the smoother your image will look, but the dim details tend to get lost. Finding the right balance between these two points requires some experimentation.

My personal preference is to use the automatic histogram stretches in MaxIm DL, CCDOPS, CCDSoft, and other image processing tools only to check the overall quality of my data. If you save the image in TIFF format for editing in Photoshop or another image editing program, you'll have to set the black point and white point all over again anyway. The TIFF format does not store these settings.

Histogram Changes in CCDSoft

CCDSoft version 5 includes a powerful and flexible Histogram tool that allows you to make histogram changes quickly and accurately. Figure 8.1.13 shows an image of M42. The brilliant nebulosity for which M42, the Great Nebula in Orion, is known is nearly absent. The information is in there, but you can't see it. This section will show you how to use the CCDSoft Histogram tool to reveal the hidden data. To download:

http://www.newastro.com/newastro/booknew/samples/c8_m42.zip

The vertical streaks are bloomed stars. The stars and the blooming spikes are all extremely bright, and the pixels in these areas have values close to 55,530. The nebulosity near the center of the image is also very bright, and that is why a very small part of it is visible. But most of the image is at the dim end of the 55,530 brightness values.

FIGURE 8.1.13. AN IMAGE OF M42 SHOWING THE FULL RANGE OF ORIGINAL DATA (16-BIT).

You can determine how many pixels have a given brightness value by examining the histogram. Right click on an image and choose Histogram.

Figure 8.1.14 shows the CCDSoft Histogram tool with the histogram for the image in figure 8.1.13. The histogram is a graph of the number of brightness values in an image. Dim values are at left and bright values at right. The taller the graph is at any given point, the greater the number of pixels that have that value.

The histogram tells us several things:

- Most of the data is in the dim values. The highest peak is at left. Unlike previous examples, the curve slopes gradually to the right, and there are multiple peaks at the right end due to blooming.
- The peak at the far left of the histogram represents pixels that make up the black background in the image.

FIGURE 8.1.14. THE HISTOGRAM FOR THE IMAGE IN FIGURE 8.1.13.

- The broader slope just to the right of the peak occurs because M42 is a very bright nebula.
- The two peaks at the far right represent, from left to right, the brightest nebulosity, and the brightest stars (the ones that have bloomed).
- Other than the bright peaks, there aren't many pixels in the entire right half of the histogram.

FIGURE 8.1.15. ADJUSTING THE WHITE POINT IN THE HISTOGRAM.

There are two triangles just beneath the histogram. These identify the black point (left) and white point (right) of the image. In figure 8.1.14, the black point is 0, and the white point is 56,152. These values are outside the minimum and maximum brightness values (612 and 55,530, respectively), so the entire range of brightness levels is visible.

Figure 8.1.15 shows a change to the histogram. The white point has been moved to the left. The new white point is 9,493. Figure 8.1.16 shows the appearance of the M42 image with the new white point. More of the nebula is visible, but there are problems. The bright portion of the nebula is burned out (pure white), and the dimmest portions of the nebula are still not visible.

Drag the triangles for the black and white points to make changes. In figure 8.1.17, the black point has been raised to 720, and the white point lowered to 1,565. This reveals a wealth of detail in the dim portions of the nebula. NGC-1977 has become visible near the top of the image.

After these changes, the entire core of the nebula is white. Many celestial objects have a bright core and dim extended regions. You'll learn how to avoid burning out the core later in this chapter, in the section on non-linear histogram adjustments. To reveal just the right amount of detail in your images, you usually set the black and white points first, and then use non-linear adjustments to reveal the dim details.

FIGURE 8.1.16. MORE OF THE NEBULOSITY AND MORE STARS ARE VISIBLE.

The histogram in figure 8.1.18 has a lot of data to the right of the white point. That is why there is such a large burned-out white area at the center of the nebula.

The changes you make to the histogram using camera control software are not destructive. The values to the left of the black point and to the right of the white point are not gone; they are hidden. If you save an image after a histogram adjustment, the black and white points are saved with the image, and will be reapplied when you reopen the image.

FIGURE 8.1.17. FURTHER ADJUSTMENTS REVEAL LOTS OF DIM DETAILS.

Image editing software, such as Photoshop or Picture Window Pro, does not retain data outside the black and white points. Histogram changes of any kind in those programs alter the image permanently.

In previous versions of CCDOPS and in CCDSoft, the black and white points were set using Background and Range. CCDSoft still offers this option if you prefer it. The Background setting is the same as the black point. The range is the distance between the black point and the white point.

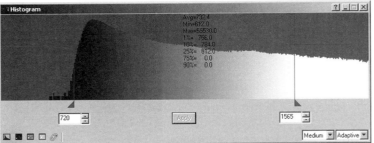
FIGURE 8.1.18. ADDITIONAL HISTOGRAM ADJUSTMENTS TO RAISE THE BLACK POINT, AND LOWER THE WHITE POINT FURTHER.

The Histogram Tool

A CCD image contains more data than your monitor can display, so histogram adjustments are the norm. You can make profound changes in the appearance of your images by altering the histogram. If an image doesn't look right, adjust the histogram to better visualize the data.

My favorite tool for setting black and white points is the Histogram tool in CCDSoft (see figure 8.1.19). I find it really easy to use. You click on the histogram to alter it. There are four zones, and clicking in each zone will change the histogram in different ways. For each zone, the further you click from the black point or white point, the greater the magnitude of the change.

Click in Zone 1 to lower the black point. This lightens the background. The lowest possible black point is zero.

Click in Zone 2 to raise the black point. This darkens the background. If you go too far, dim details will also get darker.

FIGURE 8.1.19. THE CCDSOFT HISTOGRAM TOOL.

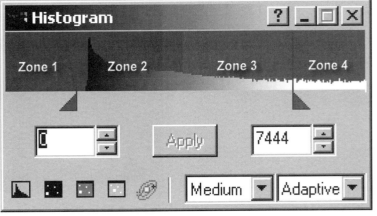

Click in Zone 3 to lower the white point. This brings out dim details, and bright areas get whiter. If you go too far, bright areas will burn out and lose detail.

Click in Zone 4 to raise the white point. This darkens the image generally, and dim details are less obvious. Star sizes get smaller.

There is a vertical bar above the black and white points. The closer you click to one of these vertical bars, the smaller the change in the histogram. This proportional response allows you to make very precise adjustments quickly. You can click the histogram presets at bottom left of the Histogram tool, or use the drop-down lists to create a custom histogram setting. The left and right lists affect the black and white points, respectively.

Histogram Changes in Photoshop

Photoshop, like many other image editors, contain tools for manipulating the image histogram. The histogram tools in most image editors are more sophisticated than what you will find in the camera control programs like CCDSoft and MaxIm DL.

FIGURE 8.1.20. NGC 1893 WITH ADJUSTED BLACK AND WHITE POINTS.

For a typical image, most of the data is at the dark (left) end of the histogram. Other than the bright stars in an image, most of the image *is* dark. It makes sense that dark pixels make up most of an image.

The problem is that for most subjects, including galaxies and nebula, the brightness difference between the background and the dim details is very small. Figure 8.1.20 shows an example. It is an image of NGC 1893. Dark and flat-field frames have been applied, and the black and white points have been set to display the nebula as clearly as possible. There isn't a lot of nebula visible, but if we lower the white point further to show more nebulosity, the image looks very grainy, and the stars begin to bloat.

Figure 8.1.21 shows the MaxIm DL Information window in action. The cursor is over a portion of the background just to the right of center.

FIGURE 8.1.21. MEASURING THE AVERAGE LEVEL OF THE BACKGROUND IN MAXIM DL.

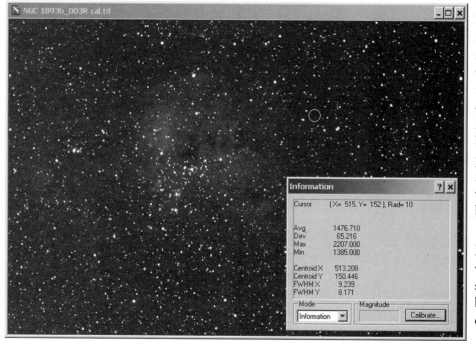

That area is as star-free as possible (a challenging task in this image because of the rich Milky Way star field). The Information window shows that the average background value in the area of the cursor is 1476. This is the background level of the image.

Passing the cursor over two other areas in the image reveals some interesting facts. The brightest area of the nebula has a value of 1782, and a more typical portion of the nebula has an average brightness of 1617.

Out of the 65,000 brightness values in this image, the difference between the background and the medium-bright portions of the nebula is less than 150 units. That means that the dim portions of the nebula are only 0.2% brighter than the background. The difference between the brightest portion of the nebula and the background is only a little over 300 units. At best, the bright portion of the nebula is 0.4% of the total brightness range.

The bright stars in the image, however, have extremely high values, from 50,000 to 65,000. Even the dim stars have values of more than 5,000 units. It is the range from the background level (1476) to the brightest portion of the nebula (1782) that is most interesting. The goal is to display as much of that range as possible without making it look grainy.

Figure 8.1.22 shows what happens when you go too far. There is lots of nebula visible, but the image doesn't look natural at all. This does reveal faint details, but you start to lose the brighter portions of the nebula; dim portions become grainy; and stars become bloated and are not as pleasing to look at. To download two versions of the NGC1893 image to test yourself:

http://www.newastro.com/newastro/book_new/samples/c8_n1893.zip

The reason why you can't find a happy balance point for this image using just the black point and the white point is that these are linear adjustments. No matter where you put the black point and the white point, the area between them is divided evenly among the available shades of gray. The shape of the histogram between the black and white points remains the same. The next step is to change the shape of the histogram.

FIGURE 8.1.22. INCREASING THE BLACK POINT AND DECREASING THE WHITE POINT TO EMPHASIZE NEBULA DETAILS.

This is called a non-linear histogram adjustment. It allows you to take that 0.04% of the data and stretch it out so that it occupies most of the available gray shades.

Altering the Histogram

Unlike CCDSoft, MaxIm DL, Mira and other camera control programs, Photoshop makes only permanent changes to the image histogram. With camera control programs, the full range of data remains, but you are free to change which portion of that data gets displayed visually. With Photoshop, and most other image editors, the image data below the black point and above the white point are discarded when you make a change. You can always use Undo features to regain the data, but once you save the file, the data is gone. Always save your image to a new file when you use an image editor to modify the image.

Let's walk through a Photoshop session to see how to change the histogram. If you are using a different image editor, it will have similar features.

You could save an image file in TIFF format in the camera control software. This would alllow you to open it later in an image editing program like Photoshop. However, Eddie Trimarchi has written a Photoshop plug-in that allows you to open .FIT and .FTS files directly in Photoshop. I usually save the file as a

FIGURE 8.1.22. FITS AND TIFF FILES DO NOT RETAIN SETTINGS FOR BLACK AND WHITE POINTS IN PHOTOSHOP.

Photoshop (.PSD) file after making changes to it. You can visit Eddie's site at the following location. Look for the FITS plug-in link on the menu:

http://users.fan.net.au/~eddiet/indexh.html

When you open a CCD image in Photoshop, any adjustments you made to establish the black point and white point in your camera control program have disappeared. You will be viewing the full range of image data, as shown in figure 8.1.23. You'll need to make major histogram adjustments just to see your image.

All you are likely to see initially are the brightest stars, since the interesting data is clustered at the dim end of the histogram. The image shows no hint of what it contains, other than the very brightest stars. You can change the histogram's black point, white point, and mid tones using Photoshop's Image | Adjust | Levels menu item. Figure 8.1.24 shows the unaltered histogram for this image; it is extremely compact, with a small spike of values at the left and a lot of space to the right.

Things to note about this histogram:

- The background is not actually black, as evidenced by the small blank range at the far left of the histogram. This is due to sky glow.
- The data we are interested in is in a narrow range at the dim end of the histogram.
- The current midpoint (the middle triangle below the histogram) is far away from the values we are interested in.

The black and white points you set in the image processing software mean nothing to Photoshop, so the first job is to re-establish them. Photoshop does not use the same numerical scale found in most camera control software. Camera control programs typically use the full 16-bit range of values, while Photoshop uses a scale from zero to 255. This is true even if you load a 16-bit image into Photoshop. As a result, I usually adjust the black and white points in several steps.

Drag the small triangles at the left and right ends of the histogram inward to set the black and white points. The idea is to drag them in so that they enclose the data of interest, as shown in figure 8.1.25. This retains the linearity of the data, but unlike MaxIm DL and CCDSoft, the changes are permanent.

If you have a very small range of useful values, as in this example, you should set the black and white points in stages to make sure you don't accidentally lose some brightness levels that you want to keep. Figure 8.1.26 shows the result of three stages of black and white point adjustments. Although the white point could clearly be lowered more to show more detail, I usually stop at this point. It is almost always better to finish up the histogram adjustments using the non-linear methods described in the next section. You need to leave the

FIGURE 8.1.23. THE HISTOGRAM FOR THE NGC1893 IMAGE.

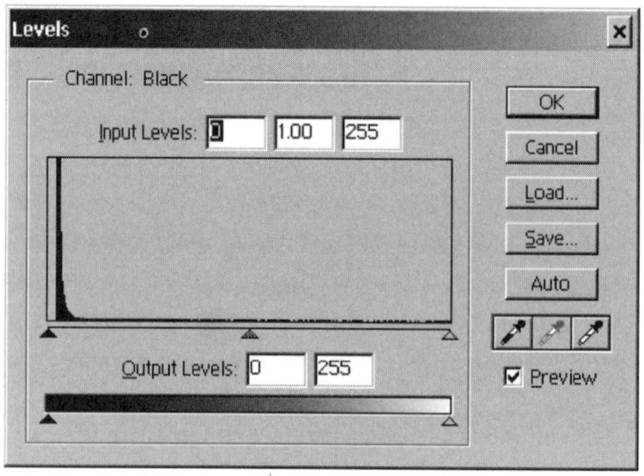

SECTION 1: SELECTING DATA TO DISPLAY

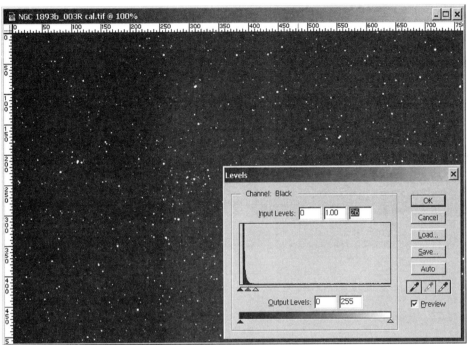

FIGURE 8.1.22. USING PHOTOSHOP LEVELS TO CHANGE THE BLACK AND WHITE POINTS.

white point somewhat higher than optimal in order to get good results with your non-linear adjustments. I usually set the white point so that the subject (galaxy, nebula, etc.) just begins to show up in the image.

For best results, leave about 60 to 90 percent of the histogram flat as shown in figure 8.1.26. Experiment to get a feel for what works best for a particular image. If the image has an especially bright core, and you don't want to burn it out, go more toward 90%. If the image is uniformly dim, and burning out is not as much of a concern, then go closer to having about 60% of the histogram being relatively flat on the right-hand side. Then make your non-linear changes. If you get too much contrast from the non-linear changes, use undo and go back and use a higher white point.

Notice that there is a third triangle below the histogram. It is the midpoint adjustment. Unlike setting the black point and white point, changing the midpoint slider takes your data down a non-linear road.

You can use the midpoint adjustment to make improvements to your image. This isn't as powerful as the types of non-linear adjustment covered later in this chapter, but it's still a very useful tool. Generally speaking, adjusting the midpoint so that it sits within the data of interest helps to emphasize details in the image, as shown in figure 8.1.27.

FIGURE 8.1.23. ADJUSTING THE BLACK POINT IN PHOTOSHOP.

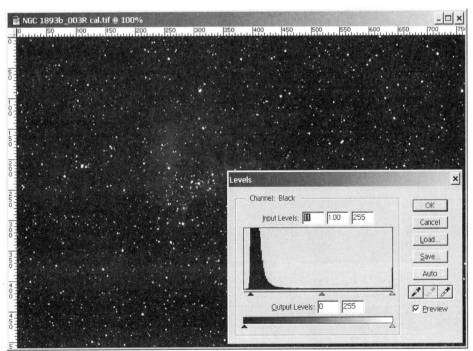

THE NEW CCD ASTRONOMY 379

This adjustment is called a gamma adjustment, and it skews the histogram. A gamma correction causes the image data to be displayed so that 50% of the available brightness values are on one side of the midpoint, and 50% are on the other side. This is a non-linear change because the displayed brightness values no longer have a one-to-one relationship with the actual brightness values. In figure 8.1.27, moving the midpoint to the left brightens the image overall, and this emphasizes the dimmer levels of the image.

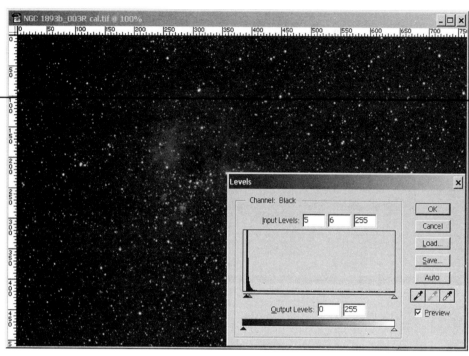

FIGURE 8.1.22. ADJUSTING THE MIDPOINT.

TIP: Even when the image data is in 16-bit format, with 65,000 possible values, Photoshop only shows adjustment values from 0 to 255. You can use fractional values to get precise results. For example, you could set a black point of 5.13, and a mid level of 5.96, instead of the values I chose for the example.

In this example, a gamma (midpoint) setting of 6 is applied to the image. The gamma of any curve is its slope, expressed as the ratio of the logs of the output to input values. In plain English, gamma establishes how the input brightness values are mapped into output brightness values. A gamma of 1.0 means nothing changes. A gamma of 2 lightens the image; a gamma of .5 darkens it. A gamma of 6, as in figure 8.1.27, is an extreme change. For non-astronomical images, gamma changes outside the range of 0.6 to 2.0 are rarely used. For astronomical images, very large gamma changes are needed to bring out dim details in most images.

Even a gamma of 6, which is an extreme midpoint adjustment, doesn't fully solve the problems presented by this image. Figure 8.1.28 shows the shape of the histogram after the gamma change. Note that the data of interest now occupies much more of the histogram.

The nebula is visible after the gamma change (see figure 8.1.27), but it is still dim and the faint details are obscure. Moving the

FIGURE 8.1.23. THE HISTOGRAM SHAPE AFTER A GAMMA ADJUSTMENT.

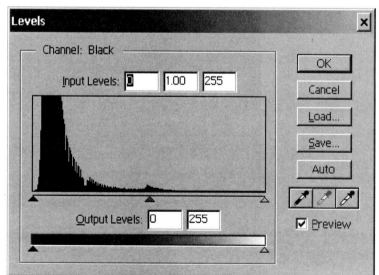

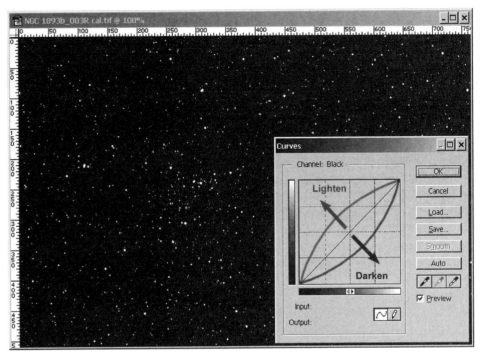

FIGURE 8.1.22. WORKING WITH THE CURVES DIALOG.

Non-linear Histogram Adjustments: Curves

Some programs allow you to edit the shape of the histogram directly. Some, like Photoshop, provide a curve that you can manipulate to change the data. Direct manipulation of the histogram is one of the most powerful and effective ways to modify your images. I use it as the backbone of almost every image processing session because it gives me complete control over many aspects of an image's appearance.

Photoshop's Curves dialog is one of the most powerful tools available to tease out subtle data in your images. Picture Window Pro also contains some useful tools you can use, and products such as Image Plus and Paint Shop Pro also give you similar tools.

Use the Image | Adjust | Curves menu item to open the Curves dialog. Figure 8.1.29 shows the general appearance of the dialog. There is a thin black line running diagonally from lower left to upper right. I have marked up the dialog to show how dragging the line will lighten or darken an image. You manipulate the line by clicking on it to create points. You can then drag a point to change the shape of the line/curve. Now you know why this is called the Curves dialog. You alter the overall shape of the curve, and then Photoshop uses the curve to alter the histogram. The straight diagonal line represents no changes.

> **TIP:** Note the grayscale bars at left and below the graph in the Curves dialog. As shown in figure 8.1.29, the darker ends of the bars are at lower left. If the bars are reversed, click the double-headed arrow at the center of the bottom bar to change them. This will make it easier to follow along with the tutorials in this and later chapters.

midpoint can be effective for brighter objects, but for a dim nebula like this one, more extreme processing is required. If moving the midpoint alone won't solve the problem, back off on the amount of the gamma adjustment. Use a setting that just begins to pull the galaxy or nebula out of the blackness, or use the white point alone to just reveal the object of interest. Then use the non-linear techniques described in the next section.

For any given image, experiment with gamma changes and black point/white point changes to see which method gives you the best results. If you use both methods, always adjust the black and white points first. Linear changes, like white point and black point selection, come first.

The midpoint adjustment offers significant power, but it is not the most advanced in Photoshop's histogram toolbox. What might seem at first like raw power is really just a taste of what Photoshop has to offer in this area.

If you can't get the results you want with gamma adjustments, you can gain finer control over the process using yet another Photoshop tool: Curves.

FIGURE 8.1.22. TOP: LIGHTENING THE MIDTONES.

FIGURE 8.1.22. MIDDLE: DARKENING MIDTONES.

FIGURE 8.1.22. BOTTOM: USING MORE COMPLEX CURVES TO ACHIEVE SPECIAL EFFECTS.

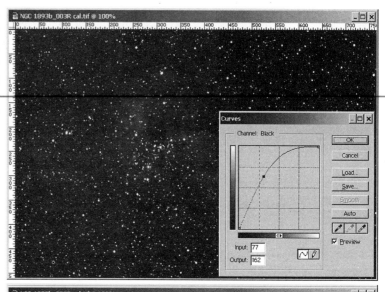

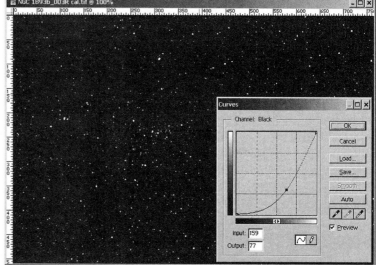

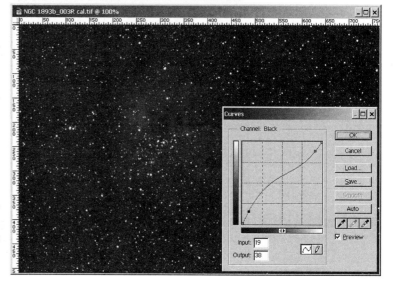

In figure 8.1.29, the only changes that have been made prior to opening the Curves dialog are black point and white point adjustments. This is the ideal starting point for using Curves. If you make the nebula or galaxy too bright by setting the white point too low, you won't get as pleasing a result when you adjust the image with the Curves dialog. Figure 8.1.30 shows what happens when you click on the line and drag the point to the left: the image brightens. Because you are using a curve to the adjusting, the middle values in the image are affected the most. In this example, that's effective at bringing out more detail in the nebula.

Figure 8.1.31 shows what happens then you drag the curve down and to the right. The image gets darker. It's rare to go in this direction for astronomical images because you are usually trying to brighten dim objects. However, in some cases, a downward tweak at the end of your image processing can give you a pleasing dark background.

You can also set more than one point on the curve to make more complex changes to the image. Figure 8.1.32 shows a curve that will brighten the dim areas of the image, and darken the bright areas. This softens the effect of the adjustment, and gives a look more like film than CCD. However, you will need long exposures to use this technique. For average images, it won't give you the brightness needed to display dim extended objects.

You can create extremely complex curves with Photoshop, using as many points as you need on the curve. However, for a natural appearance, one or two points will usually do it. Additional points can be useful, however, to anchor a portion of the curve that you do not want to move when you adjust some other part of the curve.

For example, figure 8.1.33 shows two points on the curve. The point at lower left brightens the dimmer portions of the nebula. The point higher up on the curve has been used to pull the upper part of the curve down a bit. This limits the amount of brightening in the rest of the image. The additional point serves as an anchor.

Figure 8.1.34 shows what happens without the anchor point. The curve is so steep that the right half of it is all scrunched up against the top of the graph. This is called clipping. The curve has been cut off, or clipped, and data is lost. The image may look sort of OK, but the stars have gotten much brighter relative to the nebula, and the brightest stars have become fully-saturated disks that will not look good in the completed image. The overall image can also get out of balance and become difficult to manage.

As with the Levels dialog, you will get the best results by moving in steps rather than trying to reach your goal all at once. It is better to click OK when you

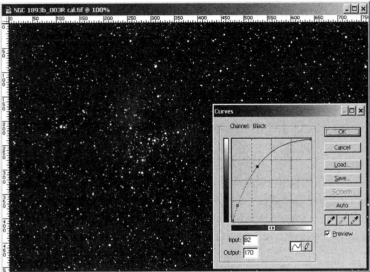

FIGURE 8.1.22. USING A SECOND POINT TO ANCHOR A CURVE.

get to the stage shown in figure 8.1.33 than to try to do it all at one time. During the first few uses of the Curves dialog, keep an eye on the brightest portions of the image. You may need to use an anchor point to prevent them from becoming too bright while you are boosting the dim areas.

Figure 8.1.33 shows a good first application of Curves. Figure 8.1.35 shows a reasonable second application of Curves to the image. When you re-open the dialog, the curve is again a straight diagonal line. Each time you apply Curves, the straight line will be your starting point.

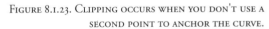

FIGURE 8.1.23. CLIPPING OCCURS WHEN YOU DON'T USE A SECOND POINT TO ANCHOR THE CURVE.

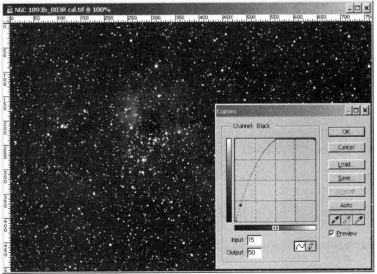

The curve in figure 8.1.35 is similar to the one shown in figure 8.1.33, but it is different in some important ways. The lower-left point is not set as aggressively. The slope of the line at lower left is closer to the original diagonal line. This means that the increase in brightness in the dim areas will be less. The numbers at lower left show the nature of the change. At the lower left point, the input value is 20, and the output value is 29. This is about a 30% increase in brightness. The more you move a point away from the diagonal line, the more aggressive the change you are making. Use Alt + Tab to move from point to point for editing the numeric values. The numbers give you another way to fine-tune you changes. For example, if you felt that the lower left point was giving you just a little too much brightening in the dim areas, you could change the output to 28.

The lower left point does brighten the darkest areas of the image, but it also serves double duty as an anchor point. The second point is increasing the brightness of the middle and upper-middle tones. NGC 1893 has very small changes in brightness, and this curve helps bring out brightness differences more effectively. If you make sure that the Preview checkbox is checked, you get instant feedback on the effectiveness of your Curves settings.

The curve shown in figure 8.1.35 has the effect of slightly brightening the dimmest portions of the nebula (lower left in the dialog), and making the brightest por-

FIGURE 8.1.22. USING A COMPLEX CURVE TO INCREASE CONTRAST.

tions of the nebula stand out a little better from the medium-bright portions. This reveals the structure of the nebula more effectively.

I wasn't quite satisfied with the results after saving the changes shown in figure 8.1.35. The nebula was still a little too dark overall. This is due in large part to the short exposures. To get a really good result, I needed a longer exposure (or more images to combine). But you will often find yourself in the same situation, trying to get the most out of the data you have. For this image, my goal was to show the structure in the nebula. This requires as much contrast as possible between the dim and bright portions of the nebula. Figure 8.1.36 shows a curve that darkens the very dimmest pixels in the image and brightens the very brightest pixels. This increases the contrast in the image in a more gradual way than a simple contrast increase would. This brings out even more structure in the nebula, but at a price. The dimmest portions of the nebula are barely visible, and the brightest portions are competing a little bit with the

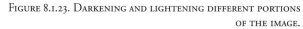

FIGURE 8.1.23. DARKENING AND LIGHTENING DIFFERENT PORTIONS OF THE IMAGE.

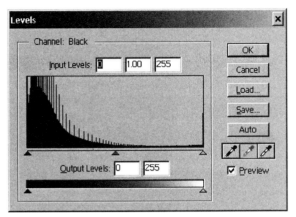

FIGURE 8.1.22. THE IMAGE HISTOGRAM FOLLOWING NON-LINEAR ADJUSTMENTS.

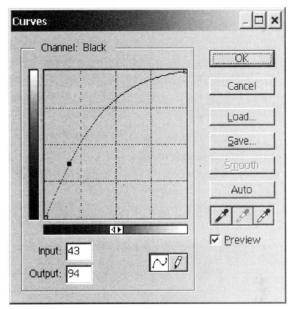

FIGURE 8.1.22. A SINGLE-POINT ADJUSTMENT.

stars in the same area. You need to be careful when applying double curves like the one in figure 8.1.36. Most of the time, a very slight double curve will be all you need.

Figure 8.1.37 shows how the changes through figure 8.1.36 have affected the histogram of the image. Compare the shape of the histogram to the way it looks in figure 8.1.26, prior to all the non-linear changes. The data peak is more spread out, which shows that the data of interest -- the nebula -- now occupies a larger portion of the visible image. Compare the histogram curve to figure 8.1.28, the end result of the gamma stretch, and the image to figure 8.1.27, which shows the appearance of the image after the gamma stretch. For this image, Curves provides more control over the changes to the image and allows you to bring out more of what you are interested in.

All of the changes made so far have been focused on increasing the contrast in the nebula itself to emphasize the variations in brightness. Another goal might be to show the overall shape of the nebula. To do this, you would brighten all portions of the nebula, with a little emphasis on the dimmest portions to make sure they show up clearly. Repeated applications of curves like the one shown in figure 8.1.38 would do this. Such a curve brightens the dimmest areas a little

more than the middle tones and significantly more than the existing highlights.

The final result of this kind of adjustment, starting with the original images and with five applications of Curves similar to figure 8.1.38, is shown in figure 8.1.39. Note that the curve is gentle enough that no clipping occurs. I used the Levels dialog one time during the process. The background was getting too bright

FIGURE 8.1.23. THE HISTOGRAM THAT RESULTS FROM THE CURVE IN FIGURE 8.1.38.

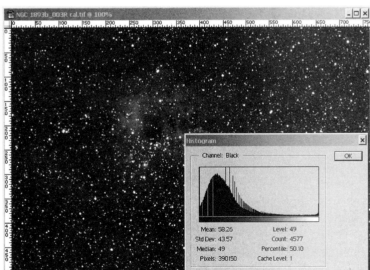

because of the repeated boosts to the dimmest levels. I used Levels and moved the left-side slider in a bit to raise the black point. This restored a black background. The revised histogram is also shown in figure 8.1.39. The brightness levels of interest occupy even more of the image, and the very dimmest portions of the nebula are clearly visible.

This image doesn't have the high level of contrast of the first example, but it does a better job of showing the dimmest, extended portions of the nebula. This is a common trade-off. Is one version better than the other? It depends on what you want to accomplish; there's no one right way to process an image.

TIP: If you want to measure the brightness of stars or perform any science, stick to linear processing so that the brightness relationships in the image remain true.

Digital Development

CCD cameras deliver images that are different from the images you get with film. Film is non-linear: the response to light is not constant over time or intensity. The beauty of CCD chips is that they have a consistent response. Not every photon will kick out an electron to be read at the end of an exposure, but if you examine the large-scale behavior of a chip, it's safe to say that for every X photons that hit, Y electrons will be generated. This is called the quantum efficiency of the chip.

The result is that if twice as many photons strike a pixel, it will be twice as bright (within the range of error for the chip due to readout noise, the quantum nature of light, and other sources of error).

Not every CCD chip is nicely linear, however. Pixels on CCD chips with anti-blooming gates (ABG) bleed off charge at some point short of their full-well capacity, and this behavior is a little more like film. Without ABG, a CCD chip will spill excess electrons into adjoining pixels along a column. This is called blooming.

The downside of a linear response is that a lot of information is squeezed into a small portion of the histogram (see figure 8.1.24).

The manipulations so far have been designed to alter the histogram so that the area of interest occupies more of the histogram.

Digital Development gives CCD images a look more like that of film. It includes a general sharpening of the image and non-linear adjustments to the image histogram. There are various interpretations of digital development in various CCD imaging programs. We'll use MaxIm DL as the first example.

TIP: You can approximate digital development using image-editing programs like Photoshop by adding sharpening to the histogram adjustments. In fact, manual digital development in Photoshop is ideal because it gives you such a hige degree of control.

Digital Development with MaxIm DL

Figure 8.1.40 shows an image of M42. It has a typical astronomical image histogram: most of the interesting image data is at the low end (left) of the histogram. You can see the image histogram at upper left of figure 8.1.40. To download this image of M42:

http://www.newastro.com/newastro/book_new/samples/c8_m42b.zip

FIGURE 8.1.24. AN IMAGE OF M42, RIPE FOR DIGITAL DEVELOPMENT.

FIGURE 8.1.41. LEFT: A LINEAR STRETCH SHOWS DETAILS, BUT BURNS OUT THE CORE.

FIGURE 8.1.42. BELOW: DIGITAL DEVELOPMENT USING DEFAULT PARAMETERS.

raised to 470, and the white point is lowered a lot, to 1040. This reveals the dim details, but now the core is burned out. What we need is a way to show the full range of detail in the gray levels available to us. As with most images of galaxies and nebulae, there is just too great a range of brightness values to display them all at one time.

Digital development can solve the problem. It is

What this image needs is a good, non-linear stretch. We could do it manually, as in the previous examples -- and for many images, a manual stretch is a good thing, for a variety of reasons that will become clear shortly. A linear stretch doesn't do the trick. Figure 8.1.40 has the black point (minimum) set at 300, and the white point (maximum) set at 10,151. That's a lot more than the 256 gray levels available, so most of the image data is still lost to our eyes.

Figure 8.1.41 shows what happens when we simply move the black and white points to reveal more details. The black point is

quick and remarkably effective, but it brings a few tricks to the party that may not be in your image's best interests. It's a quick solution, however, as suggested by the details revealed in figure 8.1.42. I let MaxIm DL set the default parameters for digital development in that example.

The image shows more dim details and more of the core, but it has a few problems that are characteristic of digital development. Digital development doesn't just alter the histogram; it also applies sharpening to the image. Sharpening may or may not be an improvement, so you need to take this into account when deciding whether to apply digital development to one of your images.

Sharpening causes several problems in this image:

- The spikes from blooming have black lines around them, so they are going to be a little harder to edit out manually. Some software, such as Mira, has features for removing blooming spikes. However you deal with them, you will usually be better off clearing up the spikes before you apply digital development.

- Some stars now have a dark halo around them. This occurs when the star is against a bright background, such as a nebula. Most types of sharpening (including unsharp masking and deconvolution) will create these halos if you sharpen too much. You can usually eliminate the halos by using a limited amount of sharpening.

- Stars that are against a black or very dark background have a light halo around them. This is also a common artifact of the sharpening process, but digital development seems to emphasize it the most. The halo around the brightest stars is usually made up of photons reflecting off of the surfaces inside the camera.

To some degree, you can control these artifacts of digital development (especially sharpening) by adjusting the parameters instead of using the default values. Figure 8.1.43 shows the MaxIm DL dialog for digital development.

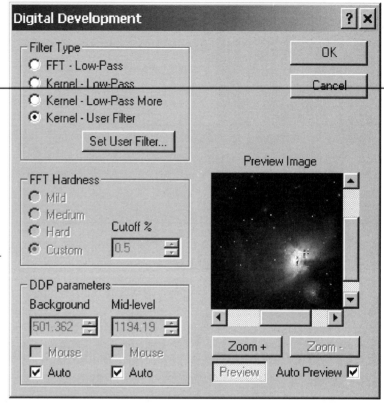

FIGURE 8.1.43. DEFAULT PARAMETERS FOR MAXIM DL DIGITAL DEVELOPMENT.

The preview image shows you the approximate result you will get. You can zoom in and out. The view you see will vary depending on what part of the image you are look at, or what level of zoom you are using. Only the area visible in the preview image is used to calculate the preview. For example, I have zoomed out to see the entire nebula in figure 8.1.43. If you zoom in to look at just a portion of the nebula, you will see a preview that is based only on what you can see. If it is a dim portion, you will see an extreme stretch. If it is a bright portion, you might not see much stretch at all. So use the preview carefully, or it won't be a good prediction of the final results. You need to include as large a brightness range in the preview as you have in the full image for best results.

You can right click and drag in the preview window to bring the important area of the image into view for the preview. If the important area is too large to show, zoom out.

The Digital Development dialog contains the following settings:

Filter Type - This is the filter that will be used for sharpening.

FFT - Low-Pass - Provides more sharpening and less histogram stretch. Useful when sharpening is your main goal. FFT stands for Fast Fourier Transform, which is a mathematical technique for performing sophisticated filtering quickly. If you choose this option, set FFT Hardness (explained below).

FFT Hardness - Active if *FFT - Low-Pass* is selected. Choose between Mild, Medium, Hard, and Custom. The preview window is especially unpredictable when used with FFT filters. You may need to apply the filter in order to find out what you get. The FFT filters are processor-intensive and may take from a few seconds to a few minutes to complete. From Mild to Hard, provides increasingly dramatic sharpening.

Kernel - Low-Pass - Provides a moderate amount of sharpening and lots of histogram stretch. A good choice for nebulae and galaxies, where too much sharpening can create undesirable halos around stars.

Kernel - Low-Pass More - This provides about the same amount of histogram stretch as the Low-Pass filter, but a higher degree of sharpening.

Kernel - User Filter - Allows you to define your own filters using numeric arrays that define how sharpening is to occur. This is an advanced feature, and requires an understanding of the theory of image sharpening.

DDP Parameters - Determines how DDP will affect the histogram. Examples follow that show you what to expect from various settings.

Background - This is the black point. All pixels dimmer than this value will be black in the final image.

Mid-Level - This is the mid point for the remaining portion of the histogram. Pixels with brightness levels between the background setting and the mid-level setting will occupy 50% of the final histogram. Pixels with brightness levels greater than the mid-level will occupy the other 50% of the histogram.

Mouse - Use the mouse to click in the source image to set the value for the corresponding parameter.

Auto - Allows MaxIm DL to set the parameter.

In figure 8.1.43, MaxIm DL picked default parameters for the M42 image:

Background: 501
Mid-level: 1194

These parameters, applied with a "Kernel - Low-Pass" filter type, resulted in the image shown in figure

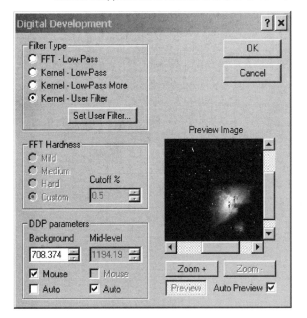

Figure 8.1.44. Raising background (black point).

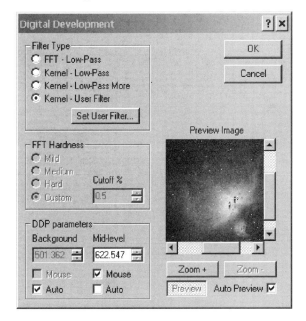

Figure 8.1.45. Lowering mid-level (more faint detail).

8.1.42. You can adjust the default settings to see if you can do a better job than the software. Figure 8.1.44 shows what happens when you increase the setting for the background from 501 to 708. You are raising the black point, so more of the image is now black. You can adjust the background setting when the default background is too light. In this case, raising the background setting doesn't help; the default was a good setting.

If the Mouse checkbox is checked, you can click in the original image (not the preview window) to set the background level. This is useful for quickly experi-

FIGURE 8.1.46. RESULTS OF CHANGING BACKGROUND AND MID-LEVEL SETTINGS.

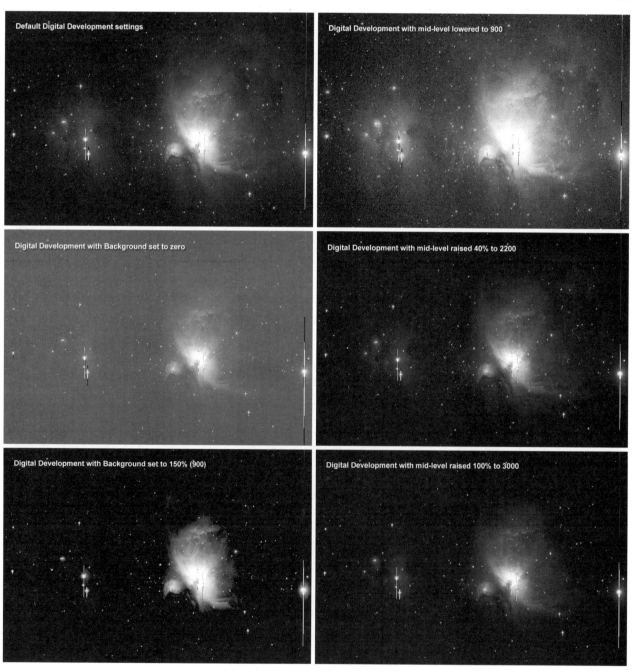

menting with different portions of the image as the black point. For example, if you have serious background problems, find a pixel that is just outside the galaxy or nebulosity you are interested in, and click on it to set a black level. You can click to find just the right spot. If the background has too much variation in it, you probably need to remove the gradient or other problems before using digital development. See chapter 6 for information on gradient removal.

If you set the background too low, you will see too much brightness in the background. If you set it too high, you will lose faint detail. MaxIm DL is pretty good at picking a good background level, so you will not need to adjust it very often.

Figure 8.1.45 shows what you can do by altering the Mid-level setting when using the "Kernel - Low-Pass" filter type. The default mid-level was 1194, and I have changed it to 622 in figure 8.1.45. The lower mid-level setting shows more detail, but it also reduces the contrast in the image. This is a trade-off that occurs when reducing the mid-level. Because of the contrast reduction, a lower mid-level will sometimes create more problems than it solves.

Increasing the mid level hides minor variations in background level. Digital development brings out dim details, and sometimes you want to choose which dim details those will be. Raising (and occasionally lowering) the mid level will help you do that.

Figure 8.1.46 shows the results of six different settings for background and mid-level. All of the images have the same contrast settings: minimum is zero; maximum is 35,000. All examples use the "Low-pass" option. This is the option I use most often with digital development, as I find the stronger sharpening from the other methods objectionable most of the time.

Careful use of digital development can be a good thing. Most of what you can do with digital development you can also do manually in Photoshop, Picture Window, and other image editors. But none of those other techniques are as quick and simple as digital development is.

Keys to figure 8.1.46: Digital Development Settings in MaxIm DL

Top left - default settings

This image of M42 was made by applying digital development using the default settings.

Middle left - background set to zero

The background is brighter; the black point is too low. Dim details are lost in the haze.

Bottom left - raise background to 150% of default setting

The black point has been set too high. The dim details are gone.

Top right - decrease mid-level to 60% of default setting

Lowering the mid-level makes more of the dim areas of the nebula visible. Despite the overall loss of contrast, the increase in dim details pays off, making this a good choice.

Middle right - increase mid-level to 140% of default setting

Many imagers like to increase the default mid-level setting to reduce the noise in the dim portions of the image. However, with a bright subject such as M42, this isn't really necessary. The 140% approach works best on dim nebulae and galaxies. Here, it just hides some of the interesting detail. If you see excessive graininess using the default setting, try raising the mid level, then fine-tune the increase to match your image.

Bottom right - increase mid-level to 200% of default setting

The dim areas are gone. The mid level is set too high, and we lose too many dim details. There is very good contrast in the brightest portions, however. You can even create two copies of the image, one with a low mid-level and one set high, and then combine them using masks in Photoshop to get the best of each in the final image. See chapter 9 for details on using Layers and Masks in Photoshop.

> **TIP:** After applying digital development, you often still need to adjust the black point and white point to show the details you are most interested in. The default settings may not show dim details, or they may show a burned-out core on nebulae or galaxies. Don't evaluate the success or failure of digital development until you have adjusted the histogram.

All of the examples so far have used the "Kernel - Low-Pass" filter. This gives you the best balance in many situations, but there are times when a different level of sharpening would be better. Grainy images will always be difficult for digital development. It enhances grain just like any other sharpening process.

Images with background problems, especially uneven background levels because of a poor flat field or no flat field, can look horrible after digital development. Images that have a lot of bright detail, such as planetary or lunar images, don't get much of a boost from digital development. If you find yourself using five-digit mid-level settings, 10,000 or more, digital development could reduce image quality rather than improve it.

The "FFT - Low Pass" filter type applies more sharpening. The "Kernel - Low-Pass More" filter type also applies a bit more sharpening, but not as much as the FFT type. Figure 8.1.47 shows four images of M42 with different filter types applied to them. Clockwise from the upper left corner:

- Default values for "Kernel - Low-Pass," same as figure 8.1.43.
- The FFT filter type, with a hardness setting of Mild. Note that the sharpening is much more pronounced, with large dark areas around bright stars and especially around blooming spikes.
- The FFT filter type, with a hardness setting of Hard. The dark areas around bright objects are much larger still, and sharpening is more intense.

FIGURE 8.1.47. COMPARING DIFFERENT DIGITAL DEVELOPMENT FILTER TYPES.

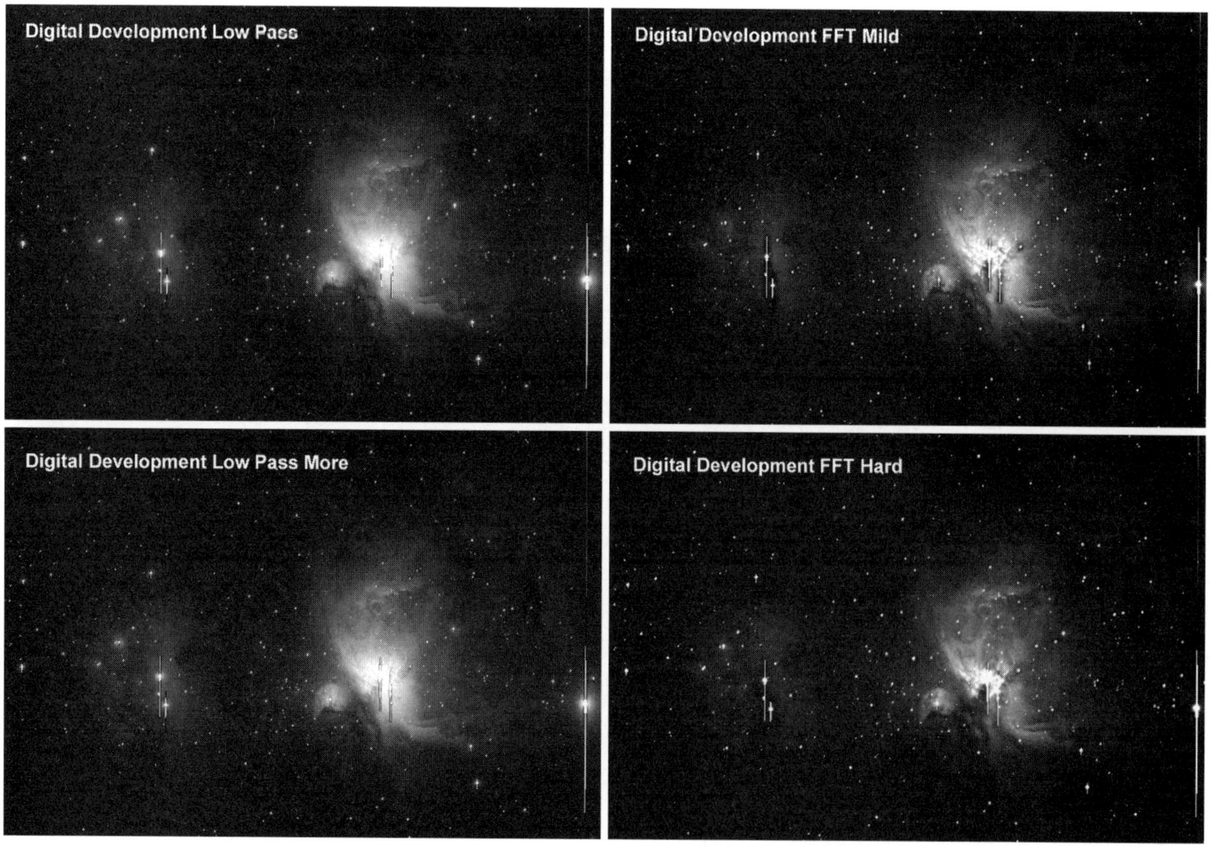

- The Low-Pass More filter type. Sharpening is a little stronger than with Low-Pass type, but not nearly as much as for the FFT types.

Generally, I use the FFT filter types when I find that I need more sharpening than I can get with the Low-Pass and Low-Pass More filters. My taste runs toward a minimum of sharpening, but you may prefer sharpening; it depends on individual tastes. There is no one right way to process an image.

The User Filter setting is for designing your own filters. Figure 8.1.48 shows an example of a sharpening filter. You can increase the amount of sharpening by decreasing the number in the center, or by going more negative with the negative numbers. You can decrease the amount of sharpening by increasing the number in the center, or make the negative numbers closer to zero. This is a fairly crude filter; large stars tend to take on the shape of a cross. You can use a 5x5 or 7x7 filter to get a finer degree of sharpening.

Figure 8.1.49 shows a smoothing filter. To increase the amount of blur, decrease the number in the center. To decrease the amount of blur, increase the center number.

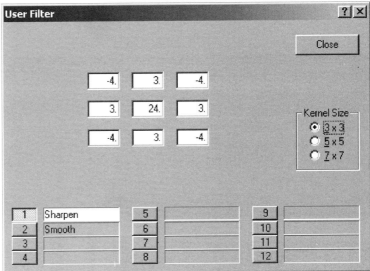

FIGURE 8.1.48. A USER-DEFINED SHARPENING FILTER.

Digital Development in Astroart

Other image processing programs handle the digital development parameters differently. Figure 8.1.50 shows the digital development dialog for Astroart 2.0. There are two settings in the Astroart dialog: Threshold (upper slider), and High Pass (lower slider). The threshold setting determines the black point for the output image, and the high pass setting determines the amount of sharpening. If you are using a different software package for digital development, you will usually find some means of controlling the histogram and the sharpness of the output image.

FIGURE 8.1.49. A USER-DEFINED SMOOTHING FILTER.

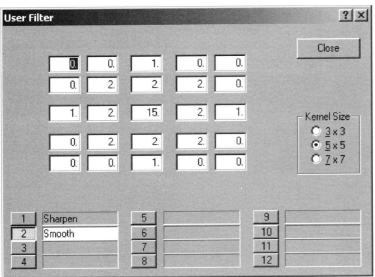

Figure 8.1.51 shows a comparison of the results you can expect with the two options available to you in the Astroart 2.0 Digital Development (DDP) dialog. The upper slider controls the black point (called "Threshold" in Astroart). The image at upper left has the threshold set way too low; the background is too light and details are washed out. The image at upper right is better. The version at lower left has a dark background. The image at lower right shows the sharpness increased to 3; this brings out a lot of details in the image.

As much as I complain about too much sharpening during digital development, the sharpening often enhances the result if it is not over-used. I like the Astroart approach to Digital Development because it gives you control over the degree of sharpening in a simple, straightforward manner.

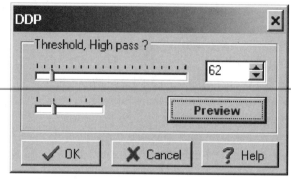

FIGURE 8.1.50. DIGITAL DEVELOPMENT IN ASTROART 2.0.

FIGURE 8.1.51. ADJUSTING DIGITAL DEVELOPMENT PARAMETERS IN ASTROART 2.0.

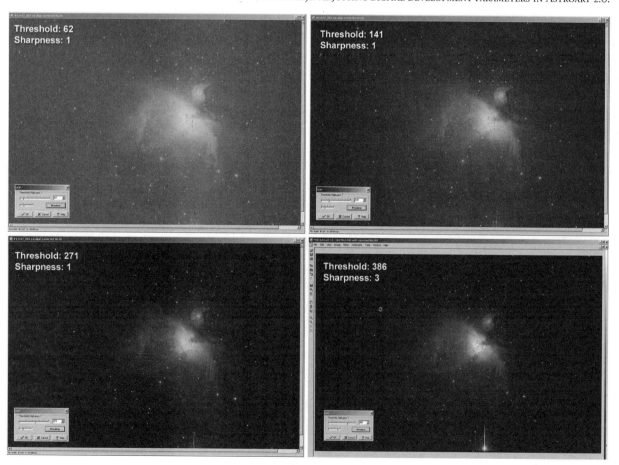

Section 2: Improving Image Clarity

Sharpening has already been discussed in the context of digital development, but it might surprise you to learn that blurring can also improve your images. Many images are noisy, and smoothing/blurring can improve the appearance at the cost of some loss of detail.

Sharpening works best with images that have low noise. Sharpening noisy data gives you a very sharp image of...noise. When you are imaging very dim subjects, it is common to wind up with portions of your image that have more noise than others. I often sharpen some parts of the image, and smooth out others.

There are various techniques for sharpening, and they work in very different ways. It is useful to have several sharpening tools available. Some tools, such as digital development, including a sharpening component. Other tools, like deconvolution, are not strictly sharpening tools, but they do sharpen up your image.

The trick when sharpening is to know how much sharpening is enough. A mild amount of sharpening can improve many images. Heavier sharpening will show some artifacts of the sharpening process. For example, a slightly brighter pixel in the background can become a false star if you sharpen too much. The amount of sharpening you can use depends on the noise level in your image.

I use blurring to smooth out the noisy areas (sky background; dim portions of a nebula or galaxy). I use sharpening on the portions of the image that have a very good signal to noise ratio (S/N). You can also use Layers and Masks (see chapter 9) to blend sharpened/deconvolved and smoothed versions of an image.

If the overall image has an excellent signal to noise ratio, blurring may be unnecessary. You can also use blurring of the background to reduce the file size of your JPG images for the web -- a blurred background reduces the amount of detail that must be compressed, and usually has no adverse impact on the appearance of the compressed image.

> **TIP:** The search for data, rather than beauty, can involve some serious sharpening. If you are looking for very faint objects, such as Bok globules in M42, you can use sharpening to enhance detail beyond the level you would normally use. You have to be careful not to create stars where they don't exist! Likewise, blurring can remove stars that do exist if carried too far.

Sharpening

Just about every image processing program has some kind of sharpening routine in it. These range from simple to fancy. For purposes of astronomical imaging, there are two types of sharpening: brute sharpening, and unsharp masking.

Brute sharpening, often called high pass filtering, is accomplished by increasing the brightness contrast between groups of adjacent pixels to enhance subtle differences. By doing this at small scales (2x2 pixels; 3x3 pixels; 5x5 pixels; etc.), the image looks sharper. Figure 8.2.1 shows an example of an image with increasing amounts of sharpening, from none at the top to severe at the bottom. The top band has no sharpening at all. The next band down has typical

FIGURE 8.2.1. THE EFFECT OF SHARPENING ON AN IMAGE.

FIGURE 8.2.2. SIMPLE, BRUTE SHARPENING BRINGS OUT DETAILS BUT MAY HAVE UNDESIRABLE EFFECTS ON STARS.

"brutish" sharpening. Note that even one application of this type of sharpening is noticeably grainy. The third band has double sharpening, and the bottom band has triple sharpening. This type of sharpening often isn't good enough to enhance a CCD image.

The background of figure 8.2.1 shows the effects of sharpening on areas with low signal. The bands with additional sharpening show harsh and rapid deterioration. The best approach is to only sharpen those portions of the image that can handle it gracefully.

This type of sharpening goes by various names in various products. In MaxIm DL, it is called a high-pass filter (the "high-pass more" filter is an extra dose of sharpening). In Photoshop it is simply called sharpening. In Astroart, you will find sharpening under the Filter menu as a high-pass filter. It is available in light, medium, heavy and degauss versions. The degauss filter is variable in effect, which reduces the noise problem. It can be a useful sharpening tool, as it looks more natural than many simple high-pass filters. In CCDSoft, the Image menu contains a Sharpening sub-menu, which includes two grades of sharpening: Sharpen, and Sharpen Gentle. Sharpen is too intense for most images, but Sharpen Gentle might work some of the time.

Figure 8.2.2 shows one of the problems with high-pass-filter sharpening. The image of M101 at left is the raw image after image reduction. The signal to noise ratio is good, but the object isn't very sharp. The middle image shows a simple high-pass filter applied in MaxIm DL. Notice that the image has better sharpness and detail, but the bright pixels representing stars have shrunk and have dark halos. This level of sharpening is too aggressive. Even if you darken the background (far right), you still have the problem of black halos when the star is in front of the galaxy.

You won't always see these halos around stars; different programs apply different amounts of sharpening, or allow you to adjust the amount of sharpening that occurs. If possible, adjust the available settings to balance sharpening with the halos -- get as much sharpening as you can without creating problems in the image.

FIGURE 8.2.3. A COMPARISON OF UNSHARP MASK SETTINGS IN PHOTOSHOP.

Generally, I prefer to use unsharp masking rather than high-pass sharpening on most images. Unsharp masking works by subtracting the blurred part of the image out, leaving behind a sharper image. The trick with unsharp masking is to know when to stop. A little experimentation can usually uncover the sweet spot.

Figure 8.2.3 shows an image of M101 with four different levels of unsharp masking in Photoshop. Many other programs include unsharp masking, including Astroart and MaxIm DL. The parameters vary for each program, but the concepts are similar.

The numbers at the left of each of four horizontal slices in figure 8.2.3 indicate the unsharp masking settings used for that slice. The area of the image to the right of the vertical bar has not been sharpened, to allow comparison to the various sharpened bands. See figure 8.2.4 for a close-up view of the boundary between the sharpened and unsharpened areas.

The numbers correspond to the settings in the Photoshop Unsharp Mask dialog. You can use the Filter | Sharpen | Unsharp Mask menu item to open this dialog. Three slider controls determine how the image will be sharpened. The numbers in figure 8.2.3 represent, from left to right, amount, radius, and threshold:

Amount - The amount of sharpening to apply to the image. You can apply a partial sharpening, such as 25 or 50%, to get a gentle effect. You can apply 100% sharpening, or even beyond this to double, triple, or more. That level of sharpening is almost always too much for astronomical images. A conservative amount of sharpening is usually the best choice, and I use 50% or less most of the time.

Radius - This controls the scope of sharpening. The larger the radius in pixels, the more obvious the sharpening will be. For most images, a radius in the range of 1-2.5 pixels will work best.

Threshold - This determines how large a difference must exist between pixels before sharpening occurs. A zero threshold allows sharpening to apply to every pixel in the image. Larger threshold settings mean that sharpening will only occur when pixels have larger differences in brightness. As you increase the threshold setting, fewer and fewer pixels are involved in the sharpening operation.

Other programs will have some or all of these settings, under various different names. You can experiment with the available settings to determine how they affect the image. Figure 8.2.3 shows one type of experiment. There are no major differences in the results. Figure 8.2.4 shows a close-up of the sharpened and unsharpened portions of the image (rotated 90 degrees). This close-up view shows what happens with different unsharp mask settings. The top row shows the unsharpened image. The bottom row from left to right matches the settings from top to bottom in figure 8.2.3.

From left to right, the bottom portion of figure 8.2.4 uses the following settings for unsharp masking.

Amount 50%, radius 2 pixels, threshold 0

These settings are the ones I use the most often. You can see a slight increase in the noise level in the background area. At 1X, this noise is generally not objectionable, and you get a nice gain in sharpness. However, for best possible results, select only the bright portions of the image for unsharp masking whenever possible. Use a smaller radius if the noise is objectionable, and a larger threshold if the image is noisy.

Amount 100%, radius 1.5 pixels, threshold 0

This setting causes more sharpening to be applied to the image, but the smaller radius decreases the harshness. The result isn't quite the same as the first

FIGURE 8.2.4. A CLOSE-UP LOOK AT THE EFFECTS OF SHARPENING.

setting, however. The noise increase is more noticeable, and gives you some idea of why I favor the 50% setting for amount. I had concerns initially that the larger radius required with the 50% amount setting would be troublesome, but in practical use the results are better overall.

Amount 100%, radius 3 pixels, threshold 3

In this example, the larger threshold setting somewhat mitigates the fact that the amount is still at 100%. There is still noise, but it is softer than in the previous example. Sharpening is applied to only larger features in the image because of the increased threshold setting.

Amount 50%, radius 6 pixels, threshold 3

This example combines the lower amount setting with a larger radius and threshold setting. The radius and threshold settings partially offset each other. The sharpening is more aggressive, but it is applied only to areas with high contrast (threshold setting). There is very little change in the background noise, but the overall sharpening is also minimal. Only large details, like the knot of star formation in the lower right, get any sharpening. This setting, or something close to it, is a good one to use when you have a fairly noisy image that desperately needs at least a little sharpening.

I encourage you to experiment with various unsharp mask settings on your images to see what works for you. A good starting point is to set amount to 50%; radius to 1.5 to 2.0, and threshold at zero. These settings are ones I worked out to suit the results I get with an ST-8E camera; a different camera with a different CCD chip in it might require different settings to get optimal results. The general rules to live by with Photoshop's unsharp mask tool are:

- Use the amount setting to control the overall amount of sharpening to apply to the image.
- Use the radius setting to control the smoothness/harshness of the sharpening, and to some degree its scale.
- Use the threshold setting to control the scale at which sharpening begins, and to some extent the harshness of the sharpening.

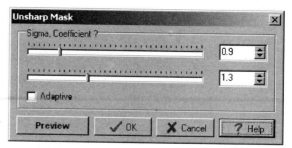

FIGURE 8.2.5. UNSHARP MASK SETTINGS IN ASTROART.

Astroart also has a very versatile unsharp masking tool. The Astroart web site has a nice tutorial on manually using unsharp masking that can give you some insights into what is actually involved in the process. The tutorial is located at **http://users.iol.it/mnico/lab/tut_en3.htm**. The example is for planetary image processing, but it is useful for understanding unsharp masking in general.

Figure 8.2.5 shows the unsharp mask dialog in Astroart. It has sliders for Sigma and Coefficient. The Sigma setting defines the amount of blur applied to the

FIGURE 8.2.6. SETTING AN UNSHARP MASK USING AN FFT - LOW-PASS FILTER.

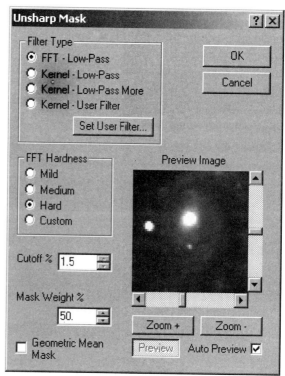

FIGURE 8.2.7. COMPARING FILTER TYPES FOR THEIR EFFECT ON STAR IMAGES.

the greatest impact; Kernel - Low-Pass has the least, and Kernel - Low-Pass More has a bit more but not nearly as much as FFT.

Figure 8.2.7 compares the preview of the FFT - Low-Pass filter (right) and the Kernel - Low-Pass filter (left). Note that there is a hint of a halo around objects in the right-hand image, but not in the left-hand image. The degree of sharpening, however, is comparable. I find that the Kernel - Low-Pass filter is the one I use most often.

Figure 8.2.8 shows a comparison of images sharpened with MaxIm DL and Astroart. The results are similar, with only minor differences between the images. A comparison of the detail near the core of M101 from each of the images underscores the similarities (see figure 8.2.9).

The heavier the unsharp masking, the more likely you are to get halos or other artifacts that spoil the image. Artifacts may show up where you least expect them, away from the object of interest. Noise in the original image, in the form of hot pixels or a short exposure, are the most likely sources of artifacts. But look for increased noise in the background, too.

image. The blurred version is subtracted from the original image, "suppressing large scale features and leaving fine details," according to the Astroart documentation. The coefficient is a multiplier that is applied to every pixel in the image. If the coefficient is too low, little sharpening will occur. If the coefficient is too high, the sharpening will be too aggressive. The coefficient seems to be similar to the Amount setting in Photoshop in its effect. The Adaptive checkbox allows the process to adjust itself to avoid any negative values in the image, and should be checked for most applications. This tends to mellow out the sharpening, which is a good thing for CCD images.

MaxIm DL also provides unsharp masking, using the Process | Unsharp Mask menu item. The Unsharp Mask dialog (see figure 8.2.6) looks a lot like the digital development dialog because the same filters are used here, but for a different purpose. In the digital development dialog, the sharpening filters are used along with a filter that alters the histogram. The filters define the type of blurring to use for the unsharp mask.

The rules for choosing filter types are similar to digital development: The FFT - Low-Pass filters have

FIGURE 8.2.8. UNSHARP MASKS IN MAXIM DL AND ASTROART ARE MORE ALIKE THAN DIFFERENT.

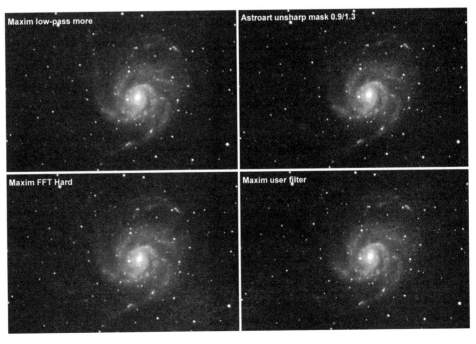

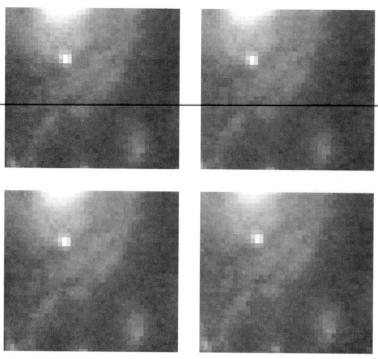

FIGURE 8.2.9. A DETAILED COMPARISON SHOWS VERY LITTLE DIFFERENCE BETWEEN THE VARIOUS UNSHARP MASKING TECHNIQUES.

The most important use of blurring is to remove graininess in portions of the image that have poor signal levels. This can occur in the areas between the brighter spiral arms of a galaxy; in the outer edges of an elliptical galaxy, and just about anywhere in a nebula. You can do several things to increase signal when you are taking images, of course, including combining multiple images and taking longer exposures. But sometimes there are portions of an object that are so dim that they remain noisy/grainy even after you've taken steps to limit the noise. And sometimes you just don't have enough data to get a great S/N ratio, due to light pollution or time restrictions. And even in a long exposure, the outer edges of galaxies and nebulae can have high noise and low signal.

Blurring can smooth out the grain that is characteristic of noise. The general rule of thumb is to sharpen when you have excellent signal, leave it alone if the signal is only good (since sharpening would increase the noise), and to blur where the noise becomes noticeable. Many times, a single image will present all three situations. You can use masking and/or selections in Photoshop or other image editors to select specific portions of the image to be treated by each technique.

For example, in the image with the MaxIm DL FFT Hard filter the outlying bright stars are surrounded by a large but subtle dark halo. This doesn't matter when zoomed in on M101, but for a wide field shot that shows the dimmest details in M101, it would be more of a problem.

TIP: The backgrounds of the M101 images show artifacts from unsharp masking. This argues in favor of applying unsharp masking in programs like Photoshop and Picture Window Pro, where you can select the portion of the image you wish to sharpen.

Smoothing

If you aren't familiar with image processing, it would be easy to dismiss smoothing (blurring) techniques as useless. After all, why in the world would you want to blur a perfectly good image? Well, the answer is that you wouldn't want to do that. But not many images are perfect to start with, and blurring can help make the image more presentable.

FIGURE 8.2.10. AN IMAGE OF A FAINT NEBULA, FIVE MINUTES WITH AN ST-8E BINNED 2X2 AND AN FSQ-106.

FIGURE 8.2.11. DIGITAL DEVELOPMENT DOESN'T HELP THIS IMAGE.

Consider figure 8.2.10, which shows an image of the area around the Cone Nebula (the nebula is barely visible). There is actually quite a bit of nebulosity in this image. You just can't see it yet. The blooming is present because I had to take a long exposure with a non-anti-blooming camera. The blooming spikes can be edited out later. The image is noisy due to a short exposure. Let's see if we can work a little magic and bring out the nebulosity while controlling the noise.

The first instinct is often to apply digital development to bring out dim details, but the sharpening component of digital development makes a mess of a noisy image, as shown in figure 8.2.11. The sharpening increases the overall grain of the image and is not the best way to approach this image. Even sharpening without digital development introduces too much grain.

A histogram adjustment with the tool of your choice (I used MaxIm for figure 8.2.12), brings out the dim nebular detail. It also brings out noise (grain) as shown in the inset area of figure 8.2.12. This is less true in the bright portions of the nebula, but all areas suffer to some extent from noise. If only a *portion* of the image had this problem, you could select it as shown later and apply the blur only where it is needed. In this example, the entire image needs help.

I applied a low-pass filter in MaxIm DL (see figure 8.2.13). The filter has fuzzed out the bright portions of the image. The blooming spikes are wider than they are in figure 8.2.12. The stars are a little softer. The dim stars show the softening most clearly. But the nebulosity looks smoother, especially in the enlarged detail. This is what smoothing does for noisy images.

FIGURE 8.2.12. A HISTOGRAM STRETCH MAKES THE GRAIN IN THE IMAGE STAND OUT.

FIGURE 8.2.13. A LOW-PASS FILTER IN MAXIM DL REDUCES THE GRAININESS OF THE IMAGE.

CHAPTER 8: IMAGE PROCESSING FUNDAMENTALS

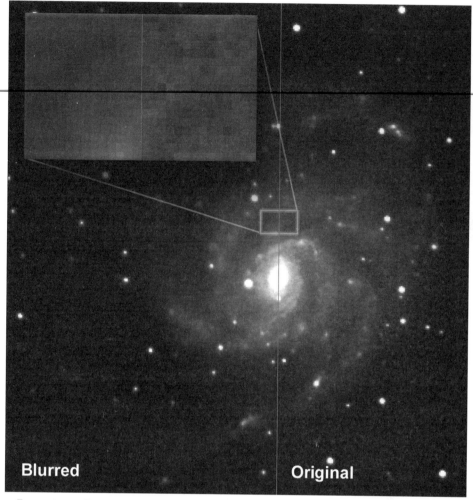

the dim areas. Use the Select | Color Range menu item, and then shift + click in the background and dim areas to select them. Try setting the Selection Preview to "Quick Mask" so the unselected areas will be easily visible. The Quick Mask hides the unselected areas with a red mask (see figure 8.2.15).

It is a good idea to examine your image to see which portions are noisy, and which are not. The greater the grain, the greater the noise. Look for a brightness level in the image where the graininess more or less disappears. Blurring will be effective for everything darker than the level where the graininess disappears.

FIGURE 8.2.14. COMPARING BLURRED AND UNBLURRED VERSIONS OF M101.

FIGURE 8.2.15. MASKING THE BRIGHT AREAS SO YOU CAN BLUR THE DIM AREAS.

You can achieve similar results in Photoshop using the Filter | Blur | Gaussian menu item. Picture Window includes a grain reduction filter that is also very effective, and most image editing software will include some kind of ability to blur the image, or selected portions of the image. Figure 8.2.14 shows the same M101 image that we applied sharpening to treated with blurring instead. Notice how the left side, which has had three stages of blurring applied, has much smoother fine detail.

Blurring is fine for the dim areas of the image, but the core of M101 is fairly bright and can stand some sharpening. You can use Color Range in Photoshop to select the dim or bright areas of an image. Figure 8.2.15 shows a selection I created to allow blurring of

THE NEW CCD ASTRONOMY

FIGURE 8.2.16. THE IMAGE ON THE LEFT HAS HAD ONLY A HISTOGRAM STRETCH. THE IMAGE ON THE RIGHT HAS HAD THE DIM AREAS BLURRED, AND THE BRIGHT AREAS SHARPENED.

Deconvolution

Deconvolution is magic, but it's dangerous magic. It can take a so-so image, and turn it into something special. It can also take a perfectly mediocre image and turn it into a real mess if you aren't careful. Not every image is a good candidate for deconvolution, and deconvolution isn't always (or even often) the answer to the problems in your image.

Deconvolution is mathematical magic. Between seeing, any imperfections of focus, and limitations of your optical system, the image you get is blurred. The light from a star, for example, is spread out into a circle, even though a star is nearly a perfect point source. Deconvolution will cleverly dig out the sharper image hiding inside of your blurred image and bring it to the surface. It does this using a Point Spread Function (PSF). The PSF defines how the light from a point source has been spread out. The deconvolution algorithm works backward from the PSF, and attempts to sharpen the image by removing the blurring effects.

The key to success with deconvolution is taking long exposures. Noise tends to make a mess of deconvolution. Longer images have a better signal level and less noise (remember that signal increases faster than noise during long exposures). A noisy image isn't a good candidate for deconvolution. Just as with conventional sharpening, deconvolution will emphasize noise as well as signal.

There are significant differences in the quality of the deconvolution you get with various image processing tools. MaxIm DL is OK, but not great. Astroart has a really friendly and effective implementation of deconvolution. CCDSharp and CCDSoft have simple but effective implementations. How you use deconvolution, and whether you use it on a given image, depends on both the noise in your image and the image processing software you are using. I'll give you some examples

> **TIP:** You can blur in stages for more control. Apply blur more heavily in the dark areas, and more lightly in bright areas. For example, select the background and apply a 0.4 Gaussian blur. Then select the background and some of the dimmest areas of the arms and apply a 0.3 Gaussian blur. Depending on the noise level, you might then try another blur selecting more of the object. This is what I did for the M101 images above.

Figure 8.2.16 shows a comparison of M101 images before (left) and after (right) sharpen/blur treatment. The blurring reduced the noise in the dim areas, allowing me to show more dim detail using a histogram stretch. Contrary to what you might expect, you can sometimes use blurring to *increase* the amount of detail you can show in an image!

There is a price to paid for any processing, however. The dim areas, while they are clean and show up more clearly in the right-hand image, are devoid of fine details. The brightest stars show just a hint of a dark halo from the sharpening, and the stars are more pronounced as a result of sharpening. This is why processing is really such a personal decision -- there is no one right way to process your images. For this image, you might consider applying your sharpening only to the bright areas of the arms, and not to the stars. You can add to and subtract from selections in Photoshop.

of the good and the bad of deconvolution, and show you what to watch out for when you are using it.

Like many sharpening tools, going too far with deconvolution can create artifacts. The dark halo effect happens more easily with deconvolution. Depending on the noise level, you can also get background artifacts or false features and details. You'll see examples that illustrate all of these situations.

Properly used, deconvolution can work wonders on an image. Figure 8.2.17 shows four versions of an image of M57, the Ring Nebula. The raw image at top left is a little disappointing because it is blurry and there is a slight elongation of the star images top to bottom (probably from a slight guiding error). The image at top right has had a Lucy-Richardson deconvolution applied. This is one of several types of deconvolution you will find in various software. The deconvolution has sharpened the image, revealing more detail including knots of structure in the ring. Star images are smaller and rounder. The image at bottom left shows the result of applying a Maximum Entropy deconvolution. The character of the image is slightly different, but the increased sharpness and detail is also clear to see. The image at bottom right is the result of an unsharp mask. It is improved over the raw image, but the deconvolution routines are clearly more effective. Both deconvolution routines bring out more detail than sharpening does. In particular, sharpening does nothing to help the out-of-round star images. You can think of deconvolution as a smart sharpening routine.

Which algorithm you use will vary with your preferences and your images. When in doubt, try both of them and compare. You will also find other flavors of deconvolution in other software packages, such as Van Cittert in AIP for Windows.

Two things are important to deconvolution: the point spread function (PSF), and the number of iterations you use. Unlike sharpening, which is normally applied one time to an image, deconvolution proceeds through a number of iterations. A few iterations mean a small improvement in the image. Many iterations means major changes in the image. Choosing the right

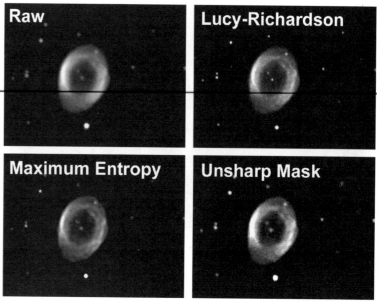

FIGURE 8.2.17. EXAMPLES OF WHAT DECONVOLUTION CAN DO.

number of iterations is not something you can always get right on the first attempt. MaxIm DL allows you to back out of the most recent iterations. This allows you to try again different values. With most software packages, you need to evaluate the final results using a given number of iterations. If you don't like the result, you have to start over again with different numbers.

Figure 8.2.18 shows the results of deconvolution in MaxIm DL. The top left image is the original image of the Eskimo Nebula. Not a bad image, certainly, but I was curious to see what MaxIm's Maximum Entropy deconvolution could do to improve the image.

The first few iterations look like steps backward, and they are. Determine the number of iterations that will show improvement by trial and error. Iteration 5 is a little blurrier than the original; iteration 6 is a little sharper, so 6 iterations would be the minimum number to work with for this particular image. The number will vary from image to image.

Iterations 7, 8, and 9 don't really show a lot of difference, but there are some very small improvements. The most noticeable is the shrinking of the central star, and the darkening of the area around it. The large version of the Eskimo at lower right is the result after 15 iterations. The outer areas are a tiny bit sharper, but the center is mostly unchanged by the additional iterations.

The large image at lower left is 15 iterations with digital development to bring out the dim portions of the nebula. This loses some contrast and detail in the core, but is much more effective at showing details in the outer regions of the nebula. You may or may not find that digital development helps after deconvolution, and you can try any of the other tools in your arsenal if you think it might help after deconvolution.

That part about "after deconvolution" is important. To work effectively, deconvolution should be applied to the image before other tools. After you apply your bias/dark/flat-field frames, the next step is deconvolution if you are going to use it. On objects like the Eskimo, deconvolution is a wonderful tool. It's not always so wonderful on images that contain nebulosity or dimly illuminated areas, such as many galaxies. The noisier the image, the less likely it is to benefit from deconvolution. Figure 8.2.19 shows why. Depending on the noise level of your image, you can wind up with artifacts. In figure 8.2.19, these include excessive graininess and small dark halos around the stars in front of the galaxy arms.

FIGURE 8.2.18. MAXIMUM ENTROPY DECONVOLUTION IN MAXIM DL.

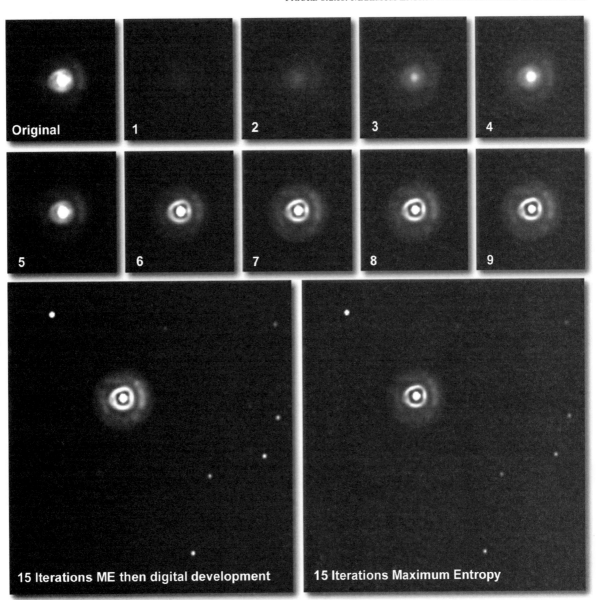

FIGURE 8.2.19. DECONVOLUTION IS NOT ALWAYS THE BEST CHOICE FOR AN IMAGE.

However, there are times when you will want to set your own parameters.

Figure 8.2.20 shows the start of the manual deconvolution process: selecting a star to create a point spread function (PSF). There are a variety of ways to set the point spread function. Usually, the best option is to pick a star in the image and let MaxIm DL set the PSF based on the star. The star is a record of the way in which a point source has been spread out by whatever happened during the exposure. If the star is slightly elongated because of a minor guide error, that will be part of the PSF. If the seeing was poor, and the star spread out, the PSF will reflect that. Make sure the star is not saturated.

TIP: A small point spread function can give better results on nebulas and galaxies. The PSF for this image was automatically selected by MaxIm DL by scanning the image. If you see the heavy mottling evident in this image after deconvolution, you might get better results by manually choosing a smaller star (make sure it is not saturated) instead of letting MaxIm DL do the choosing. You can generate a PSF using a Gaussian function in MaxIm DL or Astroart. If you see mottling, try a smaller number.

You can also have MaxIm DL construct a PSF mathematically. You can do this if there are no stars present, such as for a planetary image. Or you can experiment with generated PSFs to see if you can get a

FIGURE 8.2.20. SETTING THE POINT SPREAD FUNCTION (PSF) IN MAXIM DL.

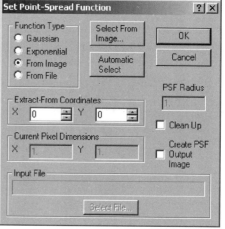

MaxIm DL includes a Deconvolution Wizard, which will automatically select everything you need to do a deconvolution.

Figure 8.2.21. Left: An excellent, clean PSF.

Figure 8.2.22. Right: Automatic selection of the star for the PSF doesn't always give useful results.

Figure 8.2.23. Below right: If the background is noisy, the PSF can be a mess.

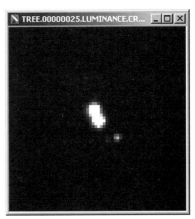

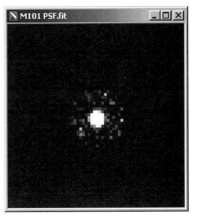

better result than when using a star. If you want to see the PSF that MaxIm DL has extracted from the image or created mathematically, check the "Create PDF Output Image" checkbox. Figure 8.2.21 shows an enlargement of the PSF for the Eskimo Nebula image, which is based on the star at upper left in figure 8.2.20.

TIP: Pick a bright star for the PSF when choosing one manually. If the star is too dim, MaxIm won't generate a useful PSF. But you also do not want a star that is saturated or bloomed. Such a star will not reflect the way that other, non-bloomed stars' light has been spread, and the PSF from such a star can lead to really bad results. One of the reasons the Eskimo deconvolution went so well is that the star used for the PSF was very noise-free, as was the image.

I always ask MaxIm DL to create an image of the PSF so I can verify that it's OK. Figure 8.2.22 shows why. MaxIm DL sometimes gets a little too creative when picking the star for the PSF. In this case, it has picked two stars close to each other, with a third dim star and some random pixels thrown in for good measure.

Most of the time, you will also want to check the "Clean Up" checkbox. Otherwise, you can wind up with a mess of pixels that will only serve to create chaos in your deconvolved image. Figure 8.2.23 shows a real-life example of a PSF created without Clean Up turned on. If your image is exceptionally clean, you might get ever so slightly better results by leaving Clean Up turned off, but most of the time leave it turned on.

I recommend letting MaxIm DL automatically select the star for the PSF (except on planets). On planets, you will most often get the best results letting MaxIm DL generate a Gaussian PSF. Depending on the quality of your optics, the image scale, and the seeing conditions, create a PSF using a value between 0.5 and 3.0. For most planetary images, a deconvolution with a Gaussian PSF of 1.0 to 1.5 will be a good starting point. If you don't see much sharpening, increase the size of the Gaussian PSF. If you see too much mottling (like in the background of the M101 image in figure 8.2.19), reduce the size of the Gaussian PSF. It takes a little experimentation to find the right balance point for a given image.

TIP: Although deconvolution can be a powerful tool for planetary image processing, sometimes it just doesn't help much. Or you may find that unsharp masking does a better job than deconvolution on planets. Unsharp masking is often the best choice for improving planetary images. Deconvolution can result in pleasing planetary images, but unsharp masking will usually be more accurate. Unsharp masking uses the data in the planetary image to enhance contrast and reveal features. Deconvolution is more likely to create artifacts because it is designed to manipulate point sources like stars.

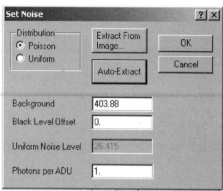

FIGURE 8.2.24. THE SET NOISE DIALOG.

Once you have set the PSF, the next step is to have MaxIm DL assess the noise level and background level in the image. Figure 8.2.24 shows the Set Noise dialog where these steps take place. MaxIm does this step very accurately, so I always choose Auto-Extract. Click the Auto-Extract button, and MaxIm will set the background average level and the noise level. An accurate noise analysis is critical to success with deconvolution. Some software does this step automatically.

After you set the PSF and the noise level, you are ready to deconvolve. Figure 8.2.25 shows the deconvolution dialog.

Once you have input the deconvolution parameters, follow these steps to perform the deconvolution:

FIGURE 8.2.25. PERFORMING A MAXIMUM ENTROPY DECONVOLUTION.

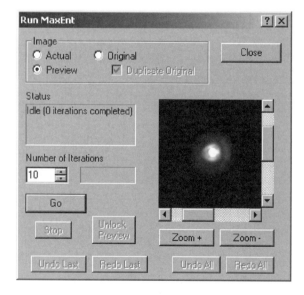

1. Verify the "Preview" radio button is selected.
2. Put the portion of the image you are most interested in into the preview window. Zoom in/out to show the region of interest and some background. If you don't include some background, the preview won't match the deconvolution in the actual image.
3. Choose a number of iterations; 7-10 is a good starting number for most images.
4. Click the Go button. Depending on how far you zoomed in or out, the preview may take a while to show results.
5. After the requested number of iterations is complete, check the preview to se if you want more or fewer iterations. If you want more, enter the number of additional iterations and click Go. If you want less, click Undo All or Undo Last, and enter a smaller number of iterations. Repeat until you determine the right number of iterations. Not all software allows successive iterations; check the documentation. With some software, you may need to start over if you want to change the number of iterations. You may want to save the various versions that you create for comparison. I typically include the settings for the deconvolution in the filename so I can adjust or recreate the same outcome at a later time. For example, a filename might be: Eskimo_MaxEnt_12iter_Gauss1.2.fit.
6. Click the "Actual" radio button. This makes sure that deconvolution applies to the source image, not the preview window.
7. The number of iterations you applied to the preview should now appear. If not, go ahead and change the number to be what it should. Click Go. The deconvolution of the actual image will take longer than the preview. For very large images, such as with an ST-8E camera, deconvolution may take a long time.

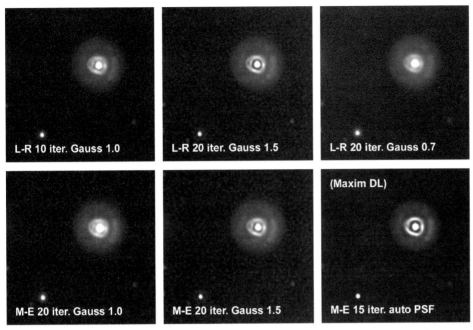

FIGURE 8.2.24. A COMPARISON OF VARIOUS DECONVOLUTION SETTINGS IN ASTROART, WITH A COMPARISON TO MAXIM DL AT LOWER RIGHT.

then increase. For example, increase the size of the PSF until you see undesired artifacts, such as mottling. Increase the number of iterations but stop when you get artifacts like dark halos.

You can also create a custom PSF from a star in the image. Before you open the deconvolution dialog, click and drag a small box around the star you want to use to generate the PSF. Use a medium-sized star, and avoid any star with saturated pixels. Keep the box fairly tight around the star. Zoom in on the image as needed to keep the box as tight as possible. Then use the Filters | Richardson-Lucy/Maximum Entropy menu item to open the dialog shown in figure 8.2.27.

Astroart has one of the faster implementations of deconvolution, and the flexibility and quality are both excellent. I find myself turning to the Astroart deconvolution routines more often than any of the others. It includes two different algorithms: Lucy-Richardson and Maximum Entropy. I opened the same image of the Eskimo Nebula in Astroart and tried both algorithms on it. I applied various settings using both Lucy-Richardson (L-R) and Maximum Entropy (M-E), as shown in figure 8.2.26. The first number (e.g., 20 iter.) refers to the number of iterations. In general, Astroart needs more iterations for a given image than MaxIm DL does, but it is also much faster so there is a net gain in speed. The second number (e.g., Gauss 1.0) refers to the size of the Gaussian PSF generated for the image. As you can see, the Astroart algorithms are better at bringing out faint details, and they don't soften the edges of high-contrast portions of the images.

The number of iterations you use, and the size of the Gaussian PSF, will vary from image to image and can only be determined by experimentation. In general, start low and

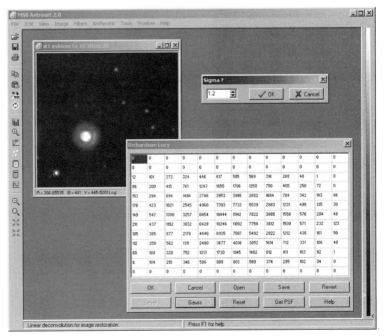

FIGURE 8.2.25. DECONVOLVING IN ASTROART BY CREATING A PSF FROM A STAR IN THE IMAGE.

The PSF is defined by the brightness (ADU) values of the pixels in the box you draw around the star. The box can be no larger than the matrix shown in figure 8.2.27, which is why you should make it tight to the star and drag it out carefully. "Tight" doesn't mean close to just the bright portion of the star. It means close to the pixels where the star emerges from the background. Initially, the matrix is filled with zeroes except for a value of 1 at the center (the ultimate in small PSF, I guess!). To generate a PSF for the selected star, click the "Get PSF" button. The matrix will fill with numbers as shown in figure 8.2.27.

> **TIP:** If you are going to subtract from the background to get it in the range of 50-100 (or 25-35 for 12-bit cameras), do so before you create the PSF. This will give you a cleaner background for the PSF.

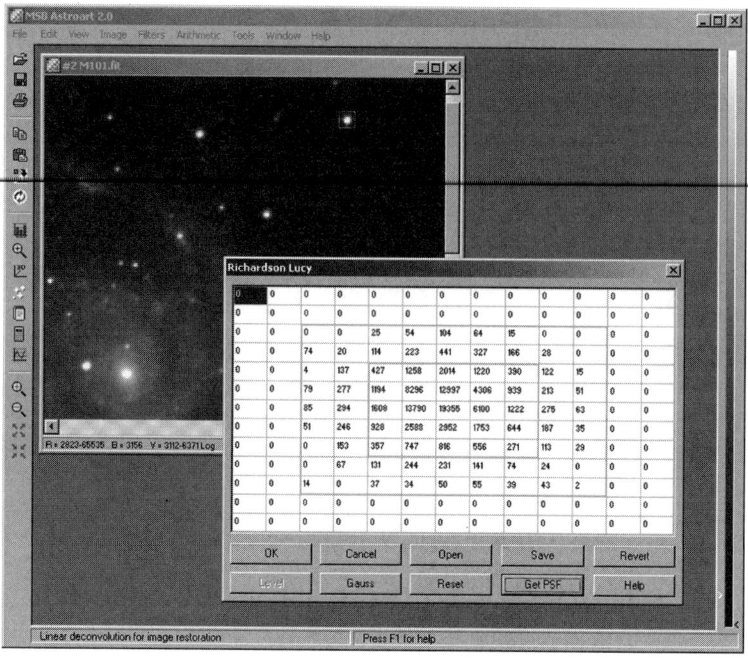

FIGURE 8.2.24. AN EXAMPLE OF A BETTER PSF.

The PSF shown in figure 8.2.27 is actually a poor one. Several things are going on here that can cause problems:

FIGURE 8.2.25. SETTING A PSF USING A GAUSSIAN FUNCTION WITH SIGMA OF 1.5.

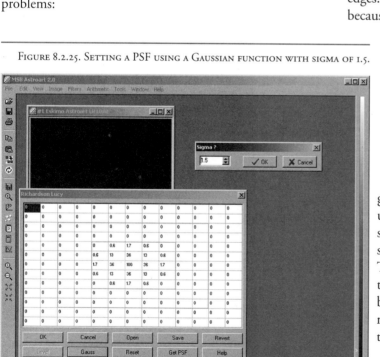

- The box for the PSF is a little too wide. Note that the numbers are still large near the left and right edges. Selecting a dimmer star will solve this because a dimmer star will by physically smaller as well.

- The PSF has a hard edge. Note that the numbers at the top and bottom are quite large, then drop suddenly to zero in the last row or two. This happens when the background levels are too high, or when the box is too tightly placed around the star.

You can try a Gaussian PSF if you can't get a good PSF from a star in the image. Figure 8.2.28 shows a much better PSF from a star in an image. Notice the box around the star at upper right in the image of M101. There is just enough room around the edges to enclose the star where it emerges from the background. The numeric values in the matrix vary appropriately and the transition to the background is smooth.

To generate a PSF using a Gaussian function, click on the Gauss button in the deconvolution dialog. A small window will open, titled "Sigma." This number defines an adaptive radius for the generated PSF. The radius will adjust automatically for both small and large star images, so that the same apparent sharpening occurs for both.

The default Sigma is 0.7, and this is too small for the average image. Use a sigma of 0.7 only when seeing is superb, guiding is perfect, and the image has excellent signal to noise ratio. A sigma of 1.0 to 1.2 can be used on a good image, and a sigma of 1.5 is good for the average image. If your seeing is worse than average, you may find that deconvolution can't solve the problem adequately, but it never hurts to try. As with all things deconvolution, experiment with different settings to see what works best. Figure 8.2.29 shows the PSF that results from a Sigma setting of 1.5. Notice how much smaller a Gaussian PSF is compared to the PSFs generated from stars.

To deconvolved the image with the PSF (however you created it), simply click the OK button in the deconvolution dialog. Yes, I know, this isn't very intuitive, but that is generally what you will find with Astroart: excellent speed and function, but the user interface has some rough edges. On the plus side, the procedure is the same whether you are doing M-E or L-R deconvolution.

Figure 8.2.30 shows the next step in the process: determining the number of iterations to use. Astroart's iterations are different from MaxIm DL's. In MaxIm, the first few iterations make the image look worse, then it starts to look better once you get to about 3-6 iterations. With Astroart, even a single iteration is likely to show some level of improvement. I have also had excellent results with as many as 20 and 30 iterations. Astroart has a large range of useful iteration values.

Generally speaking, the noisier your image, the fewer iterations you will be able to use. On the flip side, at some point you will not see any further benefit from additional iterations. If the PSF is too large, however, you will see artifacts in Astroart, particularly mottling and dark halos. Reduce the size of the PSF and try again.

TIP: If you are using the Maximum Entropy method, the window in figure 8.2.30 allows you to check a box to "Cut sky background." This cut operation will make the background black. In most cases, this creates excessive contrast, and is not often visually pleasing on galaxies and nebulae. It is more likely to be useful when working with very bright objects like planets.

FIGURE 8.2.26. SETTING THE NUMBER OF ITERATIONS.

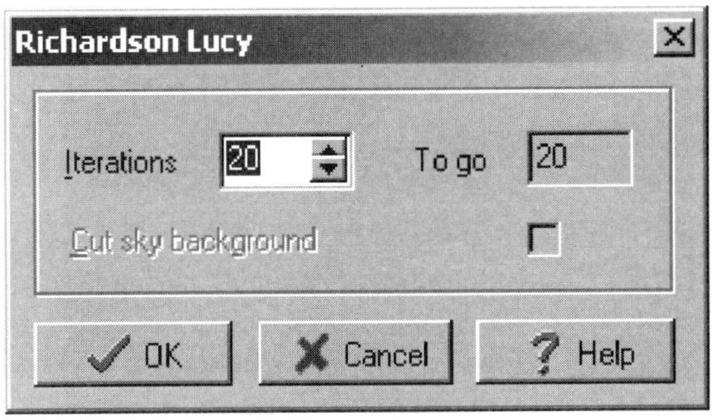

9 Image Processing for Celestial Objects

Our nearest celestial neighbors, the sun, moon, and planets, provide some of the most fascinating subjects for astronomical imaging.

Most solar system objects offer a wealth of detail. Our planet's atmosphere is often turbulent, making it difficult to capture that detail. But when you put everythnig together to get good results, the images can be breathtakingly beautiful.

Section 1: Processing Sun, Moon, and Planets

FIGURE 9.1.1. A PROCESSED IMAGE OF MARS.

Figure 9.1.1 is an image of Mars that I took with Adam Block at Kitt Peak as part of the Visitor Observing Program. Good seeing (steady air) and high magnification are the essence of planetary imaging.

One way to image these objects is to watch for nights or partial nights when the seeing is better than average. The better the seeing is, the better your chances are to get good results.

The brightness of these objects also helps. Because they are so bright, short exposures are possible. Short exposures freeze the action and give you sharp detail.

One problem with short exposures is that freezing bad seeing won't give you a good image. You need good seeing, short exposures, high magnification, good optics, *and* good technique to capture details on solar system objects. Sharpening is main tool you will use for processing sun, moon, and planetary images.

Note: Many of the techniques and descriptions in this section apply to any high-resolution image. The sun, moon, and planets are especially appealing targets for high-resolution imaging, and many people image them when the seeing becomes superb. But you can apply the same high-resolution techniques to other objects, too. Because other objects are not as bright, however, you may need extremely long exposures to compensate for the high magnification involved in high-resolution imaging. High magnification often comes at the cost of slower focal ratios, which means that exposure times must be greatly increased. It's one thing to increase your exposure from 5 to 20 milliseconds to image Jupiter at f/40; It's quite another to increase your exposure of M16 from one hour to four hours! The alternative is to have a fast scope with a really large aperture, such as a 24" f/4 Newtonian. Then you have both a fast focal ratio and high magnification.

Seeing Conditions

There are a lot of ways that seeing conditions can get in the way of acquiring good, sharp images of the sun, moon, and planets. Turbulence is the worst problem. The movement of the air will swamp details even with very short (millisecond) exposures. But a bigger problem is how the seeing distorts the image. It doesn't just blur it; the image is actually moving and changing shape at a very high rate of speed when the seeing is poor to average. You can view at low powers, but planetary imaging is all about high magnification.

Distortion can also occur when the seeing conditions are good, but you are more likely to be able to freeze and get good results at such times. Even in good seeing, slow atmospheric movements can distort the geometry of a planetary image and make combining images impossible. The first link below shows how typical seeing conditions mess with a planetary image. The second shows this in slow motion, emphasizing geometric distortions:

http://www.newastro.com/newastro/book_new/samples/saturn_motion.gif

http://www.newastro.com/newastro/book_new/samples/saturn_motion2.gif

The following sections offer some suggestions for various types of objects.

All solar system objects

View the object at the greatest possible elevation in the sky. The lower the object is, the greater the distance its light must pass through the atmosphere. This can result in additional turbulence, differential diffraction (splitting the object's colors), color changes, and other nasty effects. The moon is highest in the winter; the sun is highest in the summer. Planets vary in elevation with each apparition. Elevation tends to be high in one hemisphere and low in the other. For example, in 2001, Mars was quite low for northern hemisphere observers, but high and well placed for observers in the southern hemisphere.

Use an appropriate optic. The larger the aperture you use, the greater the potential for resolution. Since high-resolution imaging must take place when the seeing is excellent, you can use larger apertures without fear of seeing-induced problems.

Good contrast transfer is essential for maximum detail. Refractors have no central obstructions, and APO refractors tend to have excellent baffling and superb contrast. However, aperture is limited. You may find the best detail in a 20" Dob or some other large instrument, as long as the optics are of exceptional quality. The short exposure times mean that you can image without tracking, so large Dobs are in the running for best scope for planetary imaging.

If you must use a scope with a large central obstruction, you may lose some contrast but you can boost contrast during image processing. This won't recover all of the contrast, but it will help you get the most out of your instrument.

The most important consideration in getting good contrast is optical quality. Quality in this case means extremely smooth optics with excellent figure (accuracy). A smooth optic that is ground to exact curves will always be the best planetary instrument at a given aperture, even if it has a central obstruction. The obstruction diameter should, however, be no more than about 20% of the aperture diameter.

CCD isn't the only way to image. Video is fun, even if you don't save the images to tape for later capture. Digital cameras have short exposure times, and eyepiece projection (technically, afocal projection) with digital cameras can yield high magnification.

Sun

The sun creates daytime turbulence. The sun throws off a lot of heat, even in the winter. This heating causes air currents to swirl, and the swirling increases as the heating increases. Imaging the sun in the midmorning will help you eliminate the worst of the heat-induced atmospheric motion. Look for the sweet spot where the sun is at a high enough elevation to provide good images, but low enough that it hasn't caused major heating yet.

You will get reduced cooling. Speaking of heat, you won't be able to cool your CCD camera as effectively when imaging the sun because the ambient daytime temperature is much warmer than it is at night. The short exposures and high brightness level will hide most of the noise, however, so this is not a critical problem. You may well be able to get excellent images without even using a dark frame.

Focusing solar images is very challenging. Do your initial focusing using the edge of the solar disk. Use a sunspot and/or faculae to do final focusing. If you simply can't find a good focus point, it's likely that turbulence is too severe for high-resolution imaging anyway.

Don't use so-called photographic solar filters for CCD-based solar imaging. They are intended for film photography. Film requires more light than visual observing, and these filters admit considerably more light. So much more, in fact, that they are not safe for visual observing. For CCD imaging, use a visual-grade solar filter. Depending on the shortest exposure available in your camera, you may need additional filtering. Use a moon filter or a polarizing filter on the nosepiece of your camera if that's the case.

Moon

The moon is the easiest of the nearby celestial objects to image. Even if you aren't enjoying a night of excellent seeing, you still have surface detail that you can image -- craters and seas. When the seeing is very good or better, you will start to see fascinating details on the lunar surface. You don't have to wait for a perfect night to image the moon, but your images on those perfect nights will really be something special.

You will probably need filters to image the moon. A simple neutral density moon filter, often used for visual observing, may do the trick. If not, consider a polarizing filter that uses two elements, one of which rotates. By rotating one element you change the amount of light that passes through the filter. This lets you tune the amount of light striking the CCD chip to suit your camera's capabilities.

Planets

Planets require magnification for meaningful imaging. This makes them the toughest solar system objects to image. With many celestial objects, imaging at 2-3 arcseconds per pixel gives good results. That image scale will barely show the planets as a disc, so you've got to magnify to the point of 1, 0.5 or even 0.25 arcseconds per pixel to get sufficient detail. This makes planetary imaging the most seeing-dependent type of imaging. If you don't have good seeing, you won't have good planetary images.

Barlows are the simplest way to get more magnification. Assuming you are using a standard 2X Barlow, you get 2x by putting the Barlow just in front of the camera. To get approximately 3X, use a spacer between the camera and Barlow. About two to three inches should be adequate. The 4X Powermate from TeleVue gives you a whole lot of magnification in one gulp, and this item is frequently found in the toolkit of serious planetary imagers.

Sharpening is the entire game when it comes to planets. Unsharp Masking and Deconvolution are the image-processing weapons of choice. Most of the time, Unsharp Masking will be the best choice. To get good results with sharpening, you have to start with a very high signal to noise ratio. The simplest way to do this is to combine multiple images. Imagers in search of the best possible quality will take not just dozens, but hundreds of images in a night. You then sift through them, looking for the best quality images, which you then combine and sharpen.

Jupiter is a special case. It rotates very quickly, and you will need to work very efficiently to capture enough images for combining. If you take too long to capture a series of images, key features will move and create false colors when you combine red, green, and blue images. Use a subframe if at all possible when imaging Jupiter to minimize your download times. Try to acquire all of your images in less than 10 minutes for best results. This includes all of your color and luminance images that you intend to combine at one time.

Sharpening

Once you have obtained the best possible images, use sharpening to bring out details. It takes some experience to determine how much, and what type, of sharpening to apply to a given image. If you go too far with sharpening, you will create artifacts in the image (false details). If you don't go far enough, you'll miss out on finding the details that are hidden in the image.

There are two basic kinds of sharpening:

Unsharp masking - Uses a blurred version of the image to perform a subtraction that removes blurriness and leaves behind the sharp features in the image.

Deconvolution - Uses a sophisticated algorithm to repeatedly transform the image into a sharper version of the original. Deconvolution uses a point spread function that defines the shape and extent of the blur, and then attempts to remove that blur from the image.

Both techniques are covered in detail in chapter 8. Which one to use for a given image of the sun, moon, or a planet is not a simple choice. A little trial and error is your best option. Try each technique, and variations of each technique, to find out what works best. Figure 9.1.2 shows a set of images taken through red, green, and blue filters, then processed to create a sharp, detailed image of Mars. The three images at the top are individual red, green, and blue light images. I acquired ten images through each color filter.

The second row of images shows the results of a median combine of each group of ten images. Note how even this single processing step yields additional detail in each of the color-filtered images. As I've said many times throughout the book, combining images increases signal and decreases noise. This is yet another good example of the benefits of combining images.

The third row shows the results of applying a Lucy-Richardson deconvolution in Astroart to each of the combined images. The parameters used were a 1.2 pixel Gaussian point spread function (PSF) with 20 iterations. I determined these parameters by trial and error, as each processing session tends to be unique.

Despite my success with deconvolution, most (but clearly not all) planetary images will benefit more from unsharp masking. It never hurts to try every tool at your disposal on planetary images, however.

The color image at the bottom of figure 9.1.2 is the result of combining the deconvolved images in MaxIm DL with some additional histogram adjustments and a final light pass with unsharp masking in Photoshop.

As pointed out in chapter 8, sharpening works best when the image you are attempting to sharpen has a very good signal to noise ratio. Sharpening a single image has limited benefit. Sharpening combined images is more effective. Figure 9.1.3 shows a variety of sharpening techniques and parameters using a red-fil-

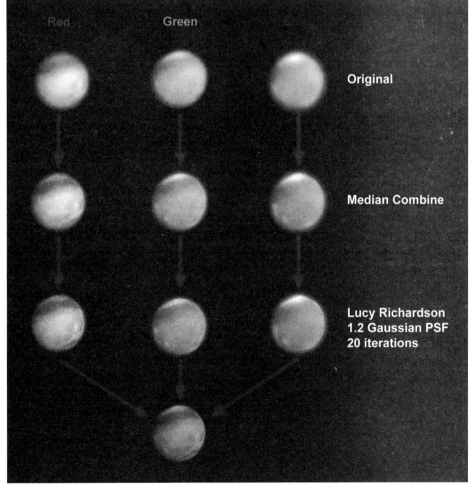

FIGURE 9.1.2. APPLYING SHARPENING FOR 30 IMAGES OF MARS. EACH COLUMN SHOWS THE PROCESSING STEPS FOR A SINGLE COLOR. EACH ROW SHOWS THE RESULTS OF ONE STEP IN THE PROCESS. THE FINAL IMAGE AT THE BOTTOM IS THE RESULT OF COLOR-COMBINING THE FULLY-PROCESSED RED, GREEN, AND BLUE IMAGES.

tered Mars image. The left side shows samples of different settings for a Lucy-Richardson deconvolution in Astroart. The right side shows three variations in settings for the Astroart unsharp mask tool. For ease of reference, each image in figure 9.1.3 has a number associated with it.

Image 1 is a single image through a red filter. Images 2 through 6 have had some level of Lucy-Richardson deconvolution applied. Images 7-9 have had an unsharp mask applied using different parameters.

There are two numbers below each Lucy-Richardson deconvolution image. The first number is the size, in pixels, of the Gaussian point spread function (PSF) used to deconvolve the image. (See chapter 8 for information about the Astroart deconvolution filters.) The second number is the number of iterations. For example, "Combine, LR 1.2 20" means that the combined image (10 individual images median combined in Maxim DL) was deconvolved with a Gaussian point spread function of 1.2 pixels, and that 20 iterations of deconvolution were used.

There are also two numbers associated with the unsharp masked images. The first number is the sigma, which defines how blurry the unsharp mask will be. Larger numbers generate more sharpening. The second number is the coefficient. Larger values intensify the sharpening effect.

I find Astroart's implementations of deconvolution and unsharp masking to be among the best available. CCDSharp's Lucy Richardson implementation is also very good. This same algorithm is also used in CCD-Soft version 5. A good unsharp mask tool should allow

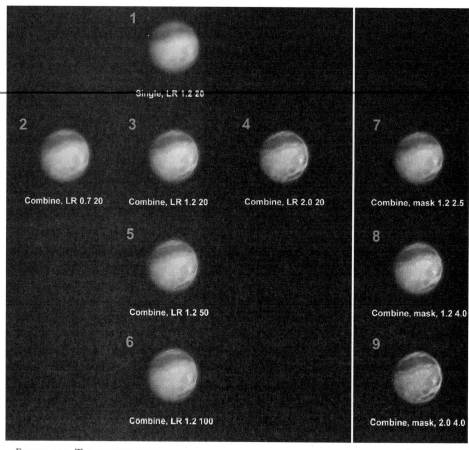

FIGURE 9.1.3. THE RESULTS OF USING VARIOUS SETTINGS FOR DECONVOLUTION AND UNSHARP MASKING.

a very fine degree of control over the parameters, down to one-tenth units at the very least. Astroart's unsharp mask is excellent, and Photoshop's unsharp masking is pretty good for an image editor.

Here are some tips for sharpening images:

- A combined image can be sharpened much more effectively than a single image.

- If you are making a color image, apply sharpening to the color components before you combine them. **This runs counter to my advice for all other types of images.** For planets, sharpening is so important that it can come first. Early sharpening may upset the color balance, requiring a lot of hard work on your part to restore it.

- Sharpening may saturate some pixels. If this happens, back off on the sharpening so you can do a proper color combine. Saturated pixels will throw off your color balance every time.

- Finding the right size for the PSF is critical. If it is too small, you don't get enough sharpening. If it is too large, you get so much sharpening that false details are created. There is no rule for the size of the PSF; trial and error is the way to go. Planetary image processing requires patience.

- Additional iterations during deconvolution don't necessarily show more useful detail. Image 3 is fairly smooth, and the details are mostly real. Image 5 has 50 iterations, and noise has begun to dominate -- no more detail is revealed, and some details are now hidden by the artifacts of deconvolution. Image 6 has extensive noise. See the close-up of these three images in figure 9.1.4.

- Although there is no hard and fast rule for what values to use for the parameters of your sharpening tools, the appearance of noise tells you when you've gone too far. When noise becomes obvious, cut back a bit and you should be at the sweet spot for getting the best sharpening possible.

- Unsharp masking behaves like deconvolution, with additional noise showing up as you increase the amount of sharpening. However, the actual appearance of unsharp masked images is different, and for planets in particular, unsharp masking tends to do the best job at revealing details (see figure 9.1.5, and compare to figure 9.1.4).

Whatever you use to sharpen high-resolution images, it all starts with good seeing. Good seeing allows you to tease details out of fuzzy original images. Don't hesitate to take large numbers of images. When you select the best and combine, you give yourself the best possible chance at a really good image.

FIGURE 9.1.4. ABOVE: A CLOSE-UP COMPARISON OF IMAGE 3 (LEFT), 5 (CENTER), AND 6 (RIGHT). NOTE THE INCREASING NOISE FROM INCREASING ITERATIONS (20, 50, AND 100, RESPECTIVELY).

FIGURE 9.1.5. BELOW: A CLOSE-UP COMPARISON OF IMAGES 7-9 (LEFT TO RIGHT). NOTE THE INCREASING NOISE FROM LEFT TO RIGHT AS A RESULT OF MORE SEVERE UNSHARP MASKING.

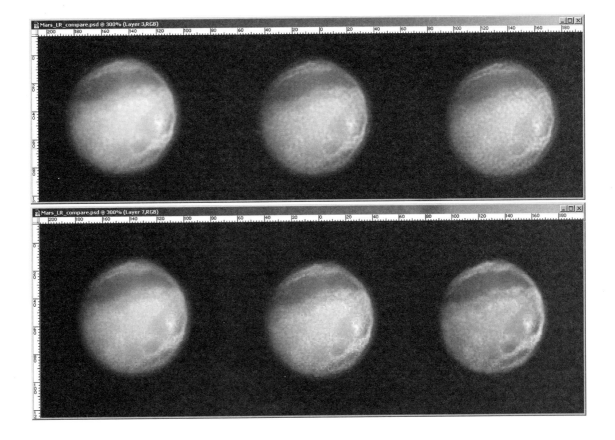

The same techniques that work on the Mars image will also work on other images. If you image the sun and moon under less than ideal conditions, try a very light sharpening, either with a minimal unsharp mask, or use conventional high-pass sharpening in Photoshop, Picture Window, etc.

Figure 9.1.6 shows an example of a sharpened moon image. Figure 9.1.7 shows a solar image that has been enhanced with unsharp masking.

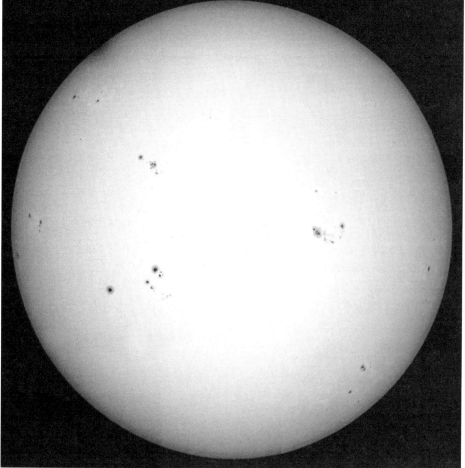

FIGURE 9.1.6. TOP: AN IMAGE OF THE MOON THAT HAS BEEN SHARPENED.

FIGURE 9.1.7. LEFT: AN IMAGE OF THE SUN SHARPENED TO SHOW FINE DETAIL.

Section 2: Globular Clusters

FIGURE 9.2.1. AN IMAGE OF M13 THAT SHOWS DETAIL IN THE CORE, AS WELL AS THE DIM STARS OUT AWAY FROM THE CORE.

Globular clusters are fascinating: huge collections of stars massed together into a giant ball. They are bright enough to image even when the moon is out, yet they are complex enough to benefit from long exposures. There is a lot of interesting science being done with globulars, and imaging them enables you to see what all the fuss is about. Figure 9.2.1, for example, is a color image that shows both the red and blue stars in the globular cluster M13.

If you look at a just-downloaded image of a globular cluster in your camera control software, you are likely to see one of two things: a small group of stars that represents just the very core of the cluster, or a big white blob. Neither is the best way to view the image of a globular cluster. For best results, a globular cluster requires a non-linear histogram stretch to reveal the full range of stars in the cluster.

You can quickly test a globular image to see how extensive the cluster is, and check to see if the core is burned out or not. A high white point makes the image show the core more clearly, as in the left half of figure 9.2.2. See chapter 8 for details on white points, black points, and histogram adjustments. A low white point shows the true size and extent of the globular, as shown in the right half of figure 9.2.2.

It is easy to over-expose a globular image and wind up with a burned-out core. The best exposure time is long enough to capture the dim outer members of the cluster, but short enough to prevent the core from burning out. You can take multiple exposures to show more of the dim stars without burning out the core.

The images in figure 9.2.2 are both of the same globular, M13. When the processing tricks described in this section are applied to the image, you can see the full range of stars all at one time. To download globular images for following along with this section:

http://www.newastro.com/newastro/book_new/samples/c9_globs.zip

One quick approach to showing the full range of details in a globular is to use digital development. Figure 9.2.3 shows the result of using digital development in MaxIm DL. Since digital development does both histogram changes and sharpening at the same time, I prefer to process my globulars manually. The sharpening from digital development is often more than is really needed to enhance an image. Over-sharpening of the core can make a mess, for example. Processing globulars manually allows you to get every last little bit of detail, and gives you complete control over the final result. The techniques described here are also applicable to many other bright but extended objects, including open clusters and galaxies with bright cores.

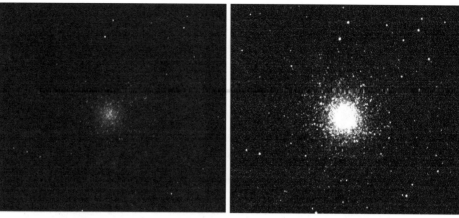

FIGURE 9.2.2. THE LEFT IMAGE SHOWS WHAT A GLOBULAR OFTEN LOOKS LIKE WITH DEFAULT HISTOGRAM SETTINGS. THE RIGHT IMAGE SHOWS HOW LARGE THE GLOBULAR REALLY IS.

Resolution Requirements

You will get the best results with globulars if you have good seeing. All those stars look best when they are small pinpoints of light. Poor focus, lousy seeing, or any tracking problems will quickly ruin an image of a globular cluster. Globs, as they are affectionately known, are a challenge to image for this reason. Figure 9.2.4 shows examples of globs shot under windy conditions (left) and under steady conditions (right).

Focus is especially critical. If you are even slightly out of focus, the quality of the image will suffer greatly. See chapter 2 for information about obtaining the best possible focus.

This emphasis on resolution doesn't mean you need high magnifications to image globs. As figure 9.2.1 shows, the quality of your images is more important than their scale. You can remedy problems to some extent with sharpening, but see the last portion of this section for some thoughts on sharpening globulars.

FIGURE 9.2.3. M13 AFTER DIGITAL DEVELOPMENT.

Digital Development in Maxim DL

The basic problem with imaging and processing a globular is that the full brightness range is larger than you can reproduce on paper or a computer monitor. There are some very bright giants in the population of a globular, and there are small stars down to the size of white dwarfs. The solution is to stretch the brightness values (histogram) so that the middle ranges are compressed, and the dim and bright ranges are emphasized. This makes dim stars visible, and the bright core retains detail.

SECTION 2: GLOBULAR CLUSTERS

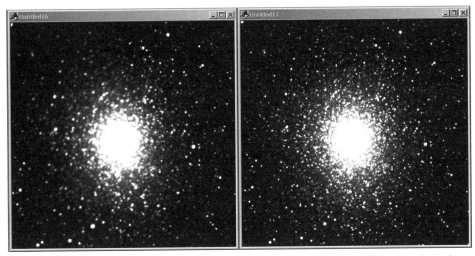

FIGURE 9.2.4. GLOBULAR IMAGES REQUIRE EXCELLENT SEEING AND STEADY CONDITIONS (RIGHT), OR POOR RESOLUTION (LEFT) WILL RESULT.

For most globulars, the default values for digital development will do a decent job. Digital development includes some sharpening. I prefer a minimal amount of sharpening on globulars. This will vary with personal taste. As always, there is no one right way to process any object.

Use the Process | Digital Development menu item in MaxIm DL to open the dialog shown in figure 9.2.5. (See chapter 8 for more information about digital development.) For globulars, the Kernel - Low-Pass filter usually offers the best balance of sharpening and histogram scaling. Make sure that the Auto checkboxes are checked. Figure 9.2.6 shows the preview for the FFT - Low-Pass filter with a Mild setting. Note the very different character of the core of the globular that results from the additional sharpening and smaller histogram adjustment.

The simplest way to display the full range of stars in a globular is to use digital development (DDP). This applies sharpening and a non-linear histogram stretch. This reveals dim details without sacrificing clarity in the bright portions of the image. The weakness in this approach is that you can wind up with dark circles around the stars if the sharpening is too aggressive.

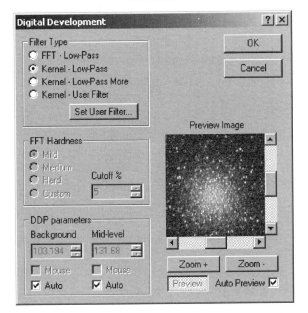

FIGURE 9.2.5. PERFORMING DIGITAL DEVELOPMENT ON A GLOBULAR IMAGE USING MaxIm DL.

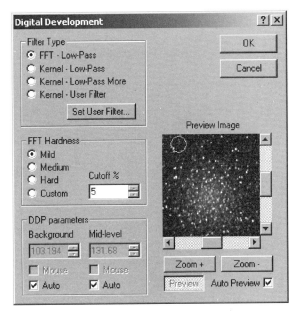

FIGURE 9.2.6. THE FFT - LOW PASS FILTER OVER-SHARPENS MOST GLOBULAR CLUSTER IMAGES.

THE NEW CCD ASTRONOMY

Levels and Curves in Photoshop

Photoshop and other image editing programs such as Picture Window Pro provide useful tools for processing images of globulars. I use traditional image editing tools to get the most out of my own images. These tools give you a very high degree of control over the appearance of the final image. My favorite way to process globulars is to use the Levels and Curves features of Photoshop to adjust the histogram, and then to do any additional processing that might be required for a given image. Compared to digital development, this separates histogram adjustments from sharpening (or anything else you care to use), and allows a finer degree of control over the final appearance of the image.

Digital development limits your options and doesn't provide adequate control. Using Levels and Curves for histogram adjustments in Photoshop (or the equivalent in other image editors), you gain complete control over the presentation of the image data.

Figure 9.2.7 shows a typical raw image of a globular, with just the core stars showing. A good way to move such an image into Photoshop is to save it as a 16-bit TIF file with your camera control program. Don't con-

FIGURE 9.2.7. A TYPICAL RAW GLOBULAR IMAGE.

vert it to 8-bits yet. You'll need all 16 bits for best results with histogram adjustments.

When you load the image into Photoshop, the globular will be dim, or even invisible. The Image | Adjust | Levels menu item displays the Levels dialog (figure 9.2.8). Use it to lower the white point (see chapter 8). This reveals the core stars. Lower the white point as far as you need to. You can apply the Levels adjustment more than once if necessary. To move the white point, click on the right-most triangle below the histogram, and drag it to the left as shown in figure 9.2.8. Keep an eye on the image so you can see what effect lowering the white point has on it. Your goal is to get to the point where the core stars are visible, roughly as shown in figure 9.2.7. Don't go so far that the core stars start blending together. Stop at the point where they are still individual stars.

Use the Image | Adjust | Curves menu item and adjust as shown in figure 9.2.9. This will brighten the image overall, and will make it easier to see both dim and bright

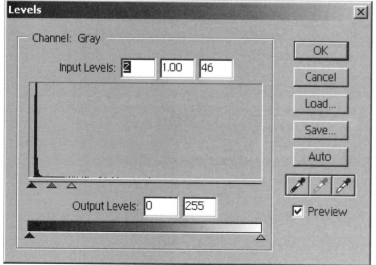

FIGURE 9.2.8. LOWERING THE WHITE POINT WITH THE LEVELS DIALOG.

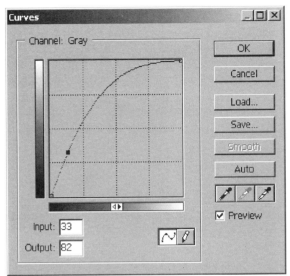

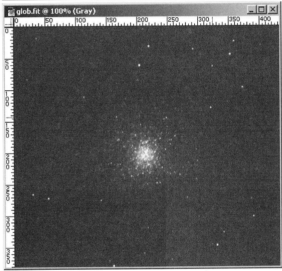

FIGURE 9.2.9. ABOVE: A CURVE THAT BRIGHTENS.

FIGURE 9.2.10. BELOW: THE RESULTS OF THE CURVE.

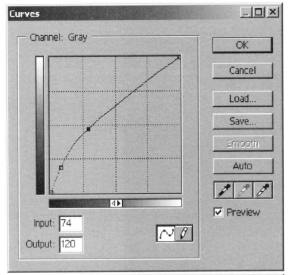

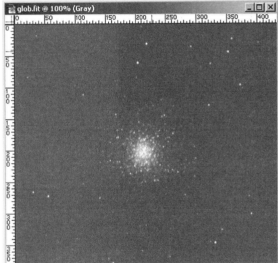

FIGURE 9.2.11. ABOVE: A MODIFIED BRIGHTENING CURVE.

FIGURE 9.2.12. BELOW: THE RESULTS OF THE CURVE.

areas at the same time. Don't allow the top right area of the curve to go straight. Use the Curves tool more than one time if necessary (see hints below on multiple use of Curves).

Figure 9.2.10 shows that the core and periperhal stars are brighter, as is the background. There is a better balance between the core and the outlying stars. The curve shape shown in figure 9.2.9 is good for generally brightening an image. Applying this type of curve multiple times leads to an over-bright core.

The curve in figure 9.2.11 has two points. Add the lower left point first, then add the upper point to flatten out the right side of the curve. This curve brightens more strongly in the dim areas, and less so in the already bright areas. The second point reduces image contrast of the image, so use this type of curve sparingly. It will lead to a washed-out image if overused.

After a few applications of the Curves tool, the background of the image may brighten considerably as shown in figure 9.2.12. The brighter your sky glow, the

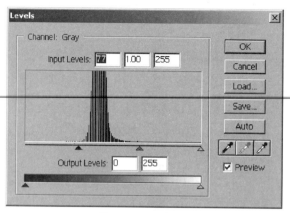

FIGURE 9.2.13. RAISING THE BLACK POINT

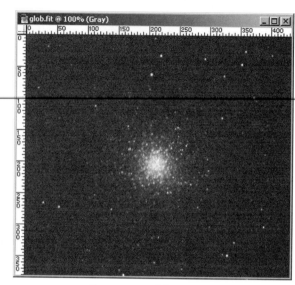

FIGURE 9.2.14. MORE STARS ARE NOW VISIBLE.

more likely it is that this will happen. A crescent moon is responsible for the skyglow in these images. This single image has faint horizontal noise. Combining multiple images would reduce the noise.

It's easy to clean up the excess brightness in the background. Open the Levels dialog again, and move the black point to the right (the black triangle at the left of the histogram). Figure 9.2.13 shows the black point raised just to the left of the main peak of data. This is a typical position for the black point while you are still making changes to an image. To make the background very dark, you'll need to move the black point closer to the midpoint of that peak.

If there is a gradient in the image the slope on the left side of the peak will not be as abrupt as in figure 9.2.13. If you see a gradual slope on the left side of the main peak, an unusually broad peak with a sloping top, or if you see multiple peaks, a gradient is often the reason. See chapter 6 for gradient removal tips.

Figure 9.2.14 shows the result of reducing the background brightness. I generally don't go after the background aggressively until I'm done making adjustments using Curves. This prevents accidentally setting the black point too high early in the process. Your last adjustment can be to raise the black point to get a good-looking background. In this example, the background is noisy, and the black point will need to be raised to hide the noise. You can also hide the noise using Photoshop's Curves function, but it's much easier to do precisely with Levels.

You can continue to apply Curves until you reveal all of the dim stars in the globular. When additional use of Curves either makes the image look too forced, or fails to reveal any additional stars in the globular, you are done. Use Levels as needed to raise the black point to keep the background reasonably

FIGURE 9.2.15. FINAL ADJUSTMENT TO LEVELS TO DARKEN BACKGROUND.

SECTION 2: GLOBULAR CLUSTERS

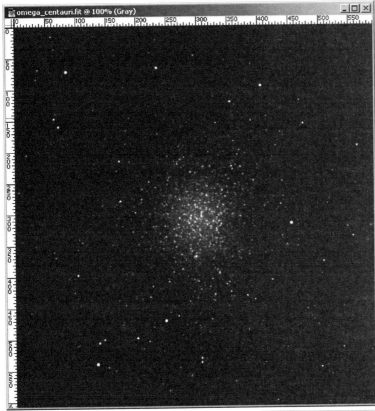

FIGURE 9.2.16. OMEGA CENTAURI WITH DEFAULT HISTOGRAM SETTINGS.

Figure 9.2.16 shows Omega Centauri, a large and famous southern hemisphere globular. The image is at the same scale as the M13 image, but the larger size of this glob yields a larger initial core. Despite the difference in size and brightness, the problem remains the same: if you set the black and white points to show the core, most of the stars in the globular remain invisible. They are there, in the image data, but you can't see them.

Figure 9.2.17 shows the result of applying Levels and Curves to bring out the additional stars. This technique is useful for bringing out dim details in a wide variety of extended objects with bright cores, including galaxies and nebulae.

dark. You can also use Curves to do the same thing, but this can also affect the midtones and bright areas in the image at the same time, and is therefore harder to control. However, it may be the best solution for a very noisy background.

Figure 9.2.15 shows the result of simply raising the black point to overcome the noise in the image. Some of the very dimmest stars in the image become marginally visible -- but that is the price of not having a superb signal to noise ratio in this image. Combining images, or going to longer exposures (as long as you don't burn out the core) are ultimately the best solutions to the noise problem in images like this one.

FIGURE 9.2.17. OMEGA CENTAURI WITH A FULL RANGE OF STARS VISIBLE.

How Much Sharpening?

None of the images of globular clusters presented so far in this section have been sharpened. The images were well-focused, and taken at image scales that reduced the impact of seeing conditions. The image scale for all of the images was 3.51 arcseconds per pixel. Images taken at smaller image scales, particularly under about 2.25 arcseconds per pixel, will often show softer stars. Deconvolution and unsharp masking are available to sharpen up the image. Even an image like figure 9.2.17 can benefit from a small amount of sharpening.

Sharpening can both improve and harm an image, but this is especially true when working with images of globular clusters. Sharpening can help by making the stars smaller and clearer, but it can also worsen the image by creating artifacts or creating detail where there really isn't any. The worst sharpening problem for a globular are the dark halos that many sharpening routines create around stars. Because globulars are so densely packed with stars, the halos obscure other stars and create a false appearance quickly.

The most useful sharpening tool for globular images is a mild unsharp mask. In Photoshop, you can set the amount of the unsharp mask to 50% or less, and use a radius under 1.5 pixels. You will usually set the Threshold to zero to allow sharpening on even the tiniest stars.

FIGURE 9.2.18. SHARPENING IN PHOTOSHOP USING SETTINGS OF 50%, RADIUS 1.5, THRESHOLD 0.

If the image is noisy, you can use a threshold of 1 or 2 to avoid sharpening the noise.

These settings provide a small amount of sharpening, as shown in figure 9.2.18. Compared to figure 9.2.17, there is slightly better definition, especially in the core.

How much sharpening to use is a matter of personal preference. Figure 9.2.19 shows two examples of harsher sharpening. The left image is about as far as you can go and still maintain a reasonably natural appearance. The right image loses the balance between a bright core and dimmer peripheral stars.

Globulars seem like simple structures, but very small adjustments to your sharpening routine can have a major impact on the final results.

FIGURE 9.2.19. LEFT: MILD SHARPENING. RIGHT: TOO MUCH SHARPENING MAKES STARS LOOK UNNATURAL.

Section 3: Galaxies

Galaxies have several characteristics that can get you in trouble when you try to image them:

- Many, but not all, galaxies have a very bright core.
- There are bright details, such as HII regions and groups of hot young stars in the spiral arms.
- There are dim areas between the arms and often well away from the core. Dust lanes, which can occur anywhere, vary greatly in brightness.

Galaxies represent a huge range in brightness, and there is both very small-scale and very large-scale detail to be resolved. In many ways, acquiring a good image of a galaxy is among the hardest tasks an imager faces.

Even so, you can resolve hints of galaxy structure with short exposures. Figure 9.3.1 shows four galaxy images taken with an ST-7E camera during my first year of CCD imaging. At upper left is an image of NGC 7441, a galaxy in Aquarius. This is a fairly dim galaxy, and this single two-minute exposure was taken with a Mewlon 210. Two minutes is just enough to show a strong central bar and hints of the spiral structure. The dim portions of the galaxy are grainy, indicating that the signal to noise ratio is very low. A much longer exposure would be needed with the f/11 Mewlon to reduce the noise.

NGC 772, at upper right, was imaged with the same setup for 10 minutes. A hint of a long spiral arm can be seen at lower right of the galaxy. The signal to noise ratio is good, and the graininess of the NGC 7441 image is nowhere in evidence. NGC 7640, at lower left, was imaged for five minutes, and shows a level of detail and noise about midway between the first two examples. This is reasonable; the exposure is also midway between the others. NGC 514, at lower right, is a ten-minute exposure that shows lots of graininess because it is a dim galaxy. No single exposure time is

FIGURE 9.3.1. VARIOUS DIM GALAXIES.

ideal for all galaxies. The dimmer the galaxy, the longer the exposure you need to get a good image.

The examples in figure 9.3.1 are all small, dim galaxies. Even a ten-minute exposure is too short to show a lot of detail. Many of the bright Messier galaxies can be imaged successfully with shorter exposures. Figure 9.3.2 shows a bright galaxy (M51) that has plenty of exposure time as a result of combining multiple exposures to help reduce noise. The image, taken with an FSQ-106 refractor and an ST-8E camera, has been processed to show both faint and bright details clearly.

At the other extreme, insufficient exposure on galaxies can be extremely frustrating because the image only hints at what is really there. Figure 9.3.3, also an image of M51, shows very little of the detail in the galaxy, and the image is very noisy. The difference between figure 9.3.2 and figure 9.3.3 is exposure time.

If you want the best possible galactic images, here are the things you need to do:

Take long exposures. Your maximum exposure length for a single exposure will be limited by things like blooming and skyglow. For the ultimate galaxy image, take an hour or two total exposure. Camera type (ABG

FIGURE 9.3.2. LONG OR COMBINED EXPOSURES REVEAL A LOT OF DETAIL IN THIS IMAGE OF M51.

or NABG), pixel size, and focal ratio will affect this. The darker your skies, the shorter your exposure can be. If your camera has a small dynamic range or limited bit depth, such as the ST-237, you will need to take a larger number of short individual exposures.

Take multiple exposures. Three is the absolute minimum, and five, seven, or even more exposures help to reduce noise when combined. Fewer images have noticeably more noise, while more images decrease the amount of noise in the final image.

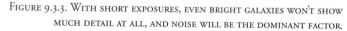

FIGURE 9.3.3. WITH SHORT EXPOSURES, EVEN BRIGHT GALAXIES WON'T SHOW MUCH DETAIL AT ALL, AND NOISE WILL BE THE DOMINANT FACTOR.

Process your images to hide the defects and bring out the good stuff. I usually apply a smoothing (blur) filter to the background and dim areas of the galaxy to reduce grain. You can sharpen the bright core region, as well as spiral arms and HII regions. Digital development is one option for galaxy images, but you can probably get the best balance with manual histogram adjustments.

When working with galaxies, you'll want to hone your histogram skills. The large brightness range of most galaxies requires creative use of non-linear histogram stretches.

Exposure Guidelines

As described in chapter 4, exposure is dependent on the focal ratio of your telescope. For a given CCD camera, no matter what your aperture is, the right exposure length is determined by the focal ratio alone.

If you switch to a different camera, of course, that camera's quantum efficiency, pixel size, and well depth may be different. That would affect your exposure time. Since galaxies almost always have some extremely dim areas and some extremely bright areas, the issue is balance. You want exposures that are long enough to show the dim details, yet short enough not to burn out the core of the galaxy.

Other issues can limit maximum exposure time:

- Skyglow can cause saturation on cameras with a limited well depth or dynamic range. The faster the focal ratio, the more likely you are to run into this problem. For example, with an ST-237 on an f/2 Fastar, a two-minute exposure can saturate the chip from a suburban location.
- If you have a non-antiblooming CCD detector, you need to decide how much blooming, if any, you want to deal with. There are processing tricks for removing blooming spikes, but you can't recover the data lost in the areas where blooming occurs. You can edit out the bloom to a neutral or matching background. If you have the right software, such as Registar or StellaImage 3, you can rotate the camera between exposures and combine the rotated images to remove blooming.
- Saturation of bright stars can lead to false or missing color in stars. You need to make a choice: do you go with long single exposures for the sake of dim detail, or do you limit exposure length (and decrease convenience) to help preserve star color?
- Longer exposures entail more risk. Examples include cosmic ray hits, clouds moving in, satellites, etc. You could trip on a tripod leg, or a tree branch might come into the field of view. A neighbor might turn on a porch light. Airplanes could fly through the field of view, not to mention the risks of meteors and fireflies! If you don't mind taking an extra 30-minute exposure now and then, then this isn't an issue.

Despite all the hazards, there is one key benefit of long exposures: better signal to noise ratio. The gain will sometimes be a small one, but when it comes to faint detail in galaxies, the reward goes to the longer exposure. If you take many shorter exposures, you can get excellent signal to noise if you take enough of them. However, many long exposures will still win out.

In other words, if you compare a combination of 12 five-minute exposures with a combination of two 30-minute exposures, you will find less noise in the image that uses the longer exposures. On the other hand, if an airplane flying through the heart of your galaxy ruins one of those 30-minute exposures, and one of the five-minute exposures is similarly ruined, the 11 five-minute exposures will have better signal to noise than the single 30-minute exposure.

So the real trick for imaging galaxies is to determine what your longest possible exposure is for a given galaxy. You need to weigh the factors that apply to your situation:

- At what exposure length does the core start to burn out? If the galaxy is dim enough, and your focal ratio is slow enough, this may not occur often.
- At what exposure length does the skyglow saturate the detector? This will vary with sky conditions, pixel size, quantum efficiency, focal ratio, etc.
- When and where does blooming occur, if any, and how willing are you to deal with it?
- How much time and energy are you willing to spend imaging the galaxy?

You can then create an imaging plan for your galaxy of choice on any given evening. Consider the following example. Suppose your goal is to image the galaxy M81. M81 has a bright core, and the spiral arms are much dimmer. Imaging with a 5" f/8 refractor, there really isn't enough light to burn out the core when using an NABG camera. Blooming will occur in the bright nearby stars in 5 to 8 minutes. With an NABG camera, that would be the limit of your single exposures. With an ABG camera, you could use 30-minute exposures, and face the risk of possibly burning out the core of the galaxy. Some test exposures of varying duration would help you define the exact trade-offs and enable you to make a decision on the exposure lengths that will work best.

Getting Better Signal-to-Noise Ratios

Figure 9.3.5 shows a two-minute exposure of M81 using a 5" f/6 refractor with an ST-7E camera. The image squeezes every possible amount of detail out of a single two-minute image. The core isn't burned out; stars aren't bloomed. And there is some detail in the spiral arms, but considerable noise, too. A bit of dust lane detail can just be made out near the core.

Now let's look at a 10-minute exposure at f/8 (see figure 9.3.4). Despite the difference in focal ratio, the result is still instructive. The longer focal ratio means that we don't quite have five times the exposure, but the exposure is still several times the level of that in figure 9.3.5. Both images have been processed in Photoshop using the techniques described later in this section.

The differences are striking. The dust lane detail near the core is easier to see in the 10-minute image. Detail in the spiral arms is much clearer, and the image overall has better contrast. The noise level is also lower. One star at lower left has bloomed.

Enlargements of similar sections of these images show the differences. Figures 9.3.6-7 shows 3x blow-ups of the images. Three arrows point to the same features of M81 in both enlargements.

The arrows point to, from left to right:

- Faint area of hot young stars in a spiral arm
- HII region
- A dust lane

Compare the appearance of these features in the enlargements. The area of hot young stars is just visible in the ten-minute image, but it can't be found in the two-minute image The HII knot is visible in both, but is

FIGURE 9.3.4. ABOVE: A TEN-MINUTE EXPOSURE OF M81 WITH A 5" REFRACTOR AT F/8.

FIGURE 9.3.5. BELOW: A TWO-MINUTE EXPOSURE OF M81 WITH A 5" REFRACTOR AT F/6.

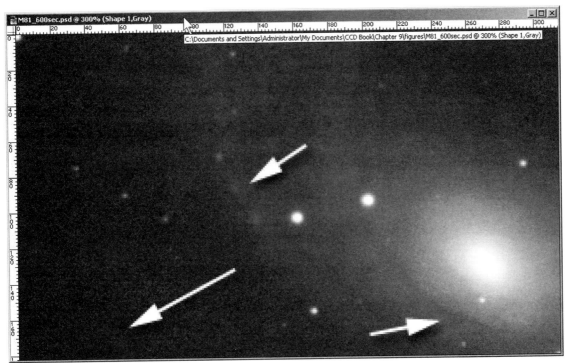

FIGURE 9.3.6. A DETAIL OF THE TEN-MINUTE M81 IMAGE.

obscured by noise in the shorter exposure. The dust lane is crisp in the ten-minute image, but buried in noise in the short exposure.

Although the 10-minute image shows lower noise, noise is still readily apparent. I know that when I first started imaging, it was a big deal to take exposures for a full minute. It was a little disillusioning to learn that even five and ten-minute exposures have limitations. That's where combining images comes into play. What can you accomplish by combining images? How much improvement can you realize from combining a bunch of shorter exposures? Let's take a look.

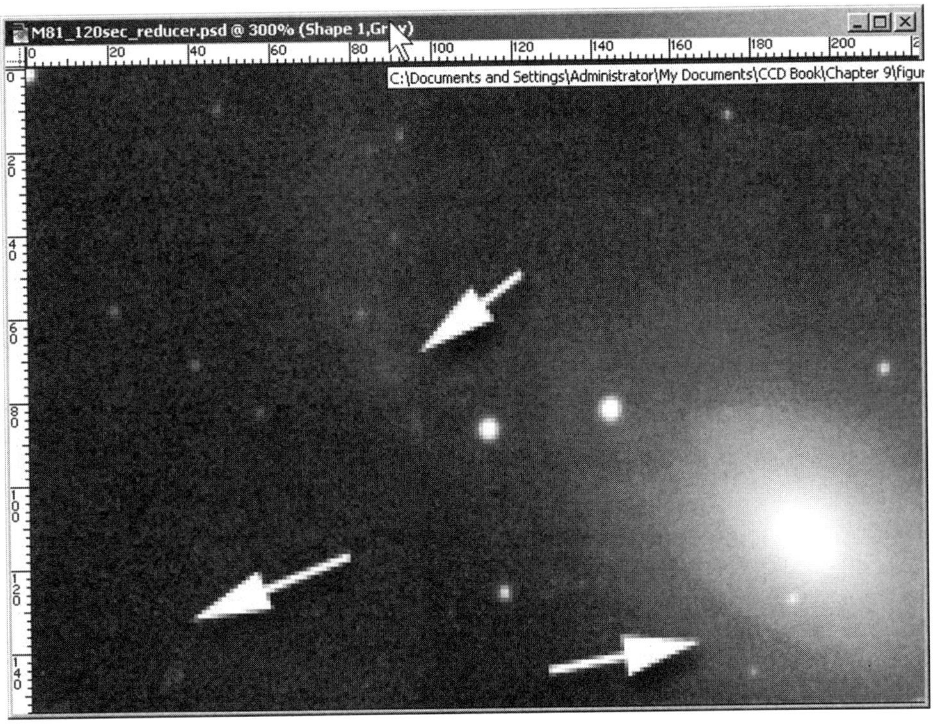

Figure 9.3.8 shows a detail from an image that is

FIGURE 9.3.7. A DETAIL OF THE TWO-MINUTE M81 IMAGE.

a combination of ten two-minute exposures (total exposure time: 20 minutes). There are definite improvements over figure 9.3.7. The faint galaxy arm at lower left is just barely visible. The HII area is better defined and the dust lane is less noisy. The entire image is less noisy than the single ten-minute exposure in figure 9.3.6. The bottom line is that combining images is very effective at reducing noise. But figure 9.3.6 still has a bit more contrast, especially in the dust lanes. I give the edge to the combined image in the dimmer areas, however.

Let's take this to its logical conclusion: what happens when we combine some ten-minute images? We get the best of all possible CCD worlds: high signal level, low noise level, and better detail in the dim areas. Figure 9.3.9 shows that the galaxy arm at lower left is clearer. The HII regions have visible structure and detail. And the dust lanes aren't just present; they show additional detail that hasn't been visible in the other images. In addition, fine structural detail is now visible in the arms of the galaxy. Look at the area above the core, which shows smaller knots of stars. The bright arm to the left of the core now shows fine detail as well.

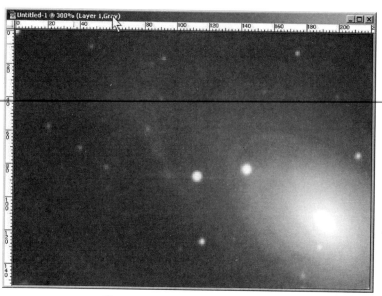

FIGURE 9.3.8. COMBINATION OF 10 TWO-MINUTE EXPOSURES.

When you have an excellent signal to noise ratio, you can process your image more effectively. Image processing amplifies both signal and noise. The better your signal, and the lower your overall noise, the more effectively you can process your images. A noisy image is all too easy to over-process. An image with a high signal to noise ratio, on the other hand, takes processing well and you can push a little harder with your processing techniques.

Figure 9.3.10 shows the effect of applying an unsharp mask to the image in figure 9.3.9. The sharpening brings out additional dust lane details. Detail in the arms is even crisper that it was before sharpening. Dim areas, however, are noisier.

Figure 9.3.11 shows the combined image sharpened. The improvements are even more obvious, demonstrating that a combination of long images is the best possible input for processing.

FIGURE 9.3.9. THE COMBINED 10-MINUTE IMAGES SHOW BETTER CONTRAST AND DETAIL.

FIGURE 9.3.10. APPLYING AN UNSHARP MASK BRINGS OUT EVEN MORE DETAIL.

FIGURE 9.3.11. THE COMBINED IMAGES SHOW LOTS OF DETAIL WHEN SHARPENED.

Dynamic Range

Dynamic range determines the number of brightness levels that your CCD camera can record with respect to the noise levels. A camera with a large dynamic range can simultaneously record useful detail across a wide range of brightness levels, with small steps (defined by the noise level) that provide low noise and greater precision. It also means that the camera can clearly show extremes of darkness and brightness. A camera with limited dynamic range will typically burn out the bright areas unless you shorten your exposures. The greater the dynamic range of your camera, the greater the range of brightness levels you can record, and the greater the precision with which you can record them.

TIP: Correcting a blooming star can be tedious work, but it gets easier once you learn the basic moves. I have created a GIF animation that shows blooming correction in action. The GIF, and a detailed explanation of how I manually corrected the blooming of the brighter star at lower left of the combined ten-minute M81 image, is located on the book web site:

http://www.newastro.com/newastro/book_new/anim/debloom.asp

This range of brightness levels is not the same as the saturation level of your camera. Two cameras can have the same saturation level, but very different dynamic ranges. Dynamic range is measured in steps. The noise level in the camera defines the step size. For example, if your camera has 10,000 brightness levels available, and the noise is +/-3 brightness levels, then the dynamic range is 10,000 divided by 6, or 1,666.

The technical definition of dynamic range is the signal divided by the noise. According to Michael Barber of SBIG, in general you can estimate dynamic range by taking the full well capacity of the CCD detector and dividing it by the read noise. This assumes that bias, dark, and flat-field frames are applied correctly. You can obtain the read noise from the camera specification on the manufacturer's web site. Poor sky conditions, of course, can reduce the dynamic range by adding noise. In any real-world situation you would need to take sky conditions and any other sources of noise into account to determine your actual dynamic range for a given imaging session.

Skyglow is a major factor in reducing dynamic range. From a dark-sky site, you might have a skyglow of a few hundred ADU for a given exposure duration. From a suburban location near a streetlight, you might have skyglow of tens of thousands for the same exposure length. If 80% of your well depth is eaten up by sky glow, you should use 20% of your well depth and then divide that by the read noise to get a more realistic number for your dynamic range.

Let's look at an example. Consider the ST-7E non-ABG camera from SBIG. It has a full well capacity of 100,000 electrons. The read noise is 15 electrons RMS. The dynamic range is a healthy 6,666. If skyglow uses up 50% of the available brightness levels, then the dynamic range is 50,000/15, or 3,333. The reduction in dynamic range will affect the quality of your images. It won't stop you from imaging as it would with film, but it will limit the level of quality you can obtain for a given expsure length.

The ST-237 has a full well capacity of 20,000 electrons, and a read noise of 17 electrons RMS. This yields a smaller dynamic range of 1,176. Skyglow that is at 50% of saturation level would lower the dynamic range to hlaf of that, a very small 588. This is why it is so much harder to get good S/N when imaging under a bright sky. There is no substitute for dark skies. However, a large dynamic range gives you the best results.

> **TIP:** The number associated with the dynamic range is a ratio. For example, the ST-237 has a dynamic range that is 1,176 times larger than it's read noise. This number is not the same as the number of brightness levels.

From these numbers, we can reasonably expect to see more detail in the core of a galaxy when taking long exposures with an ST-7E camera compared to the ST-237. The ST-237 will be more likely to burn out the core, and thus shorter exposure times will be necessary to preserve bright details. If skyglow is present, the dynamic range can shrink to the point where it is no longer possible to capture the full range of brightness levels in some objects.

Galaxy Processing Tips

I do most of my Galaxy image processing in Photoshop. However, I will often perform either a quick histogram stretch (lower the white point) or non-linear stretch (digital development) in a camera control program to get an idea of what the galaxy might look like after processing. Digital development is a quick and dirty test for data quality. Most of the time, I can get a smoother, more eye-pleasing result by processing the image without digital development. The following section shows how I processed a raw image of M81 (see figure 9.3.12). The basic processing steps apply to any galaxy, but the numeric values depend on the noise level and brightness levels in the image. These techniques also work well on bright nebulae.

Prior to processing the image in Photoshop, I saved it as a 16-bit TIFF file. I prefer to start processing initially with 16-bit files, even though some features of Photoshop are unavailable when working with 16-bit files. The most critical features are available, and you can switch to 8-bit mode for further processing after the critical steps are completed. Photoshop 6.0 has more 16-bit tools than previous versions, and this makes it a good upgrade for astronomical imagers.

Photoshop Processing

I like to do as much of my image processing as possible in Photoshop. I used versions 5.5 and 6.0 while writing this book, and I highly recommend Photoshop 6.0. Not only does Photoshop have a lot of very useful tools, it also has tools that are very convenient to use. It's clear that a lot of thought when into the user interface. When you are making adjustments to your image in Photoshop, you have a high degree of control over the outcome.

FIGURE 9.3.12. OPENING THE TIFF FILE IN PHOTOSHOP.

Photoshop has zillions of options, however, so it takes a long time to learn all of the ins and outs. It's not an overstatement to say that it takes about a year to get really comfortable with Photoshop.

The basic steps in this tutorial include:

1. Adjust black point, white point, and perform a histogram stretch
2. Smooth the background and the dim/noisy portions of the image
3. Sharpen the areas that have sufficiently good S/N (signal to noise ratio)

If you want to try these steps yourself, you can find the original TIFF file on the book web site here:

http://www.newastro.com/newastro/book_new/samples/c9_m81_sum.tif

Figure 9.3.12 shows the starting point in the process: the raw TIFF file. The background/range (or min/max in MaxIm DL) of the image are not used when you open the image in Photoshop. Your first tasks are to set the black and white points, and to adjust the histogram to show the data to best advantage.

TIP: If you don't see the glow immediately around the core of the galaxy, use the Image | Adjust | Levels menu item to lower the white point. The more glow you get using Levels, the higher the contrast will be in the final image and the dimmer the darker details will be. If you start early with Curves, the contrast will be lower (the image will look flatter), but dim details will be seen more easily.

If you aren't familiar with the Photoshop Curves tool, please see the introductory material in chapter 8 before you follow the rest of this tutorial. Curves is a non-linear histogram stretch and it is a surprisingly sophisticated tool. You can make major changes to the quality of your images with Curves, for better and for worse! You will see some new curves here that expand on and add to the curves covered in chapter 8.

Figure 9.3.13 shows the first step in bringing out some detail. I used the Image | Adjust | Curves menu item to open the Curves dialog. Take note of the orientation of the grayscale bars at the left and bottom of the grid. If your grayscale bars are the opposite of what is shown, click on the double-headed arrow in the middle of the lower bar to make yours the same as in figure 9.3.13. This will make it easier to follow along.

I seldom make changes to Curves using the numeric values. I evaluate the image by eye as I move the curve. Make sure Preview is checked so you can see the effect of the changes as you make them. The Input/Output values are not critical. The general idea at the start is to boost the dim areas more than the bright ones. Keeping the point on the line in the left quarter of the grid, as shown in figure 9.3.13, should do the trick nicely.

Click OK to save your change. Unlike many histogram changes in camera control programs, this is a permanent change. The data in the image is altered. If you make a mistake, use the Undo feature to get rid of the change.

After making the change to the curve, the background is brighter, the core is brighter, and the area immediately surrounding the core shows some detail. You may see a hint of some arm detail, depending on the shape of your curve. The area outside the core is the main target of this exercise.

Typically, I do 2-5 rounds of Curve adjustments. If the background gets too bright, I adjust Levels as needed. The first adjustment (figure 9.3.13) helps, but there is more detail to come. Perform another Curves adjustment (see figure 9.3.14). This curve looks a lot like the one from figure 9.3.13 because the goal is the same. The background

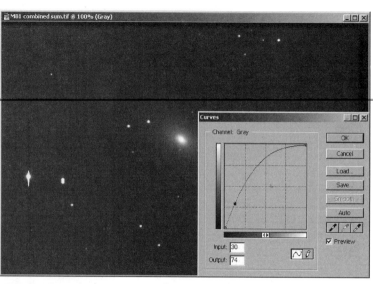

FIGURE 9.3.13. TOP: FIRST ADJUSTMENT TO THE HISTOGRAM CURVE.

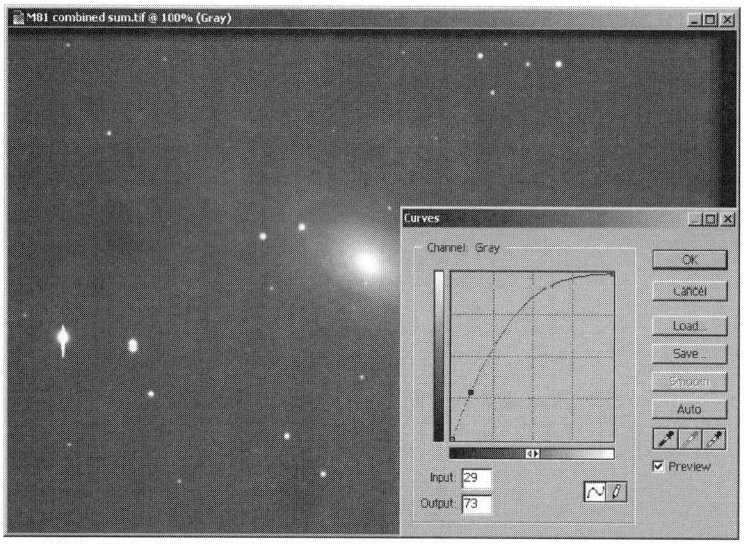

is very bright, so it's hard to tell how much detail there really is. Raise the black point with Image | Adjust | Levels to darken the background (see figure 9.3.15). Leave the new black point a little to the left of the main peak. The small peaks reflect differences in background brightness, and they are common when you combine images. I set the black point by eye, using both histogram and image to determine the right setting.

With the background cleaned up and contrast restored, the level of detail visible in the galaxy is easier to evaluate. There's no magic number that tells you when to stop using Curves. If the image starts to develop fully-white, burned-out areas, you've gone too far. Whenever possible, continue using Curves until the noise in the dim areas begins to be objectionable. You'll learn when to stop based on experience with these tools.

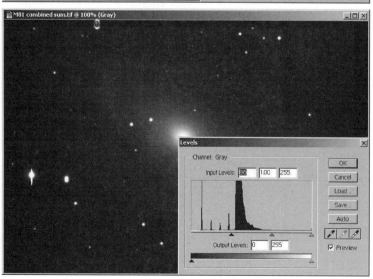

FIGURE 9.3.14. ABOVE LEFT: SECOND ADJUSTMENT TO THE CURVE.

FIGURE 9.3.15. LEFT: LOWERING THE BACKGROUND BY RAISING THE BLACK POINT.

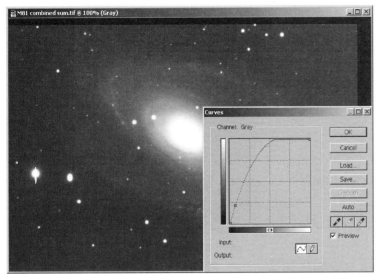

dim areas are aggressively brightened. A side effect of this second point is that the overall contrast of the image is lower. There is only so much control you can achieve with this technique. If you overuse it, the image contrast will be too low to be attractive.

The aggressive brightening has also brought up the background. This calls for another Levels adjustment. Figure 9.3.18 shows the new black point (36). Use your eye, rather than the number, to get close to the base of the large black area in the histogram. At this point, we can see lots of dim detail, and the image is much improved.

FIGURE 9.3.16. TOP: THIRD ADJUSTMENT, PART ONE.

TIP: Your best bet is to go too far. You can still get an excellent result by darkening the image later. By going a little too far, you learn where the noise level is. This allows you to apply smoothing more accurately.

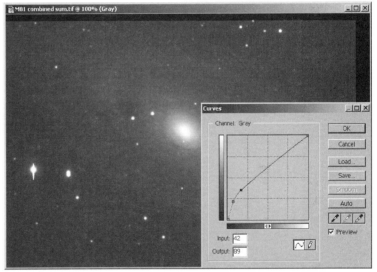

Figure 9.3.16 shows a point added to the curve that forces it up against the top boundary of the grid. This is an aggressive attempt to brighten the dimmest areas of the image. The flat part of the curve is clipped, which wil burn out the bright core. Compare the size of the core in figures 9.3.15-16. Note how much larger it is in figure 9.3.16. The lesson here is that if the right-hand side of the curve gets too high, you will eventually lose data in the bright portions of the image.

A simple change will tame the curve: add a second point. Click on the curve above and to the right of the first point to create a new point. Drag the new point downward a bit to flatten the top of the curve, as shown in figure 9.3.17. The core retains detail, and the

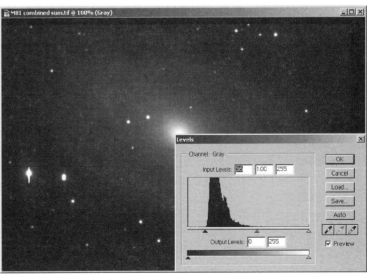

FIGURE 9.3.17. ABOVE RIGHT: THIRD ADJUSTMENT, PART TWO.

FIGURE 9.3.18. ANOTHER BLACK POINT CHANGE.

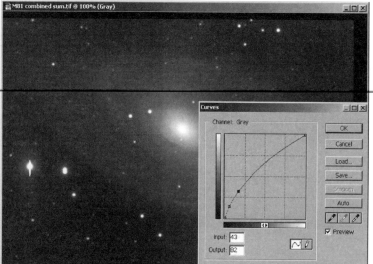

FIGURE 9.3.19. FOURTH CURVE ADJUSTMENT (TWO POINTS).

FIGURE 9.3.20. THIRD BLACK POINT ADJUSTMENT.

dim to be useful in the image. This increases the portion of the image devoted to stuff we are interested in. The black data peak is where most of the useful data is. The widening of this area as the changes proceed is a good thing. It gets easier to find the exact right black and white points for the image.

The galaxy image as it appears in figure 9.3.20 is just about right. Now comes the time to push it a little too far and reveal the noise. A Gaussian blur is then just the ticket for smoothing out the noise.

TIP: If going too far makes you uncomfortable, and there is a small cost in lost data, there is an alternative. In Photoshop 6.0, take a Snapshot before you "go too far." Examine the noise. Make your decision about how much smoothing to apply. Finally, go back to your snapshot and apply the smoothing.

Figure 9.3.21 shows one approach to taking the image too far. It minimizes the harm and adds a boost to the midrange at the same time. The Curves adjustment uses three points, and the shape of the curve is different

There is still more data to mine in this image. A less aggressive two-point adjustment (see figure 9.3.19) will do the trick. Technically, you could set a large number of points and carefully adjust them to perform the histogram changes in one step. However, you get much more control over the process by doing it in small steps.

The Curves adjustment shown in figure 9.3.19 again increases the brightness of the background, so yet another black point adjustment is called for, as shown in figure 9.3.20. Note that the data peak in the histogram is spreading out. Now that the adjustments are nearly done, it is time to start pushing the black point into the data peak. The adjustments have removed more and more values that are either too bright or too

than what we've been using so far. The very dimmest areas (leftmost point) are significantly reduced in brightness. The background gets much darker, but not black. This pretty much eliminates the signal and the noise in the background. The second point, just above the first one, slightly brightens dim areas of the galaxy. The combination of these two points increases the contrast between the dim galaxy areas and the background, and is the most likely to reveal noise. The third point, at top right, brightens the core to maintain a visually appealing relationship between the core and the rest of the galaxy. The net effect of these points is to create a bulge in the middle of the curve. This broadly brightens most of the galaxy.

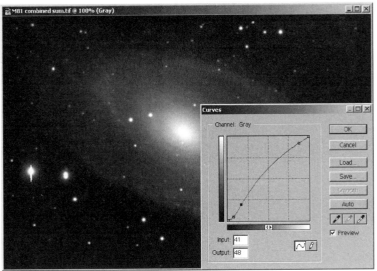

FIGURE 9.3.21. FIFTH CURVE ADJUSTMENT (THREE POINTS).

Figure 9.3.21 shows that the entire galaxy is bright. It's too bright, but that's the desired result for now.

This is the time to convert from the original 16-bit format to 8-bit. We need to use some tools that are not available when working with a 16-bit image. To convert the image to 8-bit, use the "Image | Mode | 8-bits per Channel" menu item. You won't see any change, but the internal precision of the image has dropped from 16-bit to 8-bit.

TIP: I always wait until I have completed the major histogram changes before changing to 8-bits per channel. Since the conversion represents a loss of precision, I prefer to wait until I have removed as much extraneous data from the image as possible, leaving just the good stuff. The cost of lost precision is lowest if you wait until the last possible moment.

Figure 9.3.22 shows the noise in the image. This is an enlargement of a dim area of the galaxy. There is a fair amount of grain, and one hot pixel as well. This hot pixel can be cleaned up at any point by using the Clone Stamping tool with the smallest available brush.

Examine the image by eye to see where the noise is a problem. Enlarge the view (press the Ctrl and Plus keys simultaneously) to see the grain more clearly. The dimmest areas are the noisiest. As you examine progressively brighter portions of the image, the contribution from noise becomes proportionally less There will be a certain brightness level where the signal is so strong that noise ceases to be a problem. Examine the image (enlarge as much as necessary) to see what this brightness level is. Since galaxies generally have a bright core and get dimmer further away from the core, you can fix your eye on a certain radial distance from the core where the noise ceases to be a factor.

Use the Select | Color Range menu item to open the Color Range dialog (see figure 9.3.23). A fuzziness setting of 25 is a good starting point. Click in the darkest area of the background to select it. Hold down the shift key and click in brighter and brighter areas until the mask (the red area) covers just the dim areas that are troubled by noise. This is a judgment call, and you will get better at determining the noisy areas with experience.

FIGURE 9.3.22. SAMPLE OF NOISE LEVEL IN THE IMAGE PRIOR TO SMOOTHING.

TIP: How do you evaluate whether you found the right brightness level where the noise ends? If smoothing/blurring eliminates too much detail, you've selected too much of the image. If objectionable noise remains after the smoothing, then you didn't go far enough. You'll develop a feel for the right spot with some practice. If you don't see a mask as shown in figure 9.3.23, use the Selection Preview drop-down at the bottom of the dialog to choose "Quick Mask."

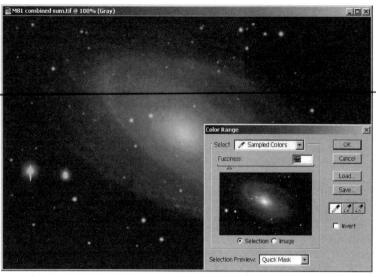

FIGURE 9.3.23. USING THE SELECT | COLOR RANGE TOOL TO SELECT JUST THE DIM PORTIONS OF THE IMAGE.

Figure 9.3.24 shows the selection that results from this maneuver. You can save the selection if you think you might want to operate on the dim areas at a later time. For example, you might want to invert the selection later to apply sharpening. Or you might want to do some additional blurring at some point. You can either save the selection (Select | Save Selection), or you can create a mask following these steps:

- Click on the Windows | View Channels menu item
- Click on the circled arrow at top right of the Channels palette, and then choose New Channel.
- Name the channel, e.g. Dim Areas. Click OK.

- The new channel becomes the active channel. It is completely black. This is normal.
- Use the Edit | Fill menu to open the Fill dialog.
- Select White in the Use drop-down list.
- Select Normal as the Mode, and 100% as the Opacity. Click OK.
- The channel fills with white based on the selection. Use it as you would use any mask in Photoshop.

If you just want to re-use the selection, save it as a selection. If you think you might have reason to modify the selection, save it as a mask. You can then paint on the mask or use tools to alter it before re-using it.

TIP: To avoid blurring the edges of stars, use the Select | Modify | Contract menu item to contract the selection by either 1 or 2 pixels.

I typically hide the selection boundary before working with a tool. This allows me to observe subtle effects at the boundary of the selected and unselected areas. If I don't like the way the boundary looks, I will undo the action, feather the selection as needed, and then try again.

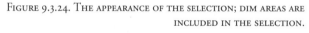

FIGURE 9.3.24. THE APPEARANCE OF THE SELECTION; DIM AREAS ARE INCLUDED IN THE SELECTION.

FIGURE 9.3.25. SETTING THE GAUSSIAN BLUR TO A 0.8 RADIUS.

The simplest smoothing tool in Photoshop for astronomical imaging is the Gaussian Blur. Use the Filter | Blur | Gaussian Blur menu item to access it. Figure 9.3.25 shows the results of a Gaussian Blur with a radius of 0.8 applied to the selected portion of the image. No detail is lost in the core because it is not inside the selection. There is less noise (grain) in the dim areas. A setting of 0.8 is a fairly strong blur, and you should try various blur settings from 0.3 to 1.0 to get a feel for what they look like.

TIP: You can also use the Despeckle tool to blur, but you can't vary the amount. Despeckling is useful if the noise level is too high for a Gaussian blur. The Dust & Scratches Filter is sometimes also a good choice, but if you go too far it will eliminate stars! Use Dust & Scratches with caution.

Figure 9.3.26 shows the noise reduction in the same area as figure 9.3.22. I used a 0.8 radius in this example, but you may need more or less smoothing to suit your image. Images with very little noise can be smoothed with a setting as low as 0.3. If the image is noisy enough, no amount of smoothing will be enough to solve the problem.

TIP: You can also apply smoothing in stages. For example, you can select just the darkest background and apply a light Gaussian blur, such as 0.3. You can then select the background and the dim areas of M81, and apply a second blur of 0.3. The background now gets the full blur (0.3 + 0.3 = 0.6), but the dim areas in the arms of the galaxy are more lightly blurred. I suggest you try this two-stage blur and compare it to the single-stage blur to see if you like it. You can also use three or more stages if appropriate.

The next step is to sharpen the brightest portions of the image. I prefer to avoid sharpening the stars at the same time. Create a selection using the Lasso or Polygonal Lasso tool as shown in figure 9.3.27. This encompasses the brighter areas of the galaxy, but none of the stars outside the object. A few stars covering the galaxy are included, but there is no way around that at this stage. If a selection exists before you open the Color Range dialog, the output selection is limited to the boundaries of the existing selection.

FIGURE 9.3.26. THE APPEARANCES OF THE SAME AREA AS FIGURE 9.3.22 AFTER SMOOTHING HAS BEEN APPLIED.

FIGURE 9.3.27. TOP: CREATING A SELECTION THAT ENCOMPASSES THE BRIGHTER AREAS IN THE GALAXY. SELECTION CREATED WITH LASSO TOOL.

FIGURE 9.3.28. MIDDLE: SELECTING THE BRIGHTER PORTIONS OF THE IMAGE.

FIGURE 9.3.29. BELOW: APPLYING AN UNSHARP MASK TO THE IMAGE.

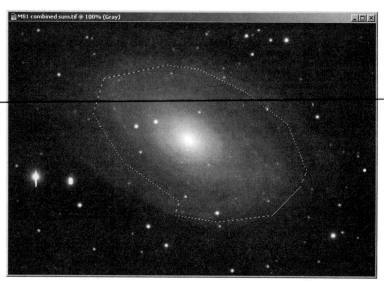

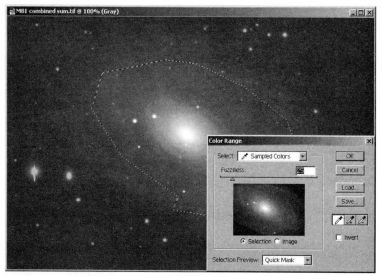

Use Color Range to select just the bright portions of the image. Figure 9.3.28 shows the range I chose. I started by clicking in the brightest part of the core. Then I held down the shift key and clicked outward, increasing the selection until the bright areas with very good signal to noise ratio were selected. Alternatively, if there is no detail to sharpen in the bright areas, you can select only the middle brightness levels. This will avoid sharpening stars, and allow you more flexibility is sharpening subtle details. If you see too many artifacts of sharpening, you've selected too much. If you don't get as much sharpening as you could, then you didn't go far enough.

Figure 9.3.29 shows the sharpening in progress. I almost always use Unsharp Masking (Filter | Sharpen | Unsharp Mask) instead of the other sharpening tools available in Photoshop. It provides a much finer degree

of control. I normally use an amount in the range of 40% to 50%, which avoids the worst sharpening artifacts. The radius is generally around 0.7 to 2.5 pixels. A noisy image will require a smaller radius. I like to use a threshold of 0 levels, but this has to be determined based on the quality of the image. A noisy image may require a threshold of 2 or 3 to get clean sharpening, but the higher the level, the less overall sharpening you will see. This is consistent with the effects of noise: the more noise you have, the less sharpening you can use.

Figure 9.3.30 shows what can happen if you try to sharpen too far. Note that some of

FIGURE 9.3.30. TOO MUCH SHARPENING CREATES SUBTLE DARK AREAS AROUND SOME STARS.

the stars over the galaxy have slight dark halos around them. The effect can be subtle, but it should be avoided as it creates artifacts that spoil the image. There is one star just to the left of the core that has an actual dark patch behind it, a dust lane in the galaxy. Ignore the dark area around that star; it's real! You may also start to see graininess if you over-sharpen, but this depends on the noise level of the image. An image with superb signal to noise ratio can take an amazing amount of sharpening if there are no stars against a bright background. In figure 9.3.30, the amount has been set to 80%, and the radius is 4 pixels. That's really heavy-duty sharpening, and I have yet to see an image that could handle that much.

You may recall that I said I like to go a little too far in brightening the image before smoothing and sharpening. The time has come to back away from those settings and establish the final brightness values for the image. Figure 9.3.31 shows a typical Curves adjustment that I use at the end of processing. It has the effect of dimming the dim areas a bit, and brightening the bright areas a bit to establish good visual contrast. It also tucks the middle brightness levels, where you just fixed the noise problems, back into a

more appropriate relationship with the bright and dim portions of the image. I don't always add the point at top right. It is sometimes too much, and other times it adds a nice dramatic touch to the image. Use your own judgment on each image to determine if helps or hurts. The dimming at lower left is almost always going to give you a change for the better because it hides noise. If done just right, you get a very clean edge between the object and the background, but without it being so abrupt as to look false. I seldom push the background all the way to black; the eye sees dim details better if the contrast isn't too abrupt.

The bright stars in the image have a minor bright halo around them. It is most pronounced around the bloomed star. This is usually due to light reflecting inside the camera. The further you push the histogram with Curves, the more likely you are to see this effect. Depending on your tastes, you may choose not to push the histogram so far as to reveal these halos. With very bright stars, reflection is so strong that there is no way to avoid the problem.

You can also select the star's halo using the Color Range tool, and then feather it with an appropriate

FIGURE 9.3.31. APPLYING A FINAL HISTOGRAM ADJUSTMENT TO CREATE A MORE NATURAL BOUNDARY BETWEEN THE BACKGROUND AND THE GALAXY.

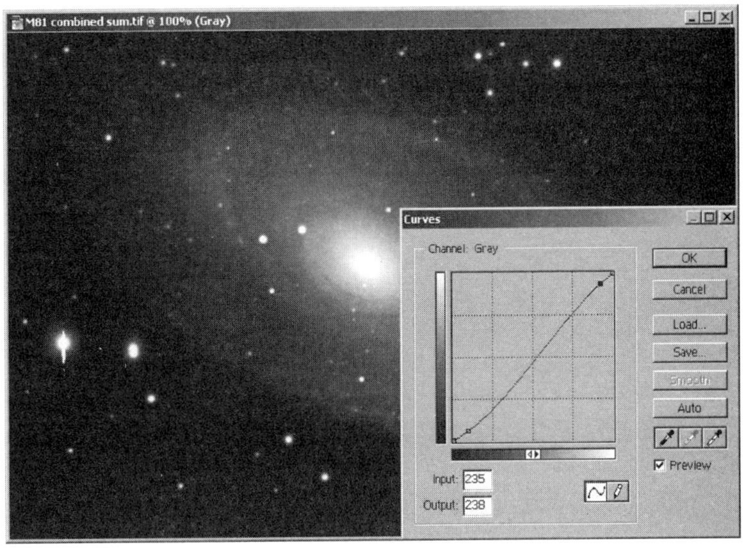

Figure 9.3.32. Top: Selecting the halo around a bright star.

Figure 9.3.33. Middle: Reducing the brightness of the halo.

Figure 9.3.34. Below: Completed M81 image.

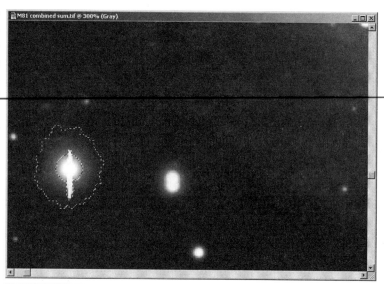

number of pixels. Finally, darken it with Levels or Curves. Be careful not to get too aggressive, or you will wind up with a dark halo around the star. Figure 9.3.32 shows a typical selection around a bright star, ready for darkening. The selection must be feathered to avoid abrupt edges. The amount of feather depends on the overall size of the selection. Turn Rulers on so you can measure the width of the selection at its widest point. Feathering that is about one-half to one-third of the radius is usually effective.

Figure 9.3.33 shows one method for darkening the halo. It does not remove the halo entirely. Some imagers may prefer a more complete removal. That requires heavy editing with the Clone Stamping tool.

There are several approaches you can use to removing blooms. One technique involves rotating the camera between images. When you combine the images, use a tool like Registar that allows you to combine using the

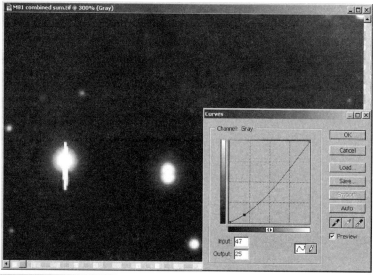

minimum pixel values. The bloomed pixels are bright. When the combine chooses the minimum pixel value, it always chooses a pixel from an image without blooming.

You can also remove the bloom using the Clone Stamping tool, or similar tools in other programs. Use the smallest available soft-edged brush. Pick up from an area that matches what should be under the bloomed area, and stamp over the bloom one small area at a time. Figure 9.3.34 shows the result of removing the blooming. This is a time-consuming approach, but if you have a steady hand you can get good results.

I have put an animated tutorial on manual bloom removal on the book web site. To view the animation, visit this link:

http://www.newastro.com/newastro/book_new/ anim/debloom.asp

The simplest method for removing blooms is the Dust & Scratches tool. By drawing a selection of a specific shape around the bloom, you can use the Dust & Scratches tool to do 90% or more of the work of removing most blooms. Figure 9.3.35 shows a close-up of the bloom at lower left in the original image of M81.

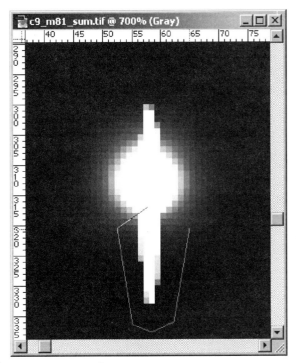

FIGURE 9.3.36. SELECTING ONE OF THE BLOOMING SPIKES.

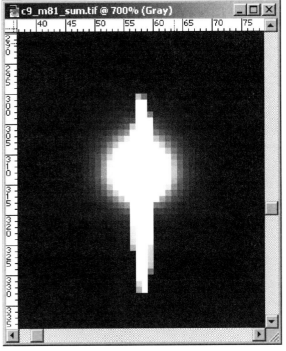

FIGURE 9.3.35. AN EXAMPLE OF A BLOOMED STAR.

Use the Polygonal Lasso tool to create a selection of the shape shown in figure 9.3.36. The exact point where the selection intersects the circular portion of the star image isn't critical, but there's not a lot of leeway, either. In this example, the selection crosses the point where the bloom and the circular portion of the star intersect. You can also try a position one or two pixels more inside the star, but you won't have good success if you create a selection away from the body of the star.

The exact shape of the selection is also not critical, but the tapered selection you see here has been the most successful in my tests. The actual appearance of the selection will vary with the extent, width, and nature of the bloom. Figure 9.3.37 shows the appearance of the selection boundary in this example.

Use the Filters | Noise | Dust & Scratches menu item to open the dialog shown in figure 9.3.38. The exact settings you use for radius and threshold will depend on the width and length of the bloom. A larger radius goes deeper toward the body of the star, while a shorter radius will not remove all of the bloom. You can

FIGURE 9.3.37. SELECTION OF THE SPIKE COMPLETED.

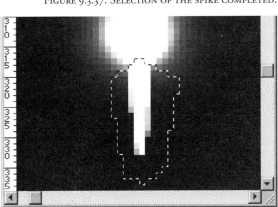

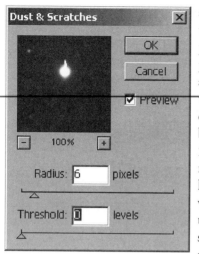

FIGURE 9.3.38. REMOVING BLOOMING WITH THE DUST & SCRATCHES FILTER.

also try feathering the selection by 1-3 pixels, with a one-pixel feather being appropriate most of the time. I usually enlarge the bloomed area by 500-800% to get a really close look at how the settings are working. The threshold has to be small enough to allow the filter to work, but not so large as to over-correct and eat away a portion of the star. A good approach is to start with a radius that removes most of the bloom, and then to adjust the threshold to get a smooth transition.

Figure 9.3.39 shows the bloom mostly removed. There are a few rough spots that can be cleaned up with the Clone Stamping tool. Pick up bright white from the core of the star, and apply it to the interior of the star to round it back out. The defects in the halo around the star are a little more challenging, but you will find that they are so slight that many times you can simply leave them without ill effects.

If your selection intrudes too far into the star, or if your radius is too large or the threshold too low, you may wind up with more significant damage (see figure 9.3.40). This is much harder to fix, and your best bet is to undo and try a smaller radius and/or a higher threshold, or simply move the selection a pixel or two further away from the center of the star.

FIGURE 9.3.39. THE BLOOM IS ALMOST COMPLETELY REMOVED.

FIGURE 9.3.40. OVERLY AGGRESSIVE REMOVAL OF A BLOOMING SPIKE.

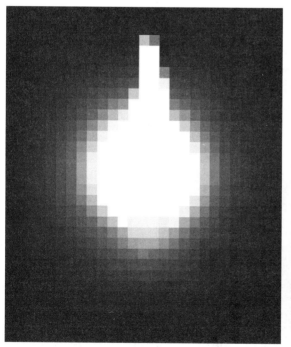
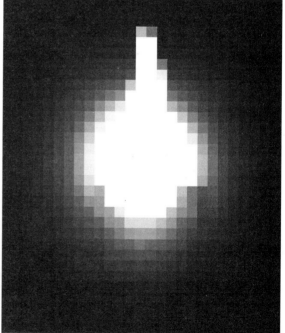

Digital Development

You can also use Digital Development (In MaxIm DL, that would be the Process | Digital Development menu item) to quickly bring out dim details in galaxy images. However, you will not have as much control as shown with the steps outlined above. As you become more proficient with the tools in Photoshop or the image editor of your choice, you will find that you can get another 10 or 20 percent better definition by careful manual manipulation of the image.

You can also run into other problems with digital development. Any flaws in the image, such as hot and cold pixels, will become significantly more pronounced because of the sharpneing included in digital development In MaxIm DL, you can set a blank user filter to eliminate the sharpening.

Digital development will sometimes give you a completely useable result, however. And it is a good tool to help you visualize the hidden data in your images. A digitally developed image can give you ideas about what you might be able to accomplish manually. Figure 9.3.41 shows digital development of the original M81 image using MaxIm DL. The result is close to the final version in figure 9.3.34, but the contrast isn't quite as good, and the sharpening isn't quite as appropriate. Worse, there are some sharpened cold pixels just right of and below the core that almost look like a false dust lane.

Still, considering that it takes about 10 seconds to perform a digital development with the default parameters, you can't beat the efficiency factor! Digital development is a great place to start out processing your galaxies, and you can use the digitally developed image as a yardstick against which you can measure your progress in mastering the skills you learned in this chapter.

FIGURE 9.3.41. DIGITAL DEVELOPMENT BOTH SHARPENS AND SHOWS DIM DETAILS.

Section 4: Nebulae

Nebulae come in two basic types: those that are generally dim, and those that contain both bright and dim areas. The processing for dim nebulae requires that you take long enough exposures to get good signal. The processing for nebulae with both bright and dim areas is a lot like that for galaxies. In this section, you'll learn another way to deal with a large brightness range. This new method works well on any object with very bright and very dim areas, not just on nebulae. The more extreme the brightness range of an object, the better suited it is to this technique, which I call Layers and Masks.

Nebulae that are generally dim can best be imaged by taking long exposures and by stacking individual images to improve the signal to noise ratio. Processing for such nebulae is identical to that for low surface brightness galaxies like M33 and M101: use a non-linear histogram stretch to emphasize dim details, and consider using a smoothing (blur) filter to reduce graininess in the dimmest areas.

Exposure Guidelines

It's easy to recommend an exposure for nebulae: make the exposure as long as possible, and take a bunch of exposures. Even the brightest nebulae, such as M42 and the Lagoon, contain extensive dim nebulosity, and only a long total exposure time will bring out all the detail. The brightness of the nebula's core determines your maximum exposure time. The central stars in M42, for example, will start to bloom at 10-20 seconds or less with most non-antiblooming cameras at f/8 or faster. A little blooming can be edited by hand. More important from an image-processing standpoint is the need to avoid saturating the bright core of the nebula. A bloomed star is editable. A bloomed nebula or galaxy core is not easily salvageable.

If the nebula does not have a bright core, you can use very long exposures and get excellent results. Blooming of bright stars with NABG cameras remains a concern, of course, but if you have an antiblooming camera, you can use exposures as long as your sky glow permits. Do test exposures with an NABG camera to determine how long you can expose without excessive blooming.

Let's look at some examples. The Trifid Nebula has bright stars, bright nebulosity, and some very dim nebulosity. It's a good example of the kind of nebula that is difficult to image with a non-antiblooming camera. Figure 9.4.1 shows a single luminance image of the Trifid taken through the clear filter of a CFW-8 with an ST-8E NABG camera. The telescope was the 16" Meade at Kitt Peak, and the image was taken as part of the Observer's Program.

FIGURE 9.4.1. A SINGLE 30-SECOND IMAGE OF M20.

FIGURE 9.4.2. A MEDIAN COMBINE OF 12 IMAGES OF M20.

is better, and the details are clearer. The blooming has been edited by hand in Photoshop; see the Galaxy section in this chapter for information about methods for cleaning up blooming.

Figure 9.4.3 shows a closer look at the differences between the images in figure 9.4.1 and figure 9.4.2. An area near the core of the nebula is enlarged 4x. The extremely grainy detail on the left is from the single image. The smooth image on the right is from the median combine using 12 images. This difference shows why it's important to get a long total exposure time. Long exposures are especially important with nebulae because of the large amount of dim detail usually present in the image. In a short exposure, the subtle variations in brightness will be lost in the noise.

Blooming has just started to occur in the image in figure 9.4.1. The exposure time is just 30 seconds, which results in a very noisy single image. The image in figure 9.4.2 is a median combine of 12 30-second images. Note how the noise is greatly reduced, contrast

FIGURE 9.4.3. DETAIL SHOWING THE HEAVY GRAIN (LEFT) OF A SINGLE EXPOSURE, AND THE SMOOTHNESS (RIGHT) OF A MEDIAN COMBINE.

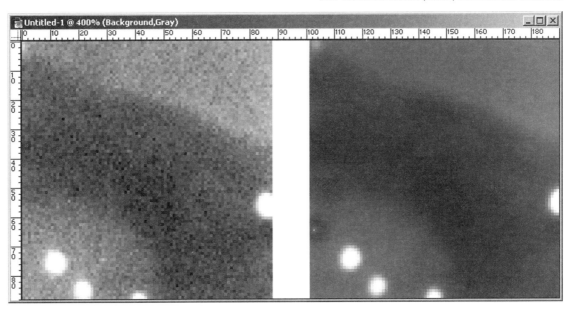

An antiblooming camera is not limited by blooming stars, and you can take very long single exposures even if there are bright stars in the nebula. Figure 9.4.4 is an image that illustrates this advantage. Despite the fact that it is a single exposure, it has very low noise. The bright star that blooms in figure 9.4.2 does not bloom in figure 9.4.4. This exposure was 15 minutes with an antiblooming camera using an f/5 refractor. It shows the very dim nebulosity surrounding the main Trifid very clearly.

The bottom line on nebulae (this applies to any extended object): take the longest practical exposures, and combine as many multiple images as you can.

FIGURE 9.4.4. A 15-MINUTE IMAGE OF M20 WITH AN ANTI-BLOOMING CAMERA SHOWS MUCH BETTER DETAIL THAN THE SINGLE IMAGE FROM A NON-ANTIBLOOMING CAMERA.

Throwing Out the Pixel Laws

The conventional wisdom states that you should choose a camera and telescope combination that will yield about 2 arcseconds per pixel as the image scale. In other words, each pixel should cover approximately a 2x2 arcsecond square area of sky. This recommendation is based on a variety of factors, and it does have the advantage of matching your system's resolution to the typical seeing conditions.

However, you are not limited at all by this recommendation. You can get a camera with smaller pixels, allowing you to get better resolution. You can also get a camera with larger pixels, and say the heck with resolution. After all, there are nights when the seeing is poor. But the real reasons to consider larger pixel cameras are:

- You can capture nebulosity with shorter exposures
- The greater well depth of the camera typically yields a better dynamic range, allowing you to image both bright areas and dim areas effectively at the same time.

These advantages come at a price, however: reduced resolution. The good news is that for most nebulae, the advantages outweigh the disadvantages for modest increases in pixel size.

For example, if nebulae are very important to you, and you need to choose between cameras with 13 and 16-micron pixels, the 16-micron camera is likely to do a better job for you, as long as the image scale doesn't go beyond 3-4 arcseconds per pixel. See chapter 4 for information about calculating your image scale.

This doesn't mean you should disregard resolution. If you plan to use the camera for imaging other types of objects, or if you are willing to take the longer exposures required by smaller pixels, resolution may then be the dominant factor in choosing a camera.

Another factor to consider is the effects of seeing on your imaging. You can image nebulae at low resolution on nights when high-resoltuion imaging would not even be possible. You may be wondering how big a price you pay in resolution with different image scales. When it comes to nebulae, the price is very small. The continuous-tone nature of nebulae shows details surprisingly well across a wide range of image scales.

FIGURE 9.4.5. COMPARING IMAGE SCALES. TOP: 0.8 ARCSECONDS PER PIXEL. BOTTOM: 3.5 ARCSECONDS PER PIXEL.

Figure 9.4.5 shows two images of the Trifid Nebula. Can you tell which half of the image is at an image scale of 0.8 arcseconds per pixel, and which is at 3.5 arcseconds per pixel? It's not as easy as you would expect, given such a huge difference in image scale. The upper image is a taken at an image scale of 0.8 arcseconds per pixel. The stars are a little sharper. But there is only a small difference in nebula detail between the two images for such a huge difference in image scale.

Other types of images, such as galaxies and planets, are not as forgiving when it comes to image scale. If you plan to image a wide variety of objects, your image scale should be suitable for those objects. If you lean heavily toward nebula, an image scale in the range of 2-4 arcseconds per pixel will work well. An antiblooming camera makes sense for imaging nebulae because you have a freedom in choosing the exposure duration.

If you want all the detail you can get, imaging nebulae with 0.5 to 2 arcseconds per pixel will work. Just take longer exposures to get the job done.

Nebula Processing Tips

The most difficult nebula image to process is one in which there an extreme brightness range. It is challenging to get a good representation of both the dim and bright areas in such an image. If you are careful to preserve detail in the bright portions of the image, the dim stuff gets lost. If you take pains to show the dim detail, the bright stuff is too bright. The most common solution to this problem is to simply allow the brighter areas to burn out, and concentrate on the dim details. But many galaxies and nebulae have interesting details in their bright cores.

Non-linear histogram stretches are one way to deal with this problem, but this approach has its limits. If the brightness range is too great, details will still be lost.

The most effective way to handle a large brightness range involves use of layers and masks in Photoshop. You either make copies of the same image, or take different images with different exposures, and place them in stacked layers. You process each layer differently to display different features. Masks allow you to choose which portions of which layers are ultimately included in the final image. Other image editing programs may contain similar features; check the documentation.

For example, in the simplest situation, you might have two versions of an image: one for the bright core of a nebula, and one for the dim areas. You apply different histogram settings to each version appropriate to the portion of the nebula you want to emphasize. You would typically apply a very light histogram adjustment to the bright area, and a much larger adjustment to the dim areas.

When you are done with your changes, erase the burnt-out core of the dim image to allow the core of the other image to show through. In this section, you'll walk through making a two-layer version of M42.

You can obtain a copy of the images here if you would like to follow along:

http://www.newastro.com/newastro/book_new/ samples/c9_M42.zip

For this exercise, the assumption is that you have two different exposures of M42. One is 10 seconds

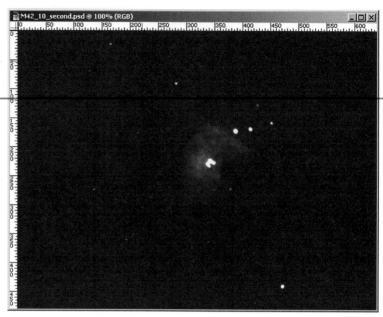

FIGURE 9.4.6. A 10-SECOND IMAGE OF M42.

long, and it has been exposed to clearly show the Trapezium (see figure 9.4.6). The second is 30 seconds long, and has been exposed to show more nebula detail (see figure 9.4.7). Both images are supplied courtesy of Kevin Dixon, and were taken by Kevin with his Fastar C8 and an ST-237 camera. The core of the 10-second image preserves the Trapezium and core detail, but the outer portion of the nebula is too dark. The 30-second image has a burnt out core, but many outer nebula details are visible. By using two layers, you can make a final version that clearly shows both the Trapezium from the shorter image and the dim details from the longer image.

The overall process of making a two-layer version of the image includes these steps:

1. Copy and paste the images into a single file as layers, with the longer exposure layer above.

2. Modify the histograms of both images to show the detail you want to show in each.

3. Create a selection that includes just what you want to show from the bottom layer.

4. Erase the selection in the top layer allowing the bottom layer to show through.

FIGURE 9.4.7. A 30-SECOND IMAGE OF M42.

There are a few interesting twists along the way, but the basic procedure outlined above will do the job. You can have more than two layers. Not too many objects have a sufficiently wide range of brightness to need such treatment. A wider field of view on M42 would be one such case, as would some galaxies with extremely active, bright cores.

Paste the Images

Photoshop supports layers, which allows you to stack images one above the other. A layer above some other layer hides whatever is in the lower layer. If you erase a portion of a layer, lower layers will show through. This is the technique that allows us to show different versions of an image selectively.

Note: It is not necessary to have multiple exposures, one long and one short, to use the Layers and Masks technique. If your image contains both the bright and dim details clearly, without blooming or a burned out core, you can also create two copies of the same image and process them differently.

To begin, open both images in Photoshop. Copy the 10-second image to the clipboard, and then create a new image (File | New). The size of the new image will automatically match the size of the image you copied to the clipboard, so you can accept the default values. Just click OK or press Enter to create the new image.

Paste the 10-second image from the clipboard into the new image. By default, it will be Layer 1. Now copy the 30-second image to the clipboard, and paste it into the new image as well. By default, it will be Layer 2, and will be located above Layer 1. All you can see is Layer 2, the 30-second image. It covers the lower layer. You can show or hide a given layer by clicking on the eyeball icon that appears next to the layer in the Layers palette (see figure 9.4.8). If the Layers palette isn't visible, use the Window | Show Layers menu item to display it.

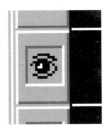

FIGURE 9.4.8. CLICK ON THE EYEBALL AT THE LEFT OF THE LAYERS PALETTE TO SHOW AND HIDE LAYERS.

Modify Histograms

Because the images have different exposure durations, you need to make adjustments to the histograms independently. For the 10-second image, the goal is to get good clarity in the bright areas of the image, which includes the Trapezium and the bright central portion of the nebula. In M42, there is a clear break between the bright portion and the dimmer portion; not all nebulae are as cooperative. If there isn't a clear break between bright and dim zones, you may have to experiment to find the sweet spot where you should establish your boundary between the two (or more) images.

For the 30-second image, the goal is to reveal the details in the dim areas away from the core. Since the core will be erased in this layer, you do not need to be concerned about burning it out. You are free to adjust the histogram to reveal dim details.

That said, you shouldn't treat the two images completely independently. They still have to blend into each other at the end of the histogram adjustments. The core should be bright enough to look natural in the final image, but it must also be dim enough to blend with the dimmer image. There is a fine line between success and disaster with this approach. Photoshop's History feature allows you to undo one or more steps. This will come in handy and help you get a feel for how to match the histogram adjustments in the two images.

The following images will give you an idea of what to aim for with your adjustments. Because the core does have a reasonably definite boundary, you have some leeway in setting the brightness relationship between the two images.

My usual procedure is to begin adjusting the histogram of the layer that will show the faintest details. This layer typically requires the most attention and second-guessing, and it will set the limits for what you can do with the other layer(s).

In this example, the 30-second image is the one I would start with. Figure 9.4.9 shows approximately how far you should go with Levels and Curves to emphasize dim details. Use your own taste to determine how far to go. If you prefer to keep a very dark background, then sacrifice some detail to get it. If you prefer to bring out the details as much as possible,

FIGURE 9.4.9. RIGHT: USE LEVELS AND CURVES TO EMPHASIZE DIM DETAILS IN THE 30-SECOND IMAGE. COMPARE TO FIGURE 9.4.7.

FIGURE 9.4.10. BELOW: USE SMALL AMOUNTS OF THE LEVELS AND CURVES TOOLS IN PHOTOSHOP TO REVEAL ADDITIONAL DETAIL IN THE CORE. COMPARE TO FIGURE 9.4.6.

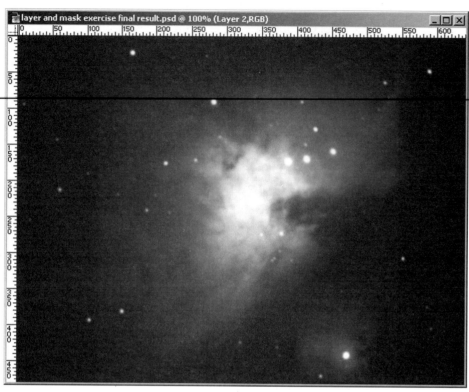

you'll wind up with a lighter background. There is no one right approach. You could emphasize more outer detail, or you could make the background darker. If you don't have experience with using Levels and Curves to emphasize dim details, read the section on galaxy processing earlier in this chapter.

The next step is to process the core image so that it will show details in the bright core, yet merge reasonably well into the 30-second layer. Click on the eyeball for the 30-second layer in the Layers palette to hide it, revealing the 10-second layer. Figure 9.4.10 shows how I processed the 10-second image. I used a very light touch with Levels and Curves. The trick is to approximately match the brightness of the 30-second layer. You will use a blend to combine the two layers, so an exact match is neither needed nor desirable. You just want to get into the same brightness ballpark.

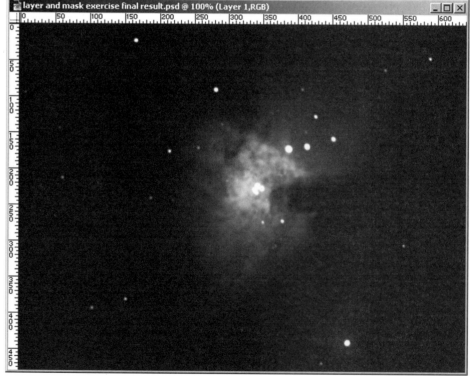

You could sharpen the core at this stage to bring out details. Or you could wait until later to do any

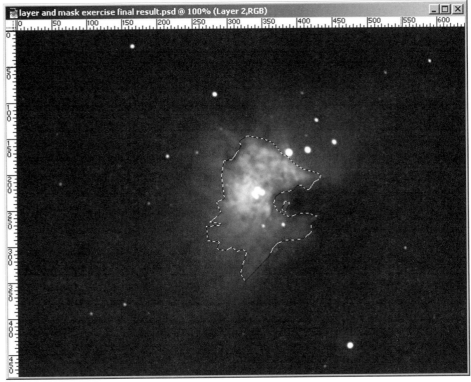

FIGURE 9.4.11. USE THE POLYGONAL LASSO TO DEFINE THE SELECTION.

feathering. That way, when you apply the feathering, the edge will be a bit further out than the edge we just defined. This may seem convoluted, but it works!

To expand the selection, use the Select | Modify | Expand menu item. Enter a number that is one-third of the amount of feathering you intend to use. This is a judgment call, experience will teach you how large a feather to use. This example calls for a 60-pixel feather. The narrowest portion of the selection is about 120 pixels, and I usually choose a feather that is about half of that distance. This one-half rule works often, but not every time. As I mentioned earlier, the undo feature will come in handy as

sharpening or other processing so you can compare the result with the two images blended. That's what is so great about working with separate layers. You can adjust them independently.

Create a Selection or Mask

The next step is a tricky one, as it requires judgment that only comes with experience. You need to define the selection that will be used for the blending of the two images. Figure 9.4.11 shows how I defined the selection for this particular case. I drew a selection boundary using the Polygonal Lasso tool. I attempted to follow a more or less uniform brightness level around the bright core of the nebula. The selection is just slightly outside the bright portions of the core, and excludes any and all dim portions of the nebula.

The next step is to feather the selection. A feather fades the selection inside *and* outside the selection boundary, so first expand the selection by one-third of the width of the

FIGURE 9.4.12. THE SELECTION HAS BEEN EXPANDED BY 20 PIXELS.

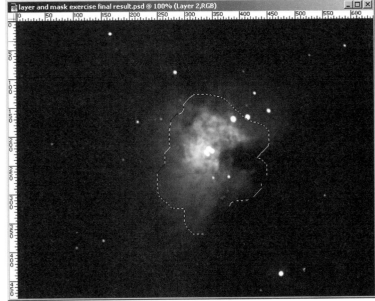

you get a feel for this part of the procedure.

One third of 60 is 20; use that as the expand amount. Figure 9.4.12 shows the expanded selection.

Now feather the selection by three times the amount you used for expansion (3 * 20 = 60 pixels). Use the Select | Feather menu item to open the Feather dialog. Figure 9.4.13 shows the result of feathering the selection. The selection doesn't appear to match the boundary now, but it's actually OK. If you want finer control over the selection, I suggest that you save it as a mask. You can then modify the mask using the various painting tools, filters, etc. to get precise control over the blending of the two images.

Erase to Reveal Details

The selection now has a soft, feathered boundary. Click on the 30-second layer to activate it. Make sure that the 30-second layer is the active layer. Use the Edit | Clear menu item or the delete key to remove the area inside the selection boundary. This allows the 10-second image to show through from below.

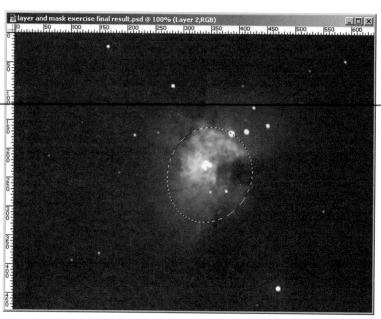

FIGURE 9.4.13. THE RESULT OF FEATHERING THE SELECTION.

Figure 9.4.14 shows the result with the 10-second layer showing through the hole you created in the 30-second layer. The brightness levels match well enough that the two images blend effectively. If the brightness levels don't match, you can either use the Undo feature to try again, or simply adjust the brightness of one layer or the other to get a closer match.

I suggest using Curves to do the matching. Keep in mind that lowering the white point removes information from the image.

FIGURE 9.4.14. THE FINAL RESULT OF COMBINING BOTH IMAGES USING LAYERS AND MASKS.

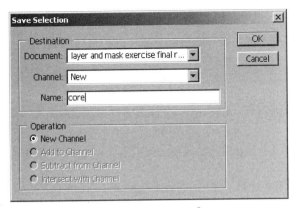

Figure 9.4.15. Saving a selection.

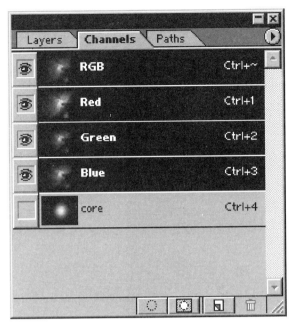

Figure 9.4.16. Looking at saved selections using the Channels tab.

The final result shows both the very dimmest details and the very brightest ones.

A Little Theory

Let's take a look at some tricks that will help you gain control over the blending process. Save the selection using the Select | Save Selection menu item (see figure 9.4.15). Give it a meaningful name, such as "core" and click OK to save it.

Click on the Channels tab in the Layers palette to show the image channels (see figure 9.4.16). Notice that the selection you saved shows up as a channel.

Click on the new channel ("core") to activate it. Notice that the window changes to show a grayscale image that defines the selection with shades of gray. The area inside the selection is shown as lighter shades; the area outside the selection shows up as darker shades. Figure 9.4.17 shows what the channel looks like.

The nice thing about this feature is that you can edit the channel to make changes to it. For example, if there was some detail in the image that wasn't showing up because the feathering was too soft in one area, you could use the Airbrush tool to paint with white and increase the brightness of that area. You can use almost any Photoshop tool to alter the channel, in fact, so you have tremendous freedom to customize your saved selections.

You can view your selection as a mask at any time by clicking on the Quick Mask button on the Photoshop main toolbar (see figure 9.4.18). To view a selection normally again, click the Selection button to the left of the Quick Mask button.

Figure 9.4.17. The contents of the channel "core."

Figure 9.4.19 shows what a Quick Mask looks like. The simple "crawling ants" selection boundary doesn't give you any information about the degree or location of feathering, (look back at figure 9.4.12 to see what I mean), but the Quick Mask shows this kind of information very clearly. You can use the Quick Mask to evaluate the effectiveness of your selection. In this case, the mask shows which portions of the layer will be removed when you use the Edit | Clear menu item.

If you don't like the mask, you can edit the channel to revise it. For example, there is a dark area to the right of the core of M42. You might choose not to include this area in your selection. To remove it, go to the Channels tab in the Layers palette. Click on the "core" channel to make it active. Also click on the eyeball for the RGB channel. Both channels are visible. Click the Quick Mask button. Set the color to black, and use the Airbrush tool to paint in the "core" channel to subtract from the selection. With the "core" channel and the RGB image visible, you can see the underlying image through the mask (see figure 9.4.19), showing you where to paint.

FIGURE 9.4.18. CLICK THE QUICK MASK BUTTON TO VIEW THE CURRENT SELECTION AS A MASK.

If this sounds complex, it will become easier as you do it. There are a few things to keep in mind:

- Paint with black to mask an area (that is, to remove it from the selection).
- Paint with white to add to the selection.

You can also paint with shades of gray, which is like feathering. Experiment with the various options on your own images to get a feel for how they work.

Narrow-Band Filters

You can take a completely different approach to imaging nebulae by using narrow-band filters. The most commonly used narrow-band filter is hydrogen-alpha.

Hydrogen-alpha is the wavelength of light that a hydrogen atom emits when its electron makes a specific jump between energy levels. It is in the deep red portion of the spectrum. Hydrogen atoms absorb photons from nearby stars, and then emit a photon with a wavelength of 656.3nm. There are also other emissions (Hydrogen-beta, Hydrogen-delta, etc.). These are at progressively shorter (bluer) wavelengths. Filters are available for some of these, including Hydrogen-beta, another common emission wavelength in the blue portion of the spectrum. Other elements also have interesting emissions, such as triply-ionized Oxygen (OIII; 500.7nm), for which there are filters.

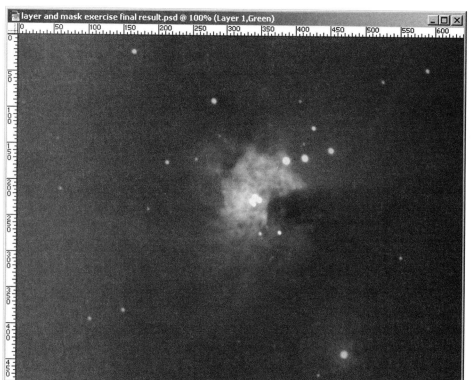

FIGURE 9.4.19. VIEWING THE SELECTION AS A MASK.

Hydrogen-alpha filter

FIGURE 9.4.20. LEFT: AN IMAGE OF THE HORSEHEAD AND FLAME NEBULAE TAKEN WITH A HYDROGEN-ALPHA FILTER.

FIGURE 9.4.21. MIDDLE: AN IMAGE OF THE SAME AREA AS FIGURE 9.4.20 TAKEN THROUGH A RED CONTINUUM FILTER. NOTE BLACK BODY RADIATION FROM STARS, FLAME NEBULA, AND NGC2023.

FIGURE 9.4.22. BOTTOM: RESULT OF SUBTRACTING CONTINUUM IMAGE FROM H-ALPHA IMAGE. NOTE THAT MOST STARS ARE PRACTICALLY REMOVED.

Most high-quality narrow-band filters have a pass band that is *very* narrow: just 3 nanometers. This restricts transmission to only the wavelength of interest. However, everything in the universe emits *black-body* radiation. This is commonly seen on earth as heat (infrared), but out in space objects at a huge variety of temperatures are radiating. Many objects you image through a h-alpha filter are also emitting black-body radiation at the h-alpha wavelength. You can use a continuum filter to capture the black-body radition in a separate exposure. You can then subtract the continuum image from the h-alpha image, leaving you with just the h-alpha data.

The continuum filter works its magic because it is also a narrow passband filter, centered on a wavelength slightly different than the desired wavelength. For example, a red continuum filter that is designed to remove black-body radiation from h-alpha images is centered on a wavelength of 645nm.

The h-alpha filter tends to reduce the size of stars because most of their energy is in other wavelengths. Applying the continuum image will nearly remove stars from the image.

All filters used for these examples were provided courtesy of Custom Scientific.

Continuum filter

Subtract Continuum filter

Section 5: Making Mosaics

All but the very largest CCD chips are small in relation to the size of most film formats. The vast majority of commonly available CCD chips are much smaller than even a 35mm negative. Since a smaller footprint results in covering a smaller area of sky, this means that astrophotographers using film can easily image wider fields of view. By comparison, the field of view available to a CCD imager is comparatively narrow.

Nonetheless, there are some wide-field imaging options. The most straightforward approach is to use a telescope or camera lens with a very short focal length. Something in the range of 200-400mm will provide a wide field of view. This could be a camera lens or a short focal length refractor with a focal reducer. Small-pixel cameras such as the ST-237, ST-8E, and ST-10E provide reasonable resolution at these short focal lengths.

Another approach is to take a number of overlapping images and assemble them into a mosaic. Figure 9.5.1 shows a mosaic of the Lagoon/Trifid/M21 area. The mosaic is made from five separate images. The images were carefully aligned and matched by hand in Photoshop. Products such as Registar can make it easier to align images, but the blending, smoothing, and histogram tools of Photoshop and other image editors are indispensable in matching brightness levels. The best approach would be to use Registar for alignment, and Photoshop or a similar image editor to do the actually assembly and blending. MaxIm DL 2.11 also includes a built-in mosaic feature, but it is limited in functionality.

The mosaic in figure 9.5.1 is haphazard, with the images being at rather odd angles to each other. You can also take a more rigorous approach to taking your mosaic and plan the location of each component image.

FIGURE 9.5.1. A MOSAIC OF THE TRIFID AND LAGOON NEBULAE.

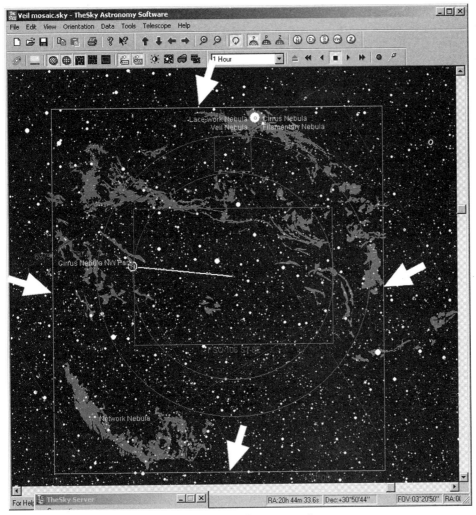

FIGURE 9.5.2. SELECTING THE AREA OF THE MOSAIC.

The best mosaics have two things in common: the individual images that make up the mosaic provide seamless and complete coverage of the area, and the brightness and colors in the images match very closely. The best program for achieving both objectives at one time is Photoshop, and that is what I have used for the sample mosaic in this section.

Taking the Images

TheSky from Software Bisque is an excellent tool for taking a mosaic. For imaging, you'll need level IV of TheSky. It contains a Mosaic feature that defines the size and position of the individual images you take for the mosaic.

To use the mosaic feature, you need a goto mount or a mount with digital setting circles connected to TheSky. A high-precision goto mount will be the easiest to use for making mosaics, as it will point your telescope more accurately at the center of each frame. If your mount does not point accurately enough to automate the taking of the images, you can check the positioning for each image by taking test exposures at the highest available bin mode. To get a seamless result, you will need to make sure that the images overlap.

You will need to have a field of view indicator for the camera and telescope you use for the mosaic. The indicator need not be visible when you are setting up the mosaic in TheSky, but it must be available to create the mosaic.

To define a mosaic, zoom in to approximately the area where you want to take the mosaic. Make sure that the orientation is correct; use the North handle (center of figure 9.5.2) to rotate the view to match the orientation of your camera. This is an essential step. Failure to match rotation will result in mosaic frames that do not match up with the images you will take, and you won't be able to assemble a mosaic with full coverage.

Click and drag out a rectangle that covers the area you want included in the mosaic. See figure 9.5.2 for an example mosaic of the Veil Nebula. The four large arrows in figure 9.5.2 point to the sides of the box I used for a Veil mosaic. The inner rectangle shows the field of view.

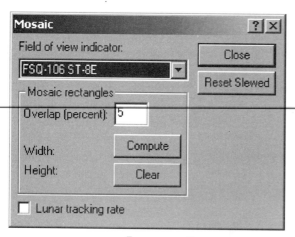

Figure 9.5.3. Setting up a mosaic.

Do not click in the program window, or you will lose the rectangle. Use the Tools | Mosaic menu item to open the Mosaic dialog (see figure 9.5.3). Choose the Field of View Indicator for your camera-telescope combination from the list of available options at the top of the dialog. Only field of view indicators you have already created will appear.

Set the amount of overlap as a percentage. The default value is 5%. This is not sufficiently reliable for some automatic image registration programs. If the overlap percentage is too small, you run the risk of having to do image alignment manually. This isn't all that difficult, but automatic alignment is always easier. The following example shows you how to do a manual alignemnt. Try a setting of 10% if you plan to use a program like Registar to align the images. If you have a high degree of mount pointing accuracy and a superb polar alignment, this should be fine for manual image alignment. If you are using digital setting circles, or if your pointing accuracy is not high, use a larger percentage in the range of 10-20%.

To generate the mosaic, click the Compute button. TheSky draws a series of numbered rectangles the same size as your camera's field of view. The rectangles fill the area that you originally drew. The actual coverage may be larger than your original rectangle. If you don't like the arrangement, click the Clear button, and start over with a new rectangle. Figure 9.5.4 shows a typical mosaic, in this case for the Veil Nebula. If the number of stars is too large to see the mosaic boxes clearly, reduce the number of stars visible (View | Filters menu item) by changing the value for the dimmest magnitude displayed.

Once the mosaic is created in TheSky, you can close the Mosaic dialog and the outlines will remain. There is a small cross at the center of each field of view. Right click on the first cross and choose Slew if you have a goto mount, or use your digital setting circles to move to that position. Take your image(s) of each rectangle in turn. It may take more than one night to take all of the images, especially if you are taking color images or if you are taking multiple images at each location. If you are using digital setting circles, or if your goto mount doesn't point with high precision, take a short, binned image to verify your position and adjust as necessary. Digital setting circles have a lower precision than most goto mounts, and require more testing and adjusting prior to taking each image.

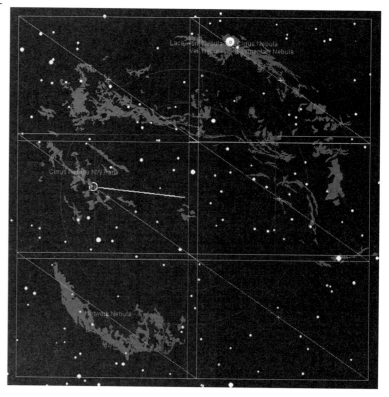

Figure 9.5.4. TheSky creates purple rectangles for the images that make up the mosaic.

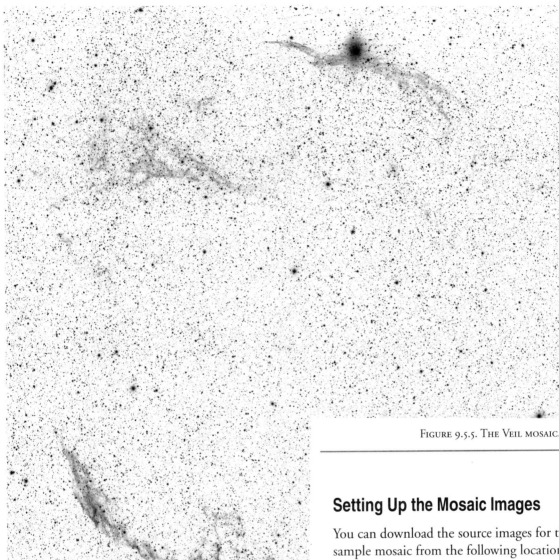

FIGURE 9.5.5. THE VEIL MOSAIC.

Figure 9.5.5 shows the completed mosaic of the Veil nebula as a negative. A negative image often shows extremely dim details more clearly than a positive image does. Only five of the six images were actually taken. The mosaic isn't seamless; note that there is a slight darkening near the top of some of the component images. This comes from an inadequate flat field for this series of images. To correct this, create a selection that roughly matches the dark area and feather it. Use the Levels and Curves tools to adjust the selection to match the rest of the image.

Setting Up the Mosaic Images

You can download the source images for the following sample mosaic from the following location:

http://www.newastro.com/newastro/book_new/samples/c9_mosaic.zip

The above file is large (4MB); for a 1MB version:

http://www.newastro.com/newastro/book_new/samples/c9_mosaic2.zip

Before you can combine the images in Photoshop, make basic histogram adjustments. Other processing should be done on the assembled image. Do enough pre-mosaic contrast adjustment to show details in the images, but leave the white point a little high to allow room for final adjustments. The approach is to do only the contrast adjustment (Levels and Curves) needed to reveal enough detail for assembly; convert to 8-bit for-

Figure 9.5.6. Applying initial histogram adjustments to the individual images.

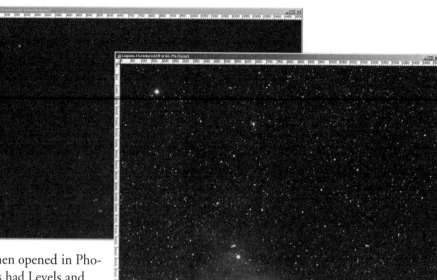

mat; paste all of the images into one file; and then align them and perform final image processing on the completed image.

Figure 9.5.6 shows a before-and-after comparison. The background image is the original .FIT or .TIF file when opened in Photoshop. The foreground image has had Levels and Curves treatment to bring out the nebulosity. You do not need to do exactly the same processing on each image. You can try, but you'll find that even with identical exposure and processing there will still be variations between the images.

Even if you are very careful to match exposure times while taking images, there will still be variations from image to image. For example, the object moves across the sky while you are imaging, and could move into or out of light pollution zones. Unless you are at an extremely pristine dark-sky site, there will be background differences between images. You will need to do different processing on each image to get them to match. It is easiest to match contrast when the images are beside in each other in the mosaic. However, final processing may reveal additional issues later, and you can make final contrast adjustments at that time.

It takes a keen eye and patience to match brightness levels. Make initial rough changes with all images open at one time. Make all Levels changes first. The black and white points should be at the same relative places on each image histogram. Verify this by comparing the images side-by-side. You cannot necessarily use the same numeric values for Levels changes, as the skyglow or transparency varies from one image to the next.

Next, make your rough Curves adjustments. Here you *can* use identical numeric values for the changes, with perhaps a small adjustment at the end to each image just before you make a conversion to 8-bit color (Image | Mode | 8-bits per channel menu item). You can't copy and paste 16-bit images in Photoshop, so you need to delay the conversion until after the last Curves adjustment.

To create the blank image into which you will paste the individual images, use the File | New menu item. Enter a width and height sufficient to contain the number of images you are inserting into the mosaic. (You can change size later using the Image | Canvas Size menu item.) Set the type as RGB (24-bit color).

Figure 9.5.7. Paste a copy of one image into the new file.

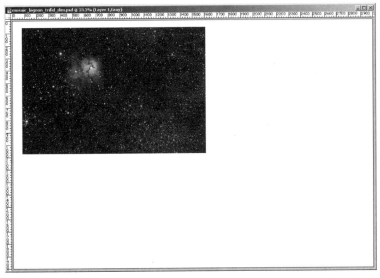

FIGURE 9.5.8. POSITION THE IMAGE IN ITS APPROXIMATE FINAL LOCATION.

Aligning Images

Open the individual images in Photoshop, and visually confirm that the brightness and contrast of the images is at least close. Then copy one of the images to the clipboard, and paste it into the new, empty image (see figure 9.5.7).

Hold down the Control key and drag the image to its approximate location in the final image. For convenience, I usually do the image at upper left first (see figure 9.5.8).

I then fill the background layer with black. A white background makes it harder to see subtle brightness variations between images. Paste an adjoining image into the mosaic (see figure 9.5.9). If the new image is not in a layer above the first image, drag the second image's layer above that of the first image in the Layers palette (Window | Show Layers menu item). Note that in this example, although the images are close in brightness, there is still a noticeable difference between them. Your ability to match levels will grow with experience using these techniques.

Set the blending percentage of the top layer to 50%. To set the blending percentage, right click on the layer in the Layers palette, and choose Blending Options. Move the Opacity slider in the General Blending section to a value of 50%. In Photoshop 6, you can set the blending percentage directly.

Select the layer, and use the Blending drop-down list on the Layers palette to adjust the blend percentage. These methods are identical, and you can use whichever one you prefer.

FIGURE 9.5.9. ADDING THE SECOND IMAGE.

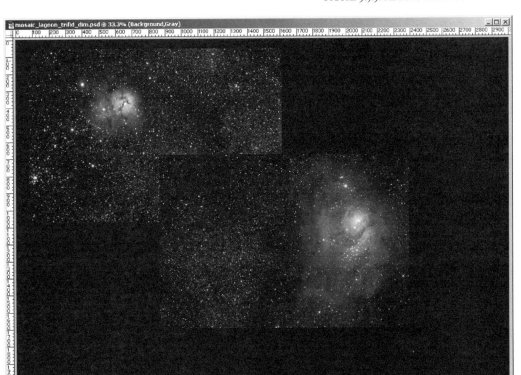

With the blending set to 50%, the layer becomes semi-transparent, as shown in figure 9.5.10. Because you can see through the top layer to the layer beneath it, you can align on stars that you can see in both images.

Hold down the control key to drag the semi-transparent layer around until it matches up reasonably well with the lower layer. Often, there will be some mis-rotation between the two images. Ignore the rotation problem for the moment. Pick a star roughly in the middle of the overlapping area, and align the two images on this star carefully. If the rotational misalignment is small, you will see the single star that you aligned at the center of circular trails (see figure 9.5.11). If there is no rotational misalignment, you can align the two images on all of the stars in the overlapping area.

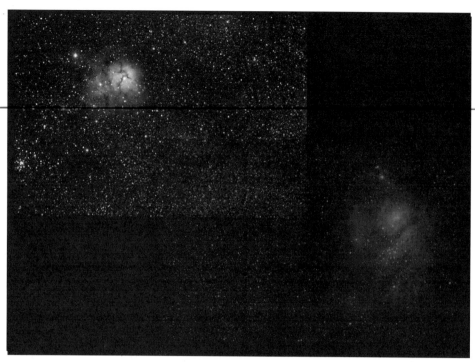

FIGURE 9.5.10. CHANGING THE BLENDING OPTIONS.

If you need to rotate the semi-transparent layer to achieve alignment, use the Edit | Transform | Rotate menu item (Photoshop 6; see figure 9.5.12). For best results in previous versions of Photoshop, use the Numeric menu item to rotate, rather than trying to use the mouse to align the images. Numeric input is always available when rotating in version 6. Numeric values allow finer control over the amount of rotation. For most images, precision to the nearest 10th of a degree will be adequate.

Once the images are aligned (with or without rotation, as appropriate), set the blending back to 100% (see figure 9.5.13). When you have added all of the images, use the Image | Crop menu item to remove any extra space in the image.

FIGURE 9.5.11. ALIGN ON AT LEAST ONE STAR.

FIGURE 9.5.12. ABOVE: ROTATING THE IMAGE TO ALIGN IT.

FIGURE 9.5.13. BELOW: BLENDING HAS BEEN SET BACK TO 100%.

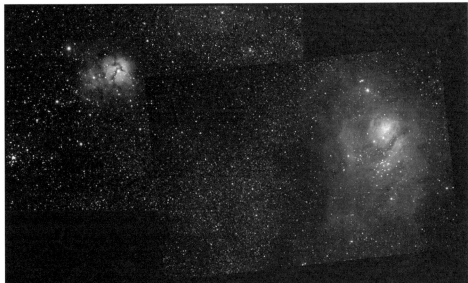

Matching Brightness and Contrast

If there are any brightness and contrast differences between the two images, as in figure 9.5.13, use Levels and/or Curves to adjust the images so they match (see figure 9.5.14). See chapters 8 and 9 for many examples illustrating use of these tools. Although portions of the images match contrast in figure 9.5.14, other portions of the boundary do not match. This can be caused by background gradients due to light pollution, or a not-quite-accurate flat field. To recover from this problem, blend the edges of the top image.

There are many tools in Photoshop that will allow you to do the blending. For example, you could use the Eraser tool in Airbrush mode. Click on the Eraser tool to activate it, and then choose Airbrush as the type of eraser. Choose a large brush size, in the range of 50-200 pixels depending on your overlap. The goal is to get a smooth blend. In general, use the largest brush size you can. Make sure the top image is active in the Layers palette, and then erase the edge using a very light airbrush setting (15% usually works well). Use your judgment to determine how much erasing to do, and be very careful not to erase in areas where there is no underlying image! When you have blended the edges of the upper layer, it will merge smoothly with the underlying image, as shown in figure 9.5.15.

You can also create selections with the Polygonal Lasso tool, and feather the selection before erasing or using the Edit | Clear menu item.

Continue to add additional images in the same manner until you have built the entire mosaic. If the images are taken on different nights, you will need to spend some extra time matching brightness and blending edges.

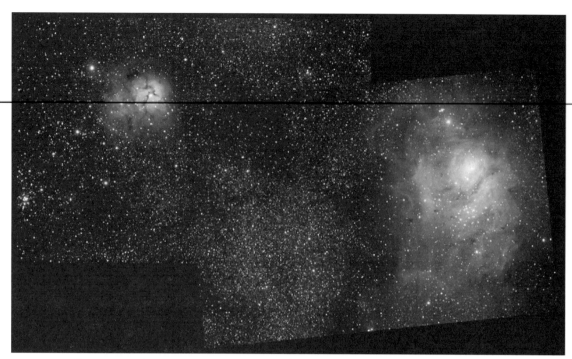

Figure 9.5.14. Adjusting with the Levels and Curves tools to balance the brightness of the images.

Figure 9.5.15. The completed two-image mosaic.

INDEX

Symbols

@Focus (CCDSoft) 38–39, 68–77
  parameters 72–75
  requirements 69
  sharpness 75
  step size 73
  step size tips 75

Numerics

16-bit versus 8-bit 436

A

A/D converter 264
aberrations 83
ABG 30, 152–160, 230, 452
ADU 355, 367, 436
aggressiveness 208
Airy disk 35
airy disk 34
aligning images 287–295
antiblooming camera
  see ABG
aperture 24
aperture mask 95–96
API4Win 165
Apogee 161–162
artifacts 428
Astroart 164
Astromart link 174
astrometry 112
Astro-Physics 132–133
AstroPIX 165
AutoAstrometry 115–117
AutoDark 14
autoguiding
  see guiding

B

background 352
background and range 16–17, 371
backlash 73, 86–88, 136
  compensation 87–88, 208
  measuring 87–88
Barlow 101, 416
bias frames 14, 229, 245–247, 282–284
binning 11–12, 43–44, 158, 326–327, 332, 347–348, 359, 431

black point
  see histogram
blooming 16, 95, 152–155, 157, 446
  manual removal 435
  removal 447–448
Blue Snowball 167
blurring
  see smoothing
book web site 4
brightness and contrast settings
  see histogram
brightness level 280
Bubble Nebula 61, 193

C

calibration
  scaling 203
calibration errors 209
calibration times 207–208
calibration tips 213
California Nebula 16–17, 359, 363
camera
  self-guiding 194–197
CCD
  typical session 4–5
CCD Calculator 168–171
CCD chip 230
  amplifier glow 236
  flaws 243
  full well 263, 366, 436
  reading 231–232
CCD comparisons 3
CCDSoft 164
central hot spot 240, 243, 459
collimation 33, 43, 82–86, 151
color 103, 322–323, 325–329, 334–363, 417–418
  alignment 361
  automated imaging 330–333
  balancing 325, 351, 353, 363
  blue extinction 324
  blurring RGB 363
  CCDSoft 330–332, 336, 339–347
  CMY 7, 328–329
  combining 323, 338, 344–348, 350–355, 357–363
  exposure 333
  false 326
  filters 330–333
  guidelines 336–337
  LRGB 7, 326–327, 343–347, 350, 359–362

color *(cont.)*
    MaxIm DL 332–333, 336, 347–348, 350–352, 354–355
    one-shot camera 161
    Photoshop 353, 355–363
    RGB 7, 323, 325, 328–329, 339–343, 347, 353, 357–359
    UBVRI 329
color combining 339–343
color correction 25
coma 25, 43
combining images 92–93, 110, 158, 267, 287, 290–291, 294–296, 334, 417, 430–435, 451
    union 294
Cone Nebula 91, 401
contrast 23, 25
contrast transfer 415
cooling 3, 228, 250, 415
cosmic ray hits 235, 257, 431
Crescent Nebula 2, 35
critical focus zone 36–38, 74
    formula 39

D

dark frames 14, 229, 232–237, 248–251, 253–262
    applying 256–262
    scaling 245–247, 282–284
    temperature 249
    variations over time 252, 255
dark peak 369
data reduction
    see reduction
data transfer 230
DDP
    see digital development
deconvolution 403–411, 416–419
    Astroart 409–411
    Lucy-Richardson 404, 409, 411, 417
    MaxIm DL 406–408
    maximum entropy 404–405, 408–409, 411
delay 44
diffraction 33–35
    spikes 58–64
diffusion 265, 275
digital development 352, 386–394, 401, 424, 449
    Astroart 393–394
    MaxIm DL 386–393, 422–423
Dobsonian 134–135, 146–147
downloads
    CCD Calculator 168
    CCDCalc 12
    focuser control 67

drift 108
drift alignment
    camera assisted 224
    manual 223
dust motes
    see dust shadows
dust shadows 238–242
dynamic range 435–436

E

elevation 415
equatorial platform 134–135
equipment
    cheap 174
    over $10,000 176
    under $10,000 176
    under $3,000 175
    under $5,000 175
Eskimo Nebula 405
exposure
    CCDSoft 10–12, 14–15
    guided 8–9
    guidelines 431
    length 3, 8–9, 12, 14, 20–22, 28–30, 43–44, 89–93, 98, 107, 109, 120, 155–157, 159–160, 192, 250–251, 333–334, 403, 414, 429–433
    short versus long 6–7
    suggested times 13
    unguided 8–10
exposure and focal ratio 8–9
eyepiece projection 105–106

F

Fastar 172
field rotation 108
filter wheel
    see filters
filters 15, 22, 94–97, 235, 300–302, 322–328, 418
    CMY 328
    continuum 461
    filter wheel 330–333
    high pass 394, 396–398, 403
    hydrogen-alpha 96, 326, 460–461
    IR (infrared) blocking 14, 111, 302, 322
    light pollution 160–161, 300–302
    low pass 389, 391–392, 401
    moon 13
    narrow-band 326, 460–461
    polarizing 13
    PSF 403
    solar 13, 94–95, 416
Finger Lakes (FLI) 163

FITS plug-in 355, 377–378
flat-field frames 14, 238–243, 263–281, 297
   applying 275–281
   brightness level 264
   comparison 271–273
   diffusion flat 265
   dome flat 266
   evaluation 270–273
   sky flat 266–267
   T-shirt flat 268–270
flatness of field 25
flexure 27
flip mirror 78–80
focal length 20–21, 27, 137–141, 154, 219
focal plane 79, 82–83
focal ratio 20–21, 27–29, 37–39, 159–160, 171–173, 334, 414, 432
focus
   comparison 36
   evaluation 42, 44, 49, 52, 54–57
   examples 36
   thermal drift 214
focus shift 8–9
focuser 65
   backlash 73
   motorized 65–67
focuser quality 68–70
focuser types 70
focusing 32–44, 46–53, 54–80
   automated 39, 65–77
   brightest pixel 48–49, 52, 54
   CCDOPS 53, 59–61
   dim stars 56–57
   examples 41
   FWHM 50–52
   mask 61–64
   masking tape 58–61
   solar images 415
focusing knobs 65
formulas
   critical focus zone size 39
   image scale 12, 115, 208
   minimum calibration time 207
   minor planet exposure delay 119
FOV indicator 204–206
full width at half maximum
   see FWHM
FWHM 50–52
   plot 51

G

galaxies 13, 429–431, 436–441, 443–444
   Photoshop 437
gamma 380–381
Gaussian blur 402
GEM
   see German equatorial mount
German equatorial mount 130–131'
globular clusters 13, 421–428
good-weather setup 180
gradients 111, 270, 297–319, 352
   removal 303–313
   subtract blurred image 314–317
   Veil Nebula 305–313
grain
   see noise
grain surgery 105
gudiing 218
guide errors 217–218
guide star 202–206, 216
guidelines for processing 362
guiders 198–200
guidescope 198
guiding 188–191, 194–197, 213, 215–216, 223–225
   aggressiveness 211
   corrections 108, 216
   errors 108
   oscillations 192
   resume after download 215
   spreadsheet analysis 224–225
   typical session 215

H

histogram 16–18, 98–101, 309–313, 341–343, 351–352, 355–357, 363, 367–386, 401, 421, 426–427, 437–441, 443, 455, 466, 469
   CCDOPS 371–373
   CCDSoft 373–376
   MaxIm DL 368, 370–371
   non-linear changes 381
   Photoshop 369, 376–386
Horsehead Nebula 262, 297–298, 461
hot pixel 233

I

IDAS 301
image processing 366–367
   globular clusters 421
   solar system 414–417
image reduction
   see reduction

image scale 11–12, 119, 140–141, 166–168, 171–173, 219, 452–453
imaging
  casual 177
  remote control 181–183
  setup 177–182, 184–185
infrared (IR) 324
internal reflections 238
IRIS 165
iterations 405, 408, 411, 419

J

Jupiter 13, 102–104

L

Lagoon Nebula 107, 150, 335, 462, 470
Layers & Masks 454–460
light box 274–275
light curve 113, 124–127
light frames 14
light leak 233, 235–236
light pollution 20–23, 110–111, 160, 297–299, 324, 358
light pollution suppression filter 300–302
LISAA guider 200
Losmandy 133
luminance 326–327, 343

M

M101 13, 109–110, 143, 151, 329, 334, 395–396, 403
M13 13, 36, 421
M16 40, 347, 353, 363
M27 20–21, 25, 141, 322, 327
M31 13, 146, 337
  widefield 166
M33 165
M42 3, 11, 13, 28, 30, 111, 139, 141
M46 28
M5 150
M51 28–30, 40, 142, 151, 299, 430
M57 140–141, 404
M65 7
M81 150, 298, 335, 431–432, 446, 449
M82 28–29, 147, 298
M92 36
Maksutov-Cassegrain 46, 149
Maksutov-Newtonian 23–24, 147
Mars 13, 414, 417–419
mask
  see focusing, mask

MaxIm DL 164
maximum 368, 371
Meade 163
MegaFix 165
mesh 27, 131
midpoint 380–381
minimum 368, 371
minor planet search 113, 119–124
Mira AP 164
Mira AP background correction 304
moasics 462
Moon 13, 40, 98–101, 150, 416, 420
morning after, the 177
MOS technology 230
mosaic
  manual setup 114
mosaics 463–470
  aligning 467–468
  blending 468
  defining 463–465
  matching brightness 469–470
  matching brightness and contrast 466
  versus widefield 462
mount 26–27, 107
  goto 109
  tuning 221–222
  visual versus CCD 26
mount calibration 201–213
  MaxIm DL 210–212
mount quality 26–27, 161
mounts 191
  fork 135–136
  types 130

N

NABG 8–9, 11, 13, 30, 152–160, 431
nebula 13
Nebulae 450–461
  processing tips 453
negative image 465
Newtonian 23–25, 41, 146
NGC 1977 153
NGC 2174 343
NGC 3190 149
NGC 7331 26
NGF-S 39, 46, 65–66, 198, 214
noise 14, 89–93, 154–156 158, 193, 228–229, 287, 296, 325, 334–335, 395–403, 405–406, 408, 432–436, 440–443, 451–452
  defined 89
  reducing 91
  system 229, 249

non-antiblooming camera
    see NABG
normalizing backgrounds 340–341, 349–351, 354–355
North American Nebula 40

O

observatory setup 185
off-axis guider 78–80
off-axis guiding 198
Omega Centauri 427
one night stand 178–179
Open clusters 13
Optec TCF-S 66
optical quality 3, 23–26, 33–34
overlay 289

P

Paracorr 25
parfocal rings 78

PE
    see periodic error
PEC 190
periodic error 131, 190, 220
photometry 112, 127
Photoshop
    Apply Image 316
    Channels 459
    Channels palette 308, 357–358, 442, 459
    Color Range 442
    color range 444
    Curves 99, 310, 312–313, 316, 357–359, 363, 381–385, 424–425, 427, 437–441, 445–446, 456
    Despeckle 443
    Dust & Scratches filter 317–319, 448
    Gaussian blur 443
    Layers 309, 319, 458
    Layers palette 360, 362, 454–455, 469
    Levels 310–312, 356–357, 369, 378–380, 385, 424, 426–427, 437–441, 454–455
    masking 402
    masks 460
    Merge Channels 357
    Quick Mask 459–460
    rotation 360–361, 469
    selections 442, 444, 446–448, 457–459
    sharpening 396–398
pixel math 355
planetary imaging 101
planets 13, 416

point spread function
    see PSF
polar alignment 3, 10, 47, 107–108, 190, 222–225
primary mirror
    moving 46, 214
PSF 404–406, 409, 411
    Gaussian 406, 410
    samples 407
    star 410

Q

quantum efficiency 366

R

random errors 220
recalibration 217
recalibration frequency 201
reduction 15, 47, 228–229, 282–286, 405
    groups 282–286
reflectors 146–147
refractor 47, 109, 214
    achromat 143
    APO 23, 26, 142, 145
Registar 291–295
    calibration control 294
    Combine tool 294
register shifting 230–232
remote control 181–183
resizing 348
resolution 167, 422
response curves 324
rich field imaging 144
Ritchey-Chretien 151
road show setup 183
RoboFocus 39, 67, 214
Rosette Nebula 13

S

S/N
    see Noise
saturation 95, 264–265, 435
Saturn 13, 104–105, 242
SBIG 162–163, 196
Schmidt-Cassegrain
    see SCT
scope reviews
    links 174
screen stretch 368
SCT 25, 27, 41, 46, 54, 65, 82–86, 95, 109, 148–149, 214
Sculptor Galaxy 147

seeing 19–21, 33, 35, 44, 57, 69, 102, 414
self-guiding
   see guiding
seminconductor 230
SExtractor 121–122
sharpen 419
Sharpening 24
sharpening 24, 97, 103–105, 352, 388–401, 416, 418, 420, 443–445
   Astroart 398
   globular clusters 428
   MaxIm DL 398–400
   Photoshop 398
shift 289
sigma 411
signal to noise ratio
   see noise
sky glow 70, 436
SkySensor 2000 PC 134
smoothing 105, 363, 393, 395, 401–403, 441, 443
   Photoshop 402
Software Bisque 133
SSC Astronomy 165
stacking
   see combining images
star halos 144, 388–389, 445
star trails 108
STAR2000 196–197
Starlight XPress 162, 196
StellaImage3 165
step size 74
Stephan's Quintet 192, 229, 368
STV 80, 147, 199–200
subframe 53, 242
Sun 13, 94–97, 415
SuperFix 165
supernova search 113, 124

T

Takahashi 133
Takahashi Epsilon 148
Takahashi Mewlon 150
TCF-S 39, 66–67, 198, 214
telescope
   selecting 137
   types 142–146, 148–151
TheSky 113, 204–206
   mosaics 114, 463–464
threshold 394
telescope
   types 147
Tom Osypowski 134

Track & Accumulate 193, 295–296
   flat fields 264
Trifid Nebula 150, 158, 323, 450, 462, 470
turbulence
   see seeing

U

unguided imaging 109
unsharp masking 396–400, 404–405, 416, 418–419, 428, 434, 444–445
UV 324

V

Veil Nebula 148
   mosaic 464–465
Venus 13
vignetting 238, 240, 243, 459
Virgo Cluster 155–157

W

white point
   see histogram
whoops! 294
worm 27, 131
worm gear 27, 131

Y

Yahoo groups links 174